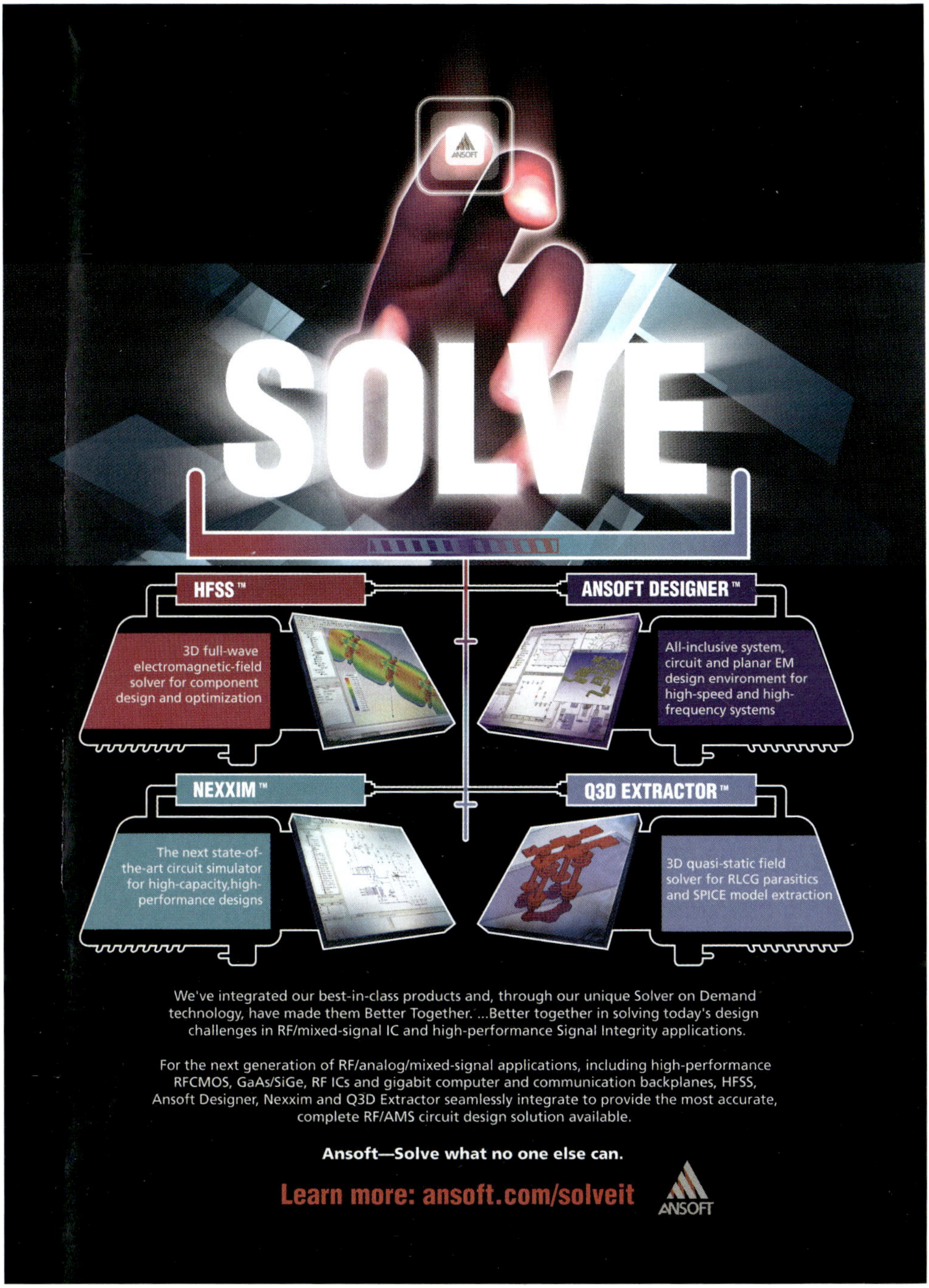

Klaus Kupfer (Ed.)

Electromagnetic Aquametry

Klaus Kupfer (Ed.)

Electromagnetic Aquametry

Electromagnetic Wave Interaction
with Water and Moist Substances

With 379 figures

 Springer

Dr.-Ing. Klaus Kupfer
Materialforschungs- und -prüfanstalt
an der Bauhaus-Universität Weimar
Amalienstr. 13
D-99423 Weimar
Germany
klaus.kupfer@mfpa.de

ISBN 3-540-22222-7 Springer Berlin Heidelberg New York

Library of Congress Control Number: 2004109775

Springer. Part of Springer Science+Business Media
springeronline.com

© Springer-Verlag Berlin Heidelberg 2005
Printed in Germany

Typesetting: data delivered by editor
Cover design: deblik Berlin
Printed on acid free paper   62/3020/M - 5 4 3 2 1 0

Water

Water – the vital element exists

- *as solid ice – in snow crystals, collected in glaciers, and maritime ice of polar regions*
- *as liquid – in rain, collected in rivers, lakes, and oceans*
- *as humidity – evaporated from oceans or agricultural areas, transpired from plants and collected in clouds*
- *as moisture – bound to surfaces or volumes of nearly all substances*

The water molecule consists of an oxygen and two hydrogen atoms. Extremely high dipole forces are the basis of its absorptions in the microwave and infrared ranges.

Its high permittivity and strong dielectric losses enable the determination of water in the composite of other substances using different dielectric measuring methods.

Numerous hydrogen nuclei in moist materials possess high magnetic moments. In strong magnetic fields they are detectable using nuclear magnetic resonance.

By the measurement of moisture

- *for the estimation of quality characteristics,*
- *for energy savings and water dosage,*
- *for reducing environmental damage.*

Thus, we

- *investigate the properties and effects of this life donating substance,*
- *learn to understand the world around us, and*
- *appreciate nature as our partner.*

Klaus Kupfer

Foreword

Information about a material can be gathered from its interaction with electromagnetic waves. The information may be stored in the amplitude, the phase, the polarisation, the angular distribution of energy transportation or the spectral characteristics. When retrieved from the wave, certain material properties may thus be determined indirectly. Compared on the one hand to direct material analysis, an indirect method requires calibration and is prone to interference from undesired sources. On the other hand, however, it permits the determination of features inaccessible by direct methods, such as non-destructive material interrogation, high measurement speed, or deep penetration depth. However, being a physical method, the use of electromagnetic waves is still handicapped by the lack of acceptance by many chemists, who are used to applying direct approaches.

Historically, the first application of electromagnetic wave interaction with matter involved measurement of amplitude changes at a single frequency caused by material properties, and it is still used today by some systems. This approach was soon supplemented by single frequency phase measurements, in order to avoid distortions through amplitude instabilities or parasitic reflections. Such single parameter measurements of course require dependence only on one variable in the measured process and sufficient stability of all other ancillary conditions. If that is not the case, the single parameter measurement fails.

Single parameter electromagnetic measurements can be complemented by other methods, to remove disturbances by undesired sources. The effect of temperature is frequently eliminated by an additional temperature sensor. However, a much greater effort is often deployed. For example, in microwave moisture determination, the influence of varying mass density is sometimes removed by applying radiometric gauges in order to determine the density. In order to simplify the measurement set-ups, an electromagnetic two parameter measurement can be applied and is now state of the art. Amplitude and phase changes due to a material are determined simultaneously at a single frequency. Instead of acquiring amplitude and phase, the permittivity and loss factor of a material may also be determined by measuring the shift of the resonant frequency and deterioration of the quality factor of a resonator. Hence for example the simultaneous determination of material density or mass and moisture content becomes possible. This has led to today's two most important applications of electromagnetic material measurements in process control. One is density independent moisture measurement. This is important, because the moisture content often has to be determined within an accuracy of fractions of a percent. The other is moisture independent mass determination. Thus widespread radiometric mass gauges, which require specialized personnel for maintenance, can often be eliminated, thus avoiding any undesired adverse influence of the radiation on materials such as foodstuffs.

Microwave two parameter measurements hitherto work well only in environments and for materials, where the conductivity is sufficiently small and where dielectric losses dominate. This is because the effective permittivity and the dielectric loss factor already constitute a set of two parameters, which can be determined by a two parameter microwave measurement. If for example an ionic conductivity is included in the effective loss factor, as it is often the case in practise, one additional microwave measurement is required. Such a three parameter determination is not yet standard. The third parameter could for example be acquired by adding a measurement at another frequency. Since the dielectric properties change quite slowly with frequency, care has to be taken to choose a frequency sufficiently distinct from the original one.

In real applications, it is not only the conductivity of a material which interferes but there are also contributions from many other constituents, which also affect the permittivity spectrum. Dielectric multi-parameter measurements are then required in order to take account of that situation and allow the simultaneous determination of a larger set of unknowns. The question then arises of how to correlate the direct measurement variables like amplitudes and phases at various frequencies and the originally desired information about constituents or properties of the material and thus calibrate the instrument.

One solution could consist in a physical model of the interrogated material, which would allow the direct calculation of the desired relations. Such models have been developed in the past for some idealized compositions of constituents by applying exact electromagnetic analysis. In addition empirical approaches exist, but very often it is experienced that a model which works satisfactorily for one situation fails for another even though it only differs marginally from the first. This is because in nature, many materials are too complex to describe them sufficiently accurately with such necessarily simplified models.

Hence another solution has been recently proposed which evaluates the statistical properties of the interrogated matter. Experience has shown that the dielectric spectrum, when recorded across a wide bandwidth at a limited number of sampling frequencies, contains a vast amount of information, although in a subtle manner, which is not yet understood through physical models. The measurement instrument has to undergo a learning phase, where it is brought into contact with samples of materials having exactly known properties. The recorded measurement values represent a section of the dielectric spectrum, and a calibration curve can be extracted by applying suitable multivariate statistics. Then unknown materials can be evaluated using that calibration.

The latter approach has been demonstrated to deliver excellent results not only with measurement values recorded in the frequency domain, but with time domain measurements as well. Its utilization is just at the beginning in the radio frequency and microwave domain, and a great potential can be seen for a large number of applications. It offers the opportunity to apply and take advantage of electromagnetic multi-parameter measurements in cases where a concise physical model for a measurement object still does not exist.

For the researcher working in the field, exciting and challenging problems have yet to be solved in order to make multi-parameter electromagnetic measurements work. Benefits will accrue for potential users both from superior manufacturing equipment and the resultant improved products.

Kiel, June 2004   *Reinhard Knöchel*

Preface

Material investigation and moisture measurement using electromagnetic waves in a wide frequency spectrum are useful for quality assessment in many branches of industry, civil engineering, agriculture, and commerce, but also for foodstuffs, e.g., quality detection of meat, fruits, coffee, and so on. Moisture damage to buildings requires restoration expenditure of approximately one billion (10^9) dollars every year in Germany alone.

Electromagnetic Aquametry is a widespread application area of the measurement of water in solids and liquids using methods of frequency domain, time domain, and nuclear magnetic resonance. This special branch addresses all problems relevant to both physical concepts and technological aspects of the practical implementation of electromagnetic measurement techniques for determining the electromagnetic properties of materials, and the amount of water in moist substances.

The research of moist materials and development of measuring devices are absolutely necessary. Because this complex area is too complicated and time consuming for large industry, experts are relatively rare. Their cooperation around the world helps to solve quality problems in the framework of ISO 9000, to use water and energy effectively, and to reduce environmental damages.

Conferences about Electromagnetic Aquametry are titled *"Electromagnetic Wave Interaction with Water and Moist Substances"*. The first workshops were held in Atlanta in 1993, in San Francisco in 1996, in Athens, GA, in April 1999 and were organised by Dr. Kraszewski. The conferences, and also the foundation of the *International Society of Electromagnetic Aquametry (ISEMA)* should help to support the cooperation of scientists and users of material and moisture measurements by using microwave and dielectric measuring methods.

The Material Research and Testing Institute (MFPA) in Weimar was host of the *Fourth International Conference on Electromagnetic Wave Interaction with Water and Moist Substances* in May 2001. This conference was intended as a forum of theory and practice, where more than 130 scientists, manufacturers, and users from 24 countries met each other in order to exchange information, and the latest findings, to offer solutions and devices of research, development, and application, and to establish contacts as well as acquaintances for the future.

The newest results of research and development in Electromagnetic Aquametry were presented in 70 contributions. The contributors came from 24 countries: Australia, Belarus, Brazil, Canada, China, Czech Republic, Denmark, Spain, France, Germany, Greece, Hungary, India, Israel, Italy, Japan, Korea, Malaysia, New Zealand, Norway, Russia, The Netherlands, UK, and USA.

An exhibition accompanied the scientific conference, which demonstrated the basis for practical applications. Exhibitors came from Belarus, New Zealand, USA, and Germany.

The measuring instruments presented in oral presentations, posters, and exhibits operated in the frequency or time domains. The whole frequency spectrum from

0.01 Hz – 300 GHz was used for dielectric measurement methods, but the micro-wave range up to 10 GHz dominated. Multiparameter methods as well as methods of principal components and artificial neural networks were used to carry out density independent measurements. Increasingly the imaging of moisture distributions were realised by using time domain methods, ground penetrating radar and microwave instruments connected with various methods of data processing.

The interdisciplinary contributions of the conference covered 21 fields such as physics, technology, electrotechnical engineering, biological and forest engineering, environmental physics, food engineering, civil-, geotechnical-, geological-, and geophysical engineering, oil-, coal-, and mining industry, hydrology, chemistry and biochemistry, medicine, and pharmacy.

The participating authors work at 17 different institutes e.g. for horticulture, soils and bio-meteorology, computing and biometrics, biomaterials, agriculture, meteorology and climate research, material research, geo-ecological research, bio-diversity and ecosystems dynamics, physical geography and soil science, soil and rock mechanics, agrophysics, building and climatology, and non-destructive testing.

The ISEMA is glad to get support and cooperation from so many disciplines. A new society of international scientists, engineers, and manufacturers should not be limited to moisture measurement alone. Investigations of electromagnetic material properties, cancer detection in medicine and in veterinary medicine, water and air pollution, determination of physical properties such as, density, mass, consistency, concentration of composites, quality, authenticity of foodstuffs, and standardization of these measurement methods present a very wide work area.

The editor of the book and Springer Verlag, represented by Dr. Merkle, agreed to publish a collection of actual papers given at the conference which were selected and revised to represent the state of the art. The book contains 21 chapters written by well-known experts in the field of Electromagnetic Aquametry from Finland, France, Germany, Greece, Hungary, New Zealand, Norway, Russia, The Netherlands, UK, and USA.

The scope of the book covers all aspects of Electromagnetic Aquametry, which summarizes the broad area of metrology, including science and technology, applied in electromagnetic sensing of moist materials, foods, and other dielectrics. The book is divided into five parts.

The first part is devoted to the physical properties of water in various degrees of binding in moist substances in the electromagnetic field promoted by model systems.

The second part contains measurement methods and sensors in the frequency domain. It covers the presentation of different methods of density independent measurements, resonator sensors, microstrip transmission- and reflection-type sensors in different applications of industry. The use of resonators for water content determination in oil but also the application of small band frequency domain spectroscopy in the low RF range will be demonstrated during the on-line process monitoring in soil, and concrete strength.

The third part contains TDR-techniques for environmental problems, such as the measurement of water in waste deposits, the moisture content determination in soil and snow to prevent mud streams, and avalanches, and the estimation of filling of artificial lakes and the power-generation, as well as the prevention of water ingress in waste deposits which are located in salt mines.

With the introduction of the High Frequncy Structure Simulator HFSS, complete new applications were introduced for the development and optimization of sensors and their calibration.

Methods and sensors for quality assessment of biological substances, such as grain, palm oil, timber and foodstuffs are included in the fourth part.

The nuclear magnetic resonance shown in part five will be applied not only for foods and pharmaceuticals, but also for brickwork, for density and moisture content determination in wood panels, and for monitoring concrete hardening.

I want to thank all the authors, the MFPA Weimar, and Springer Verlag for their cooperation during the editing and revision of this book. I thank the company IMKO for the cover picture.

I greatly appreciate the help of Prof. Kummer and my daughter Heike as well as the extensive work done by my wife Helga in making corrections and reproducing many figures.

Weimar, June 2004   *Klaus Kupfer*

Table of Contents

Measurement Methods and Sensors in Frequency Domain

Measurement Methods and Sensors in Time Domain

Application of Nuclear Magnetic Resonance

List of Abbreviations

Abbreviation	Explanation	Page
ADC	analog to digital converter	281
AIA	active integrated antenna	244, 250
ANN	artificial neural network	256, 455
A-Φ	attenuation-phase shift	151, 152, 155
BPSK	binary phase shift keying	247
CBCPW	Conductor-Backed Coplanar Wave-guide	483
CFR	Cylindrical Fin Resonator	229
CW	continuous wave	249
DDS	direct digital synthesis	339
DMA	dynamic mechanical analysis	51
DRC	**Dry Rubber Content**	467
DRS	dielectric relaxation spectroscopy	40,42
DS	dielectric spectroscopy	39, 42
DSC	differential scanning calorimetry	41
DSP	digital signal processor	249
DWCM	Downhole WaterCut Meter	238
EC	electrical conductivity	278
EMA	effective medium approximation	88
EMC	environmental moisture content	485
EMP	evanescent microwave probe	243
FD	frequency domain method	278
FDR	frequency domain reflectometry	278
FDS	frequency domain spectroscopy	278
FFA	free fatty acid	467
FID	free induction decay, abbreviated to	494
FMCW	frequency-modulated continuous wave	407
FRF	frequency response function	389
FSA	feedback self-oscillating amplifier method	240
FSK	frequency shift keying	246
HFSS	High Frequency Structure Simulator	262
HN	Havriliak–Negami	43
IMA	integrated microstrip antenna	251
IPNs	interpenetrating polymer networks	47, 58
I-Q	in-phase, quadrature phase	253, 254
IRF	impulse response function	387
ISM	industrial-, scientific-, medical-	249
LPF	low pass filter	247, 281
MBT	modulated backscatter technology	249
MC	moisture content	249, 467
MDF	medium density fibreboard	126
MiMo	multiple-input – multiple-output	386
MLFF	multi-layer feed forward	455
MLR	multiple linear regression	447
MMM	microwave moisture measurement	243
MRI	magnetic resonance imaging	491

Abbreviation	Explanation	Page
MSA	microstrip antenna	249
MUT	material under test	217
MW	microwave	139
MWS	Maxwell–Wagner–Sillars	45, 55
NDT	non-destructive testing	245
NMR	nuclear magnetic resonance	41, 491
OOK	on-off keying (modulation)	247
OSA-NMR	One-Sided Access NMR	498
PALS	positron annihilation lifetime spectroscopy	41
PAs	polyamides	51
PBG	photonic band gap (material)	250
PCA	principal components analysis	452
PCR	principal components regression	452
PDB	passive detector/backscatter	247
PEO	poly(ethylene oxide)	40, 49
PHEA	poly(hydroxyethyl acrylate)	40, 47
PHS	phase shifter	248
PLSR	Partial least squares regression	453
PSF	point spread function	258
PUs	polyurethanes	41
RAO	reflection-,aperiodic-, open-	248
RCS	radar cross-section	247
RF	radio-frequency	169,419
FRF	frequency response function	389
RMS	root mean square value	395
RMSA	rectangular microstrip antenna	245
$RMSE_c$	root mean square error calibration group	448,452
$RMSE_v$	root mean square error validation	448,452
RX	microwave receiver	249
SiSo sensors	single-input – single-output	410, 411
SMA	sub-miniature, A-type (connector)	252
SMP	micro-machined patch (antenna)	250
SNR	signal-to-noise ratio	394
SRF	step response function	389
TDS	time-domain spectroscopy/ spectrometry	21,42,277
TDR	time-domain reflectometry	42, 277,317
TE	transverse electric	222
TEM	transverse electromagnetic	173
TFR	time – frequency representation	414
TM	transverse magnetic	220
TSC	Total Solid Content	467
TSDC	thermally stimulated depolarisation currents	42, 45
TX	microwave transmitter	249
UWB	ultra-wideband	383
VCO	voltage controlled oscillator	397
WSA	water, solid, air	87

1 Recent Developments in Electromagnetic Aquametry

Andrzej W. Kraszewski

Balion-Milopotamos, 74057 Panormos, Crete; Greece

1.1 Introduction

The term "Aquametry" is used here as a synonym of: "measurement of moisture content in solid and liquid materials," analogous to "hygrometry" which is a well established branch of metrology devoted to "measurement of water vapor content in gases, mainly in air." The adjectives "microwave" or "electromagnetic" in the title of this book indicates that it will be concerned with moisture content measurement of solids and liquids using electromagnetic methods and instrumentation derived from classical microwave techniques (resonant cavities, waveguide, transmission line, free-space measurements). The subject of interest for microwave aquametry is searching solids of different form and structure, as well as liquids containing water, for identification of their properties when placed in electromagnetic fields of radio and microwave frequencies (attenuation, reflection, phase angle, shift of resonant frequency, etc.). In this aspect electromagnetic aquametry utilizes some physical theories on dielectric mixtures and bound water. But electromagnetic aquametry has also strictly defined practical objectives, namely quantitative measurements of water content in materials, which are important from an economic point of view. Since water occurs in most materials in nature as a natural component of the material or is introduced during technological processes, it is quite obvious that measurement and control of moisture content have great economic and technical importance. The domination of practical aims over cognitive purposes, influenced the development of microwave aquametry in the past and also had a serious impact on its present state. The purpose of this chapter is to present in some detail the recent developments in the field, and to provide a sketch of the overall field of electromagnetic aquametry without delving too deeply into the more complicated components.

Typical nondestructive techniques for determining moisture content in material consist of measuring the electrical properties of the material in a sample holder and relating these properties to the moisture content. These techniques have their roots at the beginning of the twentieth century when the possibility of rapid determination of moisture content in grain by measuring the dc resistance between two metal electrodes inserted into the grain sample was established [1]. This resistance was found to vary with moisture content. Later, samples of wet materials were placed in the path of an electromagnetic wave between two horn antennas and the simple relationship between the propagation constant and the amount of water was easily determined. Both the simplicity of the measuring arrangement and practical utility of the results were fascinating. Because of the particular properties of microwave radiation, (frequencies between 1 and 100 GHz), the new method appeared to surpass all other previous methods for measuring moisture content in solids, such as chemical methods, methods using radiofrequencies (several MHz), infrared and ionizing radiation. The following advantages were obvious since the early experiments:

(a) Contrary to lower frequencies, the dc conductivity effects on material properties can be neglected.

(b) Penetration depth is much larger than that of infrared radiation and permits the probing of a significant volume of material being transported on a conveyor or in a pipe.

(c) Physical contact between the equipment and the material under test is not required, allowing on-line continuous and remote moisture sensing.

(d) In contrast to infrared radiation, it is relatively insensitive to environmental conditions, thus dust and water vapor in industrial facilities do not affect the measurement.

(e) In contrast to ionizing radiation, microwave methods are much safer and faster.

(f) Water reacts specifically with certain frequencies in the microwave region (relaxation) allowing even small amounts of water to be detected.

(g) Contrary to chemical methods, it does not alter or contaminate the test material, thus the measurement is nondestructive.

These features combined with great potential savings in fuel, energy, manpower and improvement of the quality of products resulting from the application of moisture content measurement and control, created a powerful incentive for research and equipment development all over the world.

In the mid-sixties several manufacturers of microwave moisture meters were established on the market. Among them were Scanpro AB in Sweden. AEI and Rank Precision Industries Ltd. in Great Britain, Uniplan/Wilmer in Poland, Kay-Ray in the U.S. (who expanded its line of nuclear radiation meters to include microwave instruments), and Compur AG in Germany (who produced meters based on research by Bayer AG). There were new companies being created and old ones bought by others; successful projects were developed and others closed and forgotten; but fascination with the potential of the technology has lasted with varying intensity to this day. The state of knowledge on the subject was summarized several times during the years [2-6]. The first meeting devoted to the exchange of ideas on the subject took place in 1980. The list of papers published at that time exceeded 400 [5]. Professional meetings took place more frequently in the late eighties and a tradition of annual meetings (Feuchtetage) was established in Germany [7, 8]. Later another meeting of more international character was established, namely the IEEE International Microwave Symposium Workshop on Electromagnetic Wave Interaction with Water and Moist Substances in 1993 in

Atlanta [9], in 1996 in San Francisco [10, 12], in 1999 in Athens, Georgia [11], and recently as the International Conferences on Electromagnetic Wave Interaction with Water and Moist Substances, organized by the International Society for Electromagnetic Aquametry (ISEMA) in Weimar, Germany, in 2001 [13] and Rotorua, New Zealand in 2003 [14].

The total number of microwave moisture meters manufactured during the last forty years throughout the world is unknown. The total investment in research on the adaptation of microwave techniques to aquametric purposes and the number of unsuccessful projects also remain unknown. But the bibliography of the subject, on both physical background and practical application contains well over one thousand entries [5, 8, 9, 12] and it does not cover internal reports of proprietary character nor contributions to closed or semiclosed conferences and seminars (the full text of which were not published in the generally accessible literature). A recent survey indicated the existence of over thirty companies involved in manufacturing and applying moisture meters based on the measurement of microwave parameters [12]. Eleven countries on three continents are represented and materials involved extend from grain and soil to living fish. This list provides evidence that microwave aquametry is not only a subject of academic discussions and dissertations, but also an accepted tool in the field of non-destructive moisture monitoring and control in modern factories and laboratories.

1.2 Principles and Definitions

The moisture content of material may be defined on a wet basis (w.b.) as a ratio of the mass of water, m_w, to the mass of the moist material, m_m,

$$\xi = \frac{m_w}{m_m} = \frac{m_w}{m_w + m_d} \tag{1.1}$$

or, on a dry basis (d.b.), as a ratio of the mass of water in the material to the mass of dry material, m_d,

$$\eta = \frac{m_w}{m_d} = \frac{m_m - m_d}{m_d} \tag{1.2}$$

Most often the quantities ξ and η are expressed in percentage. The definition expressed by Eq. (1.1) is most frequently used in practice, and when the concept of moisture content is related to a certain volume of material, v, it can be rewritten as follows:

$$\xi = \frac{m_w / v}{m_w / v + m_d / v} = \frac{k}{k + g} = \frac{k}{\rho} \tag{1.3}$$

where k is the partial density of water, g is the partial density of dry material, and ρ is the density of moist material. Other relationships resulting from Eqs. (1.1-3) are:

$$\xi = \frac{\eta}{1+\eta}, \qquad \eta = \frac{\xi}{1-\xi}, \qquad k = \frac{m_w}{\upsilon} = \rho\xi, \tag{1.4}$$

$$g = \frac{m_d}{\upsilon} = \rho(1-\xi) = \frac{\rho}{1+\eta}$$

There are many parameters of materials that can be correlated with the density of water in the material, k, but from Eq. (1.3) it is obvious that fluctuations in the material density, ρ, have as much influence on moisture content as the variation in k. This observation is universal, because this disturbing effect of material density does not depend on the electrical method applied for moisture content determination. Thus, when k is determined from electrical measurement, determination of moisture content from Eq. (1.1) or Eq. (1.3) requires that ρ be known. This information can be obtained by keeping the mass of moist material in the measuring space constant during the calibration as well as during the measuring procedure; or by performing separate density measurements, for example by weighing a sample of given volume, or by using γ-ray density gauge. A third approach is to use a *density-independent* function, an expression relating the moisture content with electrical properties of the material independent of density and to eliminate, or seriously limit, the density effect in moisture content measurements. It might be interesting to note that in many cases the dry basis moisture content is linearly related to the measured electromagnetic quantities, while the wet basis moisture content exhibits quite nonlinear relationship. In such cases it is wise to calibrate the system for dry basis moisture content and then transform the results of measurement to the moisture content on the wet basis required in many branches of industry.

Standard methods of moisture content determination are *direct* methods, based on the definitions of Eq. (1.1) or Eq. (1.2) and performed in laboratories according to procedures described in formal documents of national or international character. The most often used method involves weighing a sample of moist material, removing water by evaporation and reweighing the remaining dry material; another (the Karl Fischer method) uses extraction and chemical titration. The whole procedure is precisely described, giving time and temperature of drying, exact amount of chemicals to be used, etc. These methods are accurate but do not provide rapid results. Drying for up to three days is required in some cases. For rapid moisture content determination and monitoring, indirect methods calibrated against the standard methods have been used, and the method using measurement of microwave properties of moist material is one of them.

Interaction of an electromagnetic wave with moist material can be expressed in terms of the complex value of the propagation constant of the wave in a dielectric medium as

$$\gamma = \alpha + j\beta = j\frac{2\pi}{\lambda}\sqrt{\varepsilon - p} \tag{1.5}$$

where $\varepsilon = \varepsilon' - j\varepsilon''$ is the relative permittivity of the medium, where ε' is the dielectric constant and ε'' is the loss factor, and $p=(\lambda/\lambda_c)^2$, where λ and λ_c denote free-space and waveguide cut-off wavelengths, respectively. Eq. (1.5) may be solved for two components of the propagation constant being expressed as:

$$\alpha = \frac{2\pi}{\lambda}\sqrt{\frac{\varepsilon'-p}{2}\sqrt{1+\left(\frac{\varepsilon''}{\varepsilon'-p}\right)^2}-1} \qquad [\text{Np/m}] \qquad (1.6)$$

for the attenuation constant

and for the phase constant

$$\beta = \frac{2\pi}{\lambda}\sqrt{\frac{\varepsilon'-p}{2}\sqrt{1+\left(\frac{\varepsilon''}{\varepsilon'-p}\right)^2}+1} \qquad [\text{rad/m}]. \qquad (1.7)$$

In free space, where $p = 0$, the following approximate expressions can be used to relate the electromagnetic wave propagation to the properties of moist materials, assuming that $\varepsilon'^2 \gg \varepsilon''^2$ which is valid in most practical situations,

$$\alpha \cong \frac{\pi}{\lambda}\frac{\varepsilon''}{\sqrt{\varepsilon'}}, \qquad (1.8) \qquad \beta \cong \frac{2\pi}{\lambda}\sqrt{\varepsilon'} \qquad \text{and} \qquad (1.9)$$

$$|\Gamma| \cong \frac{\sqrt{\varepsilon'}-1}{\sqrt{\varepsilon'}+1} \qquad (1.10)$$

where Γ is the voltage reflection coefficient from the surface of the moist material. Thus, by measuring the more practical quantities,

$$A = 20\log|\tau| = 8{,}68\,\alpha d \qquad [\text{dB}] \qquad (1.11)$$

and

$$\phi = (\beta - \beta_o)d = \frac{2\pi}{\lambda}\left(\sqrt{\varepsilon'}-1\right)+360n \qquad [\text{deg}] \qquad (1.12)$$

where A is the attenuation of the material sample in decibels and ϕ is the phase shift in degrees; β_0 is the phase constant in free space; n is an integer to be determined when the thickness d of the material layer is greater than the wavelength in the material, and the transmission coefficient $|\tau| = \exp(-\alpha d)$. The integer n can be found by repeating the measurement with samples of different thickness or by taking the measurements at two frequencies [15]. It is clear from the above that the parameters of the electromagnetic wave are affected by the material relative permittivity which in turn is related to the water content in the material. It is true, however, that the relative permittivity also depends on material temperature, density, shape and dimensions of its particles, chemical composition, etc. [16]. This is where the real troubles start and the science of aquametry begins.

1.3 Instrumentation

All instruments manufactured recently contain modern microwave integrated circuitry, high speed signal processors and efficient power supplies, and are equipped with modern microcomputers (often operating on the Windows platform), with modems, high capacity data storage and other gadgets typical of modern measuring instruments. Because of general progress in microwave integrated circuit (IC) technology during recent years, the price of microwave components has recently been quite comparable with the price of components (mixers, amplifiers, oscillators, filters, etc.) operating at much lower frequencies. This is another advantage, since very often in the past, application of microwave meters was restrained because of their higher costs. In addition to the progress which has been typical for other measuring instruments, there are certain developments and recent enhancements specific for microwave moisture meters. Some of them are briefly reviewed below.

1.3.1 Metrological Enhancements

It may be observed from Eqs. (1.8, 1.9) that the components of the propagation constant, α and β, are dependent upon the relative permittivity of the moist material. Since the relative permittivity in turn depends on moisture content ξ, density ρ, and temperature τ, the components can be written in a general form:

$$\alpha = \psi_1(\xi,\rho,T) \quad \text{and} \quad \beta = \psi_2(\xi,\rho,T) \tag{1.13}$$

According to the definition of Eq. (1.3) and expression in Eqs. (1.11, 1.12) and Eq. (1.13), the components of the propagation constant can be easily expressed in terms of measured variables A and ϕ, as

$$A = \Phi_1(k,g,T) \quad \text{and} \quad \phi = \Phi_2(k,g,T) \tag{1.14}$$

These two equations can be solved to express the partial densities of water and dry material in terms of measured variables:

$$k = \Psi_1(A,\phi,T) \quad \text{and} \quad g = \Psi_2(A,\phi,T) \tag{1.15}$$

In general, this operation known as an *inverse problem* can be very complex and uncertain, but in the case of moisture content in most materials, it can be quite simple. Thus, the moisture content can now be expressed as:

$$\xi = \frac{\Psi_1(\alpha,\phi,T)}{\Psi_1(A,\phi,T)+\Psi_2(A,\phi,T)} \qquad (1.16)$$

which contains only the wave variables, A and ϕ, and temperature T, determined experimentally. Also the density of the wet material

$$\rho = \Psi_1(A,\phi,T)+\Psi_2(A,\phi,T) \qquad (1.17)$$

can be determined at the same time. Thus, the density of moist material is no longer a disturbing factor in the moisture content measurement, but it can be determined during this measurement and used for other purposes in a technological process. Identification of the relationships in Eqs. (1.16) and (1.17) is called the *calibration* of the measuring system. It has been suggested recently [17] that carrying out the measurements of two wave parameters at two different frequencies should allow determination of *four variables*, for example, moisture content ξ, bulk density ρ, temperature of the material T and material layer thickness d. Selection of two appropriate frequencies remains to be determined for any given material.

In a search for more efficient and accurate ways of instrument calibration, other approaches were also explored. Artificial neutral networks seem to be especially useful [18]. An artificial neutral network is a collection of simple interconnected analog signal processors, providing a mathematical structure that can be trained to map a set of inputs to a set of outputs. The inputs are the measured data of A and ϕ, and the output is the value of moisture content, ξ. For experimental data taken in free space at eight frequencies between 10.3 and 18,0 GHz for wheat in the moisture content range from 10% to 19% and at temperatures between -1°C and 42°C [19], the standard error of calibration was 0.135% moisture. When the network was trained using only the amplitude of the transmission coefficient measurements as the inputs, the value increased to 0.219%. This is an important observation, because eliminating the need for the phase measurements greatly reduces the complexity of the hardware required to make the measurements. Application of principal component analysis [20] and partial least squares regression [21] to the same set of experimental data provided standard errors of performance, of 0.232% and 0.210%, respectively. One should keep in mind, however, that the average spread of moisture content in triplicate 10-gram samples determined by the standard oven method for wheat (130°C for 19 hr) was 0.176% moisture with standard deviation of 0.077% moisture [22].

For some time now, the concept of a density-independent function has been considered a vital way of limiting the density effect in moisture content measurement [23]. The idea is to find an expression or function, X, which can be correlated with moisture content, ξ, in a form

$$\xi = a_1 + bX \qquad \text{or} \qquad \xi = a_2 + b_2\sqrt{X} + b_3 X \qquad (1.18)$$

providing elimination or serious limitation of the density effect. Originally, the ratio of the two measured quantities, A and ϕ, was considered to be such a function. Later, a permittivity related function was proposed [24, 25] in the form

$$X = \varepsilon'' / (\varepsilon' - 1) \tag{1.19}$$

It has been shown [26], that Eq. (1.19) is a part of the original ratio which in turn is a linear function of the material loss tangent and can be expressed as

$$A / \phi = c \tan \delta \sqrt{\varepsilon'} / \left(\sqrt{\varepsilon'} - 1 \right) \tag{1.20}$$

where $\tan \delta = \varepsilon'' / \varepsilon'$, and c is a constant. Because both measured variables of Eqs. (1.11, 1.12) are directly proportional to the layer thickness, the density-independent function. Eq. (1.16), can be correlated with the material moisture content without regard to fluctuations in the material layer thickness d. This is often a valuable feature.

A recently established density-independent function [27, 28] is based on the observation that in the complex plane, the normalized variables ε'/ρ and ε''/ρ for all temperatures and moisture contents can be expressed by the linear equation

$$\varepsilon'' / \rho = a_f \left(\varepsilon' / \rho - b_o \right) \tag{1.21}$$

where a_f is the slope of the line, which depends only upon the operating frequency, and b_0 is the intercept constant, which, for a given material, has the same value at all frequencies and corresponds to the density-normalized zero-moisture material permittivity or to the density-normalized permittivity of the material at very low temperature. Graphical representation of Eq. (1.21) is sometimes called an Argand diagram. The density of the material can be calculated from Eq. (1.21); thus, the procedure allows simultaneous determination of the material density and moisture content. The density-independent function for moisture determination can be written as

$$X = \sqrt{\frac{\tan \delta}{\rho}} = \sqrt{\frac{a_f b_o \tan \delta}{a_f \varepsilon' - \varepsilon''}} \tag{1.22}$$

The interesting feature of the above relationship is that as more and more experimental data for grain have become available (for corn, soybeans, oats, etc.), all variables in Eq. (1.22) have been found to have similar values, and since at a given frequency $a_f b_0$ is a constant, one can consider it a *universal* function. It must be stressed that the mentioned commodities have pronounced differences in kernel dimensions, shape, bulk densities and composition, and the potential for a common calibration equation for moisture content determination in all of them should motivate further research.

1.3.2 New Sensors and Transducers

There have been recent developments in microwave moisture sensing devices related to progress achieved in flat, microstrip, patch antennas [29-35], as well as to wide application of microwave resonators (cavity and microstrip) [36-49]. Sensors based on the principle of the reflected wave measurements are easy to use as they allow one-sided sensing and their robust construction permits operation at the bottom of concrete mixers, in walls of chutes, etc. Several open-ended sensors are shown in Figure 1.1 together with their simplified equivalent circuit.

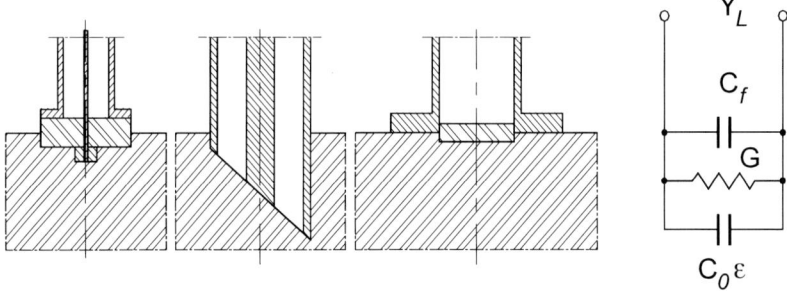

Fig. 1.1. Various open-ended transmission line structures used as sensors in microwave moisture content measurements. From left, microstrip line, coaxial line, cylindrical wave guide and their simple equivalent circuit.

Theory related to the operation of the conical-tip open-ended probe was presented [39], as well as a comprehensive study of various probe types [40]. The input admittance of the sensor is related to the permittivity of the material in which the line is immersed, and, in turn, the reflection coefficient is a function of the admittance expressed as:

$$\Gamma = \frac{1 - Y_L}{1 + Y_L} \quad \text{where} \quad Y_L = G + j\omega\left(C_f + \varepsilon C_0\right) \tag{1.23}$$

and G is the conductance and C_f and C_0 are capacitances as shown in Figure 1.1. One example of a reflection sensor is a microstrip open-ended probe used for moisture sensing in the production of curd cheese [41]. Changes of moisture content produce changes in the resonant frequency, as shown in Figure 1.2, where the reflection loss (Γ expressed in decibels as $R_L = -20\log |\Gamma|$) is presented as a function of frequency for various moisture contents in cheese.

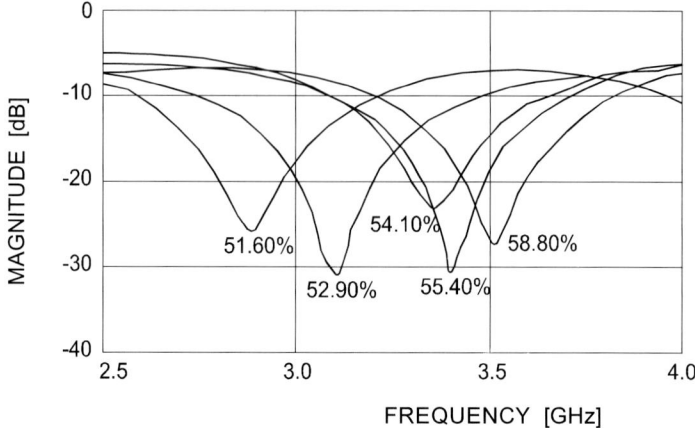

Fig. 1.2. Return loss of open microstrip sensor immersed in curd cheese, as a function of frequency for various curd moisture content [41]

Sensitivity of moisture measurement can be significantly increased by using resonant structures as sensors. A microwave resonator is a metallic chamber resonating when the operating wavelength exactly matches its dimensions. Inserting a dielectric object into the cavity changes its electrical dimensions and the change can be correlated with the object permittivity and then with its moisture content. This principle has been used for moisture and mass determination of single kernels and seeds [42], as well as in bulk materials up to 5 liters in volume (sugar, pharmaceutical products, grain, cigarettes, margarine, etc.) measured in laboratory and industrial conditions [43-49]. A simple cylindrical resonator coupled through loops with external coaxial lines is shown in Figure 1.3.

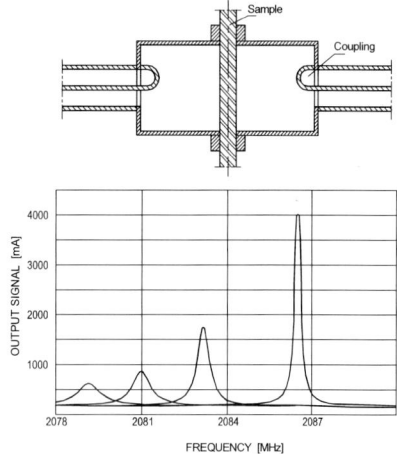

Fig. 1.3. Cylindrical resonator and several resonant curves with amplitude and frequency decreasing with high moisture content [45]

The output signal vs. frequency has the shape of a resonant curve, magnitude of which decreases for increasing moisture content in the material. The potential accuracy of frequency measurement has been used advantageously by coupling a pipe conducting liquid material (e.g., crude oil) into a microwave oscillator circuit in such a way that variation in water content changes the operating frequency of the system [49, 50].

Some more interesting concepts should be mentioned here. First, the study on retrieval of the moisture profile in materials of non-uniform moisture distribution [51-53]. In many cases the interesting material property is not only a global moisture content, but also its distribution along one or even two axises of the material. Another concept is based on using the effect of the relative permittivity of the medium on the cut-off wavelength of a waveguide (see Eq. (1.5)). For rectangular waveguide this can be written as $\lambda_c = c/2a(\varepsilon')^{1/2}$, where c is the speed of light and a the wider dimension of the waveguide [54]. Thus, when wet material is flowing through the waveguide, its moisture can be correlated to the attenuation measured along the waveguide. The other concept is applying time-domain reflectometry (TDR), the technique well established for *in situ* determination of water content in soil with two-conductor line, to other granular materials such as grain [55]. These convenient two-wire sensors in combination with sharp pulse generating circuitry and precise time measuring devices can be a valuable complement to the family of high-frequency and microwave moisture meters.

1.4 Summary

This review does not claim to be complete in covering all problems facing microwave aquametry. It was the intention of the author to show that microwave aquametry is a dynamic developing branch of metrology, full of practical potential and many needs for research in various fields of physics, chemistry and metrology. The truth is that a successful development requires not only an adequate microwave technology, but also specific knowledge concerning the material of interest. While the microwave performance of a sensor can be expressed in terms of accuracy of amplitude and phase measurement, predicting the final accuracy in terms of moisture content dramatically depends on the properties of the material to be monitored and the particular details of the technological process. As a consequence, there is significant work to be done before making an effective microwave sensor for a given application.

References

1. Briggs LJ (1908) An electrical resistance method for the rapid determination of the moisture content in grain. Bureau Plant Industry Circular No. 20, U.S. Department of Agriculture.
2. Watson A (1965) Measurement and control of moisture content by microwave absorption. In: Wexler A (ed) Humidity and Moisture vol 4, Winn PN (ed) pp. 87-93
3. Kraszewski A (1973) Microwave instrumentation for moisture content measurement. Jour Microwave Power 8: 323-336
4. Benzar VK (1974) Microwave techniques of moisture content measurement. Izdat. Vysheyshaya Shkola (University Publishers): Minsk (In Russian)
5. Kraszewski A (ed) (1980) Microwave Aquametry. Jour Microwave Power 15: 207-310
6. Pyper JW, Buettner HM, Cerjan CJ, Hallam JS, King RJ (1985) The measurement of bound and free moisture in organic materials by microwave methods. In: International symposium on moisture and humidity, Washington DC, pp 909-917
7. Kupfer K (ed) (1997) Material moisture measurements (in German). Expert Verlag, Renningen-Malmsheim
8. Kupfer K (ed) (1997) Proceedings of 9th meeting on moisture measurement (in German: 9 Feuchtetag), MFPA University of Weimar, Weimar
9. Kraszewski A (ed) (1996) Microwave aquametry. IEEE Press, Piscataway MJ
10. Kraszewski A (ed) (1996) Workshop on electromagnetic wave interaction with water and moist substances - summaries of papers. IEEE Internatl. Microwave Symp., San Francisco CA
11. Kraszewski A, Lawrence KC (eds) (1999) Collection of papers for the third workshop on electromagnetic wave interaction with water and moist substances. U.S. Dept. Agriculture, Athens GA
12. Kupfer K, Kraszewski A, Knochel R (eds) (2000) RF & microwave sensing of moist materials. Sensor updates 7, Wiley VCH Verlag, Weinheim
13. Kupfer K (ed) (2001) Proceedings of the 4th international conference on electromagnetic wave interaction with water and moist substances. MFPA an der Bauhaus-Uni Weimar
14. Thakur K (ed) (2003) Proceedings of the 5th international conference on electromagnetic wave interaction with water and moist substances. Industrial Research Ltd, Auckland, New Zealand
15. Trabelsi S, Kraszewski A, Nelson SO (1997) Phase-shift ambiguity in microwave dielecrtic properties measurements. IEEE Trans Instrum Meas 49: 56-60
16. Nelson SO (1981) Review of factors influencing the dielectric properties of cereal grains. Cereal Chem 58: 487-492
17. Kraszewski A, Trabelsi S, Nelson SO (1997) Moisture content determination in grain by measuring microwave parameters. Meas Sci Technol 8: 857-863. Also (1998): Addendum, *ibidem* 9: 543-544
18. Bartley PG, McClendon RW, Nelson SO, Trabelsi S (1998) Determining moisture content of wheat with an artificial network from microwave transmission measurements. IEEE Trans Instrum Meas 47: 123-125
19. Kraszewski AW, Trabelsi S, Nelson SO (1996) Wheat permittivity measurement in free space. Jour Microwave Power & EE 31: 135-141
20. Archibald DD, Trabelsi S, Kraszewski AW, Nelson SO (1998) Regression analysis of microwave spectra for temperatura-compensated and density-independent determination of wheat moisture content, Appl Spectroscopy 52: 1435-1446

21. Ben Slina M, Morawski RZ, Kraszewski AW, Barwicz A, Nelson SO (1999) Calibration of micowave system for measuring grain moisture content. IEEE Trans Instrum Meas 48: 778-782
22. Kraszewski AW, Trabelsi S, Nelson SO (1998) Simple grain moisture content determination from microwave measurements. Trans Am Soc Agric Engrs 41: 129-134
23. Kraszewski A, Kulinski S (1976) An improved microwave method of moisture content measurement and control. IEEE Trans Indust Electron and Control Instrum IECI 23: 364-370
24. Meyer W, Schilz W (1980) A microwave method of density independent determination of moisture content in solids. Jour Phys D 13: 1823-1830
25. Kraszewski AW, Trabelsi S, Nelson SO (1998) Comparison of density independent expressions for moisture content determination in wheat at microwave frequencies. Jour Agric Engng Research 71:227-237
26. Trabelsi S, Kraszewski AW, Nelson SO (1997) Simultaneous determination of density and water content in particulate materials by microwave sensors. Electronics Letters 33: 874-876
27. Trabelsi S, Kraszewski AW, Nelson SO (1999) Unified calibration method for nondestructive dielectric sensing of moisture content in granular materials. Electronics Letters 35: 1346-1347
28. Trabelsi S, Kraszewski AW, Nelson SO (2001) Optimizing universal calibration for industrial microwave moisture sensor. In [13]: 117-124
29. Pozar DM, Schaubert DH (1995) Microstrip antennas. IEEE Press, Piscataway NJ
30. Volgyi F (1993) Microstrip antenna array application for microwave heating. In: Proceedings 23rd European microwave conference, Madrid: 412-415
31. Volgyi F (2000) Monitoring of particleboard production using microwave sensors. In [12]: 249-274
32. Volgyi F (2001) Microstrip sensors used in microwave aquametry. In [13]: 135-142
33. Volgyi F, Shephardson R, Burrows J (2003) New microwave moisture sensors for use in bins of raw materials and in concrete mixers. In [14]: 138-145
34. Daschner F, Knoechel R (2003) A new transmission-line sensor for measuring the composition of foodstuffs using microwaves. In [14]: 24-31
35. Okamura S, Zhang Y (2003) High moisture content measurement using microstrip transmission line. In [14]: 85-90
36. King RJ (1992). Microwave sensors in process control. Part II: Open resonator sensors. Sensors 9:(10): 25-30
37. Knoechel R (2000) Technology and signal processing of dielectrometric microwave sensors for industrial applications. In [12]: 65-105
38. King RJ (2000) On-line industrial applications of microwave moisture sensors. In [12]: 109-170
39. Keam RB, Holdem JR (1997) Permittivity measurement using coaxial-line conical-tip probe. Electronics Letters 33: 353-355
40. Kim SW, Cho YS, Huyn YS, Kim SY (2001) A comparative study on the stability of four different conversion models of the open-ended coaxial probes. In [13]: 185-192
41. Ball JAR, Horsfield B, Holdem JR, Keam RB, Holmes WS, Green A (1996) Cheese curd permittivity and moisture content measurement using six-port reflectometer. In: Proceedings Asia-Pacific Microwave Conf, New Delhi, 2: 479-482
42. Kraszewski AW, Nelson SO (1996) Moisture content determination in single kernels and seeds with microwave resonant sensors. In [9]: 177-203
43. Fischer M, Vainikainen P, Nyfors E (1995) Design aspects of stripline resonant sensors for industrial applications. Jour Microwave Power & EE 30: 246-257

44. Sarabandi K, Li ES (1997) Microstrip ring resonators for soil moisture measurements. IEEE Trans Geosci & Remote Sensing 35: 1223-1231
45. Hermann R, Sikora J (1997) Moisture content measuring with microwave resonators. In [8]: 291-310
46. Avitabile G, Sottani N, Salvador C, Biffi Gentili G (2001) A slot-coupled ring resonator for moisture measurement in an industrial environment. In [13] 308-315
47. Tsentsiper B (2001) One-sided microwave moisture sensors. In [13]: 156-164
48. Gallone G, Lucardesi P, Martinelli M, Rolla PA (1996) A fast and precise method for measurement of the dielectric permittivity at microwave frequencies. Jour Microwave Power & EE 31: 158-164
49. Nyfors E (2001) Permanent downhole microwave sensor for the local measurement of the water content of the fluid being produced in an oil well. In [13]: 293-300
50. Scott BN, Cregger BB, Shortes SR (1993) Technology for full-range water-cut measurements. In: Proc 25th Offshore Technology Conf, Houston TX: 279-286
51. Glay D, Lasri T, Mamouni A, Leroy Y (2001) Free space moisture profile measurement. In [13]: 235-242
52. Glay D, Lasri T (2003) Microwave sensing of moisture profiles in layered materials. In [14]: 71-78
53. Thakur K, Chan KL, Holmes WS (2003) Microwave measurement of layered dielectric from microwave reflection spectroscopy using an inverse technique. In [14]: 118-128
54. Jean B (2000) Guided microwave spectroscopy for on-line moisture measurement of flowable materials. In [12]: 171-184
55. Stacheder M, Blume P, Fudinger R, Koehler K (1999) Grain moisture measurements with time domain reflectometry. In [11]: 133-137

Dielectric Properties of Water and Moist Substances

2 Electromagnetic Wave Interactions with Water and Aqueous Solutions

Udo Kaatze

Drittes Physikalisches Institut, Bürgerstr. 42-44; D-37073 Göttingen; Germany

2.1 Introduction: Water, the Omnipresent Liquid

Water is the elixir of life on our planet. Molecular processes in the biosphere proceed almost exclusively in aqueous reaction media. The fact that water was present long before the evolution of life on earth suggests that its unique properties have strongly conditioned life as we know it. The water content of an adult human is as high as 65–70%. Generally, the content of water in living organisms ranges from about 96% in some marine invertebrates to somewhat less than 50% in bacterial spores [1]. Water does not just serve as a filling material in biological systems. It promotes the formation of biological structures, enables biological hydrolysis, and acts as a solvent distributing nutrients and removing metabolism products. The multiple functions of water in living organisms is also established by their inability to survive without a minimum supply of water. It is well known, for example, that dehydration of DNA leads to denaturation of this biopolymer.

Due to the ubiquity of water in our environment, this extraordinary chemical plays a key role in a variety of further aspects of human beings. About 95% of water available on our planet is contained in the large oceans, holding $1.3 \cdot 10^{21}$l. The Antarctic ice cap amounts to about 5% of this volume, namely $2.7 \cdot 10^{19}$l. The process of hydrological cycle mainly consists of evaporation from the oceans, subsequent precipitation and drain off back into the oceans. Within this cycle, around $1.3 \cdot 10^{16}$l of water are contained in the lower 11 kilometers of the atmosphere. The annual turnover of water amounts to $3.5 \cdot 10^{17}$l [1], leading to a continuous exposure of geological structures to water. The climate is evidently controlled by the humidity of the air but also by the moisture content of the soil. Consequences for agriculture are obvious.

In addition to the prominent role which water plays in ecological processes, it has many influences on sociological and industrial developments as well. Much industrial production would be impossible without water, like broad fields in chemical technology, power plant operation, flotation and dyeing in textile chemistry. A multitude of other areas of production and of maintenance of products depends sensitively on the water content of material. Examples are efficient oil recovery, the strength of bricks, the consistency of cosmetics, and the storage life of foodstuff. There are thus many reasons for a better understanding of the eccentric properties of water. In view of the widespread and still increasing use of electromagnetic waves, there are particular demands for deeper insights into their interactions with this omnipresent chemical. A brief tutorial on some aspects of electromagnetic wave interactions with aqueous systems is given. Details and references to original articles are presented in recent reviews on the dielectric properties of water [2–5] and aqueous solutions [6–10].

2.2 The Architecture of the Water Molecule and the Unique Hydrogen Network

2.2.1 The Isolated Water Molecule

The unusual properties of water and its multiple functions in the biosphere and in technology are related to the architecture of the H_2O molecule. As sketched in Fig. 2.1, the water molecule can be roughly represented by a regular tetrahedron with an oxygen atom at its center, with two protons at two of its vertices, and with lone pair electrons in orbitals directed toward both other vertices. The H–O–H angle is somewhat smaller than the angle 109.5° of a tetrahedron. Values in the literature vary between 104.45° and 105.05°. The electrical charges are not uniformly distributed over the water molecule so that the vertices of the tetrahedron constitute poles of electrical charges, of which two are positive since the hydrogen nuclei are not completely screened by the binding electrons. The lone pair of electrons at the other two vertices hold the corresponding negative charges. These charges amount to about $0.17e$ and $-0.17e$, respectively, where $e = 1.602 \cdot 10^{-19}$ As denotes the elementary charge [4, 5].

Because of the particular charge distribution, the water molecule, besides its electrical polarizability $\alpha = 1.444 \cdot 10^{-30} m^3$ due to electronic and atomic displacement polarizabilities, possesses also a permanent electric dipole moment, $\mu = (1.84 \pm 0.02)$ D, resulting from the vector sum of the two H–O bond moments of 1.53D. Here 1D, the commonly used unit of the molecular electric dipole moment, corresponds with 10^{-18} esu $= 1/3 \cdot 10^{-29}$ As.

Fig. 2.1. Sketch of a water molecule as a regular tetrahedron [5]. The oxygen is shown in black. Light and dark areas show the binding orbitals to the hydrogen molecules and the lone electron pairs, respectively

The water molecule also possesses quadruple moment components. The mean quadruple moment almost vanishes. Hence it is the permanent electric dipole moment that mediates electromagnetic wave interactions with water molecules.

2.2.2 Liquid Water

Because of the positive electrical charges at the only partially shielded protons and the negative electrical charges of the lone electron pairs, water molecules interact to form hydrogen bonds. As illustrated by Fig. 2.2, the binding hydrogen atom in a water dimer forms a covalent bond to one oxygen and a hydrogen bond to the oxygen of the other water molecule. The bond strength differs by an order of magnitude. The enthalpy of the covalent H–O bond is as high as 463 kJ/mol, whereas that of the hydrogen bond is about 20 kJ/mol only. The most stable configuration of a hydrogen bond is a linear H–O-H arrangement. In ice, the distance between two hydrogen bonded oxygens is 0.276 nm with the hydrogen being 0.101 nm apart from one oxygen atom and 0.175 nm from the other one. The H-bond is largely ionic in character, with covalent parts that can be neglected on many events [4, 5].

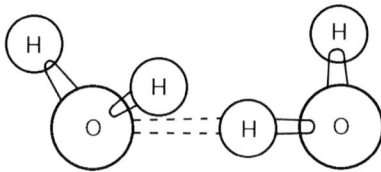

Fig. 2.2. Hydrogen bonded linear water dimer

As each water molecule is capable of four hydrogen bonds, a macroscopically percolating three-dimensional H-bonded network is formed in the condensed phases. For hexagonal ice, the ordered hydrogen network structure is illustrated by Fig. 2.3. Computer simulation studies of liquid water reveal the bond order j_b ($j_b = 0...4$) to follow a binominal distribution. Hence j_b may be considered a random quantity. Water molecules bound by more than one H-bond are prevented from reorientational motions. Consequently, only the molecules which, at a time, are non- or single-hydrogen bonded ($j_b = 0,1$) are able to rotate the direction of their permanent electric dipole moment into the direction of an external electric field and thus to contribute to the orientational polarization.

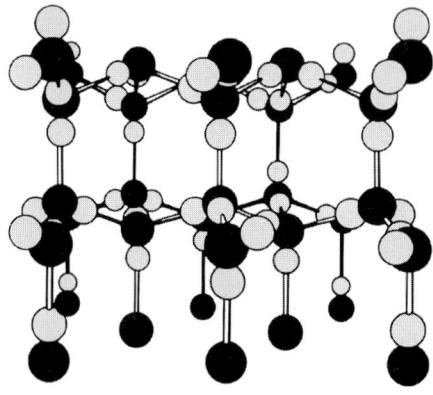

Fig. 2.3. Hexagonal structure of ice-Ih

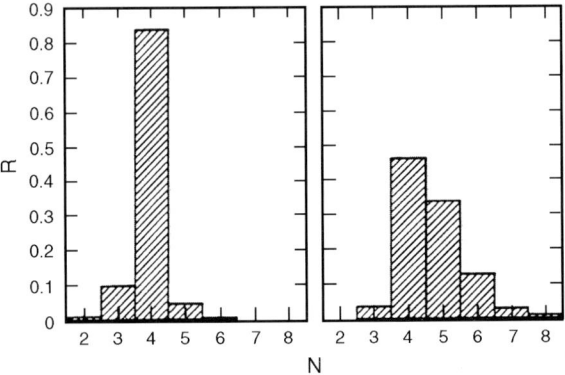

Fig. 2.4. Fraction R of molecules with N neighbors within a distance of 0.33 nm for water at 0°C and at reduced (left side, $\rho = 0.75$ g cm^{-3}) as well as normal (right side, $\rho = 1$ g cm^{-3}) pressure [4]

Computer simulations also show that, at reduced density ρ, almost perfect tetrahedral order is adopted in water. The fraction R_4 of molecules with four neighbors within a distance $r = 0.33$ nm in water at 0°C amounts to more than 80 per cent at $\rho = 0.75$g cm^{-3}. On normal conditions ($\rho = 1$g cm^{-3}) less than 50 per cent of the water molecules are fourfold coordinated and about one-third of all molecules reveals five neighbors within the $r = 0.33$ nm distance (Fig. 2.4). Hence the regular tetrahedral structure of ice, yielding a tridymite-like lattice in which the oxygen atoms form puckered six-membered rings (Fig. 2.3), is significantly disturbed in water on normal conditions. The defects in the water structure resulting thereby lead to far reaching consequences for the microdynamics of water. The much higher number of defects, for example, is the reason why the dielectric relaxation time of liquid water at 0°C ($\tau = 17.7$ps) is by six orders of magnitude smaller than the relaxation time of ice ($\tau = 20\mu$s) at the same temperature.

The fact that, within the hydrogen bond network of water, the energy of the covalent bond exceeds that of the hydrogen bond by a factor of more than twenty suggests the idea of well-defined water molecules also for the condensed phases. There exists nevertheless an autoprotolysis equilibrium $2H_2O \Leftrightarrow OH^- + H_3O^+$ in the liquid, with considerable import for chemistry and biology. The equilibrium constant of the autoprotolysis, however, is small ($pK_w = 14$, $25°C$) so that there is only a small ion concentration in pure water.

2.3 Hydrogen Network Fluctuations and Polarization Noise

Since the enthalpy of a hydrogen bond (20 kJ/mol) is on the order of the thermal energy at room temperature ($RT = 2.5$ kJ/mol, $25°C$) the bond strength of the hydrogen network of water fluctuates rapidly due to thermal agitation. Fluctuation correlation times as small as 0.1 to 1 ps have been reported. Normally, after the weakening of a bond, however, the same bond is reformed again. Reorientation of a water molecule through a significant angle and thus formation of a hydrogen bond at another site occurs only on favorable conditions. These conditions include the existence of an additional neighbor, the "fifth neighbor", in a suitable position. Such a neighbor constitutes a network defect with considerable importance to the reorientational motions of water. The additional neighbor molecule promotes the formation of a branched (bifurcated) hydrogen bond and flattens the potential energy barriers between different network fluctuations. Reorientational motions are significantly facilitated thereby, particularly as the additional neighbor offers a site for the formation of a new bond. For water at room temperature, it takes about 10 ps until a fifth neighbor molecule is present in a position that promotes reorientation. The reorientation of a molecule itself into a new direction resembles a switching process since it occurs again in a short period of about 0.1 ps. Hence the reorientational motions of water molecules in the liquid may be characterized by a wait-and-switch process in which the reorientation time is predominantly governed by the period for which a water molecule has to wait until favorable conditions for the reorientation exist. As it is essential for these conditions that an additional hydrogen bonding neighbor has to approach, the orientational motions of the water molecules are evidently controlled by the concentration of partners capable of forming H-bonds. The higher this concentration, the larger the probability for the availability of the additional neighbor – hence the smaller the reorientation time of the water molecules [5, 10].

Since the water molecules are provided with a permanent electric dipole moment, their reorientational motions will produce electrical polarization noise. In principle, this noise could be used to measure the dielectric properties of the aqueous systems under study, in particular to determine the time constants of interest. Let us, for simplicity, consider an imaginary experiment. The dipolar liquid may be contained in an ideal plane parallel-plate capacitor with the distance between the plates small as compared to their lateral dimensions. To be able to monitor all changes in the electrical charges on the plates, the capacitor is connected to a suitable instrument with vanishing internal impedance.

Thermal fluctuations in the electrical polarization $\vec{P}$ of the dipolar liquid induce electrical charges on the plates of the capacitor. For isotropic liquids the amount of charges is proportional to the amount $P = |\overline{P}|$ of the dielectric polarization. The noise signal monitored by this experiment (Fig. 2.5) displays two essentially different molecular processes. Fast changes in the signal result from electronic and atomic displacement polarization mechanisms, slower variations in the noise are due to the reorientational motions of the dipolar molecules. The details in the time-dependent properties of the noise are reflected in an obvious manner by the normalized auto-correlation function [5, 7]

$$\phi(t) = < P(t) \cdot P(0) > / < P(0) \cdot P(0) > \tag{2.1}$$

also named "dielectric decay function". For water at 25°C, as an example, the autocorrelation function of the polarization noise is shown in Fig. 2.5. Due to the fast displacement polarization mechanisms and also to a high frequency relaxation process, the decay function decreases rapidly from $\phi = 1$ at $t = 0$ to $\phi(t_0) = 0.94 \cdot \phi(0)$ at $t = t_0$. The slower decay in the autocorrelation function represents the reorientational motions of the water molecules. This part of the autocorrelation function at $t > t_0 = 2 \cdot 10^{-12}$s (=2 ps) follows almost an exponential

$$\phi(t > t_0) = \phi(t_0) \exp(-(t - t_0)/\tau_w) \tag{2.2}$$

The decay time τ_w will be named principal dielectric relaxation time of water in the following. As mentioned above, for water at room temperature τ_w is on the order of 10 ps. Notice, that the dielectric relaxation time of water corresponds with the macroscopic polarization.

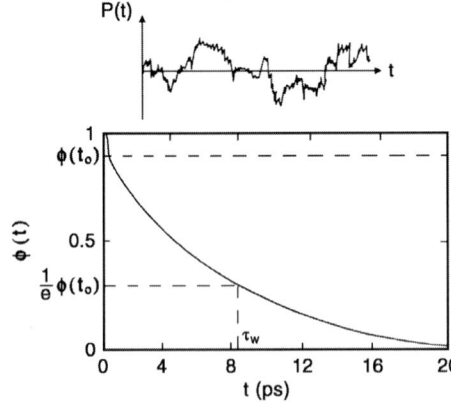

Fig. 2.5. Polarization noise (top) and autocorrelation function $\phi(t)$ (Eq. (2.1)) for water at 25°C (bottom)

Hence it is not the reorientation time of individual water molecules but rather a collective quantity. It is a measure of the period over which, at thermal equilibrium, the direction of the polarization is correlated to an originally existing direction.

2.4 The Dielectric Properties of Water

2.4.1 Complex Permittivity Spectrum

Due to the unavoidably existing noise of the measurement system itself, analysis of the noise signal of real experiments is difficult. For this reason, it is more convenient to expose the sample liquid to a disturbing electromagnetic signal. In doing so the field strength of the signal has to be sufficiently small to guarantee the sample remains in almost thermal equilibrium and thus avoid any nonlinear effects during the measurements. The advantage taken from the use of an external signal is the small preferential orientation of all dipolar molecules that is super-posed to the thermally driven reorientational motions, resulting in a substantial enhancement of the signal-to-noise ratio. Two different types of disturbances are common in measurements of the principal dielectric relaxation of water. Sequences of step-voltage pulses are used in time domain spectrometry (TDS) that probes the dielectric decay function $\phi(t)$. Alternatively, sinusoidally varying electromagnetic fields $E(v)$ are applied in frequency domain techniques. The electric polarization $P(v)$ is then not only a function of the electric field strength E but also of the frequency v. Since, due to molecular interactions, the polarization needs a finite time to establish, it cannot follow electrical field changes instantaneously. Hence there exists a dispersion in $P(v)$. In addition, a phase shift between $P(v)$ and $E(v)$ occurs in the dispersion region. Energy of the external electromagnetic field is dissipated as heat. Both effects, the dispersion in the polarization and the absorption of electromagnetic energy, are considered by a frequency dependent complex permittivity [5–7]

$$\varepsilon(v) = \varepsilon'(v) - i\varepsilon''(v) = \frac{P(v)}{\varepsilon_0 E(v)} + 1 \qquad (2.3)$$

Here, ε_0 denotes the electric field constant and $i^2 = -1$. According to linear system theory the transfer function $\varepsilon(v)$ and the step response function $\phi(t)$ are related as

$$\varepsilon(v) = (\varepsilon(0) - 1) \int_0^\infty \left[-\frac{d\phi(t)}{dt} \right] e^{-i2\pi vt} dt + 1 \qquad (2.4)$$

where $\varepsilon(0) = \lim_{v \to 0} \varepsilon'(v)$ is the low frequency ("static") permittivity of the dielectric.

As $\phi(t)$ is a real function the real part $\varepsilon'(v)$ and the negative imaginary part $\varepsilon''(v)$ of the complex permittivity are not independent from one another but are different forms of the same phenomena, the relaxation of the dielectric.

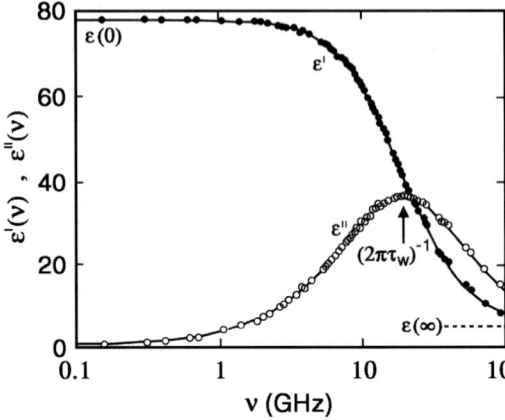

Fig. 2.6. Real part ε' and negative imaginary part ε'' of the complex permittivity spectrum of water at 25°C. Microwave permittivity data have been taken from the literature [2, 3]. Lines are graphs of Eq. (2.5) with parameter values of Table 2.1

Table 2.1. Dielectric parameters of water at different temperatures T: Static permittivity values as recommended by the IUPAC Commission on Physico-Chemical Measurements and Standards [14] and parameters of Eq. (2.5) as following from a regression analysis of microwave complex permittivity data ($v < 100$ GHz, [2, 5])

$T/°C$	$\varepsilon(0)$, [14]	$\varepsilon(0)$	$\varepsilon(\infty)$	τ_w, ps
0	87.87 ± 0.07	87.91 ± 0.2	5.7 ± 0.2	17.67 ± 0.1
0.3		87.70 ± 0.2	5.9 ± 0.2	16.42 ± 0.2
5		85.83 ± 0.2	5.8 ± 0.2	14.50 ± 0.4
10	83.91 ± 0.07	83.92 ± 0.2	5.8 ± 0.3	12.68 ± 0.1
15		82.05 ± 0.2	6.0 ± 0.2	10.84 ± 0.1
20	80.16 ± 0.05	80.21 ± 0.2	5.7 ± 0.2	9.37 ± 0.05
25	78.36 ± 0.05	78.36 ± 0.05	5.4 ± 0.2	8.28 ± 0.02
30	76.57 ± 0.05	76.56 ± 0.2	5.2 ± 0.3	7.31 ± 0.05
35		74.87 ± 0.2	5.3 ± 0.4	6.54 ± 0.1
37		74.17 ± 0.2	5.3 ± 0.2	6.27 ± 0.1
40	73.16 ± 0.04	73.18 ± 0.2	4.6 ± 0.7	5.82 ± 0.1
50	69.90 ± 0.04	69.89 ± 0.2	4.0 ± 0.5	4.75 ± 0.1
60	66.79 ± 0.04	66.70 ± 0.2	4.2 ± 0.5	4.01 ± 0.1
70	62.82 ± 0.05			
80	61.03 ± 0.05			
90	58.32 ± 0.05			
100	55.72 ± 0.06			

The exponential dielectric decay function (Eq. (2.2)) corresponds with a Debye-type relaxation spectral function, defined by the relation

$$\varepsilon(v) = \varepsilon(\infty) + \frac{\varepsilon(0) - \varepsilon(\infty)}{1 + i\omega\tau_w} \qquad (2.5)$$

with $\omega = 2\pi\nu$ and $\varepsilon(\infty) = \lim_{\nu \to \infty} \varepsilon'(\nu)$ denoting the permittivity as extrapolated to frequencies well above the relaxation frequency ($\nu \gg (2\pi\tau_w)^{-1}$). Parameter $\varepsilon(\infty)$ reflects the rapidly decaying polarization processes in $\phi(t)$, with relaxation times smaller than t_o (Fig. 2.5).

In the microwave region up to frequencies of about 100 GHz, the complex dielectric spectrum of water can be well represented by a Debye type relaxation function. As an example the permittivity spectrum at 25°C is shown in Fig. 2.6, where the meaning of the parameters of Eq. (2.5) is also indicated. It is only mentioned that another low amplitude relaxation term has been found toward higher frequencies, with relaxation time around 0.2 ps (19°C). Within the framework of the wait-and-switch model outlined above this relaxation term has been attributed to the single hydrogen bonded water molecules. The concentration of such molecules is small and thus the relaxation amplitude is also small. The content of molecules with more than one hydrogen bond, providing a suitable site to the single-hydrogen bonded water molecules for the formation of a new bond, is larger. Hence the time for which a molecule with only one H-bond has to wait until favorable conditions for reorientation occur is small and likewise small is the dielectric relaxation time [10].

Here the discussion will be restricted to the microwave region of the spectrum ($\nu \leq 100$ GHz). Hence $\varepsilon(\infty)$ means the high frequency limit of the dominating relaxation process with relaxation frequency $(2\pi\tau_w)^{-1}$ of about 20 GHz (25°C), Fig. 2.6). The parameters of the corresponding relaxation spectral function (Eq. (2.5)) are displayed in Table 2.1.

2.4.2 Static Permittivity

Due to the rather high permanent electric dipole moment $\mu = 1.84$D of the water molecule in the gaseous state, liquid water exhibits a large static permittivity. It decreases from $\varepsilon(0) = 107 \pm 2$ for supercooled water at –35°C to $\varepsilon(0) = 87.87 \pm 0.07$ at 0°C and finally to $55.62 \pm 0.02 \leq \varepsilon(0) \leq 55.72 \pm 0.06$ at 100°C and normal pressure [2, 3, 5]. Various empirical relations have been reported to analytically represent the temperature dependence of the static permittivity of water. In Fig. 2.7 a plot is given of the simple equation [7]

$$\varepsilon(0) = 87.853 \exp[-0.00457(T / K - 273.15)] \qquad (2.6)$$

to show that, in the temperature range between –25 and 100°C, it represents the experimental data within the limits of errors. Only for supercooled liquid water at even lower temperatures do the deviations between the measured data and the predictions from Eq. (2.6) somewhat exceed the experimental errors.

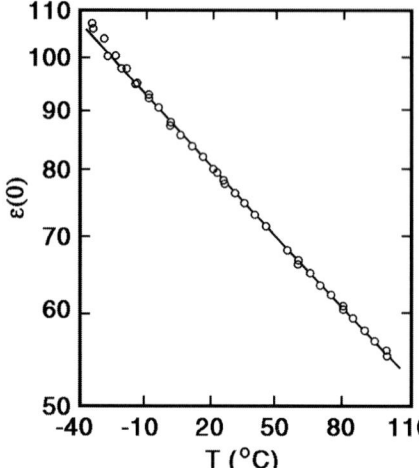

Fig. 2.7. Static permittivity $\varepsilon\,(0)$ of water on a logarithmic scale versus temperature T (Table 2.1). The line represents a simple empirical relation (Eq. (2.6))

Theoretical models relate to the static permittivity of a dipolar liquid to the dipole moment μ, to the dipole concentration c, and to the Kirkwood orientation correlation factor g [5, 7]

$$\frac{(\varepsilon(0) - \varepsilon_\infty)(2\varepsilon(0) + \varepsilon_\infty)}{\varepsilon(0)} = \frac{N_A}{9\varepsilon_0 k_B T}(\varepsilon_\infty + 2)^2 c\mu g \qquad (2.7)$$

In this equation N_A denotes Avogadro's number and k_B the Boltzmann constant. The orientation correlation factor considers the fact that preferential parallel alignment of dipole moments results in an enhanced static permittivity ($g > 1$), as evident from the dielectric properties of ferroelectrics, and that antiparallel ordering of dipole moments leads to a reduction in the static permittivity ($g < 1$). For water the situation is less clear. The reason is our insufficient knowledge of the high frequency permittivity ε_∞ ($n^2 \leq \varepsilon_\infty \leq \varepsilon\,(\infty)$) to be used in Eq. (2.7). Here n is the optical refractive index. Kirkwood using $\varepsilon_\infty = n^2 = 1.33^2$ found $g = 2.8$ whereas $\varepsilon_\infty = \varepsilon\,(\infty)$ yields $g < 1$. It has been shown that $\varepsilon_\infty = 4.3$ is in conformity with $g = 1$, which would suggest effects of orientation correlation in liquid water to be absent at all. In this context, it is interesting to notice that the aforementioned high frequency relaxation term with relaxation time on the order of 0.2 ps (Fig. 2.8) extrapolates to $\varepsilon^*(\infty) = \lim_{\nu \to \infty} \varepsilon\,'(\nu) = 3.4$ at $\nu > (2\pi\tau_w^*)^{-1}$. Despite of the still insufficient knowledge about the effect of orientation correlation in the static permittivity of water, Eq. (2.7) indicates that reorientation of permanent dipole moments occurs against thermal agitation, tending at equipartition of all states of dipole orientation.

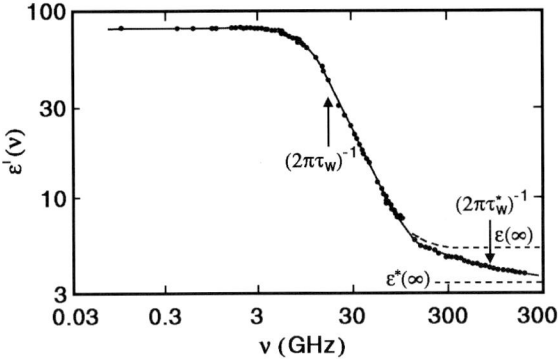

Fig. 2.8. Bilogarithmic plot of the real part of permittivity spectrum up to some THz of water at 19°C [10]

2.4.3 High Frequency Properties

In Fig. 2.9 the extrapolated high frequency permittivty $\varepsilon(\infty)$ as following from Eq. (2.5) is displayed as a function of temperature T. At low temperatures the experimental $\varepsilon(\infty)$ values significantly exceed the high frequency permittivity data that have been determined by assuming $g = 1$ and treating ε_∞ as an adjustable parameter in Eq. (2.7). At higher temperatures ($T \geq 50°C$), however, both sets of data almost agree with one another. Interesting, the $\varepsilon^*(\infty)$ data which, according to

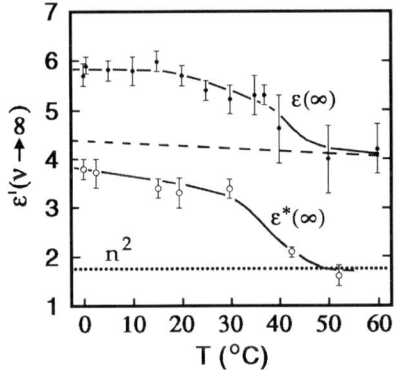

Fig. 2.9.
Extrapolated high frequency permittivities $\varepsilon(\infty)$ and $\varepsilon^*(\infty)$ (Fig. 2.8) and squared optical refractive index n^2 of water versus temperature T

$$\varepsilon(v) = \varepsilon^*(\infty) + \frac{\varepsilon(\infty) - \varepsilon^*(\infty)}{1 + i\omega\tau_w^*} + \frac{\varepsilon(0) - \varepsilon(\infty)}{1 + i\omega\tau_w} \qquad (2.8)$$

have been determined from the spectra which include complex permittivities from THz measurements (Fig. 2.8), also decrease at temperatures higher than 30°C and reach the squared optical refraction index at $T = 50°C$. Hence $\varepsilon(\infty)$ seems to just follow the trend in $\varepsilon^*(\infty)$. The relative contribution $(\varepsilon(\infty)-\varepsilon^*(\infty))/(\varepsilon(0)-\varepsilon(\infty))$ of the fast relaxation process to the static permittivity of water increases from 0.025

at 0°C to 0.034 at 50°C. This finding may be taken to support the assignment of the fast relaxation term to the reorientational motions of single hydrogen bonded water molecules, because their concentration is expected to increase with T [10]. Also in conformity with the experimental facts, however, is the assumption of the high frequency relaxation to reflect the reorientation of non-hydrogen bonded interstitial water molecules.

2.4.4 Principal Relaxation Time

The wait-and-switch model of water reorientation outlined above implies a potential barrier between two orientations of a water dipole moment as sketched in Fig. 2.10. Normally a water molecule is contained in either of the potential minima where the strength of its hydrogen bonds fluctuates rapidly with correlation times on the order of 0.12 to 1 ps, as also mentioned above. Due to thermal activation, e.g. due to collision with neighboring water molecules, a water molecule will occasionally posses a kinetic energy that is higher than the potential barrier and will thus be able to surmount it and to orientate its dipole moment in another direction [4, 5, 7, 10]. The effect from the additional neighbor in the wait-and-switch model is a reduction of the potential energy separating different dipole orientations, as indicated by the dashed curve in Fig. 2.10.

Energy

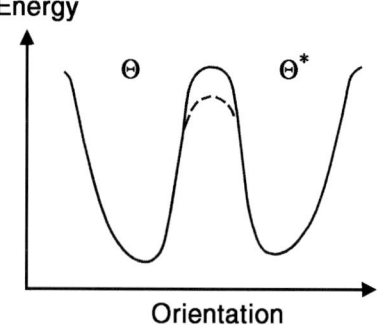

Orientation

Fig. 2.10.
Sketch of a potential energy barrier separating two directions θ and θ^* of permanent dipole moment. The dashed line shows the reduction in the energy barrier due to a suitable additional neighbor molecule

Because of the idea of an underlying thermal activation mechanism, it is an obvious attempt to assume the dielectric relaxation time τ_w of the dominating relaxation term of water to be governed by a Gibbs free energy of activation

$$\Delta G^{\#} = \Delta G^{\#} - T \Delta S^{\#} \tag{2.9}$$

and thus to be given by an Eyring relation [6]

$$\tau_w = \frac{h}{k_B T} C \exp(\Delta G^{\#} / RT) \tag{2.10}$$

Here $\Delta H^{\#}$ and $\Delta S^{\#}$ are the activation enthalpy and entropy, respectively, h is Planck's constant, C a configurational factor, and $R = k_B N_A$ denotes the gas constant. From the $\ln \tau_w$-vs-T^{-1} plot in Fig. 2.11 $\Delta H^{\#} = (16.7 \pm 0.4)$kJ/mol and $\Delta S^{\#} = (23 \pm 2)$ J/(mol K) follows. Hence the activation enthalpy is on the order of the interaction enthalpy $\Delta H = 20$ kJ/mol of hydrogen bonds in water.

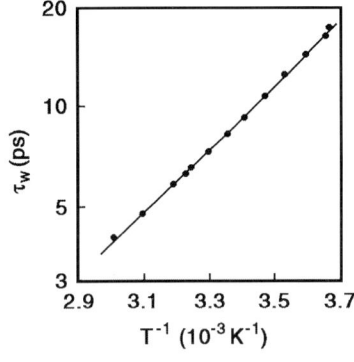

Fig. 2.11. Eyring plot of the relaxation time τ_w of water. Errors do not exceed the symbols

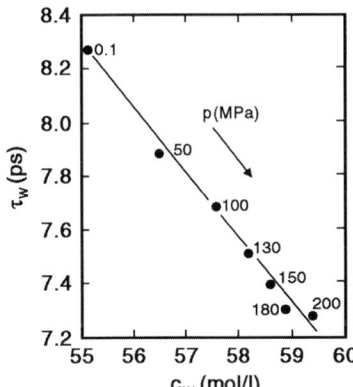

Fig. 2.12. Relaxation time τ_w of water as a function of water concentration [11]. Parameter is the hydrostatic pressure p

The quantities $\Delta H^{\#}$ and $\Delta S^{\#}$ refer to molecular rather than collective mechanisms and should thus be calculated from the dipole rotational relaxation times instead of the dielectric relaxation times τ_w. There exists, however, a nice correlation between the temperature dependencies of the proton magnetic relaxation rate $1/T_1$, reflecting the proton-around-proton reorientational motion of a water molecule, and τ_w. Therefore, the activation enthalpy from the dipole rotational correlation times almost agrees with that from the principal dielectric relaxation time τ_w.

It is well established that the principal dielectric relaxation time decreases when water is exposed to a hydrostatic pressure p. Due to the pressure the density of water increases and therefore increases the water concentration (Fig. 2.12). Hence the reduction of the dielectric relaxation time with p may be taken another confirmation of the wait-and-switch model of water reorientation.

2.5 Aqueous Solutions

2.5.1 Solute Contributions to Dielectric Spectra

In Fig. 2.13 the complex permittivity spectrum of a 1-molar aqueous solution of nondipolar quinoxaline is shown and compared to that of water at the same temperature. Due to the dilution of the dipolar water by the nonpolar solute the static permittivity of the solution is substantially smaller than that of the solvent. Additionally, because of the particular molecular interactions introduced by the solute molecules, the dispersion ($d\varepsilon'(\nu)/d\nu < 0$) and dielectric loss region extends over a broader frequency range than in water and is shifted to lower frequencies. These effects increase with solute concentration c [6, 7, 10]. The broadening of the relaxation region, which reflects a distribution of relaxation times, can be considered by a Havriliak–Negami relaxation spectral function [9]

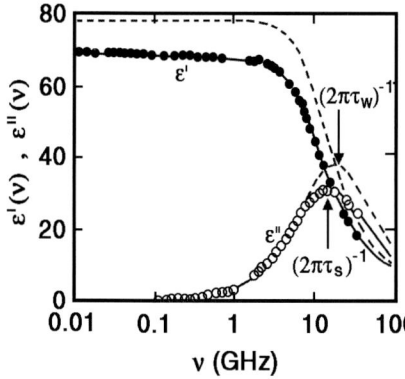

Fig. 2.13. Complex permittitivy spectrum for a 1 mol/ℓ aqueous solution of quinoxaline at 25°C [7, 9]. Dashed lines indicate the spectrum of water at the same temperature

$$\varepsilon(\nu) = \varepsilon(\infty) + \frac{\varepsilon(0) - \varepsilon(\infty)}{(1 + (i\omega\tau_s)^{1-h})^{1-b}} \tag{2.11}$$

Here, τ_s is a characteristic relaxation time of the underlying distribution function and parameters h and b control the shape and width of the relaxation time distribution. Eq. (2.11) includes some well-known and frequently used spectral functions, the Cole–Cole ($b = 0$), the Davidson–Cole ($h = 0$), and the Debye function ($b = h = 0$). With the notation $\tau_s = \tau_w$ the latter corresponds with Eq. (2.5).

A summary of results for aqueous solutions of low weight organic molecules is given in [7]. Also discussed in that article and especially in [8, 12] are spectra for aqueous solutions of low weight electrolytes and polyelectrolytes. As an example, the complex permittivity spectrum of a sodium chloride solution is shown in Fig. 2.14. The salt concentration corresponds with that of the North Sea. A particular feature of electrolyte solutions is the strong increase in the ε'' data towards low frequencies which results from the contribution of the ionic conductivity σ. With the conductivity term $-i\sigma/(\varepsilon_o\omega)$ the spectral function

$$\varepsilon(v) = \varepsilon(\infty) + \frac{\varepsilon(0) - \varepsilon(\infty)}{(1 + (i\omega\tau_s)^{1-h})^{1-b}} - \frac{i\sigma}{\varepsilon_0\omega} \qquad (2.12)$$

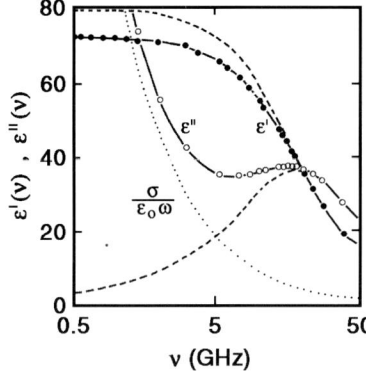

Fig. 2.14. Complex dielectric spectrum for a 0.61 mol/l aqueous solution of NaCl at 25°C [9]. The dotted curve shows the conductivity contribution to ε''. Dashed lines indicate the spectrum of water at 25°C

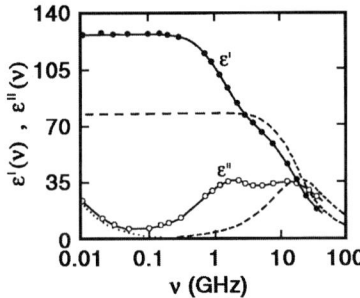

Fig. 2.15. Complex permittivity spectrum for a 1 mol/l solution of 4-aminobutyric acid in water at 25°C and at neutral pH [9]. Dashed lines represent the water spectrum at 25°C

results which at low frequencies is dominated by the σ contribution. Quite remarkably there is also a significant reduction in the static permittivity (Fig. 2.14) which, at least in parts, reflects the preferential orientation of the dipolar water molecules in the Coulombic field of small cations. This effect is normally named dielectric saturation or structure saturation.

Spectra for aqueous solutions of dipolar solutes often display two well-separated relaxation regions of which one is due to the water and the other one to the solute reorientational motions. An example is given in Fig. 2.15 for a solution of 4-aminobutyric acid in water. The large solute electric dipole moment of about 20D leads to a large amplitude in the corresponding (low frequency) relaxation term though the aminobutyric acid concentration is distinctly smaller than that of water. As not all solute molecules are zwitterionic, some ionic species contribute also a conductivity term to the spectrum. The complex permittivity may thus be analytically represented by the relaxation spectral function

$$\varepsilon(v) = \varepsilon(\infty) + \frac{\varepsilon(0) - \varepsilon(\infty)}{(1 + (i\omega\tau_s)^{1-h})^{1-b}} + \frac{\varepsilon(0)}{1 + i\omega\tau_0} - \frac{i\sigma}{\varepsilon_0\omega} \qquad (2.13)$$

where ε^* is the low frequency limit of the water dispersion as well as the high frequency limit of the solute dispersion. Parameter τ_u is the dielectric relaxation time of the solute.

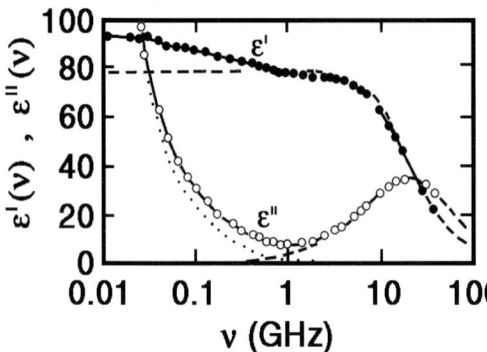

Fig. 2.16. Complex dielectric spectrum of a 0.06 mol/l aqueous solution of the cationic surfactant n-hexadecyltrimethylammonium bromide at 25°C [9]. The dotted curve shows the conductivity contribution to ε''. Dashed lines indicate the water spectrum at the same temperature

Solute contributions with relaxation characteristics may also result for limited motions of ions. Such mechanisms have been intensively discussed for many colloidal systems, including solutions of biopolymers as well as such of ionic micelles and vesicles. Fig. 2.16 shows the complex dielectric spectrum of an aqueous solution of a cationic surfactant with a conductivity contribution due to drift ions and with a solute relaxation term resulting from the limited motions of counterions on the surface of micelles. As a result of Coulombic interactions most counterions are condensed on the micellar surface where they form a diffusive layer around the micellar aggregate.

Relaxation terms with large amplitudes and relaxation times result if the limited motions extend over distances larger than molecular dimensions [12]. Such a "giant" dispersion is shown in Fig. 2.17 for an aqueous suspension of erythrocytes. Notice that in this diagram ε' and ε'' are displayed on a logarithmic scale.

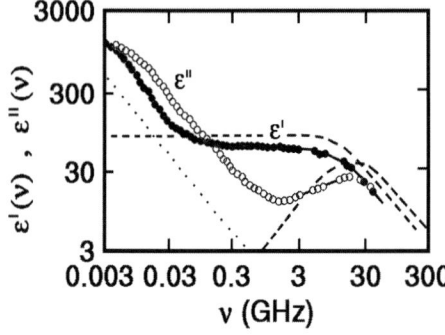

Fig. 2.17. Bilogarithmic plot of the permittivity spectrum of a solution of erythrocytes [9]. Again the dotted line represents the conductivity contribution and the dashed curves indicate the water spectrum at 25°C

Solute contributions with relaxation characteristics result also from incomplete dissociation of multivalent salts which, in many cases, is described by the Eigen-Tamm scheme [13]

$$(M^{m+})_{aq} + (A^{a-})_{aq} \Leftrightarrow (M^{m+}(H_2O)_2 A^{a-})_{aq}$$

$$(M^{m+}(H_2O)A^{a-})_{aq} \Leftrightarrow (MA)_{aq}^{m-a}$$

(2.14)

where M^{m+} is a metal cation and A^{a-} an anion. The $(M^{m+}(H_2O)_2 A^{a-})_{aq}$ complex with the anion separated from the cation by two layers of water molecules is called an "outer-outer-sphere" complex or "Bjerrum ion pair". The species $(M^{m+}(H_2O)A^{a-})_{aq}$ is called an "outer sphere" complex, and $(MA)^{m-a}$ the contact ion pair.

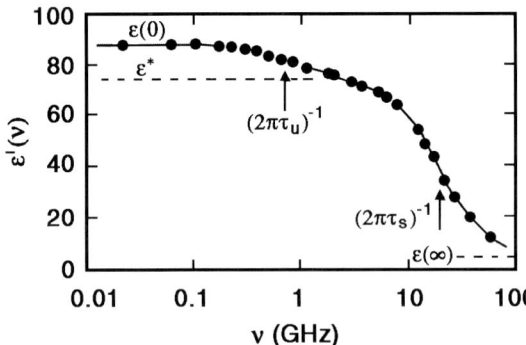

Fig. 2.18. Real part of the dielectric spectrum of a 0.1 molar aqueous solution of $Sc_2(SO_4)_3$ at 25°C [6]

The dielectric relaxation process and the effects in the ionic conductivity of the solutions which result from the complex formation are well established. An example of the real part of the permittivity spectrum for a solution of a 3:2 valent electrolyte is shown in Fig. 2.18. As another example the dielectric spectrum of a solution of 3:2 valent aluminum sulfate is compared to that of aluminum chloride in Fig. 2.19. A suggestive complex plane representation of data is given, in which the negative imaginary part of the spectrum, excluding conductivity contributions, is plotted versus the real part. For water, for which the spectrum can be well represented by a Debye type relaxation (Eq. (2.5)), the data define a semicircle with its center on the ε' axis. A semicircle with center somewhat below the ε' axis follows for the solution of 3:1 valent aluminium chloride, indicating a small distribution of relaxation times due to the disturbance of the water properties by the solute. The spectrum for the aluminium sulfate solution clearly displays two relaxation regions, of which the low frequency one reflects the reorientational motions of the dipolar ion pairs (Eq. (2.14) with $M^{m+} = Al^{3+}$ and $A^{a-} = SO_4^{2-}$). Two and three valent transition metal ions with d^{10} outer electron shell are able to form dielectrically evident ion complex structures even with monovalent ligands like the halides.

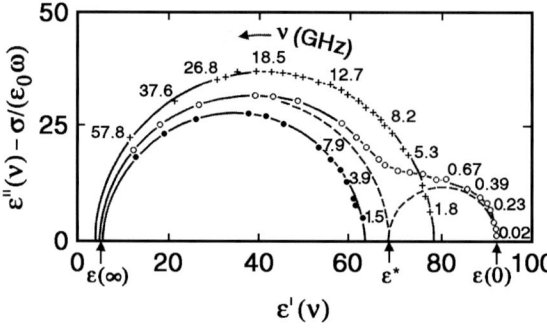

Fig. 2.19. Complex plane representation of the permittivity spectra excluding conductivity contributions for water (+), and aqueous solutions of $AlCl_3$ (0.4 mol/l, ●) and of $Al_2(SO_4)_3$ (0.15 mol/l, ○) at 25°C. Dashed lines show the subdivision of the latter into a solvent and a solute contribution

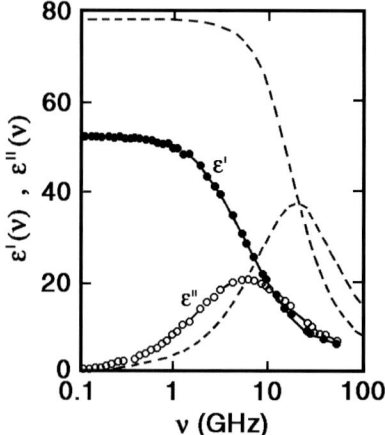

Fig. 2.20. Complex dielectric spectra for water (dashed curves) and for a 4.4 mol/l solution of *tert*-butanol in water at 25°C[9]

Many mixtures of water with dipolar solutes do not show well separated solute and solvent dielectric relaxation regions like the aqueous solutions of 4-amino-butyric acid (Fig. 2.15). Rather the spectra reflect a dielectrically homogeneous mixture with one principal relaxation, subject to a relaxation time distribution as represented, for instance, by the Havriliak–Negami spectral function (Eqs. (2.11, 2.12)). Examples are mixtures of water with alcohols, alkanediols, mono- and disaccharides, urea and its derivatives, dimethylsulfoxide, and various others. In Fig. 2.20, the real part ε' and the negative imaginary part ε'' of the complex permittivity are displayed as a function of frequency ν for a mixture of tert-butanol with water to illustrate the existence of essentially one dispersion region only. Fig. 2.21 presents a complex plane representation for the permittivity spectrum of an aqueous solution of D-glucose. This diagram shows the strong deformation of the original circular arc plot for water. With the saccharide solution there exists a broad relaxation time distribution due to the different reorientational motions of the D-glucose molecules, their dipolar side groups, as well as of the water molecules. The latter will display a relaxation time distribution by themselves because they are differently affected by the solute.

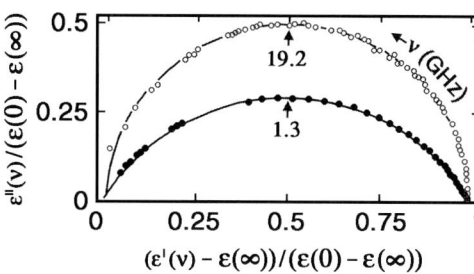

Fig. 2.21. Complex plane representation of the dielectric spectra for water (O) and for an aqueous solution of D-glucose (4.73 mol/l, •) at 25°C

2.5.2 Solvent Permittivity Contribution Aspects

The extrapolated low frequency permittivity ε^* of the water contribution to the dielectric spectra (Figs. 2.18, 2.19) offers valuable information on structural properties of the liquid. For solutions of nondipolar solutes $\varepsilon^* = \varepsilon(0)$. First of all, however, ε^* reflects the effect of dilution of the dipolar solvent, namely the reduction of the concentration c in Eq. (2.7) by the presence of the solute. Due to internal electric fields in dielectric mixtures ε^* depends also on the shape of the solute particles. This effect is sometimes used as a tool to investigate structural aspects of dielectrically heterogeneous systems like micro-emulsions. Unfortunately, however, even for homogeneous solutions of spherically shaped solute particles the effect of internal fields cannot be rigorously considered. Different theoretical approaches have lead to a multitude of mixture relations relating the resulting permittivity ε_2 of solutions of spherical solutes with permittivity ε^* in a suspending medium with permittivity ε_1 to the properties of the constituents. The graphs of two prominent mixture relations [6, 7, 9] which are frequently applied to aqueous systems are displayed in Fig. 2.22. Shown for aqueous solutions at 25°C ($\varepsilon_1 = \varepsilon(0) = 78.36$, Table 2.1; $\varepsilon_2 = 2$) are the Bruggeman formula

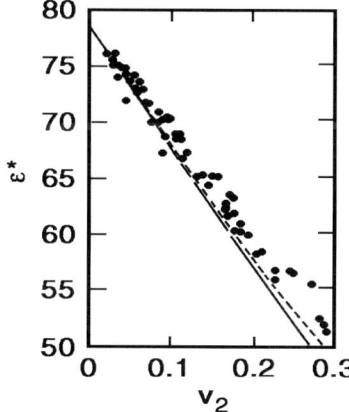

Fig. 2.22. Solvent contribution ε^* to the static permittivity versus volume fraction v_2 of solute for aqueous solutions of small organic molecules, of synthetic polymers, and of salts with large organic cations at 25°C [6, 7]. The line is the graph of the Bruggeman relation (Eq. (2.15))

$$\frac{\varepsilon^* - \varepsilon_2}{\varepsilon_1 - \varepsilon_2}\left(\frac{\varepsilon_1}{\varepsilon^*}\right)^{1/3} = 1 - v_2 \tag{2.15}$$

and a mixture relation originally derived by Maxwell and Wagner:

$$\varepsilon^* = \varepsilon_1 + \frac{3v_2(\varepsilon_2 - \varepsilon_1)}{2\varepsilon_1 + \varepsilon_2 - v_2(\varepsilon_2 - \varepsilon_1)} \tag{2.16}$$

In these formulas v_2 denotes the volume fraction of solute.

Also given in Fig. 2.22 for comparison are ε^* data for aqueous solutions of organic solutes. Interestingly, the scatter in the ε^* data for different series of solutes is comparatively small, thus indicating that the static permittivitiy of water depends only weakly on the particular interactions with the solute. The tendency in the experimental data to somewhat exceed the predictions from the mixture relations (Eqs. (2.15), (2.16)) seems to be characteristic to so-called "hydrophobic hydration" effects around largely inert molecules or ions. On the contrary, around small inorganic ions the effects from the above mentioned dielectric saturation may result in a substantial reduction of the extrapolated permittivity ε^* (Fig. 2.23). Besides the preferential orientation of water dipole moments in strong Coulombic fields, the extrapolated low frequency permittivity of electrolyte solutions may be also subject to a kinetic polarization deficiency [6, 7]. The model of kinetic depolarization proceeds from the idea that a charged particle moving through a dipolar liquid in an external electric field sets up a non-uniform hydrodynamic flow. The solvent dipole moments are turned by this flow in the direction opposed to that in which they are oriented by the electric field.

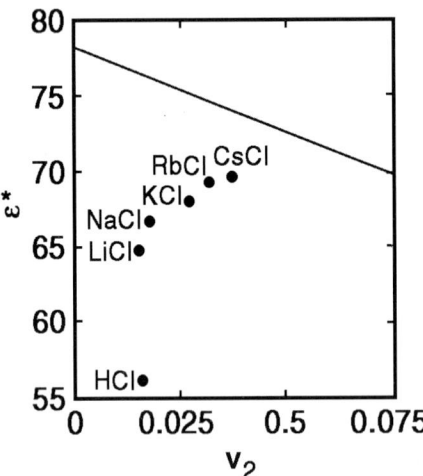

Fig. 2.23. Solvent contribution ε^* ($= \varepsilon(0)$) to the extrapolated low frequency permittivity for 1 mol/l solutions of some monovalent chlorides in water at 25°C [7]. The line is the graph of the mixture formula defined by Eq. (2.15)

It is common practice to express the effect of the solute on the principal dielectric relaxation time τ_s of the relative molal shift

$$B_d = \frac{1}{\tau_w} \lim_{m \to 0} \left(\frac{d\tau_s}{dm} \right) \tag{2.17}$$

and to assume the B_d values of electrolyte solutions to be simply given by the sum

$$B_d = \frac{m^+}{m} B_d^+ + \frac{m^-}{m} B_d^- \tag{2.18}$$

Here, B_d^+ and B_d^- denote the cationic and anionic relative molal shift and m, m^+ as well as m^- are the molal concentrations of electrolyte, cations, and anions, respectively. In Table 2.2 B_d and B_d^+ data for some series of organic molecules and ions are given. Within each series the tendency emerges for the relative shift in the relaxation time to increase with the number of aliphatic groups per solute particle. This tendency is a reflection of the aforementioned hydrophobic hydration. Within the framework of the above wait-and-switch model of dielectric relaxation the increase in the dielectric relaxation time of water around hydrophobic molecules or groups results mainly from the reduced density of hydrogen bonding sites and thus of suitable additional neighbor molecules at the water-solute interface. However, factors other than the local availability of additional hydrogen bonding partners are also important in determining the water relaxation time. These factors may include the overall size and shape of solute molecules, its flexibility with respect to the water structure, and the steric arrangement of its hydrophilic groups. Examples are the B_d–values for the stereoisomers N,N'- dimethylurean ($B_d = 0.18(\text{mol/kg})^{-1}$) and ethylurean ($B_d = 0.13(\text{mol/kg})^{-1}$).

An opposite effect, namely a reduction of the relaxation time τ_s with respect to τ_w is also found ($B_d < 0$). The ammonium ion shows indications of weak "negative hydration". A prominent example of the effects of negative hydration is the iodide ion with $B_d^+ = -0.05 \ (\text{mol/kg})^{-1}$. Around this large monovalent anion an enhanced mobility of water molecules may result from the comparatively soft electron shell, providing reorientating water molecules with transient hydrogen bond-like interactions. Hence, at least in parts, negative hydration may be also discussed in terms of the wait-and-switch model. A more detailed discussion of the dielectric properties of aqueous solutions is given in a more recent review article [7] where complex permittivity spectra are also considered in the light of a hydration model.

Table 2.2. Relative molar shifts B_d as well as cationic part B_d^+ in the principal relaxation time τ_s of water for aqueous solutions of some series of molecules and organic ions (25°C, pyridine and derivatives 20°C [6]).

Solute	B_d $(\text{mol/kg})^{-1}$	Solute	B_d $(\text{mol/kg})^{-1}$
Pyrazine	0.13	Urea	0.03
Methylpyrazine	0.19	Methylurea	0.08
2,3-Dimethylpyrazine	0.24	N,N-Dimethylurea	0.17
2,5-Dimethylpyrazine	0.27	N,N'-Dimethylurea	0.18
2,6-Dimethylpyrazine	0.25	Ethylurea	0.13
Ethylpyrazine	0.21	Trimethylurea	0.24
2,3,5-Trimethypyrazine	0.32	N-Propylurea	0.19
Quinoxaline	0.19	Tetramethylurea	0.30
2-Methylquinoxaline	0.24	N,N-Diethylurea	0,30
Pyridine	0.19	N-Butylurea	0.21
2-Methylpyridine	0.27		
3-Methylpyridine	0.22		
2,4-Diemthylpyridine	0.27		
2,6- Diemthylpyridine	0.28		
Cation	B_d^+ $(\text{mol/kg})^{-1}$	Cation	B_d^+ $(\text{mol/kg})^{-1}$
Ammonium	-0.04	Tetramethylammonium	0.17
N-Butylammonium	0.28	Tetraethylammonium	0.39
N-Hexylammonium	0.37	Tetrapropylammonium	0.73
N-Heptylammonium	0.38	Tetrabutylammonium	0.88
N-Octylammonium	0.44	5-Azoniaspiro[4,4]nonane	0.29
		6-Azoniaspiro[5,5]undecane	0.37
		7-Azoniaspiro[6,6]tridecane	0.43

2.6 Conclusions: Microwave Aquametry, an Inverse Problem

In microwave aquametry, we are normally dealing with a sophisticated inverse problem. We measure over a more or less broad frequency range the resulting permittivity $\varepsilon(v)$ of a composite dielectric and we want to calculate from it the volume fraction $v_1 = 1 - v_2$ of one of the constituents, namely the water. This is an intricate attempt, because $\varepsilon(v)$ does not just depend on v_1 but also on shape characteristics of the dielectric mixture and on the permittivities ε_1 and ε_2 of the aqueous and the non-aqueous phase. The evaluation of the experimental data is more difficult as, because of the internal polarising and depolarising electric fields, the permittivity of mixtures cannot be calculated rigorously. Hence, as already mentioned above, many different mixture relations exist even for solutions of simply shaped spherical and ellipsoidal solutes, each formula reflecting the particular assumptions made in the theoretical treatment of the problem. Irrespective of the theoretical model, the determination of the desired volume fraction v_l from the mixture permittivity $\varepsilon(v)$ requires also the most accurate knowledge of the dielectric properties of the constituents. Much progress has been made in the past decades in our understanding of the complex permittivity behavior of water as a function of frequency, temperature, and hydrostatic pressure. There exists also a large amount of data characterizing the dielectric properties of aqueous solutions, which is important since the aqueous constituent in microwave aquametry often will not just be water. In this short tutorial only some effects of solutes on the features of water have been presented and considered in the light of recent ideas about dielectric relaxation of hydrogen bonding dipolar liquids. Attention has been also directed toward solute contributions which, on various events, may dominate the complex permittivity of aqueous solutions.

References

1. Franks F (1972) Introduction – Water, the unique chemical. In: Franks F (ed) Water, a comprehensive treatise, vol 1. Plenum, New York, pp 1–20
2. Kaatze U (1989) Complex permittivity of water as a function of frequency and temperature. J. Chem. Eng. Data 34: 371–374
3. Ellison WJ, Lamkaouchi K, Moreau JM (1996) Water: a dielectric reference. J. Molec. Liquids 68: 171–279
4. Kaatze U (1996) Microwave dielectric properties of water. In: Kraszewski A (ed) Microwave aquametry. Electromagnetic wave interactions with water-containing materials. IEEE Press, New York, pp 37–53
5. Kaatze U (2000) Hydrogen network fluctuations and the microwave dielectric properties of liquid water. Subsurface Sensing Technol. Applic. 4: 377–391
6. Kaatze U (1995) Microw. dielec. properties of liquids. Rad. Phys. Chem. 45: 549–566
7. Kaatze U (1997) The dielectric properties of water in its different states of interaction. J. Solution Chem. 26: 1049–1112
8. Barthel JMG, Krienke H, Kunz W (1998) Physical chemistry of electrolyte solutions. Modern aspects. Steinkopff, Darmstadt
9. Kaatze U, Behrends R (2002) Dielektrische Eigenschaften von Wasser und wässrigen Lösungen. Technisches Messen 69: 5–11
10. Kaatze U, Behrends R, Pottel R (2002) Hydrogen network fluctuations and dielectric spectrometry of liquids. J. Non-Crystalline Solids 305: 19–28
11. Pottel R, Asselborn E, Eck R, Tresp V (1989) Dielectric relaxation rate and static dielectric permittivity of water and aqueous solutions at high pressures. Ber. Bunsenges. Phys. Chem. 93: 676–681.
12. Dukhin SS, Shilov VN (1974) Dielectric phenomena and the double layer in disperse systems and polyelectrolytes. Halsted, New York
13. Pottel R (1966) The complex dielectric constant of some aqueous electrolyte solutions in a wide frequency range. In: Conway BE, Barradas RG (eds) Chemical physics of ionic solutions. Wiley, New York
14. Marsh KN (1981) I.U.P.A.C. Recommended reference materials. Permittivity. Pure Appl. Chem. 53: 1847–1862

3 Water in Polymers and Biopolymers Studied by Dielectric Techniques

Polycarpos Pissis

Department of Applied Mathematics and Physics,
National Technical University of Athens

3.1 Introduction

Dielectric aquametry is based on the systematic variation of the dielectric properties of the material or the system considered with its water content [1–3]. Thus, one necessary condition for optimizing dielectric aquametry is to investigate and understand at the molecular level the dependence of the dielectric properties on water content for various classes of materials. This chapter deals with the systematic investigation of the relationships between dielectric properties and water content in polymers and biopolymers.

In dielectric aquametry the dielectric properties are measured and the unknown water content of the material under examination is determined on the basis of these measurements. A more fundamental concept thereby involves the independent measurement of both dielectric properties and water content and the investigation of the relationships between the measured quantities. These relationships may be described in terms of the hydration properties, a general term which refers to and includes both the way water is organised in the material and its effects on the final properties of the material itself [4–7]. Dielectric spectroscopy is just one technique (although a significant one) used to investigate the hydration properties of materials. Other techniques widely employed to this aim, in competition and/or synergy with each other, make use of other properties which change sensitively with water content. A true understanding of hydration properties is essential for several practical applications in materials science, food industry, biotechnology etc. [8–11]

Specifying the concept of hydration properties to dielectric measurements, two basic contributions are expected to the variation of the dielectric properties of a hydrated material with respect to those of a dry one: that of the polar water molecules themselves and the second one due to the modification of the various polarization and relaxation mechanisms of the matrix material itself by water [6, 12]. In the low-frequency region of measurements of interest in this chapter there is a third contribution, often ignored in work dealing with high-frequency measurements, which arises from the influence of water on conductivity and conductivity effects and often presents features of synergy of the two components [12]. If broadband dielectric relaxation spectroscopy (DRS) is considered as the main technique of measuring dielectric properties, which are then presented as complex dielectric permittivity at constant temperature, $\varepsilon(f)=\varepsilon'- i\varepsilon''$, where f is the frequency of measurements, then, at room temperature, the first contribution relaxes in the frequency region between several hundreds of MHz and a few decades of GHz, the second contribution in a broad frequency range typically below MHz and the third contribution is even slower. In this chapter we will mainly focus on dielectric measurements below 1 GHz and, thus, consider mostly effects related with the last two contributions.

In dielectric measurements for hydration studies two extreme cases can be distinguished with respect to the amount of water in the sample under investigation: aqueous solutions of the substance under study [6], if the latter can be dissolved in water, and hydrated solid samples [13]. Hydrogels, which can significantly swell but do not dissolve in water, and suspensions, that can be considered as intermediate cases, are, however, usually considered as solid samples and solutions, respectively. Both solutions and solid samples offer advantages and disadvantages. Solutions are easier to handle and to measure and for biological materials closer to practical applications. However, the dielectric response is typically dominated by the excess water, the dipolar response is often masked by conductivity and, as a result, the analysis of the experimental data in terms of hydration properties is a difficult task, often ambiguous [14]. Solid samples, on the other hand, offer the advantages of low electrical conductivity and variation of water content in small steps [13]. However, samples suitable for dielectric measurements are often difficult to prepare and handle, whereas for biological materials there is often the question of the biological significance of results obtained with such unrealistic systems. Results described here have been exclusively obtained with solid samples.

In this chapter we discuss in terms of hydration properties results of dielectric measurements obtained with polymers and biopolymers. The synthetic polymers studied include both materials with a relatively low amount of water uptake and hydrogels, i.e., hydrophilic polymeric networks which absorb large amounts of, but are not dissolved in, water. In the first class we discuss polyurethanes (PUs) and polyamides (PAs), in the second class poly(ethylene oxide) (PEO) and systems based on poly(hydroxyethyl acrylate) (PHEA). The protein casein is discussed as representative of biopolymers, together with results obtained with more complex biological systems, including seeds and plant tissue. The results presented here have been obtained by our research group in Athens in the framework of collaborations with other groups. They are discussed in relation to similar results obtained by other investigators using dielectric techniques and with results obtained by employing other techniques of studying hydration properties.

The chapter is organised as follows. In the next two sections we give some introductory remarks on hydration properties (Sect. 2) and on dielectric techniques and dielectric properties (Sect. 3). Section 4 is devoted to the overall dielectric response of the hydrated materials under investigation, whereas in the following three sections we discuss in some detail the effects of water on the local, secondary relaxations and relaxation of water molecules themselves in a separate water phase (Sect. 5), the cooperative, primary (segmental) α-relaxation and the glass transition (Sect. 6) and electrical conductivity and conductivity effects (Sect. 7).

Final conclusions are drawn in Sect. 8, where also implications of the results presented here on dielectric aquametry are briefly discussed.

3.2 Hydration Properties

The term hydration properties is used to denote both the specific organisation of water in the system under investigation and the influence of water on the structure and, in particular, the dynamics of the matrix (host) material itself. Several experimental techniques have been used to investigate the hydration properties of a variety of materials, by making use of either or both of the effects mentioned above. Next to dielectric techniques [6, 12–22], these include gravimetric water sorption/diffusion techniques [12, 23, 24], differential scanning calorimetry (DSC) [23, 25–27], equilibrium and dynamic swelling techniques [15, 28], dynamic mechanical analysis (DMA) [5, 20, 21], nuclear magnetic resonance (NMR) techniques [5, 29], ultrasonic wave attenuation measurements [30, 31], X-ray scattering techniques [7], infrared spectroscopy [32, 33], neutron scattering techniques [34], Raman spectroscopy [35] and positron annihilation lifetime spectroscopy (PALS) [5].

Two different approaches have been followed, in general, in the analysis and the interpretation of the results obtained with various hydrated materials by the aforementioned experimental techniques. The first, more common and traditional one is based on the classification of water into different classes with qualitatively different thermodynamic or dynamic properties, such as freezable and non-freezable water, free and bound water, mobile, immobile and clustered water [2, 5, 12, 16, 23, 25]. A model is frequently used to visualize such a classification, in which water molecules close to hydrophilic surfaces are bound to specific hydrophilic sites and, thus, relatively immobilized, whereas water molecules in sufficient distance from such surfaces are free, behaving like molecules in bulk water. In many cases other classes of water molecules are also considered, such as that of interfacial water, characterised by hydrophobic interaction with the matrix material, and that of loosely bound water molecules, which form clusters around water molecules bound at primary hydration sites. The classification depends, however, on the particular technique employed and, for the same technique, very often, on the method of analysing the data. Thus, the fraction of modified water measured by various techniques on the same system, e.g., non-freezable water by DSC, immobile water by gravimetric water sorption and bound water by dielectric techniques [36], is in general different [5, 15, 16, 30]. This point calls for attention when comparing with other results reported in literature on the hydration properties of materials by various techniques. In the second approach to analysing and interpreting results of hydration studies, no resource is made to the concept of various classes of water characterised by different degrees of interaction with the matrix and hence different mobilities. Thus, results obtained by DSC in hydrogels have been explained in a simple thermodynamic framework based on the phase diagram [26], whereas NMR results in hydrogels have been discussed and quantitatively interpreted by assuming the chemical exchange process between water protons and hydroxyl protons of polymer chains as the major relaxation source [37].

Molecular dynamics simulations are increasingly becoming a powerful tool for hydration studies. They show that water–water hydrogen bonds are enhanced around hydrophobic groups in hydrogels by the hydrophobic interaction and stabilised around hydrophilic groups by a severe constraint of the mutual orientation between water and polar group [38]. As a result of that, the mobility of water molecules is significantly lowered around polymer chains for both translational and rotational motions [39]. Molecular dynamics simulations in a poly(vinyl alcohol) (PVA) hydrogel with 8% water give evidence for two states of water only at a temperature below the freezing point of bulk water and show that water molecules in contact with the polymer are less mobile than free water, even below the freezing transition of bulk water [40].

3.3 Dielectric Techniques and Dielectric Properties

Dielectric techniques are a powerful tool for studying molecular mobility in various materials. Molecular mobility refers here to both dipolar reorientation and charge transport over mesoscopic and macroscopic distances. The main advantage of dielectric techniques over other techniques of measuring molecular dynamics is the extremely broad frequency range covered, which extends from about 10^{-5} to about 10^{11} Hz [41–43]. Obviously, this broad frequency range cannot be covered by a single technique.

In most cases measurements are carried out isothermally in the frequency domain and the term dielectric spectroscopy (DS) or dielectric relaxation spectroscopy (DRS) is then used. Isothermal measurements in the time domain are often used, either as a convenient tool for extending the range of measurements to low frequencies (slow time-domain spectroscopy (TDS), dc transient current method, isothermal charging–discharging current measurements) or for fast measurements corresponding to the frequency range of about 10 MHz–10 GHz (time-domain spectroscopy, TDS, or time-domain reflectometry, TDR). Finally, thermally stimulated depolarisation currents (TSDC) is a special dielectric technique in the temperature domain.

For measurements in the frequency domain, capacitance bridges, impedance analysers, frequency response analysers, radio-frequency reflectometers and network analysers are typically employed. The principle of these measurements is as follows. The sample under investigation is placed in a capacitor with empty capacitance C_0, which becomes a part of an electric circuit. A sinusoidal voltage with angular frequency ω is applied to the circuit and the complex impedance $Z(\omega)$ of the sample is measured. The complex dielectric permittivity $\varepsilon(\omega) = \varepsilon'(\omega) - i\varepsilon''(\omega)$, defined by

$$\varepsilon(\omega) = \frac{C}{C_0} , \qquad (3.1)$$

where C is the capacitance of the filled capacitor, is then obtained from

$$\varepsilon(\omega) = \frac{1}{i\omega Z(\omega)C_0} \ .$$

(3.2)

In slow TDS a voltage step V_p is applied to the sample and the polarisation or depolarisation current $I(t)$ is measured as a function of time. The time-dependent dielectric permittivity $\varepsilon(t)$ is then given by

$$\varepsilon(t) = \frac{C(t)}{C_0} \quad \text{and} \quad \frac{d\varepsilon}{dt} = \frac{I(t)}{C_0 V_p} \ .$$

(3.3)

Usually the depolarisation current is measured to avoid the dc conductivity contribution. The dielectric relaxation spectrum is then obtained by Fourier transform or approximate formulae, e.g., the Hamon approximation [41]. By carefully controlling the sample temperature and accurately measuring the depolarisation current, precision measurements of dielectric permittivity down to 10^{-6} Hz are possible [44]. In (fast) TDS or TDR a step-like pulse propagates through a coaxial line and is reflected from the sample section placed at the end of the line. The difference between the reflected and the incident pulses recorded in the time domain contains the information on the dielectric properties of the sample [45, 46].

Independently of the specific dielectric technique used, the results of dielectric measurements are usually analysed in the form of complex dielectric permittivity $\varepsilon(\omega) = \varepsilon'(\omega) - i\varepsilon''(\omega)$ at constant temperature by fitting empirical relaxation functions to $\varepsilon(\omega)$. In the examples to be given later in this chapter the two-shape-parameters Havriliak–Negami (HN) expression [47]

$$\varepsilon(\omega) - \varepsilon_\infty = \frac{\Delta\varepsilon}{[1 + (i\omega\tau)^{1-\alpha}]^\beta}$$

(3.4)

is fitted to the experimental data for a relaxation mechanism. In this equation $\Delta\varepsilon$ is the dielectric strength, $\Delta\varepsilon = \varepsilon_s - \varepsilon_\infty$, where ε_s and ε_∞ are the low- and high-frequency limits of ε', respectively, τ is the relaxation time, $\tau = 1/2\pi f_{HN}$, where f_{HN} is a characteristic frequency closely related to the loss peak frequency f_{max}, and α, β are the shape parameters describing the shape of $\varepsilon''(\omega)$ curve below and above the frequency of the peak, respectively, $0 < \alpha \leq 1$ and $0 < \beta \leq 1$. This expression becomes the single Debye form for $\alpha = 0$, $\beta = 1$, the symmetric Cole–Cole form for $\alpha \neq 0$, $\beta = 1$, and the asymmetric Cole–Davidson form for $\alpha = 0$, $\beta \neq 1$ [47, 48]. A proper sum of HN expressions is fitted to $\varepsilon(\omega)$ in the case of more than one overlapping mechanisms plus a term for the contribution of conductivity, if the latter makes a contribution at the temperature of measurements [49]. For each relaxation mechanism there are then three sources of information: the timescale of the response (τ), the dielectric strength ($\Delta\varepsilon$) and the shape of the response (α, β). By measuring $\varepsilon(\omega)$ at several temperatures, the time scale of the response is analysed in terms of the Arrhenius equation for secondary relaxations and the Vogel–Tammann–Fulcher (VTF) equation for the primary α-relaxation and valuable information on the activation parameters is obtained [50]. Examples will be given later in this chapter.

In the temperature domain the TSDC method allows for a fast characterisation of the dielectric response of the material under investigation. The method consists of measuring the thermally activated release of stored dielectric polarisation. It corresponds to measuring dielectric losses against temperature at constant low frequencies of 10^{-2}–10^{-4} Hz [51, 52]. In this method, the sample is inserted between the plates of a capacitor and polarised by the application of an electric field E_p at temperature T_p for time t_p, which is large in comparison with the relaxation time at T_p of the dielectric dispersion under investigation. With the electric field still applied, the sample is cooled to temperature T_0 (which is sufficiently low to prevent depolarisation by thermal excitation) and then is short-circuited and reheated at a constant rate b. A discharge current is generated as a function of temperature, which is measured with a sensitive electrometer. The resultant TSDC spectrum often consists of several peaks whose shape, magnitude and location provide information on the time scale and the dielectric strength of the various relaxation mechanisms present in the sample [51]. The method is characterised by high sensitivity and, owing to its low equivalent frequency [51], by high resolving power. In addition, it provides special variants to experimentally analyse complex relaxation mechanisms into approximately single responses [51–53].

3.4 Overall Behavior

Figure 3.1 shows TSDC thermograms obtained with a PU sample based on polyethylene adipate (PEA), 4,4′–diphenylmethane diisocyanate (MDI) and 1,4–butanediol at two water contents, $h = 0.002$ and 0.016 [54]. Throughout this chapter, h is defined as grams of water per gram of dry sample and is determined by weighing the sample before and after the dielectric measurement. The TSDC thermogram corresponds to measuring dielectric losses ε'' as a function of temperature T at constant frequency f in the range 10^{-2}–10^{-4} Hz. Four peaks are observed in Fig. 3.1 corresponding to two secondary relaxations γ and β in the order of increasing temperature, the primary α relaxation at about 230 K and an interfacial Maxwell–Wagner–Sillars (MWS) relaxation at higher temperatures [54]. The latter corresponds to accumulation of charges at the interfaces between phases of different conductivity and its study may provide significant information on the micromorphology of heterogeneous systems [24, 51].

The results in Fig. 3.1 indicate that the overall response increases, i.e., dielectric activity increases, with increasing water content. This result has been confirmed by measuring more PU samples at various h levels [24, 54], is characteristic for synthetic polymers [18, 55] and is often interpreted in terms of plasticising action of water on the various relaxation mechanisms present in the sample [8, 9, 11]. It is interesting to note, however, that there are distinct differences between

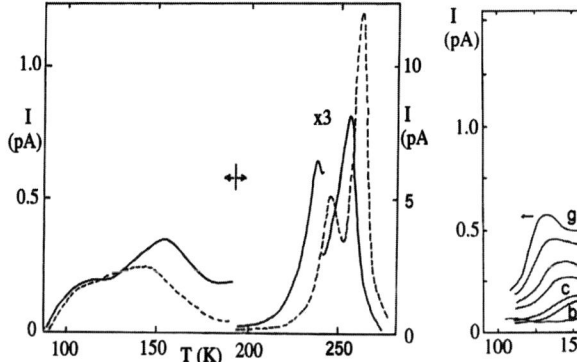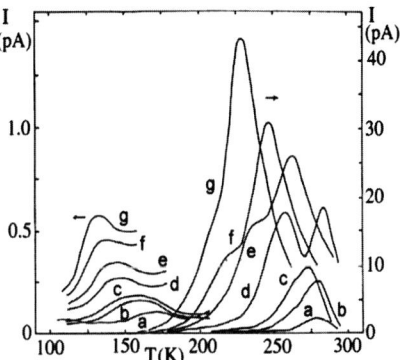

Fig. 3.1. TSDC thermograms obtained with a PU sample at h = 0.002 (--) and h = 0.016 (-). Note the change of scale at 190 K

Fig. 3.2. TSDC thermograms measured on casein samples at various water contents, h = 0.013 (a), 0.066 (b), 0.091 (c), 0.101 (d), 0.124 (e), 0.164 (f) and 0.214 (g)

the response of secondary relaxations on the one hand and primary and MWS relaxations on the other hand, i.e., between local, fast relaxations and slower relaxations of larger spatial scale. These differences, which are partly attributed to direct contribution of water molecules themselves to the dielectric response in the temperature/frequency region of the secondary relaxations, will be discussed in detail in subsequent sections of this chapter.

Typical TSDC thermograms for globular proteins are shown in Fig. 3.2. They have been obtained with casein powder supplied by Sigma (St. Louis, USA), used as received [56, 57]. The powder was compressed to cylindrical pellets of 13mm diameter and about 1 mm thickness. Prior to dielectric measurements the pellets were equilibrated at constant relative humidity and their water content h, indicated in Fig. 3.2, was determined by weighing. Similar thermograms have been recorded with lysozyme [57–59], a smaller, well-characterised globular protein, widely studied by dielectric techniques [60–62].

The thermograms in Fig. 3.2 are similar to those obtained also with other hydrated biomaterials, such as cellulose [63], DNA [64] and plant tissue [65]. Two dispersion regions are observed in these thermograms: a broad low-temperature dispersion with contributions from both secondary relaxations and loosely bound water molecules and a complex, high-temperature dispersion connected to conductivity effects [57–59]. Both dispersions shift to lower temperatures, i.e., they become faster with increasing water content, whereas their magnitude increases.

By comparing the thermograms in Figs. 3.1 and 3.2 with each other, a similarity in the overall behavior becomes obvious, despite differences in the origin of the individual relaxations to be discussed in some detail later: two dispersion regions are observed and in each of them the response increases and becomes faster with increasing water content. Measurements on more systems, a part of which will be described later, confirm that, apart from details, this behavior is typical for both synthetic polymers and biopolymers. This is true for even more complex

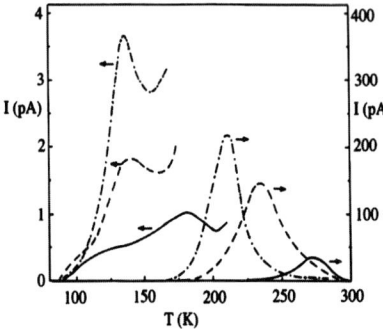 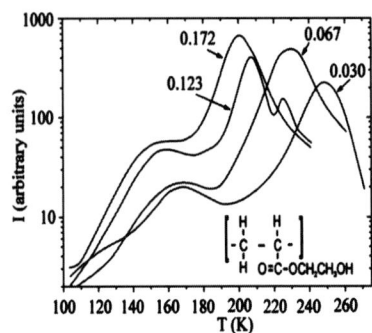

Fig. 3.3. TSDC thermograms recorded on wheat flour samples at water contents h = 0.08 (-), 0.13 (--) and 0.21 (---)

Fig. 3.4. TSDC thermograms obtained with PHEA hydrogels at several water contents indicated on the plot and the chemical structure of the repeating unit of PHEA

biological systems, such as plant tissue [65, 66] and seeds [67, 68], as suggested by Fig. 3.3, which shows TSDC thermograms obtained with wheat flour samples at various water contents [67]. At relatively high water contents the main contribution to the low-temperature dispersion in Fig. 3.3 arises from the reorientation of water molecules in water clusters and/or layers around the primary hydration sites, whereas secondary relaxations of proteins and carbohydrates dominate at lower h values. Conductivity effects and a glass-like transition [8, 69, 70] make the main contribution to the high temperature dispersion. The TSDC spectra of different plant seeds show, superimposed to an overall similar behavior, small but characteristic differences to each other, reflecting the different composition of each species with respect to protein and carbohydrate content [68]. In recent years, TSDC is being increasingly recognized as a powerful tool to investigate the hydration properties of plant tissue and seeds and to provide significant information relevant to plant physiology [71–73].

The overall TSDC response of the hydrogels is similar to that of synthetic polymers and biopolymers. With respect to interpretation of the observed relaxations at the molecular level, hydrogels take an intermediate place between synthetic polymers and biopolymers, combining the synthetic character and the structure of the former with the ability of high water uptake of the latter. Fig. 3.4 shows TSDC thermograms obtained with poly(hydroxyethyl acrylate) (PHEA) hydrogels at several water contents. The chemical structure of the repeating unit of PHEA is also shown in the figure. The hydration properties of these biocompatible hydrogels have been studied in detail over the last ten years by combining, next to dielectric techniques, mostly water sorption/diffusion measurements, density measurements, differential scanning calorimetry and electron microscopy. Next to pure PHEA homopolymer [36, 50, 53, 74–76], various combinations with a second hydrophobic component have been used as matrices to improve mechanical stability, including composites [76, 77], copolymers [78] and interpenetrating

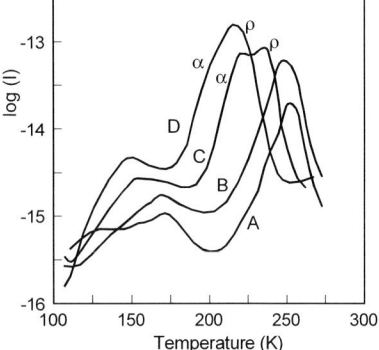

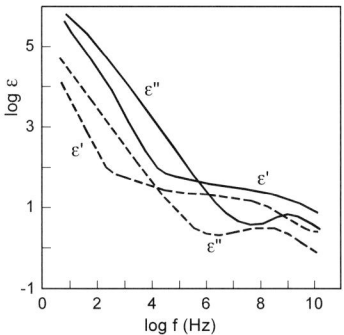

Fig. 3.5. TSDC thermograms of a PMA/PHEA IPN containing 53wt% PHEA at water contents h = 0.02 (A), 0.04 (B), 0.08 (C), and 0.12 (D)

Fig. 3.6. Log–log plot of the complex dielectric permittivity $\varepsilon = \varepsilon' - i\varepsilon''$ vs. frequency f of a PHEA hydrogel at h = 0.22 (--) and 0.42 (-) at 298 K

polymer networks (IPNs) [76, 78–80]. As an example, the overall dielectric response of an IPN of poly(methyl acrylate) (PMA) and PHEA (PMA/PHEA IPN) with 53 wt% PHEA measured by TSDC is shown in Fig. 3.5 at several water contents [80].

Similar to Fig. 3.4, the results show an overall increase of molecular mobility with increasing water content and a shift of the individual relaxations to lower temperatures. The origin of the individual relaxations will be discussed in the subsequent sections, as well as distinct differences in the behavior of the various PHEA-based hydrogels, depending on the morphology (homogeneous or microphase-separated) of the hydrogel matrix [76, 78, 80]. Such a difference is already suggested by simply comparing Figs. 3.4 and 3.5 with each other and refers to the appearance of the space charge related peak in the thermograms of the hydrogels with a microphase-separated matrix [80].

So far to the overall response measured by TSDC. Figure 3.6 shows the overall AC response of a sample similar to that of Fig. 3.4 measured by broadband dielectric relaxation spectroscopy (DRS) at 298 K. Similar to the TSDC response, the overall AC response increases with increasing water content. The high values of ε' and ε'' at low frequencies are related with conductivity effects [50]. The α-relaxation, associated to the glass transition, should appear in the kHz frequency region [78], is, however, masked by conductivity effects and cannot be studied by DRS, similar to many other polymers [47, 50]. This is an inherent difficulty of DRS, however not of TSDC, as indicated in Figs. 3.4 and 3.5, as in TSDC the steps of polarisation and depolarisation, i.e., stimulus and response, are separated from each other [53]. The relaxations at higher frequencies in the MHz–GHz frequency region correspond to the low-temperature TSDC peaks in Fig. 3.4 and are due to the secondary γ- and β_{sw}-relaxations and, at high water contents, probably also to reorientation of water molecules themselves [50],

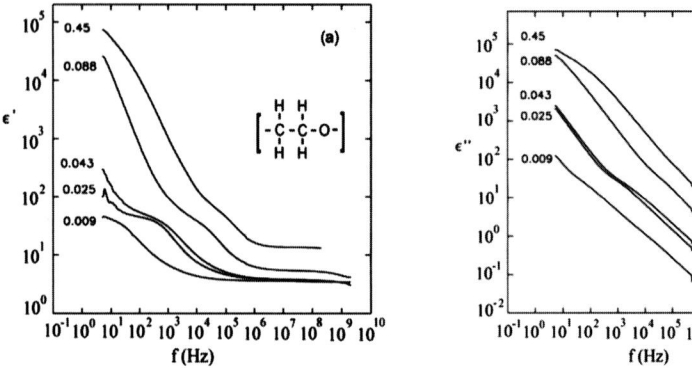

Fig. 3.7. Log–log plot of ε' (**a**) and ε'' (**b**) vs. frequency f for PEO hydrogels at T=297 K and several water contents indicated on the plot

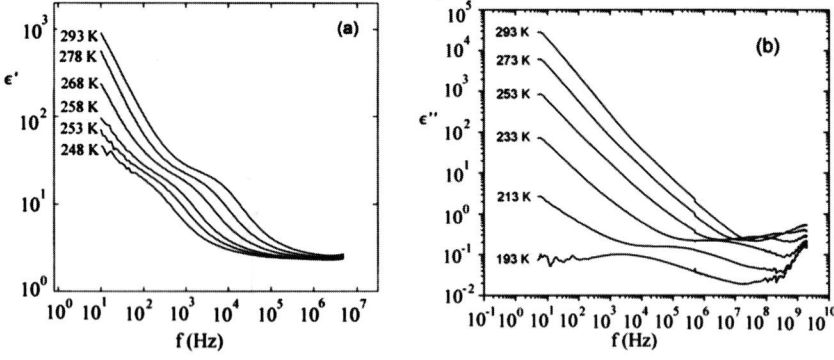

Fig. 3.8. Log–log plot of ε' (**a**) and ε'' (**b**) vs. frequency f for PEO hydrogels at h=0.052 (a) and h=0.056 (b) at several temperatures indicated on the plot

a controversial point to be discussed in the next section.

Figure 3.7 shows the overall AC response of another hydrogel, based on poly(ethylene oxide) (PEO) of the chemical structure of the repeating unit shown also in the same figure, measured by broadband DRS [81]. In addition to an overall increase of the response with increasing water content, the results suggest also the existence of critical water contents of characteristic changes of the spectra, a point which will be discussed later in relation to TSDC data too. Similar to Fig. 3.6, high values of both ε' and ε'' are observed at low frequencies, due to space charge polarisation and charge carrier motion, respectively. The decrease of the slope of $\varepsilon''(f)$ at low frequencies, 5–100 Hz, at $h = 0.45$ is due to electrode

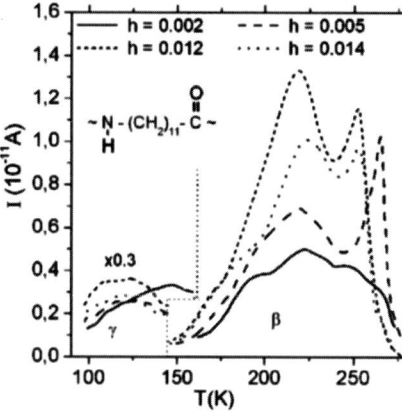

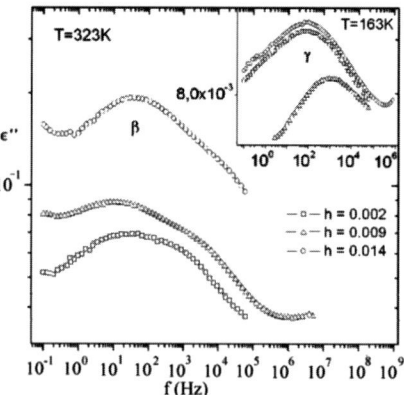

Fig. 3.9. TSDC thermograms obtained with polyamide 12 at four levels of water content indicated on the plot and the chemical structure of the repeating unit of the polymer

Fig. 3.10. Dielectric loss ε'' vs. frequency f in the region of the secondary relaxations in polyamide 12 at the temperatures and the water contents given on the plot

polarisation [81]. In contrast to the results in Fig. 3.6, a structure is observed at low frequencies in Fig. 3.7, more clearly as a step in $\varepsilon'(f)$ shifting to higher frequencies with increasing water content, attributed to the conductivity mechanism [81, 82]. Finally, the increase of ε'' at high frequencies shifting to the right with increasing water content is due to the plasticised secondary γ relaxation, studied in detail by combined DRS and TSDC analysis in Ref. 49.

It is interesting to compare the overall changes in molecular mobility induced by increase of water content h in Fig. 3.7 with those induced by increase of temperature T in Fig. 3.8: increase of T at constant h induces the same effects as the increase of h at constant T, namely an overall increase of molecular mobility. In both cases the increase of molecular mobility is reflected in both the increase of ε' and ε'' due to "free" charge motion and the shift of the secondary γ-relaxation to higher frequencies with simultaneous increase of its relaxation strength. The effects on "free" charge motion are by far more striking. Similar results have been obtained also with PHEA hydrogels [50] and plant seeds [67] and present a manifestation of the T–f–h superposition principle proposed earlier for polymers [83]. It states that the plasticising effect of increasing water content at constant temperature is equivalent to the effect of increasing temperature at constant moisture, both leading to increasing segmental mobility of the polymeric chains. The question, how the various relaxation processes are affected by temperature and humidity at the molecular level, remains thereby open.

3.5 Effects of Water on Secondary Relaxations and Relaxation in a Separate Water Phase

Figure 3.9 shows TSDC thermograms recorded in the region of the γ- and β-relaxations of polyamide 12 at four water contents. Figure 3.10 shows $\varepsilon''(f)$ plots for the same polymer and relaxations at three water contents. The γ-relaxation in polyamides is attributed to motion of $(CH_2)_n$ sequences with participation of the polar amide and carbonyl groups and the β-relaxation to associations of various dipolar groups and water [18, 55, 83–86]. The γ-relaxation is broad in dry polyamides because of the various configurations of H-bonds. When water is sorbed, H-bonds are broken and new ones are formed, NH–H_2O and CO–H_2O. Equilibrium water sorption isotherms in polyamide 11 and polyamide 12 indicate that not all NH and CO groups are accessible to water. It is also possible that H_2O makes two H-bonds with two adjacent NH or CO groups. These groups are now more heavy and participate in the β-relaxation, with the result that the magnitude of the β-relaxation increases at the expense of the γ-relaxation. The remaining non-bonded groups give a faster and narrower (more homogeneous environment) relaxation. The TSDC and DRS results in Figs. 3.9 and 3.10 are in agreement with each other and consistent with this interpretation. Moreover, the TSDC results suggest that in dry polyamide 12 γ consists of two peaks, γ_1 and γ_2 in the order of decreasing temperature, which may be attributed to two regions of different density or to CO and NH groups separately. With increasing water content the slower γ_1-relaxation diminishes, whereas γ_2 remains practically unchanged.

The β-relaxation in polyamide 12 is also double, as indicated in Figs. 3.9 and 3.10, consisting of a stronger and slower component β_1 and a faster and weaker component β_2 in relatively dry samples. These may be attributed to NH–H_2O and CO–H_2O associations, respectively. With increasing water content the response in the β-region increases, β_1 becomes slower and then faster, β_2 becomes faster. The TSDC results in Fig. 3.9 show that the magnitude of β_2 decreases again at high h values, in consistency with results in polyamide 6 obtained by TSDC and dynamic mechanical analysis (DMA) [85].

The effects of water on the secondary relaxations of polyamide 12 described briefly above are typical for polyamides [18, 55, 83–86]. Similar effects have been observed in other classes of synthetic polymers with chemical structures similar to that of polyamides [9, 84]. In particular PUs should be mentioned here, where the origin of the secondary γ and β-relaxations (Fig. 3.1) is believed to be the same as in polyamides [54]. The dielectric results give no indication for the appearance of any new relaxation in fully hydrated samples, which might be attributed to relaxation of water molecules themselves, neither in PUs [24, 54] nor in polyamides with large number of CH_2 sequences [13, 55]. This is consistent with the results of equilibrium water sorption isotherms measurements, which indicate low levels of water uptake with maximum values below 0.03 and absence of clustering of water molecules, and of differential scanning calorimetry (DSC), which show no events of freezing or melting of water [13, 24, 54].

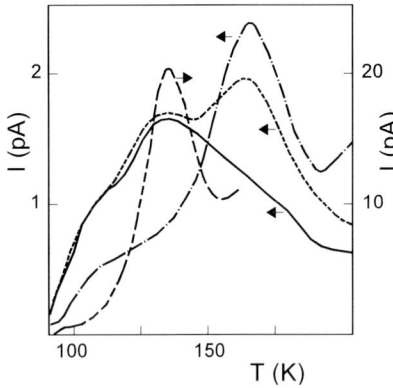

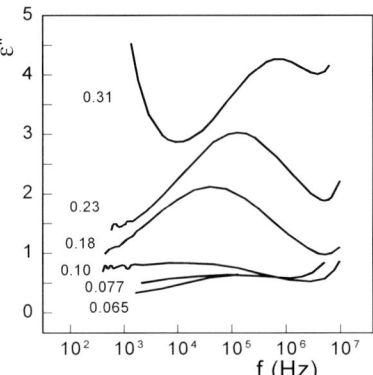

Fig. 3.11. Low temperature TSDC thermograms of PHEA hydrogels at h=0.01 (-), 0.02 (--), 0.10 (-··-) and 0.36 (····)

Fig. 3.12. Semi-logarithmic $\varepsilon''(f)$ plot for the secondary γ- and β_{sw}-relaxations measured on PHEA hydrogels at 223 K at several water contents indicated on the plot

Figure 3.11 shows TSDC thermograms in the region of the low-temperature dispersion of Fig. 3.4, corresponding to the secondary γ- and so-called β_{sw}-relaxations of PHEA hydrogels at various water contents. At low h values, $h = 0.01$ and $h = 0.02$, the dispersion consists of two peaks, γ and β_{sw} in the order of increasing temperature, well resolved because of the high peak-resolving power of the TSDC method. The γ-dispersion has been attributed to local motions within the side chain that are able to orient the dipolar moment of the hydroxyl groups, whereas β_{sw} is caused by the association of one water molecule and two hydroxyl groups of neighbour side chains [50, 53, 74, 80]. With increasing h, the magnitude of β_{sw} increases rapidly, whereas that of the γ-peak decreases, so that γ appears as a shoulder to the left of β_{sw} ($h = 0.02$ and 0.10) and finally disappears in β_{sw} ($h = 0.36$). The shift of β_{sw} to lower temperatures is due to plasticisation of the side-chain motion.

The DRS counterpart of Fig. 3.11 is shown in Fig. 3.12. As, in contrast to TSDC measurements, both temperature and frequency can be varied in DRS, measurements in Fig. 3.12 were carried out at low temperatures, in order to suppress the conductivity effects shown in Fig. 3.6 [50]. At $h \leq 0.10$ the secondary loss peak becomes slower and broader with increasing h. These results are straightforwardly explained in terms of the coexistence of γ and β_{sw}, in agreement with the (more clear) TSDC results in Fig. 3.11. At $h \geq 0.10$, the γ-loss disappears in the β_{sw}-peak and the plots show the strong plasticisation of the β_{sw}-mechanism, in agreement with the TSDC results of Fig. 3.11. Thus, from the point of view of polymer–water interactions and organization of water in the PHEA hydrogels, both DRS and TSDC results indicate the existence of strong interactions between water and the PHEA side chains, particularly at low water contents. These dielectric results are consistent with those of equilibrium water sorption isotherms measurements, which indicate the existence of strong hydrophilic groups acting as primary hydration sites at low values of relative humidity/water content and as nuclei for the formation of water clusters at higher values of relative humidity/water content [12, 36], and of dynamic water sorption measurements, which show that diffusion coefficients of water in PHEA are one order of magnitude smaller than in hydrophobic acrylates, such as poly(methyl acrylate) [80] and poly(ethyl acrylate) [79]. The molecular origin of the γ-relaxation is different in PHEA than in polyamides and in PUs and the range of water contents also. Nevertheless, there are strong analogies in the interplay between γ and β (β_{sw})-relaxations in these classes of polymers.

By measuring more water contents than in Figs. 3.11 and 3.12 and properly analysing and presenting the dielectric data, interesting effects may be revealed [50, 53, 80]. As an example, Fig. 3.13 shows the water content dependence of the peak temperature T_m and of the normalized (to the same heating rate) current maximum I_m/b of the low-temperature TSDC dispersion in PHEA hydrogels. T_m shifts systematically to lower temperatures with increasing h, for $h \leq 0.30$, and becomes then independent of h, in consistency with the results of DSC measurements showing freezing and melting events of water, suggesting the appearance of a separate water phase in the same h range. I_m/b is a measure of the number of relaxing units contributing to the TSDC dispersion. It increases practically linearly with h for $h \leq 0.30$, suggesting that new water molecules are continuously incorporated to the molecular groups responsible for this dispersion, i.e., the number of primary hydration sites increases with h [80]. The steeper increase of I_m/b with h for $h \geq 0.30$ suggests an additional contribution to the low-temperature TSDC peak, which may arise from the relaxation of water molecules in a (frozen) separate water phase. The results of equilibrium water sorption isotherms measurements indicate significant clustering of water molecules at water contents higher than about 0.2–0.3 [12, 36] and DSC measurements freezing and melting events of water in the same h range [36], in agreement with this interpretation. The results of TSDC measurements on ice [52] and on several biopolymers to be reported later in this section provide additional support for this interpretation. As for DRS measurements at room temperature, the interesting frequency range with respect to the question of relaxation of water molecules themselves is that of microwaves. The dipolar reorientation of water molecules gives rise to a loss peak at about 20 GHz at 298 K in bulk water and at lower frequencies for water bound in different states and for interfacial water [6, 11]. The loss peaks at about 0.1–0.2 GHz ($h = 0.22$) and 1 GHz ($h = 0.42$) in Fig. 3.6 may be assigned to the β_{sw}-relaxation, it cannot be excluded, however, that a part of the peak originates from the relaxation of water molecules in a separate water phase. The lower frequency of that contribution, as compared to that of bulk water, would then be explained as a result of confinement effects [12].

Johari and coworkers [87, 88] and other investigators [16] measured the relaxation of supercooled water in poly(hydroxyethyl methacrylate) (PHEMA) hydrogels in the limited temperature range of about 160–200 K in the Hz–kHz frequency range. The relaxation could not be reliably followed at higher temperatures, as supercooled water crystallises to cubic ice.

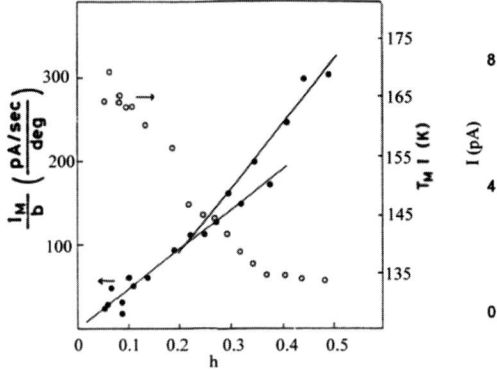

Fig. 3.13. Peak temperature T_m (o) and normalized current maximum I_m/b (•) against water content h for the low-temperature dispersion in PHEA hydrogels measured by TSDC. The lines are guides for the eye

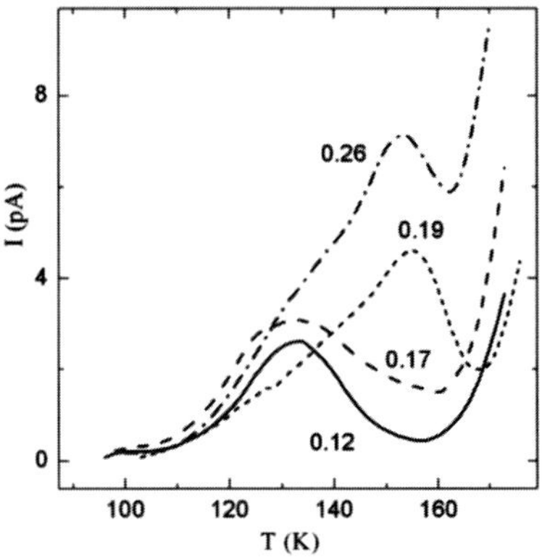

Fig. 3.14. Low-temperature TSDC thermograms of PEO hydrogels at four water contents indicated on the plot

However, it could be observed at lower temperatures by TSDC measurements. Thus, the additional contribution to the low-temperature TSDC peak for $h \geq 0.30$ suggested by Fig. 3.13 may arise from the relaxation of supercooled water. The peak temperature T_m of the low-temperature TSDC peak for $h \geq 0.30$, at around 135 K, is very close to the calorimetric glass transition temperature T_g of bulk glassy water (136 K) and of supercooled water in PHEMA hydrogels (132±4K) [87] and, thus, supports this interpretation.

With respect to plasticisation of secondary mechanisms and organisation of water in a separate phase in hydrogels, it is interesting to compare the results obtained with PHEA (Figs. 3.11–13) with those obtained with PEO hydrogels. Fig. 3.14 shows TSDC thermograms in the region of the low-temperature dispersion for PEO hydrogels at several water contents indicated on the plot. The peak at about 130 K at low water contents and the shoulder in the same temperature region at higher h values is due to the γ-relaxation, assigned to local crankshaft or kink motion in the polymer chains in the amorphous phase [49, 84]. Measurements on more water contents show strong plasticisation of the relaxation by water, providing support for the existence of strong interactions between water molecules and PEO chains [89]. For $h \geq 0.19$, a new peak appears at 152 K with approximately constant peak temperature and relaxation strength increasing linearly with increasing h, attributed to the relaxation of water in a separate water phase [49]. Interestingly, the high-temperature TSDC dispersion due to the glass transition in PEO hydrogels, not shown here, is no further plasticised for $h \geq 0.19$, in agreement with that interpretation. The peak temperature of 152 K is relatively high with respect to corresponding peaks in PHEA hydrogels (Figs. 3.11 and 3.13)

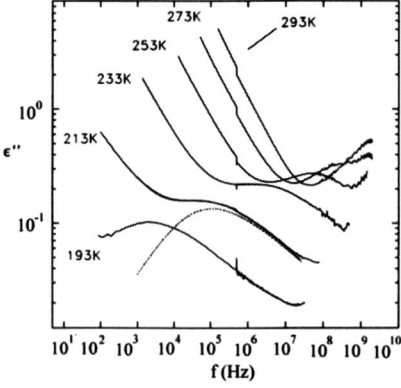 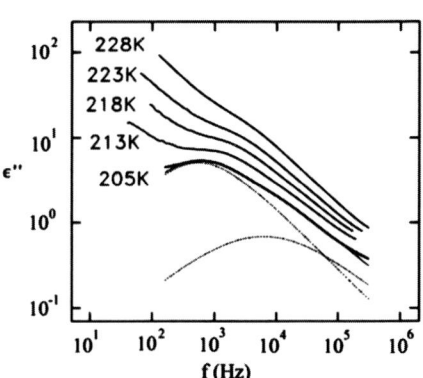

Fig. 3.15. Log–log $\varepsilon''(f)$ plot of a PEO sample with $h = 0.056$ at several temperatures indicated on the plot. The fit of a sum of a dipolar HN expression (3.4) (····) and a conductivity term to the data at 213 K is also shown (····)

Fig. 3.16. Log–log $\varepsilon''(f)$ plot of a PEO hydrogel with $h = 0.44$ at several temperatures indicated on the plot. The fit of a sum of two dipolar HN expressions (3.4) (····) and a conductivity term to the data at 205 K is also shown (-)

and in biopolymers (Figs. 3.2 and 3.3), suggesting a specific organisation of water molecules in PEO hydrogels [49, 89].

Figure 3.15 shows $\varepsilon''(f)$ spectra for a PEO sample at $h = 0.056$ over a wide frequency range and at several temperatures indicated on the plot. The secondary γ-relaxation is followed in this plot, in addition to conductivity effects at high temperatures/low frequencies. The activation energy W and the pre-exponential factor τ_0 of the relaxation are obtained from a fit of the Arrhenius equation [50, 84]

$$\tau(T) = \tau_0 \exp(W / kT) \tag{3.5}$$

to the data, a plot of the logarithm of relaxation time τ against reciprocal temperature. Measurements at several water contents give values of W decreasing with increasing h, quantifying the plasticising effect of water [49]. At higher water contents, the DRS spectra at subzero temperatures become more complex, as indicated in Fig. 3.16. A sum of two HN expressions (4), i.e., of two loss peaks, has been fitted to the $\varepsilon''(f)$ data at each temperature plus a conductivity term, an example of fitting being shown in the figure. The relaxation strength $\Delta\varepsilon$, the relaxation time τ, and the shape parameters α and β at each temperature, as well as the activation parameters W and τ_0 in the Arrhenius equation (5), were determined for each peak and the two relaxations interpreted in terms of relaxation of water molecules in a separate ice phase and of interfacial Maxwell–Wagner–Sillars (MWS) relaxation [49, 90]. In this range of water contents and temperatures, the γ-relaxation is shifted at higher frequencies out of the range of measurements in

Fig. 3.16. Thus, the dielectric relaxation of water molecules in a separate ice phase was followed in two temperature/frequency regions by DRS (Fig. 3.16) and TSDC

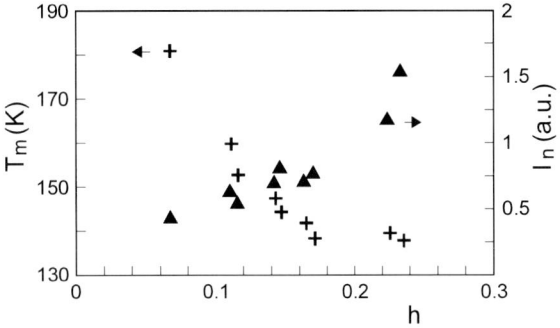

Fig. 3.17. Peak temperature T_m (+) and normalized current maximum I_n (▲) of the low–temperature TSDC dispersion measured on chickpea seeds against water content h

(Fig. 3.14), in addition to that in a separate liquid water phase, not shown here, centered at 4–5 GHz at 295 K for a PEO hydrogel with $h = 0.45$ [49].

Figure 3.17 shows, in analogy to Fig. 3.13 for the PHEA hydrogels, results of the analysis of the low-temperature TSDC dispersion in seeds (Fig. 3.3) in terms of the dependence of the peak temperature T_m and of the normalized current maximum I_n on water content h. T_m decreases continually with h for $h < 0.15$–0.20 and becomes then constant, $T_m = 130$–$135K$. I_n increases slowly with h for $h < 0.15$–0.20, whereas the increase is significant for larger h. Similar results were obtained with beans and cowpea seeds and interpreted in terms of a critical water content h_c of about 0.15–0.20, where loosely bound water molecules start to form water clusters and a separate water phase [67–68]. Similar results were obtained with other biomaterials, such as proteins [56, 57], cellulose [63], DNA [64] and plant tissue [65,66], in agreement also with other investigators [71–73, 91], h_c being a characteristic parameter for the material under investigation.

3.6 Effects of Water on Glass Transition and Primary α-Relaxation

Figure 3.18 shows isochronal (constant frequency) $\varepsilon''(T)$ plots of a model segmented PU sample based on poly(propylene glycol) (PPG) and 4,4′–diphenylmethane diisocyanate (MDI) in the region of the α-relaxation (primary or segmental relaxation) associated to the glass transition of the amorphous PPG–rich phase. The water content h of the sample was varied between 0.0003 (nearly dry sample) and 0.02 (saturated sample) and the α-relaxation was found to shift to lower temperatures (i.e., to become faster) and to increase in magnitude with increasing h. Similar results were obtained also with other PUs and interpreted in terms of plasticisation of the glass transition [24]. Measurements of equilibrium water sorption isotherms and of water diffusion coefficients on the same sample suggest that clustering of water molecules is insignificant in the whole range of water contents, in consistency with the low values of h and in agreement with the fact that plasticisation is observed in the whole h range [24].

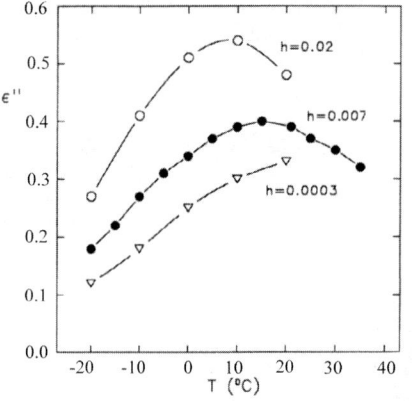

Fig. 3.18. $\varepsilon''(T)$ plot of a PU sample at a fixed frequency of 30 kHz and three water contents h given on the plot

TSDC measurements on PU samples (Fig. 3.1) and on synthetic polymers, in general, give a peak close to the calorimetric glass transition temperature T_g [12, 13, 18, 54]. This is because TSDC and DSC are characterized by similar time scales and because dielectric and thermal techniques have been found to probe the dynamics of molecular units of similar spatial scale [51–53]. The latter observation forms the basis for the convention that T_g is obtained from dielectric measurements as the temperature where τ becomes 100 s [50]. Figs. 3.19 and 3.20 show the plasticising effect of water on the glass transition in the segmented PUs of Fig. 3.1. In Fig. 3.19 the dependence of the glass transition temperature T_g determined by TSDC and by DSC, in good agreement with each other, on the weight fraction of water in the sample x_2, i.e., grams of water per gram of wet sample, is presented. The latter has been chosen as parameter instead of h to check for the validity of the Couchman–Karasz equation [92].

$$T_g = \frac{x_1 \Delta C_{p1} T_{g1} + x_2 \Delta C_{p2} T_{g2}}{x_1 \Delta C_{p1} + x_2 \Delta C_{p2}} \quad , \tag{3.6}$$

where x_1 is the weight fraction of PU in the sample, T_{g1} and T_{g2} the glass transition temperatures of dry PU and of pure water, respectively, and ΔC_{p1} and ΔC_{p2} the respective specific heat increments at T_g. For the fitting, values of T_{g2}=133 K and ΔC_{p2}=1.94 J/(gK) from the literature have been used [54]. The fit in Fig. 3.19 is satisfactory, indicating that PU and water form a homogeneous mixture [54]. In Fig. 3.20 the dependence of the normalised current maximum I_n, i.e., of the relaxation strength $\Delta\varepsilon$ [51–53], of the TSDC α peak on water content h is presented for the same PU as in Fig. 3.19. The results in Figs. 3.19 and 3.20 suggest strong plasticising effect of water. It is interesting to note that similar effects were observed on the temperature position and the magnitude of the TSDC MWS peak in the same PUs [54]. This result should be expected, as conductivity and conductivity effects in polymers are governed by the motion of the polymeric chains and, thus, reflect properties of the glass transition [12, 50].

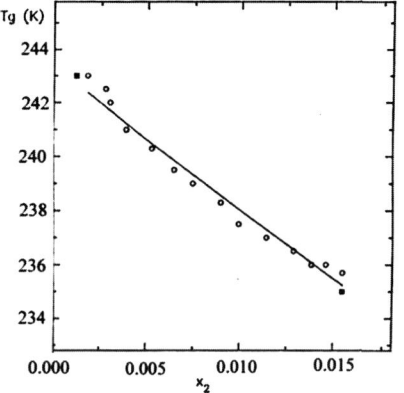
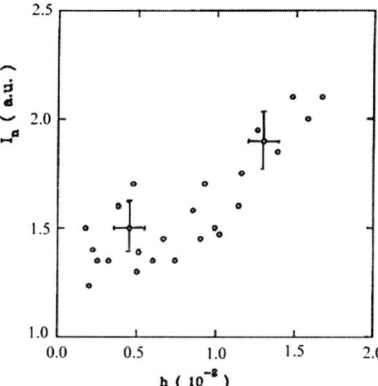

Fig. 3.19. Glass transition temperature T_g of a PU sample determined by TSDC (O) and by DSC (■) versus weight fraction of water x_2. The line is a fit of the Couchman–Karasz equation (3.6) to the data

Fig. 3.20. Normalized current maximum I_n of the TSDC α-peak of the same PU as in Fig. 3.19 versus water content h

Figure 3.21 shows results for PHEA hydrogels similar to those shown in Fig. 3.19 for PUs. The x_2 range of DSC measurements is limited to $x \leq 0.22$, due to freezing and melting events of water at higher x_2 values, which provide additional peaks in the temperature region of the glass transition and do not allow T_g to be determined unambiguously. Details of fitting and the values of the fitting parameters have been given elsewhere [12, 36]. The good fits in Fig. 3.21 support the idea that the PHEA hydrogels are homogeneous mixtures for x_2 values lower than about 0.2–0.3, whereas a separate water phase is formed at higher values. This result is in excellent agreement with results shown in Fig. 3.13 for the secondary relaxations in the same hydrogels.

The results in Figs. 3.13 and 3.21 indicate that dielectric measurements may be employed to investigate aspects of phase morphology in hydrogels. This possibility may be extended to include studies of the phase morphology of the hydrogel matrix. In most applications the hydrogel matrix is a multicomponent system, combining in the form of blends, copolymers, interpenetrating polymer networks (IPNs) or composites [77–80] a hydrophilic component to ensure high water uptakes and a hydrophobic component to improve mechanical stability. Figure 3.22 shows results for IPNs of PHEA and poly(ethyl acrylate) (PEA) for various compositions. The water content in this figure refers not to the dry IPN weight, but to the dry PHEA weight in it, and the data for the various IPNs fall on a single curve, which, moreover, coincides with the prediction of the Couchman–Karasz equation (6). The results in Fig. 3.22 show, in agreement with those of equilibrium and dynamic water sorption measurements [79], that the PHEA component in the hydrogels behaves in essentially the same manner like the pure polymer and suggest a phase–separated morphology of the hydrogel matrix at the mesoscopic level of a few nm [78, 79]. Similar results were obtained also with the PMA/PHEA IPNs of Fig. 3.5, in consistency with specific volume measurements showing volume additivity [80],

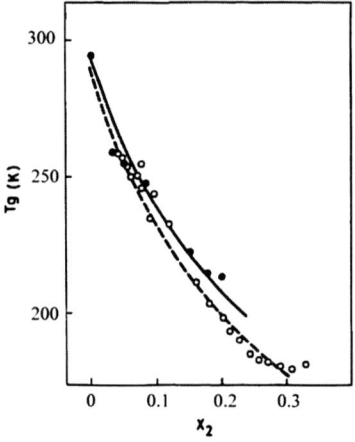

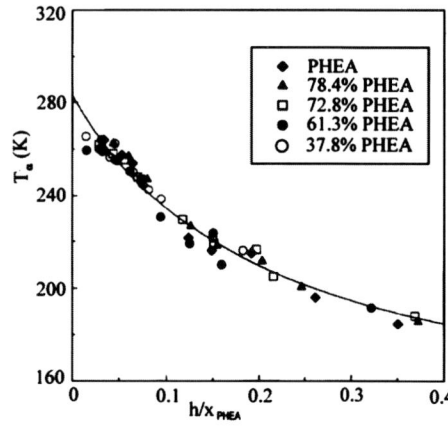

Fig. 3.21. Glass transition temperature T_g of PHEA hydrogels measured by TSDC (O) and by DSC (●) versus weight fraction of water x_2. Lines are fits of the Couchman–Karasz equation (3.6) to the DSC (-) and the TSDC data (--)

Fig. 3.22. Peak temperature T_α of the TSDC α peak of PHEA versus water content of the hydrogel referred to PHEA weight in the sample, h/x_{PHEA}, for several PEA/PHEA IPNs of the composition indicated on the plot. The line is a fit of the Couchman–Karasz equation (3.6) to the data

whereas in hydrogels based on PMA/PHEA copolymers, which are homogeneous systems characterised by a single glass transition, the peak temperature T_α of the TSDC α-peak depends sensitively not only on water content referred to PHEA content, but also on the composition of the dry hydrogel, as shown in Fig. 3.23 [78].

The phase-separated morphology of the hydrogel matrix allows to study by DRS the α-relaxation of PHEA, which in pure PHEA, like in many other polymers [84], is masked by conductivity, as indicated also in Fig. 3.6 [50]. Fig. 3.24 shows results obtained with IPNs of PHEA and poly(ethyl methacrylate) (PEMA), where PHEA has been polymerised in the pores of a sponge of PEMA network prepared by polymerisation in ethanol. At the low PHEA content of 14.6% of the IPN in Fig. 3.24, the conducting PHEA islands are isolated from each other, with the result that macroscopic conductivity of the IPN is suppressed and the α-relaxation is revealed as the strong relaxation at high temperatures shifting to higher frequencies with increasing temperature. The loss peak shifts to lower frequencies in the dry sample, in agreement with TSDC results for PHEA hydrogels shown in Figs. 3.4 and 3.5, without any significant change in the magnitude of the peak. The latter is in apparent contradiction to the TSDC results in Figs. 3.4 and

3.5, which is however resolved by considering detailed TSDC studies in pure PHEA hydrogels. These have revealed that conductivity makes a significant contribution to the TSDC high–temperature peak in Fig. 3.4 and that the magnitude of the dipolar component of that peak does not practically change with water content [53].

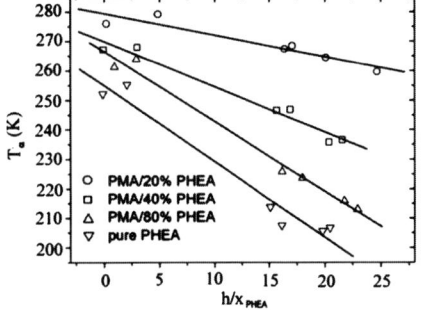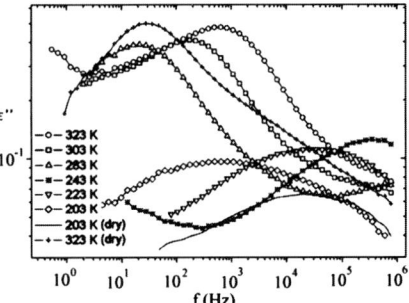

Fig. 3.23. Peak temperature T_α of the TSDC α peak in PMA/PHEA copolymers versus water content of the hydrogel referred to PHEA weight in the sample, h/x_{PHEA}. The lines are best linear fits to the data

Fig. 3.24. Log–log $\varepsilon''(f)$ spectra in PEMA/14.6% PHEA IPNs at ambient relative humidity at several temperatures indicated on the plot. For comparison two measurements on the dry sample have been included

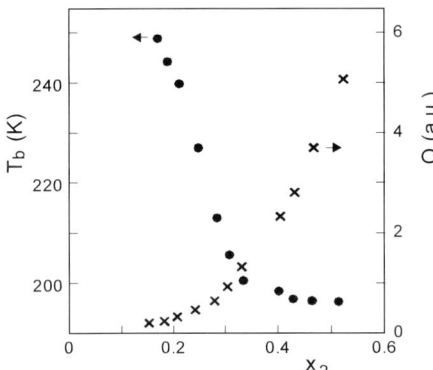

Fig. 3.25. Effective band temperature T_b ($\bullet$) and depolarization charge Q ($\times$) of the high–temperature TSDC band in Eucalyptus globulus leaves versus weight fraction of water x_2

Detailed TSDC studies in several biomaterials suggest the existence of glass-like transitions shifting to lower temperatures with increasing water content [69–73], in agreement with results obtained by other techniques [8,11]. As a result of the rather complex morphology of many biomaterials, the high–temperature TSDC dispersion of these materials is multiple, containing other components next to the α-peak associated with the glass transition, as partly indicated in Figs. 3.2 and 3.3. TSDC offers special techniques for experimentally analysing multiple dispersions into approximately single responses [51–53] and their use has provided strong evidence for glass or glass-like transitions in several biomaterials [69, 70]. This point will not be further discussed here. Instead, Fig. 3.25 shows the water content dependence of the effective band temperature T_b, defined as the geometric centre of gravity of the band, and of the depolarization charge Q, obtained

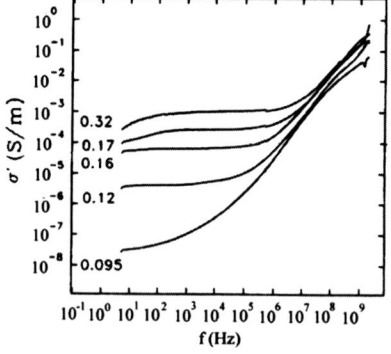 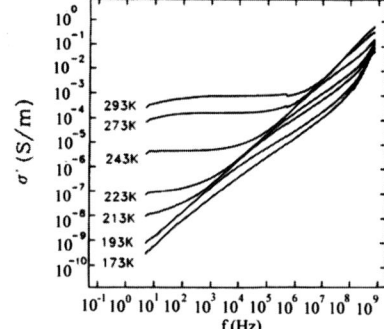

Fig. 3.26. Log–log plot of AC conductivity σ' versus frequency f for a PHEA sample at 297 K and several water contents indicated on the plot

Fig. 3.27. Log–log plot of AC conductivity σ' versus frequency f for a PHEA sample at h=0.30 and several temperatures indicated on the plot

from the area under the band, of the high-temperature TSDC band in Eucalyptus globulus leaves. Following common practice, the amount of water in the sample is characterised here by x_2, i.e., the weight fraction of water in the sample. The results show a remarkable similarity to analogous results obtained with synthetic polymers (Figs. 3.19–21), suggesting a glass or glass–like transition shifting to lower temperatures with increasing water content and becoming constant at higher water contents [69, 70].

3.7 Effects of Water on Electrical Conductivity

Electrical conductivity in polymers and biopolymers increases with increasing water content, as indicated by the results already reported for the high–temperature TSDC dispersion, e.g., in Figs. 3.1–5, and, more clearly, for the DRS response at high temperatures/low frequencies, e.g., in Figs. 3.6 and 3.7. The high protonic conductivity achieved with hydrogels, partly on addition of acids, has led to the design of novel polymeric electrolytes for various electrochemical applications [94, 95], whereas the strong dependence of conductivity on water content was utilised for designing conductimetric microsensors based on synthetic hydrogels for potential in vivo use [96]. From the fundamental point of view the interest in such studies is focused on a better understanding of the effects of water on electrical conductivity, a necessary condition for optimising the design of novel polymeric electrolytes based on synthetic hydrogels and for employing conductivity measurements for hydration studies and for dielectric aquametry [75, 81, 97, 98].

Figure 3.26 shows results for the dependence of the real part of AC conductivity σ' (protonic conductivity [75]) in wide ranges of frequency on water content h for PHEA hydrogels. σ' (in the literature very often simply AC conductivity σ) has been calculated from admittance measurements (Sect. 3) by [43, 48, 50]

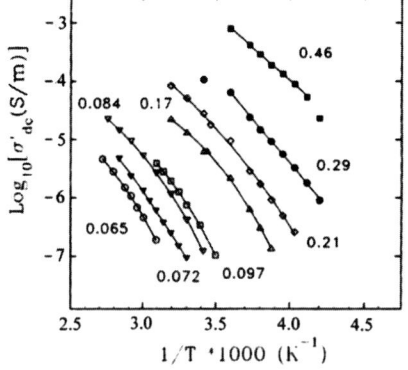

Fig. 3.28. Arrhenius plot of DC conductivity σ'_{DC} for PHEA hydrogels at several water contents indicated on the plot. The lines are best fittings of the VTF equation (3.8) at $h \leq 0.21$ and of the Arrhenius equation (3.9) at $h \geq 0.29$ to the data

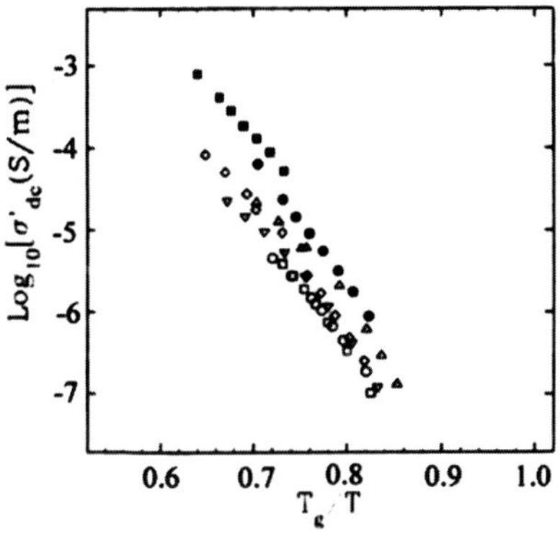

Fig. 3.29. T_g-scaled Arrhenius plot of the DC conductivity data shown in Fig. 3.28. Compare Fig. 3.28 for the symbols

$$\sigma'(f) = 2\pi f \varepsilon_0 \varepsilon''(f) \; . \tag{3.7}$$

The plateaus of σ' at low frequencies extending to higher frequencies with increasing h correspond to DC conductivity σ'_{DC}. The change from f–independent DC-conductivity to f–dependent AC conductivity, the knee in $\sigma'(f)$, corresponds to the so–called conductivity relaxation [50, 75, 81, 99]. At each h value the region to the left of the knee is where the charge carriers are mobile over long distances and the region to the right is where the charge carriers are spatially confined in their potential wells. Thus, $\sigma'(f)$ is characterized by (and its investigation reveals aspects of) the topology of conducting paths. The main result in Fig. 3.26 is the significant increase of σ_{DC} with h.

Figure 3.27 shows results for the same hydrogel at a fixed water content, $h = 0.30$, and variable temperature. Comparison with Fig. 3.26 demonstrates the validity of the T–f–h superposition principle [50, 81–83], discussed on the basis of $\varepsilon'(f)$ and $\varepsilon''(f)$ for PEO hydrogels in Figs. 3.7 and 3.8 and extended here to include conductivity phenomena in addition to dipolar ones. Peleg [100] proposed a model to quantitatively account for the effects of moisture and temperature on the stiffness of solid biomaterials at and around their glass transition.

The investigation and proper analysis of the temperature dependence of dc conductivity provides important information on the conductivity mechanism. Figure 3.28 shows the Arrhenius plot of dc conductivity of PHEA hydrogels at several water contents. Values of σ_{DC} at each temperature and water content were obtained from Z plots (plots of the imaginary against the real part of the complex impedance) [75], in very good agreement with values obtained from $\sigma'(f)$ plots. At each water content, DC conductivity appears at temperatures higher than the glass

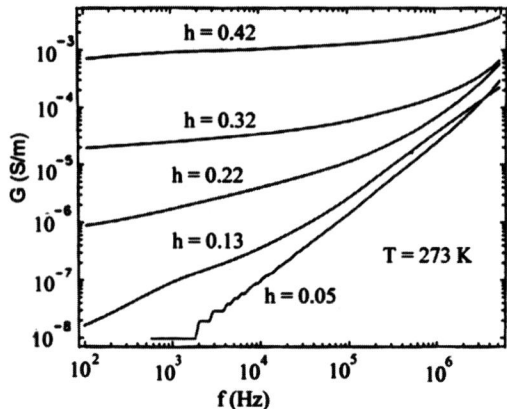

Fig. 3.30. Conductance G versus frequency f in wheat flour samples at 273 K and several water contents indicated on the plot

transition temperature T_g (Fig. 3.21). At $h \leq 0.21$, the temperature dependence of σ_{DC} is described by the Vogel–Tammann–Fulcher (VTF) equation [47, 48, 50, 84]

$$\sigma'_{dc} = \sigma'_0 \exp\left(-\frac{B}{T-T_0}\right) \ , \tag{3.8}$$

where σ'_0, B and T_0 are constants. At the highest water contents of 0.29 and 0.46, the $\sigma'_{DC}(T)$ dependence in Fig. 3.28 changes from VTF type to Arrhenius type

$$\sigma'_{dc} = \sigma'_0 \exp\left(-\frac{W}{kT}\right) \ , \tag{3.9}$$

where W is the apparent activation energy and k is Boltzmann's constant. The resulsts suggest that at $h \leq 0.21$ conductivity is governed by the motion of the polymeric chains, whereas at $h \geq 0.29$ conductivity occurs through a separate water phase [75]. Thus, with respect to effects of water on molecular mobility and to critical water contents, the results for conductivity are in very good agreement with those for the secondary (Fig. 3.13) and the primary (Fig. 3.21) relaxations.

The effects of increase of segmental mobility with increasing temperature on conductivity can be further studied on the basis of T_g-scaled Arrhenius plots [101], like the one shown in Fig. 3.29. The T_g values at the different h values were obtained from DSC and TSDC measurements [36, 75]. The results in Fig. 3.29 suggest that a significant part of the increase of σ_{DC} with increasing water content at constant temperature in Figs. 3.26 and 3.28 arises from the corresponding decrease of T_g. There is, however, a remaining part, probably due to concomitant increase of charge carrier concentration.

The results in Figs. 3.28 and 3.29 can be also discussed in terms of the strong/fragile classification scheme proposed by Angell [101], i.e., in terms of deviations from the Arrhenius behavior. They suggest that the hydrogels under investigation are fragile systems, i.e., deviation from the Arrhenius behavior is significant, and that fragility increases slightly with increasing water content [50, 75]. Interestingly, the shape of the response of the conductivity relaxation, studied within the modulus formalism [75], does not deviate much from Debye behavior of a single relaxation time. Thus, in the PHEA hydrogels, like in proteins and contrary to liquids and polymers, non–Arrhenius behavior does not correlate with non–Debye behavior [101, 102].

The plot in Fig. 3.30 shows for wheat flour samples results similar to those shown in Fig. 3.26 for PHEA hydrogels. The transition from AC-conductivity to DC-conductivity occurs for water contents between 0.13 and 0.22, i.e., in the range of critical water contents determined by low–temperature TSDC measurements (Fig. 3.17) for the completion of the primary hydration layer and the appearance of loosely bound water molecules forming clusters around hydrophilic sites and by TSDC measurements at higher, sub–zero temperatures for a glass or a glass-like transition [67–70, 103]. Careri and coworkers interpreted the appearance of DC-conductivity in hydrated seeds in terms of percolative proton transfer along threads of hydrogen-bonded water molecules on the macromolecular surface and provided evidence that the percolation threshold is close to that necessary for germination [104]. The correlation of critical water contents for physical effects, such as glass transition and percolative proton transport, to those for physiological and biological effects, such as germination for seeds and enzymatic activity for proteins [62], provides a basis for describing and understanding biological functions in physical terms [8].

3.8 Conclusions

We presented and discussed in this chapter several results on the hydration properties of synthetic polymers and biopolymers obtained with dielectric techniques. In addition to the classical dielectric relaxation spectroscopy (DRS) over the broad frequency range from 10^{-2} Hz to 20 GHz, thermally stimulated depolarisation currents (TSDC) techniques were employed, typically in the temperature range from 77 to 300 K. Information on the hydration properties of the materials under investigation were obtained by studying in detail the effects of water on (a) the local secondary relaxations present or induced by water in the materials, (b) the cooperative, primary α-relaxation associated to the glass transition, and (c) conductivity and conductivity effects. Less attention was paid here to the investigation of the dynamics of water itself, as a source of information on hydration properties, either as liquid water at temperatures above 0°C/high frequencies or as supercooled or crystallised water at subzero temperatures/low frequencies.

The results demonstrate the power of dielectric techniques for hydration studies. The main advantage of these techniques, in particular of DRS, as compared to other techniques, with respect to hydration studies, is the wide frequency range of measurements, which allows following on the same sample processes of very different time scale and, thus, spatial scale. For each of these processes information has been obtained, by proper analysis, from the time scale and the magnitude of the response. Less attention was paid here to information obtained from the shape of the response, which is related to long- and short-range interactions, this point deserving more attention in future studies.

The results show that also TSDC can be very powerful for hydration studies, mainly for other reasons than DRS: it provides a quick characterisation of the overall behavior of a material and it is characterised by high sensitivity and high resolving power. The low equivalent frequency of TSDC measurements, in the range of 10^{-2}–10^{-4} Hz, responsible for its high resolving power, also provides also the possibility to work at low, subzero temperatures, with the advantage that the water content of the sample does not change much with time. Thus, TSDC is a powerful, complementary technique to DRS and contributes much to the suitability of the combined DRS/TSDC analysis for hydration studies.

The results discussed in this chapter refer to several aspects of hydration properties, such as the organisation of water in the material under investigation, the estimation of critical water contents, the assessment of the strength of polymer–water interactions and the micromorphology developed. Similar information is also provided by various other techniques and critical comparison in this chapter has shown that, if the specifity of each technique is properly taken into account, the agreement is often good, also in quantitative terms.

Some of the results discussed in this chapter could be utilised in the future for dielectric aquametry. From the DRS results, the dependence of DC- and/or AC-conductivity on water content at room temperature, such as that presented in Fig. 3.26 for PHEA hydrogels and in Fig. 3.30 for wheat flour, could form a basis for dielectric aquametry at low frequencies. In the case of TSDC measurements, the dependence of the temperature position and/or of the intensity of the low-temperature dispersion on water content, shown in Figs. 3.13 and 3.21 for hydrogels based on PHEA and in Fig. 3.17 for seeds, appears most promising. The low temperature of measurements and, in general, temperature as a variable is discouraging. The high reproducibility of the results, however, suggests that the technique might be an interesting alternative, at least for special cases.

Acknowledgments

The author is indebted to his colleagues G. Gallego Ferrer, A.A. Konsta, A. Kyritsis, M. Monleon Pradas and J.L. Gomez Ribelles for their experimental contributions and for many stimulating discussions and to the PhD students S. Kripotou, D. Fragiadakis and G. Polizos for their assistance in the preparation of the manuscript.

References

1. Kraszewski A (ed) (1996) Microwave aquametry. IEEE, New York
2. Brandelik A, Huebner C (2000) Subsurface sensing, subsurface aquametry. Subsurface Sensing Technologies and Applications 1:365–376
3. Kraszewski A (2001) Microwave aquametry: an effective tool for nondestructive moisture sensing. Subsurface Sensing Technologies and Applications 2:347–362
4. Saenger W (1987) Structure and dynamics of water surrounding biomolecules. Ann Rev Biophys Biophys Chem 16:93–114
5. Hodge RM, Bastow TJ, Edward GH, Simon GP, Hill AJ (1996) Free volume and the mechanism of plasticization in water-swollen poly(vinyl alcohol). Macromolecules 29:8137–8143
6. Fuchs K, Kaatze U (2002) Dielectric spectra of mono- and disaccharide aqueous solutions. J Chem Phys 116:7137–7144
7. Dorbez–Sridi R, Cortes R, Mayer E, Pin S (2002) X-ray scattering study of the structure of water around myoglobin for several levels of hydration. J Chem Phys 116:7269–7275
8. Levine H, Slade L (1991) (eds) Water relationships in foods. Plenum, New York
9. Rowland SP (1980) (ed) Water in polymers. American Chemical Society, Washington, DC
10. Byrne ME, Park K, Peppas NA (2002) Molecular imprinting within hydrogels. Adv Drug Deliv Rev 54:149–161
11. Craig DQM (1995) Dielectric analysis of pharmaceutical systems. Taylor & Francis, London
12. Pissis P, Kyritsis A, Konsta AA, Daoukaki D (1999) Dielectric studies of molecular mobility in hydrogels. J Mol Struct 479:163–175
13. Neagu RM, Neagu E, Kyritsis A, Pissis P (2000) Dielectric studies of dipolar relaxation processes in nylon 11. J Phys D Appl Phys 33:1921–1931
14. Wei Y–Z, Kumbharkhane AC, Sadeghi M, Sage JT, Tian WD, Champion PM, Sridhar S (1994) Protein hydration investigations with high-frequency dielectric spectroscopy. J Phys Chem 98:6644–6651
15. Liptak J, Ilavsky M, Nedbal J (1995) Phase transition in swollen gels. 20. The dielectric behavior of swollen poly(N, N–diethylacrylamide) gels in the collapse region. Polym Networks Blends 5:55–61
16. Xu H, Vij JK, McBrierty VJ (1994) Wide-band dielectric spectroscopy of hydrated poly(hydroxyethyl methacrylate). Polymer 35:227–234
17. Jain SK, Johari GP (1998) Dielectric studies of molecular motions in the glassy states of pure and aqueous poly(vinylpyrrolidone). J Phys Chem 92:5851–5854
18. Laredo E, Hernandez MC (1997) Moisture effect on the low- and high-temperature dielectric relaxations in nylon–6. J Polym Sci Part B Polym Phys 35:2879–2888
19. Demont P, Diffalah M, Martinez–Vega JJ, Lacabanne C (1994) Study of molecular mobility at the secondary relaxations range in polyamide 66 and polyamide 66/EPR blends by thermally stimulated creep and current. J Non-Cryst Solids 172–174:978–984
20. Ceccorulli G, Pizzoli M (2001) Effect of water on the relaxation spectrum of poly(methyl–methacrylate). Polym Bull 47:283–289

21. Cendoya I, Lopez D, Alegria A, Mijangos C (2001) Dynamic mechanical and dielectrical properties of poly(vinyl alcohol) and poly(vinyl alcohol)-based nanocomposites. J Polym Sci Part B Polym Phys 39:1968–1975

22. McCrystal CB, Ford JL, He R, Craig DQM, Rajabi–Siahboomi AR (2002) Characterisation of water behavior in cellulose ether polymers using low frequency dielectric spectroscopy. Int J Pharm 243:57–69

23. Corkhill PH, Jolly AM, Ng CO, Tighe BJ (1987) Synthetic hydrogels: 1. Hydroxyalkyl acrylate and methacrylate copolymers–water binding studies. Polymer 28:1758–1765

24. Kanapitsas A, Pissis P, Gomez Ribelles JL, Monleon Pradas M, Privalko EG, Privalko VP (1999) Molecular mobility and hydration properties of segmented PUs with varying structure of soft- and hard-chain segments. J Appl Polym Sci 71:1209–1221

25. Sartor G, Hallbrucker A, Mayer E (1995) Characterizing the secondary hydration shell on hydrated myoglobin, hemoglobin, and lysozyme powders by its vitrification behavior on cooling and its calorimetric glass→liquid transition and crystallisation behavior on reheating. Biophys J 69:2679–2694

26. Rault J, Lucas A, Neffati R, Monleon Pradas M (1997) Thermal transitions in hydrogels of poly(ethyl acrylate)/poly(hydroxyethyl acrylate) interpenetrating networks. Macromolecules 30:7866–7873

27. Okoroafor EU, Newborough M, Highgate D (1998) Effects of thermal cycling on the crystallisation characteristics of water within crosslinked hydro-active polymeric structure. J Phys D Appl Phys 31:3130–3138

28. Chen WL, Shull KR, Papatheodorou T, Styrkas DA, Keddie JL (1999) Equilibrium swelling of hydrophilic polyacrylates in humid environments. Macromolecules 32:136–144

29. McConville P, Whittacker MK, Pope JM (2002) Water and polymer mobility in hydrogel biomaterials quantified by ^{1}H NMR: a simplified model describing both T_1 and T_2 relaxation. Macromolecules 35:6961–6969

30. Maffezzoli A, Luprano VA, Montagna G, Esposito F, Nicolais L (1996) Ultrasonic wave attenuation during water sorption in poly(2–hydroxyethyl methacrylate) hydrogels. Polym Eng Sci 36:1832–1838

31. Kaatze U, Pottel R (1991) Dielectric and ultrasonic spectroscopy of liquids. Comparative view for binary aqueous solutions. J Mol Liq 49:225–248

32. Librizzi F, Viappiani C, Abbruzzeti S, Cordone L (2002) Residual water modulates the dynamics of the protein and of the external matrix in "trehalose coated" MbCO: an infrared and flash-photolysis study. J Chem Phys 116: 1193–1200

33. Thouvenin M, Linossier I, Sire O, Peron J–J, Vallee–Rehel K (2002) Structural and dynamic approach of early hydration steps in erodable polymers by ATR–FTIR and fluorescence spectroscopies. Macromolecules 35: 489–498

34. Middendorf HD (1996) Neutron studies of the dynamics of biological water. Physica B 226: 113–127

35. Amorin da Costa AM, Amado AM (2001) Cation hydration in hydrogelic polyacrylamide–phosphoric acid network: a study by Raman spectroscopy. Solid State Ionics 145: 79–84

36. Kyritsis A, Pissis P, Gomez Ribelles JL, Monleon Pradas M (1995) Polymer–water interactions in poly(hydroxyethyl acrylate) hydrogels studied by dielectric, calorimetric and sorption isotherm measurements. Polym Gels Networks 3: 445–469

37. Barbieri R, Quaglia M, Delfini M, Brosio E (1998) Investigation of water dynamic behavior in poly(HEMA) and poly(HEMA–co–PHPMA) hydrogels by proton T_2 relaxation time and self-diffusion coefficient n.m.r. measurements. Polymer 39: 1059–1066

38. Tamai Y, Tanaka H, Nakanishi K (1996) Molecular dynamics study of polymer–water interaction in hydrogels. 1. Hydrogen-bond structure. Macromolecules 29: 6750–6760

39. Tamai Y, Tanaka H, Nakanishi K (1996) Molecular dynamics study of polymer–water interaction in hydrogels. 2. Hydrogen-bond dynamics. Macromolecules 29: 6761–6769

40. Mueller–Plathe F (1998) Different states of water in hydrogels? Macromolecules 31: 6721–6723

41. Daniel VN (1967) Dielectric Relaxation. Academic, London

42. Hill NE, Vaughan W, Price AH, Davies M (1969) Dielectric properties and molecular behavior. Van Nostrand, London

43. Macdonald JR (1987) (ed) Impedance spectroscopy. Wiley, New York

44. Takeishi S, Mashimo S (1992) Dielectric relaxation measurements in the ultralow frequency region. Rev Sci Instrum 53: 1155–1159

45. Nozaki S, Bose TK (1990) Broadband complex permittivity measurements by time-domain spectroscopy. IEEE Trans Instrum Meas 39: 945–951

46. Feldman Y, Andrianov A, Polygalov E, Ermolina I, Romanychev G, Zuev Y, Milgotin B (1996) Time domain dielectric spectroscopy: an advanced measuring system. Rev Sci Instrum 67: 3208–3216

47. Havriliak S Jr, Havriliak SJ (1997) Dielectric and mechanical relaxation in materials. Hanser, Munich

48. Jonscher AK (1983) Dielectric relaxation in solids. Chelsea Dielectrics, London

49. Kyritsis A, Pissis P (1997) Dielectric studies of polymer–water interactions and water organization in PEO/water systems. J Polym Sci Part B Polym Phys 35: 1545–1560

50. Kyritsis A, Pissis P, Grammatikakis J (1995) Dielectric relaxation spectroscopy in poly(hydroxyethyl acrylates)/water hydrogels. J Polym Sci Part B Polym Phys 33: 1737–1750

51. van Turnhout J (1980) Thermally stimulated discharge of electrets. In: Sessler GM (ed) Electrets Topics in Applied Physics vol 33. Springer, Berlin, pp 81–215

52. Pissis P, Anagnostopoulou–Konsta A, Apekis L, Daoukaki–Diamanti D, Christodoulides C (1991) Dielectric effects in water-containing systems. J Non-Cryst Solids 131–133: 1174–1181

53. Kyritsis A, Pissis P, Gomez Ribelles JL, Monleon Pradas M (1994) Depolarization thermocurrent studies in poly(hydroxyethyl acrylate)/water hydrogels. J Polym Sci Part B Polym Phys 32: 1001–1008

54. Pissis P, Apekis L, Christodoulides C, Niaounakis M, Kyritsis A, Nedbal J (1996) Water effects in polyurethane block copolymers. J Polym Sci Part B Polym Phys 34: 1529–1539

55. Pathmanathan K, Cavaille J–Y, Johari GP (1992) The dielectric properties of dry and water-saturated nylon–12. J Polym Sci Part B Polym Phys 30: 341–348

56. Anagnostopoulou–Konsta A, Pissis P (1987) A study of casein hydration by the thermally stimulated depolarization currents method. J Phys D Appl Phys 20: 1168–1174

57. Pissis P (1989) Dielectric Studies of protein hydration. J Mol Liq 41: 271–289

58. Pissis P, Anagnostopoulou–Konsta A (1990) Protonic percolation on hydrated lysozyme powders studied by the method of thermally stimulated depolarization currents. J Phys D Appl Phys 23: 932–939

59. Pissis P, Anagnostopoulou–Konsta A (1991) Dielectric studies of proton transport in hydrated proteins. Solid State Ionics 46: 141–145
60. Harvey SC, Hoekstra P (1972) Dielectric relaxation spectra of water adsorbed on lysozyme. J Phys Chem 76: 2987–2994
61. Bone S, Pething R (1982) Dielectric studies of the binding of water to lysozyme. J Mol Biol 157: 571–575
62. Careri G, Geraci A, Giansanti A, Rupley JA (1985) Protonic conductivity of hydrated lysozyme powders at megahertz frequencies. Proc Natl Acad Sci USA 82: 5342–5346
63. Pissis P (1985) A study of sorbed water on cellulose by the thermally stimulated depolarisation technique. J Phys D Appl Phys 18: 1897–1908
64. Anagnostopoulou–Konsta A, Daoukaki–Diamanti D, Pissis P, Sideris E (1988) Dielectric study of the interaction of DNA and water. In: Das–Gupta DK, Pattulo AW (eds) Proceedings 6th International Symposium on Electrets, IEEE, New York, pp 271–275
65. Pissis P, Anagnostopoulou–Konsta A, Apekis L (1987) A dielectric study of the state of water in plant stems. J Exp Botany 38: 1528–1540
66. Pissis P (1990) The dielectric relaxation of water in plant tissue. J Exp Botany 41: 677–684
67. Konsta AA, Pissis P, Kanapitsas A, Ratkovic S (1996) Dielectric and conductivity studies of the hydration mechanism in plant seeds. Biophys J 70: 1485–1493
68. Ratkovic S, Pissis P (1997) Water binding to biopolymers in different cereals and legumes: proton NMR relaxation, dielectric and water imbibition studies. J Mat Sci 32: 3061–3068
69. Pissis P, Anagnostopoulou–Konsta A, Apekis L, Daoukaki–Diamanti D, Christodoulides C, Sideris EG (1992) Evidence of glass transitions in biological systems from dielectric studies. IEEE Trans Electr Insul 27: 820–825
70. Pissis P (1992) Glass transitions in biological systems. In: Bountis T (ed) Proton transfer in hydrogen-bonded systems, Plenum Press, New York, pp 207–216
71. Bruni F, Leopold AC (1992) Pools of water in anhydrobiotic organisms: a thermally stimulated depolarization current study. Biophys J 63: 663–672
72. Sun WQ, Leopold AC (1993) The glassy state and accelerated aging of soybeans. Physiol Plant 89: 767–774
73. Sun WQ (2000) Dielectric relaxation of water and water-plasticized biomolecules in relation to cellular water organization, cytoplasmic viscosity, and desiccation tolerance in recalcitrant seed tissues. Plant Physiol 124: 1203–1215
74. Kyritsis A, Pissis P, Gomez Ribelles JL, Monleon Pradas M (1994) Dielectric relaxation spectroscopy in PHEA hydrogels. J Non-Cryst Solids 172–174: 1041–1046
75. Pissis P, Kyritsis A (1997) Electrical conductivity studies in hydrogels. Solid State Ionics 97: 105–113
76. Pissis P, Kyritsis A, Gallego Ferrer G, Monleon Pradas M, Gomez Ribelles JL (2000) Water in hydrogels studied by dielectric, thermal and water sorption/diffusion techniques. Subsurface Sensing Technologies and Applications 1: 417–439
77. Pissis P, Kyritsis A, Martinez Romero T, Azorin Tortosa S, Gallego Ferrer G, Monleon Pradas M, Gomez Ribelles JL (1999) Poly(hydroxyethyl acrylate)–nylon 6 nanocomposites. Dielectric and water sorption properties. In: Konsta AA, Vassilikou–Dova A, Vartzeli–Nikaki K (eds) Proceedings 10th International Symposium on Electrets. IEEE, New York, pp 561–564
78. Pissis P, Kyritsis A, Polizos G, Gallego Ferrer G, Monleon Pradas M, Gomez Ribelles JL (2001) Dielectric studies of polymer-water interactions in hydrogels. In: Kupfer K,

Huebner C (eds) Proceedings 4th International Conference on Electromagnetic Wave Interaction with Water and Moist Substances. Materialforschungs- und -Prüfanstalt, Weimar, pp 8–15

79. Gallego Ferrer G, Monleon Pradas M, Gomez Ribelles JL, Pissis P (1998) Swelling and thermally stimulated depolarization currents in hydrogels formed by interpenetrating polymer networks. J Non-Cryst Solids 235–237: 692–696

80. Gomez Ribelles JL, Monleon Pradas M, Gallego Ferrer G, Peidro Torres M, Perez Gimenez V, Pissis P, Kyritsis A (1999) Poly(methyl acrylate)/poly(hydroxymethyl acrylate) sequential interpenetrating polymer networks. Miscibility and water sorption behavior. J Polym Sci Part B Polym Phys 37: 1587–1599

81. Pissis P, Kyritsis A, Shilov VV (1999) Molecular mobility and protonic conductivity in polymers: hydrogels and ionomers. Solid State Ionics 125: 203–212

82. Kyritsis A, Pissis P (1997) Dielectric relaxation spectroscopy in poly(ethylene oxide)/water systems. Macromol Symp 119: 15–24

83. Starkweather H W Jr (1980) Water in nylon. In: Rowland SP(ed) Water in Polymers. American Chemical Society, Washington DC, pp 433–440

84. McCrumm NG, Read BE, Williams G (1967) Anelastic and dielectric effects in polymeric solids. Wiley, New York

85. Franck B, Fruebing P, Pissis P (1996) Water sorption and thermally stimulated depolarization currents in nylon–6. J Polym Sci Part B Polym Phys 34: 1853–1860

86. Yemni T, Boyd RH (1979) Dielectric relaxation in the odd-numbered polyamides: nylon 7–7 and nylon 11. J Polym Sci Polym Phys Ed 17: 741–751

87. Pathmanathan K, Johari GP (1994) Relaxation and crystallization of water in a hydrogel. J Chem Soc Faraday Trans 90: 1143–1148

88. Johari GP (1996) Water's character from dielectric relaxation above its T_g. J Chem Phys 105: 7079–7082

89. Graham NB, Zulfiqar M, Nwachuku NE, Rashid A (1990) Interaction of poly(ethylene oxide) with solvents: 4. Interaction of water with poly(ethylene oxide) crosslinked hydrogels. Polymer 31: 909–916

90. Pissis P, Apekis L, Christodoulides C, Boudouris G (1983) Dielectric study of dispersed ice microcrystals by the depolarization thermocurrent technique. J Phys Chem 87: 4034–4037

91. Capelletti R, Bridelli MG (1999) TSDC as a tool to monitor the electret state induced by water in biomolecules. In: Konsta AA, Vassilikou–Dova A, Vartzeli–Nikaki K (eds) Proceedings 10th International Symposium on Electrets. IEEE, New York, pp 159–166

92. Ellis TS, Karasz FE (1984) Interaction of epoxy resins with water: The depression of glass transition temperature. Polymer 25: 664–669

93. Mayer E (1991) Calorimetric glass transitions in the amorphous forms of water: a comparison. J Mol Struct 250: 403–411

94. Lassegues J–C (1992) Mixed inorganic–organic systems: the acid/polymer blends. In: Colomban P (ed) Proton Conductors. Cambridge University Press, Cambridge, pp 311–327

95. Kreuer KD (1997) On the development of proton conducting materials for technological applications. Solid State Ionics 97: 1–15

96. Kreuer KD (1996) Proton conductivity: Materials and applications. Chem Mater 8: 610–641

97. Stevens JR, Wieczorek W, Raducha D, Jeffrey KR (1997) Proton conducting gel/H$_3$PO$_4$ electrolytes. Solid State Ionics 97: 347–358
98. Konsta AA, Daoukaki D, Pissis P, Vartzeli K (1999) Hydration and conductivity studies of polymer–water interactions in polyacrylamide hydrogels. Solid State Ionics 125: 235–241
99. Starkweather Jr HW, Avakian P (1992) Conductivity and the electric modulus in polymers. J Polym Sci Part B Polym Phys 30: 637–641
100. Peleg M (1993) Mapping the stiffness–temperature–moisture relationship of solid biomaterials at and around their glass transition. Rheol Acta 32: 575–580
101. Angell CA (1991) Relaxation in liquids, polymer and plastic crystals – strong/ fragile patterns and problems. J Non-Cryst Solids 131–133: 13–31
102. Boehmer R, Ngai KL, Angell CA, Plazek DJ (1993) Nonexponential relaxations in strong and fragile glass formers. J Chem Phys 99: 4201–4209
103. Pissis P, Konsta AA, Ratkovic S, Todorovic S, Laudat J (1996) Temperature- and hydration-dependence of molecular mobility in seeds. J Therm Anal 47: 1463–1483
104. Bruni F, Careri G, Leopold AC (1989) Critical exponents of protonic percolation in maize seeds. Phys Rev A 40: 2803–2805

4 Thermal and Geometrical Effects on Bulk Permittivity of Porous Mixtures Containing Bound Water

Scott B. Jones and Dani Or

Department of Plants, Soils, and Biometeorology; Utah State University; Logan, UT, USA, 84322-4820 and Department of Civil and Environmental Engineering; University of Connecticut; Storrs, CT, USA, 06269-2037

4.1 Introduction

Determination of the water content of porous media (e.g., soil, powders, cereal grains) from measured bulk dielectric, ε_b, is based on the dominance of the high dielectric permittivity of liquid water relative to that of solids and air. Dielectric measurements in many minerals comprising soils for example, exhibit behavior described by a general empirical relationship [1], while measurements in high surface area materials or in particles of high aspect ratio exhibit unique and varied dielectric-water content relationships. Factors contributing to the unique dielectric-water content relation of certain porous media include water binding and dipole interactions that arise from particle geometry and water-phase configuration in addition to measurement frequency and dielectric loss. Thermal perturbation of high surface area porous media demonstrates strong interactions between bound and free water that have been shown to produce either an increase or decrease in bulk dielectric for a fixed water content. This effect is demonstrated by modeling the dielectric permittivity of bound and free water-phases, where the surface area-dependent bound water film thickness and the temperature-dependent permittivity of free water are combined. The water-phase permittivity is the least well-defined constituent and yet the most critical for predicting the bulk permittivity of a three-phase water-, solid- and air-system. A three-phase dielectric mixture model is shown to provide a physically-based approach for modeling porous media permittivity. Phase configuration in addition to estimates of particle and packing densities facilitate volume tracking with water content change.

4.1.1 Dielectric Mixing Theory

A dielectric mixture may be described in terms of the volume fractions and dielectric permittivities of constituents. A host of different mixture models have been developed for multiple phases using different methods of homogenization, including variations on constituent geometry leading to isotropic and anisotropic mixtures and combinations thereof [2]. We approximate a porous mixture as a two- or three-phase system made up of solids, water and air. The confocal ellipsoid model shown in Fig. 4.1 is an example of such a three-phase system and serves as a geometrical basis for a unit cell leading to an approximate porous mixture. We assign water-, solid- and air-phase permittivities ε_w (80), ε_a (1), ε_s (5) and volumetric fractions ϕ_w, ϕ_a, and ϕ_s, respectively [3]. As we will demonstrate later, ε_w will be subject to temperature and relaxation effects requiring further attention for improved modeling. Constituent volume fractions are related through measurable quantities of dry bulk density (ρ_b = dry solid mass per bulk volume) and solid particle density (ρ_s = solid mass per solid volume) from which we obtain total (bulk) porosity (ϕ_b), which describes the fluid-filled portion of the mixture (i.e., $\phi_b = 1 - \rho_b/\rho_s$). The air ($\phi_a$) and solid ($\phi_s$) fractions may be described as $\phi_a = \phi_b - \phi_w$ and $\phi_s = 1 - \phi_b$, respectively. Dielectric mixture theory offers the advantage of physically-based predictions of bulk permittivity based on constituent permittivities and their volumetric contributions to the mixture [4]. Predicting the dielectric permittivities of constituents in high surface area materials is complicated by the presence of bound water.

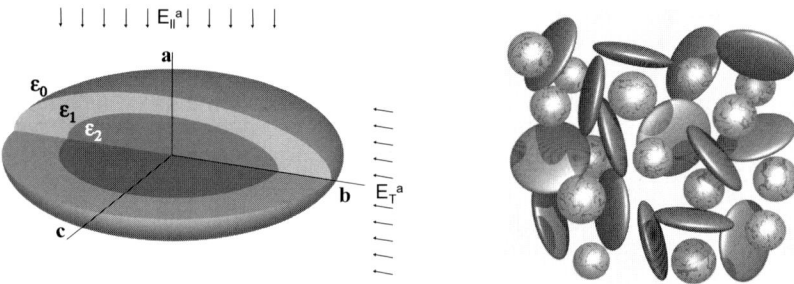

Fig. 4.1. On the *left*, a three-phase unit cell described as a confocal spheroid (core, ε_2, and shell, ε_1) contained in a background, ε_0. The orientation of the electric field (E) is either normal (E_T^a) or parallel ($E_\parallel^a$) with respect to the rotation axis, a, of the spheroid. On the *right*, a random mixture of spheroids forms an idealized porous medium

4.1.2 Geometrical Effects

Particle shape may influence electromagnetic measurements, depending on the particle geometry and orientation (random, aligned) with respect to the applied electrical field [5] as illustrated in Fig. 4.1. The shape of soil particles or portions of cereal grains from the kernel itself, to the starch granule, to the molecular components, influence the dielectric measurement due to the dipole moment each component imposes in an electromagnetic measurement of permittivity. Aspect ratio is a common descriptor for geometry, and those of clay minerals can be extreme in comparison to nearly spherical sand grains. Minimum and maximum aspect ratios for several clays are listed in Table 4.1, given based on estimates of particle planar diameter and thickness. Aspect ratios for three common cereal grains are also listed. Particle shape effects have been measured on individual kernels [6] and on mixtures of isotropic [7-9] and anisotropic particles [5].

Table 4.1. Dimensions and aspect ratios of certain clay minerals and cereal grain kernels.

Clay mineral[†]	Thickness, a [nm]	Diameter, b [µm]	Minimum a/b	Maximum a/b
Kaolinite	50	0.1 – 4	0.5	0.0125
Chlorite	10 – 100	0.1 – 2	1	0.005
Illite	5 – 30	0.1 – 2	0.3	0.0025
Montmorillonite	1 – 10	0.01 – 1	1	0.001

Kernel[‡]	Length, a [mm]	Width, b [mm]	Minimum a/b	Maximum a/b
Corn	8 – 17	5 – 15	0.5	3.4
Rice	5 – 10	1.5 – 5	1	6.7
Wheat	5 – 8	2.5 – 4.5	1.1	3.2

†[10], ‡[11].

4.1.3 Bound Water

Water that is "bound" to solid surfaces is subject to surface forces that hinder its response to an imposed electromagnetic field, resulting in both a lower relaxation frequency and in lower ε_b relative to free liquid water.

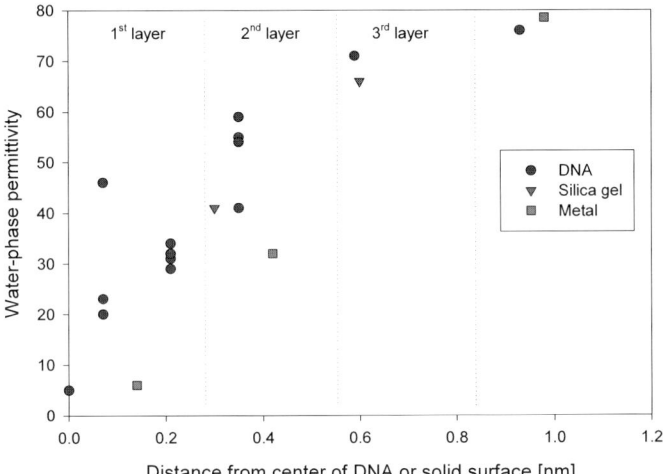

Fig. 4.2. Water-phase permittivities computed as a function of distance from the center of DNA [12] and measured thickness on the surface of silica gel [13] and metal [14]. The mono-layer water thickness is assumed to be 0.28 nm

The first monolayer of bound water closest to the solid surface is held most tightly and has the lowest measured ε_b. The bulk permittivity of successive molecular water layers increases with distance from the solid surface up to that of free water at around 3 or more molecular layers (see Fig. 4.2). The dielectric permittivity of wet materials, high surface area, (i.e., clays, peats, starches, proteins, etc.) exhibit reduced bulk dielectric permittivities compared to materials with lower surface areas at the same volumetric water content as a result of bound water and shape effects. The large surface area of minerals is often associated with plate- or needle-like particles of high aspect ratio as shown in Table 1. Assuming disk-shaped geometries, specific surface areas range from 15 to 750 $m^2\ g^{-1}$ for Kaolinite and Montorillonite, respectively. In the case of biological materials, the molecular structure is much more complex but capable of similarly large specific surface areas, only a portion of which are generally water accessible.

4.2 Theoretical Considerations

Modeling the dielectric behavior of a complex porous mixture is generally approached using a simplified representation of constituents, configurations and geometries, all of which play a role in the bulk dielectric. Here we briefly consider models describing the water–phase, the influence of shape and the configuration of constituent phases.

4.2.1 Water-Phase Permittivity

Water molecules are subject to thermal and interfacial effects that can create significant changes in the water-phase permittivity with changes in temperature and measurement frequency. In addition, the specific surface area of the solid-phase is proportional to the amount of water bound within a porous medium. Air and most solid materials, unlike water, show little change in permittivity with changes in temperature or frequency [15, 16]. In the following we will discuss water as "free" molecules in bulk water or as "bound" molecules that are hindered to some extent in their movement or rotation. The effects of temperature on the permittivity of free and bound water will also be discussed.

Free Water

A water molecule possesses both a permanent dipole moment, resulting from the structure and charges of its atomic components, and a polarizability component that is in proportion to the magnitude of an applied electric field. In the presence of strong electric fields, the polarizability component may actually reduce the dielectric permittivity of water significantly [17]. In the presence of a mild alternating electric field, water molecules exhibit dielectric dispersion leading to a frequency dependent real component of permittivity that is reduced as frequency increases. This phenomenon results from the molecules' inability to reorient in response to the changing electric field. As a consequence, the permittivity contribution of the permanent dipole orientation is diminished and eventually lost at higher frequencies. The reduction in the real permittivity component shows up in the imaginary or 'loss' component, which for free water arises in the GHz frequency band, where a phase lag develops between the applied electric field and the dipole orientation. This effect draws energy from the electric source that is dissipated as heat. For water molecules whose motion is hindered by the attractive forces of solid surfaces and adjacent molecular forces (other than neighboring free water molecules), their relaxation frequency may be reduced to within the MHz frequency range.

Permittivity measurement instruments vary both in their application (for measuring liquid, solid or gas) and in their frequency measurement range. Time domain reflectometry (TDR) instruments, for example, exhibit a frequency bandwidth, f_b, which is largely determined by the rise time (t_r) of the instrument. Instrument rise times commonly employed in soil sciences vary from 125 to 300 ps [18] corresponding to frequency bandwidths of 2.8 to 1.2 MHz, respectively (e.g., $f_b = \ln(0.9/0.1)/(2\pi t_r) \sim 0.35/t_r$ [19]). Lower frequency capacitive measurement instruments may have much lower measurement frequencies in the MHz range.

Bound Water

The influence of water binding on permittivity reduction has been measured [12-14]. The water-phase permittivity associated with attachment to different solid materials is illustrated in Fig. 4.2, where increased water layer thickness leads to an increase in permittivity. These measurements are more difficult to make than free liquid measurements owing to the molecular scale at which they occur and to differing measurement methods. It is understood that the mechanism causing the permittivity reduction is related to the hindered rotation of the water molecules in the vicinity of solid surfaces. Less well understood is the character of the bound water, its density and packing arrangement.

4.2.2 Water-Phase Temperature-Dependence

The temperature dependence of the dielectric permittivity of porous materials can be attributed largely to the effect of temperature on the water-phase. The temperature dependent permittivity of free water is described by the following expression [20].

$$\varepsilon_{fw}(T) = 78.54[1 - 4.579x10^{-3}(T - 298)$$
$$+1.19x10^{-5}(T - 298)^2 - 2.8x10^{-8}(T - 298)^3] \tag{4.1}$$

where T [K] is the temperature. The temperature dependence of the bound portion of water is more complicated, especially since the boundary between bound and free water is vague. Or and Wraith [21] derived a temperature-dependent model describing bound water content (M_{bw}), expressed in terms of the bound water layer thickness, $x(T)$ [m], the specific surface area, A_s [m^2 g^{-1}], and the bulk density, ρ_b [g m^{-3}], of the porous medium, written as

$$M_{bw} = x(T) \cdot A_S \cdot \rho_b \tag{4.2}$$

where $x(T)$ was derived from the viscosity profile of water as a function of distance from a clay surface coupled with the Debye [22] model predicting relaxation frequency of a polar liquid. Or and Wraith have used a cutoff frequency, f^* [Hz], below which bound water relaxes and thus practically does not affect a TDR signal. The resulting temperature-dependent bound water layer thickness, $x(T)$ [m], is computed as

$$x(T) = \frac{a}{-d + T \cdot \ln\left(\dfrac{k \cdot T}{8\pi^2 r^3 c \cdot f_{rel}}\right)} \tag{4.3}$$

where the constants $\alpha = 1621$ [Å K], $d = 2.047 \times 10^3$ [K], $c = 9.5 \times 10^{-7}$ [Pa s], k is the Boltzman constant (1.38062×10^{-23} [J K^{-1}]) and r [m] is the radius of the bound water molecule (~ 1.8 - 2.5 Å).

4.2.3 Particle Shape Effects

The depolarization factor accounts for the dipole effect arising from the particle shape-field alignment, which influences the measured bulk permittivity. The exact solution to the depolarization factor given by Landau and Lifshitz [23] is written in terms of field-axis orientation (e.g., $N^a = E =$ field parallel to a in Fig. 4.1)

$$N^i = \int_0^\infty \frac{(abc)du}{2(u+i^2)\sqrt{(u+a^2)(u+b^2)(u+c^2)}} \qquad i = a,b,c \tag{4.4}$$

where $N^a + N^b + N^c = 1$ and u is a scalar. Equation (4.4) was parameterized in terms of a spheroid ($b = c$) using the aspect ratio (a/b) ranging from a disk- ($a/b = 0.001$) to a needle-shape ($a/b = 1000$) [5]. The resulting depolarization factor-aspect ratio relation was defined using the following empirical expression for parallel and normal and E-field alignment with respect to a spheroid's rotation axis.

$$N^p = \frac{1}{1 + 1.6\left(\dfrac{a}{b}\right) + 0.4\left(\dfrac{a}{b}\right)^2} \qquad N^n = \frac{1 - N^p}{2} \tag{4.5}$$

where the depolarization factor for the electrical field aligned parallel to the axis of symmetry (rotation axis) is N^p and is N^n when normal to the axis of symmetry (see Fig. 4.1).

4.2.4 Three-Phase Dielectric Mixture Model

Modeling the bulk permittivity of a three-phase system consisting of inclusions comprised of a core and outer shell of confocal ellipsoids (Fig. 4.1) allows accounting for particle shape effects via the depolarization factor in addition to phase configuration effects. The core, ε_2, and the outer shell, ε_1, are contained in a background host medium of permittivity, ε_0. The effective permittivity of an isotropic three-phase confocal system of ellipsoids may be written as [4],

$$\varepsilon_b = \varepsilon_0 + \frac{\dfrac{\varepsilon_0}{3} \displaystyle\sum_{i=p,n,n} \left(\dfrac{n_v \alpha^i}{\varepsilon_0} \right)}{1 - \dfrac{1}{3} \displaystyle\sum_{i=p,n,n} N_1^{\,i} \left(\dfrac{n_v \alpha^i}{\varepsilon_0} \right)} \tag{4.6}$$

where the polarizability term in parenthesis is given as a series expansion written here for the dual, confocal ellipsoid system as

$$\frac{n_v \alpha^i}{\varepsilon_0} = \left(\phi_1 + \phi_2 \right) \cdot$$

$$\left[\frac{\left(\varepsilon_1 - \varepsilon_0\right) + \left[\varepsilon_1 + N_1^{\,i}\left(\varepsilon_0 - \varepsilon_1\right)\right] \cdot \dfrac{\left(\varepsilon_2 - \varepsilon_1\right) \dfrac{\phi_2}{\left(\phi_1 + \phi_2\right)}}{\left[\varepsilon_1 + N_2^{\,i}\left(\varepsilon_2 - \varepsilon_1\right)\right]}}{\left[\varepsilon_0 + N_1^{\,i}\left(\varepsilon_1 - \varepsilon_0\right)\right] + N_1^{\,i}\left(1 - N_1^{\,i}\right)\left(\varepsilon_1 - \varepsilon_0\right) \cdot \dfrac{\left(\varepsilon_2 - \varepsilon_1\right) \dfrac{\phi_2}{\left(\phi_1 + \phi_2\right)}}{\left[\varepsilon_1 + N_2^{\,i}\left(\varepsilon_2 - \varepsilon_1\right)\right]}} \right] \tag{4.7}$$

where ϕ_1 and ϕ_2 are the volumetric fractions and $N_1^{\,i}$ and $N_2^{\,i}$ are the depolarization factors of the shell and core ellipsoids, respectively. Various combinations of solid, liquid- and gas-phases may be assigned to the core, shell and background.

4.3 Measurements and Modeled Results

For a fixed volume fraction of water the measured bulk dielectric can be significantly different for different porous mixtures. Possible contributors to these differences are discussed here and models for describing these effects are presented, including the interplay between bound and free water, geometrical effects, specific surface area related to water binding, porosity and density and phase configuration.

4.3.1 Permittivity Measurements in Porous Mixtures

The permittivity of porous mixtures has been described using empirical relationships such as those shown in Fig. 4.3. The Topp et al. [1] equation has been widely used to describe the permittivity of mineral soils measured using TDR. The empirical relationship given by Schaap et al. [24] was fitted to measured permittivities of forest soils. The differences between these two models can be attributed to several effects, including bound water present in porous media having high surface area, effects due to particle geometry, and the configuration of the water-phase [25]. A mineral soil and a pure clay mineral from Dirksen and Dasberg [26] are well described by these two models.

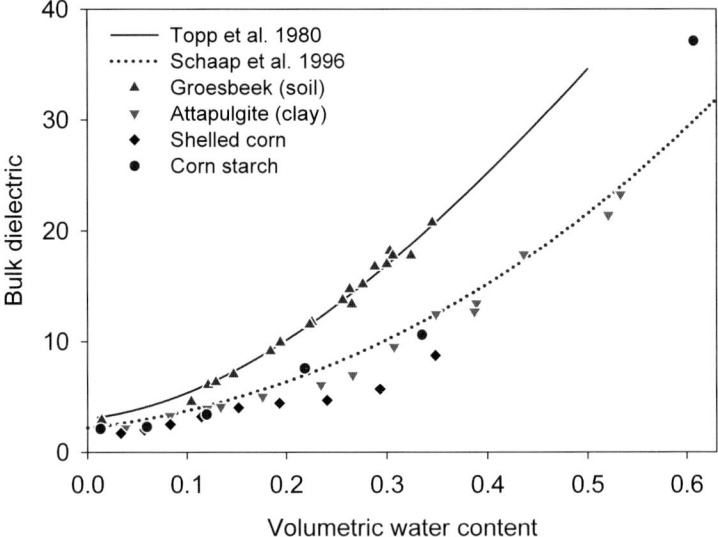

Fig. 4.3. Measured and modeled permittivities as a function of volumetric water content in different porous media. Groesbeek and Attapulgite data are from Dirksen and Dasberg (1993), the kernel and starch data were taken from Jones and Or (2002). The Topp et al. (1980) relation describes a wide range of mineral soils. The Schaap et al. (1996) model describes the permittivity of forest litter

4.3.2 Thermal Effects of Bound Water

The thermodielectric effect on the bulk permittivity of water within soil was suggested by Or and Wraith (1999) to be the result of an interplay between 1) the reduction in the permittivity of free water with increasing temperature, and 2) the increased permittivity brought about by the liberation of bound water to a less hindered state (i.e., greater permittivity illustrated in Fig. 4.4). They modeled this effect using Eq. (4.3) to represent the bound water influence on the bulk dielectric constant of sand and clay mixtures illustrated in Fig. 4.4. They found an apparent lag in the temperature response and suggested a possible correction for this effect. Surface area was also back calculated by optimizing the thermodielectric effect on permittivity measurements.

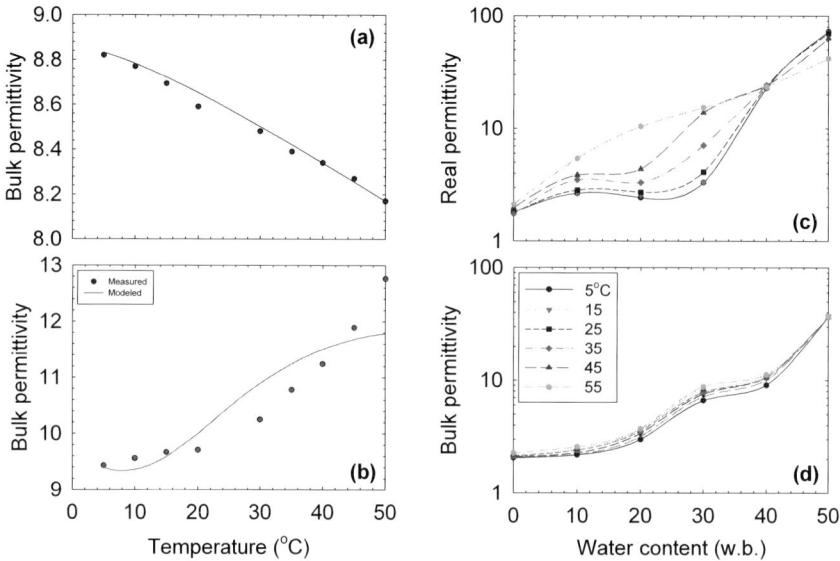

Fig. 4.4. The thermodielectric effect is illustrated in (a) for fine sand and in (b) for a sand-montmorillonite clay mixture, each at 15 % mass water content. Dominance of the opposing temperature dependence of bulk water in (a) for low surface areas, and bound water in (b) for high surface areas is demonstrated [21]. The thermodielectric effect is illustrated differently by permittivity measurements in corn starch using (c) at network analyzer at 500 MHz and (d) a time domain reflectometer [25].

The thermodielectric effect is further illustrated by bulk permittivity measurements in cornstarch shown in Fig. 4.4 and in other studies [27]. Network analyzer measurements (Fig. 4.4c) reveal a cross-over in the permittivity-temperature relationship with increasing water content at about 40 percent (wet basis) measured at a frequency of 500 MHz. TDR measured permittivities (Fig. 4.4d) also demonstrate the effect of bound water 'release' throughout the measured water content range up to 50 percent. The reversal in the temperature-permittivity trend occurring at higher water contents happens when starch granules approach water saturation, leading to a buildup of "free" water within and eventually between granules. Near saturation, temperature increase leads to reduced permittivity of the starch, similar to the reduction in free water permittivity with increasing temperature (Eq. (4.1)). The large permittivity differences at mid range moisture contents suggest an abundance of bound water which is liberated upon heating, and the non-linear nature of the starch permittivity increase with added water also suggests an evolution of the surface area with wetting or drying. Configuration of the water associated with solids combined with surface area evolution leading to changes in bound water fraction were suggested by [25] to cause the distinct plateau in the permittivity-water content curve shown in Fig. 4.4d for corn starch, and also noted in whole corn kernel measurements [28].

4.3.3 Modeling the Total Water-Phase Permittivity

Combining the temperature-dependent water-phase permittivity of bound and free water allows the use of a three-phase mixing model with shape effects. The bound water temperature-dependent expression, $x(T)$ (Eq. (4.3)), was coupled with Eq. (4.1), following an exponential increase in permittivity with water layer thickness, t_w, resulting in the following expression

$$\varepsilon_w(T) = \varepsilon_{fw}(T) \cdot \left[1 - \exp\left(\frac{-t_w}{x(T)} \right) \right] \tag{4.8}$$

where t_w [m] can be calculated with M_v where $t_w = M_v / (A_s \, \rho_b)$. Computed temperature-dependent free water and bound water-phase permittivities are shown in Fig. 4.5. The temperature effect is presented in Fig. 4.5a where the free water permittivity, ε_{fw}, at 5 and 25°C is illustrated by a reduction in the water-phase permittivity. In Fig. 4.5a the temperature increase from 5 to 25°C is described by the reduction of $x(T)$ releasing bound water to a more rotational state, which results in a higher effective permittivity. The free water permittivity is reduced with increa-sing temperature. The second phenomenon is the apparent 'release' of bound water from a lower to a higher rotational state, which occurs with increasing temperature. This bound water effect is illustrated in Fig. 4.5a by a reduction in the bound water layer thickness, $x(T)$ with increasing temperature. In Fig. 4.5b these compe-ting phenomena are illustrated as the total water-phase permittivity plotted using Eq. (4.8) ($r = 2.5 \times 10^{-10}$, $A_s = 400$ m^2 g^{-1}, $f* = 1$ GHz) and demonstrating the thermodielectric behavior as a function of water content Predicted permittivities in Fig. 4.5b are generally correlated to measurements in silica gel, metal powder and DNA shown in Fig. 4.2 and to temperature dependent measurements shown in Fig. 4.4. From Fig. 4.2 the average of the measured and simulated dielectric permittivities for mono-molecular layers 1, 2, and 3, are 27, 47, and 69 while corresponding modeled values (T = 25°C in Fig. 4.5) are 22, 49, and 63, respectively.

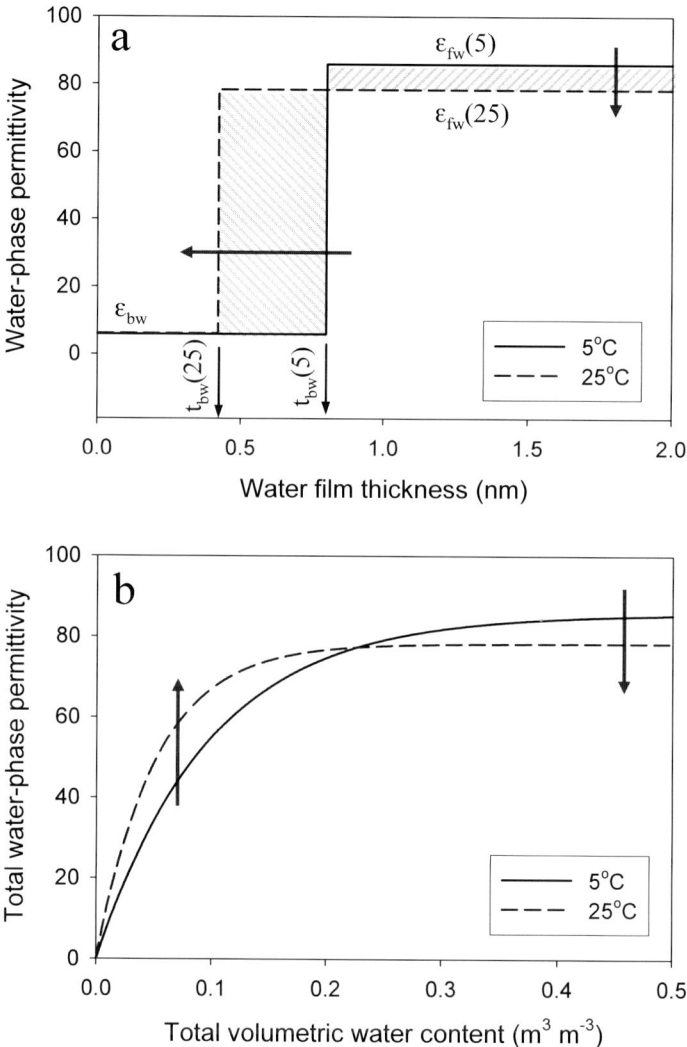

Fig. 4.5. Illustration of the thermodielectric effect described in **a**) by the temperature-dependent responses of reduced free water-phase permittivity (ε_{fw}) with increasing temperature and a reduced bound water layer thickness ($x(T)$). In **b**) these competing phenomena are illustrated as the total water-phase permittivity

4.3.4 Modeling Geometrical Effects

Particle shape effects influencing the dielectric permittivity of particulate mixtures may occur throughout a hierarchy of shapes and scales. In cereal grains, for example, this scale extends from the kernel to the starch grains and their complex internal structure, which extends down to the long chain and clustered polymers of amylose and amylopectin. In such complex structures the influence of shape on permittivity is difficult to determine. Effects of shape have been measured (using TDR) and modeled in simple systems using anisotropic packings of mica flakes where bound water is negligible [5].

Figure 4.6 illustrates measured permittivities in 0.25 mm diameter mica particles that were generally correlated to model predictions based on particle aspect ratio (geometry) and electrical field alignment as determined by probe orientation. Using Eqs. (4.5–7) and a combination of phase configurations the wet mica permittivity was modeled using an average aspect ratio of 1/25 and a porosity of 0.8. At mid- range water contents, measured data of $\varepsilon_{eff}{}^{a}$ is approximately half of the $\varepsilon_{eff}{}^{b}$ measurements. The measured permittivity of glass beads (inset) using the same system shows only a mild effect from the TDR probe (E-field) orientation. Reduced permittivity measurements in high surface area materials (e.g., clays) have been typically attributed to bound water effects. The reduction in permittivity in mica, which have very low specific surface area ($A_s < 0.06$ m^2 g^{-1}), should be attributed primarily to geometrical effects arising from the depolarizing influence of these high aspect ratio ($a/b = 1/25$) particles.

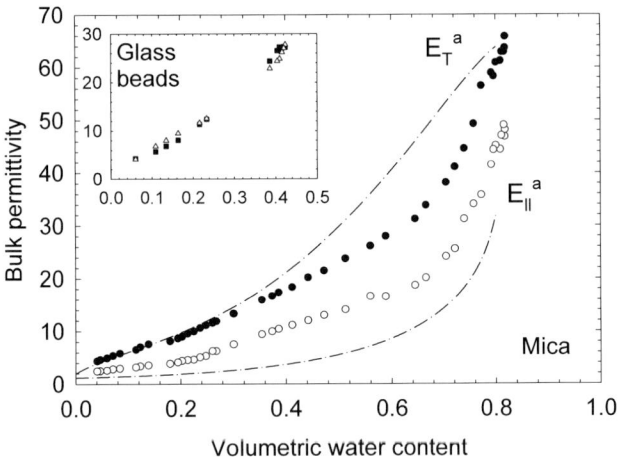

Fig. 4.6. Time domain reflectometry (TDR) measured bulk permittivity of partially saturated mica (0.25 mm diameter) and glass beads (inset). The orientation of the electric field (E) is described in Fig. 4.1 as being either normal ($E_T{}^{a}$) or parallel ($E_{\parallel}{}^{a}$) with respect to the rotation axis, a, of the mica particle assumed oblate and layered horizontally [5]

Although the binding of water is well correlated to A_s, measuring or estimating specific surface area is generally not as simple as calculating A_s based on well defined particle geometry in the simplified mica system.

Specific Surface Area Determination

Specific surface area, A_s, of a porous medium can be defined as the total surface area of solid constituents per unit mass and is intimately tied to the sizes of particles and to their shape. For example, rounded sand particles may exhibit values of A_s from 1 to 10 m^2 g^{-1} while plate-like clay particles may yield a specific surface area ranging from 100 to 1000 m^2 g^{-1}. This fundamental property of soils is correlated to important phenomena such as cation exchange, adsorption and release of chemicals, swelling, water retention and conductivity in addition to mechanical properties. Adsorption techniques are typically used to measure A_s directly [29]. Knowledge of particle size and geometry allows estimation of A_s using geometrical calculations of surface area. Another approach is based on the air-dried or hygroscopic water content, θ_h, (volume basis) as suggested by Robinson et al. [30]. They reported a linear correlation for A_s [m^2 g^{-1}] as a function of θ_h that resulted from fitting data of 42 soils, given as

$$A_s = 35.7 \cdot \theta_h \tag{4.9}$$

Various techniques have been used to determine water binding in biological materials using measurement techniques including calorimetery, dilatometry, nuclear magnetic resonance, electron spin resonance, thermally stimulated depolarization currents, and others [31, 32]. Estimating specific surface area of the kernel or starch grain may be derived from the measured mono-layered water content and an estimate of the density of the mono-layered bound water. The assumed structure of the water molecules will also influence the estimate slightly. Ryden [33] suggested a relationship for water molecule spacing, s [Å], based on water density, ρ_w [g cm^{-3}], and packing given by

$$s = \sqrt[3]{\frac{K_p}{\rho_w}} \tag{4.10}$$

where K_p is a packing constant given as 28.21 [g Å^3 cm^{-3}] for tetrahedral and 29.92 [g Å^3 cm^{-3}] for cubic packings. Assuming a water density of 1 g cm^{-3}, the resulting water molecular spacing is 3.04 Å and 3.10 Å for tetrahedral and cubic packings, respectively. These values are representative of the intermolecular spacing of free water [34]. For bound water, however, two contrasting views have been presented. One suggests that the density of water is less than the density of free water [35, 36] leading to larger spacing between water molecules. Reduced densities were suggested to arise from inter lattice spacing constraints between clay platelets which were numerically simulated, producing fluid densities both greater than and less than free water density depending on platelet separation [37]. For cereal grain constituents of protein and carbohydrate molecules, it is also conceivable that for certain structural spacing within and between molecules, reduced water densities are achievable.

Another view suggests water densities greater than that of free water [38-40], which correspond to a reduced water molecular spacing. Gur-Arieh et al. (1967) showed a constant wet flour density for moisture contents (w.b.) from 0 to 0.07 g g^{-1}, after which the wet density decreased steadily up to the final measured moisture content of 0.26. From this result they calculated a mono-layer water density, ρ_{ml}, of 1.48 g cm^{-3} and a density of the second molecular layer of water to be 1.11 g cm^{-3}. The remaining water layers were calculated to have a density of 0.967 g cm^{-3}. For a tetrahedral packing and using a mono-layer water density of 1.48 g cm^{-3} and a second layer density of 1.11 g cm^{-3}, Eq. (4.10) gives molecular spacing of 2.67 and 2.94 Å, respectively. These estimates lie on either side of spacing found in ice (hexagonal) of 2.76 Å (Robinson and Stokes, 1959) and are similar to other estimates of bound water spacing of between 2.5 Å and 2.8 Å [41]. For mono-layered water content given on a dry basis, M_{ml}, an estimate of the specific surface area, A_s [m^2 g^{-1}], is given using the estimated mono-layer water density, ρ_w [g cm^{-3}], and packing constant, K_p (Eq. (4.10)), given by

$$A_s = \frac{M_{ml}}{\rho_{ml} \cdot K_p^{1/3}} \tag{4.11}$$

Estimates based on this approach suggest A_s values of from several hundred to one thousand m^2 g^{-1} for starches and flours [32, 42, 43].

4.3.5 Modeling Porosity and Density

The physical make-up of porous mixtures is complicated by variations in texture and structure varying in size and shape. Particle geometry may take any number of forms from being plate-like to cylindrical or fiber-like to granular or blocky. Examples are found in soils where aspect ratios of certain clays (a/b = 0.033-0.025 [44]) and mica particles (a/b = 0.04, [5]) are extreme examples of oblate shapes (See Table 1). Many soil particles are generally spherical like sands (a/b = 0.46) or Tuff (a/b = 0.35) [9]. Prolate shapes are found in minerals such as hematite (a/b = 3.4, [45]) or fibers of peat (a/b = 10 to 100). As mentioned earlier, such shapes are often approximated using ellipsoidal or spheroidal geometries. From such approximations, physical characteristics of volume and surface area may be calculated. For porous mixtures comprised of spheroidal particles, an approximation of bulk porosity, ϕ_b, references the ratio of the representative particle volume to surface area [46]

$$\phi_b = 1 - k \frac{V_p^{\frac{2}{3}}}{S_p} \tag{4.12}$$

where k is an empirical scaling coefficient, V_p is the volume of the particle and S_p is the individual particle external surface area. The dimensionless volume to surface area ratio, $V_p^{2/3}/S_p$, forms a unique relationship for a spheroidal geometry and varying aspect ratio. Using approximate relationships for a spheroid of unit volume whose surface area [5] varies with aspect ratios (a/b) from 0.001 to 1000, the following empirical expression was obtained

$$\frac{V_p^{2/3}}{S_p} = \left[4.53 + 0.671 \left(\ln \frac{a}{b} \right)^2 + 0.309 \left(\frac{b}{a} \right) \right]^{-1} \tag{4.13}$$

Modeled porosities as functions of aspect ratio are plotted using Eqs. (4.12) ($k = 3$) and (4.13). The volume-surface-area relationship reproduces the general trend of increased porosity as aspect ratio deviates from that of a sphere ($a/b = 1$) for either oblate or prolate shapes. Comparison to measured or computed bulk porosities in a variety of porous media suggest these relationships provide reasonable predictions of bulk porosity illustrated in Fig. 4.7. The fitting parameter, k, simply provides vertical scaling of the curve.

Next to water content, bulk density of the porous medium is among the most varied and critical factors affecting the permittivity measurement. Bulk density is directly related to the bulk porosity through the density of solids, ρ_s, in a given volume of porous media. This relationship is of a similar form to that in Eq. (4.12) and is written as

$$\phi_b = 1 - \frac{\rho_b}{\rho_s} \tag{4.14}$$

For shrinking or swelling materials such as cereal grains or clayey soils, ρ_b varies with water content and therefore the determination of bulk density is coupled to water content determination. For cereal grains, especially, the water contained within the grain may be directly tied to the change in kernel density, which is related to a commensurate change in ρ_b. This is illustrated using an expression relating the dry solid density, ρ_s, density of water, ρ_w, and the wet basis moisture content, M_{wb}, to the kernel density, ρ_k, written as [47]

$$\rho_k = \frac{\rho_s}{1 + \left(\frac{\rho_s}{\rho_w} - 1 \right) M_{wb}} \tag{4.15}$$

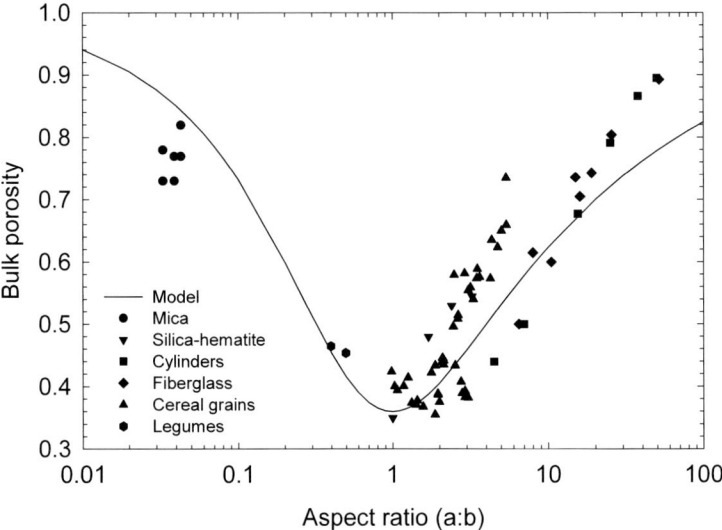

Fig. 4.7. Modeled bulk porosity of random packings of uniform spheroids as a function of particle aspect ratio. Data shown are measured porosities in (not necessarily randomly packed) mica particles [5], silica-hematite prolates [45], wooden cylinders and fiberglass rods [48], a variety of cereal grains [49], and legumes [50]

This simple expression requires only the dry solid density as a fitting parameter. For bulk density of the kernels, we consider the effect of kernel shape on packing by combining Eqs. (4.13) and (4.12) and setting them equal to Eq. (4.14) to solve in terms of bulk density. Note that for the bulk packing calculation, the solid density in Eq. (4.14) is assumed equal to the kernel density and Eq. (4.15) is substituted for ρ_s yielding

$$\rho_b = \frac{k \cdot \rho_s}{\left(4.53 + 0.671\left(\ln\frac{a}{b}\right)^2 + 0.309\left(\frac{b}{a}\right)\right) \cdot \left(1 + \left(\frac{\rho_s}{\rho_w} - 1\right) \cdot M_{wb}\right)} \qquad (4.16)$$

This expression contains effects of the additional porosity and packing going from the kernel to the bulk packing. These two models are plotted in Fig. 4.8 compared to measured kernel and bulk densities in yellow-dent field corn.

Inverse bounds on constituent volume fractions derived from complex permittivity measurements and evaluation of the structural moments and geometry [51] may provide improved measurement capability where bulk density or porosity is often an unknown and confounding factor for water content determination. Furthermore, recent work using free-space measurements of complex permittivity at

GHz frequencies in cereal grains demonstrates the potential for density-independent measurements of moisture content [52].

To summarize geometrical effects, the specific surface area is a critical parameter used to estimate bound water content in porous media. Particle shape and particle density directly affect the specific surface area, packing and arrangement of the bulk mixture and can be used to compute estimates of A_s, ρ_b and ϕ_b. Each of these characteristics plays an important role in dielectric measurements for water content determination. These parameters can be employed in modeling permittivity using applicable dielectric mixture theory.

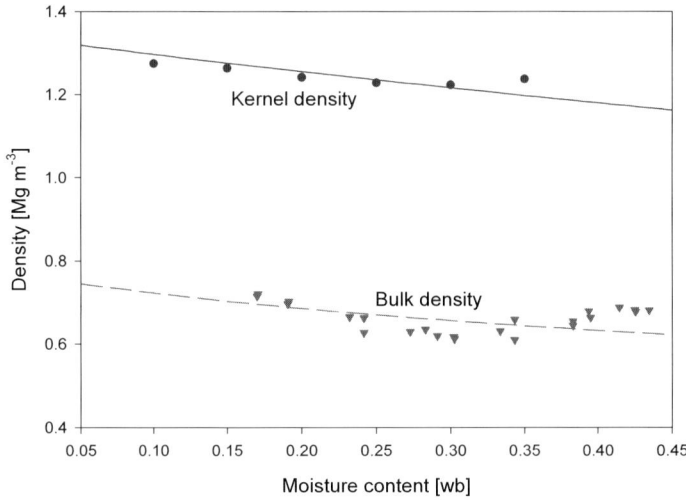

Fig. 4.8. Measured individual kernel [53] and bulk kernel densities [54] of field corn. Modeled results are obtained from Eq. (4.13) for kernel density ($\rho_s = 1.34$ Mg m^{-3}) and Eq. (4.15) for bulk density (k = 3, a/b = 2)

4.3.6 Constituent-Phase Configuration Influence on Bulk Permittivity

Friedman (1998) demonstrated the influence of constituent-phase configuration on modeled permittivities for six unique configurations of a concentric spherical model similar to the illustration in Fig. 4.1. Among the six possible combinations, water, solid, air (WSA), SWA, and ASW cover the range of permittivities exhibited by a wide range of porous media. Equation (4.6) was used to plot the configuration-dependent permittivities shown in Fig. 4.9. The combination of ASW and SWA forms an   envelope about the permittivity of common mineral soils represented by the Topp et al. (1980) equation. Reduced permittivities found in high surface area porous materials are more closely described by SWA and WSA configurations.

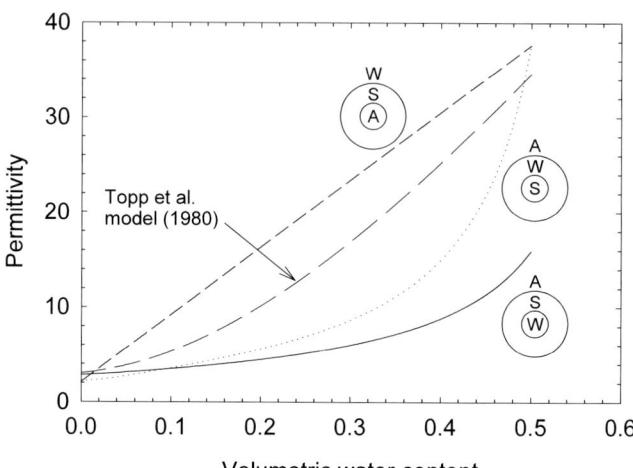

Fig. 4.9. Modeled permittivity of three different constituent-phase configurations where, for example, WSA represents water, solid and air with permittivities assuming modeled values of 80, 5, and 1, respectively. The Topp et al., 1980 expression describing mineral soils is shown for comparison

Swelling soils and plant seed constituents may expand, unfold and relax during drying and wetting, which for seeds may be accompanied by a change in shape and volume. Volume increases in soybean with hydration were found to be sub-stantially greater than the weight of water imbibed, suggesting that polymeric seed constituents unfold with hydration [55]. We suggest this is a mechanism for in-creased surface area or evolving water-phase configuration. Evidence of this is seen in the sigmoidal shape of corn starch (Fig. 4.4c and d) and corn kernel (Fig. 4.10) moisture content-permittivity curves. Measured permittivities (200 MHz, [28]) of yellow-dent field corn kernels are shown in Fig. 4.10 where a combination of SWA and WSA could describe the complex water-solid configuration in cereal grains. A sigmoidal weighting distribution describes the relative contribution of SWA and WSA configurations shown in Fig. 4.10, which resulted in the modeled curve marked EMA. Such an approach was used successfully to model the three-phase system of moist soil combining configurations of ASW and SWA with an effective medium approximation (EMA) and estimates of bound water based on surface area [56]. Our justification for this approach is linked to kernel swelling or shrinking that occurs throughout the hierarchy of geometries contained in the starch grain. A second attempt to model the corn data was based on the surface area of the corn, which was increased 10 times at a critical water content to approximate a surface area increase for additional water binding (curve marked $A_s \times 10$). The effects of water binding in cereal grains and swelling soils are com-plicated by the volume change accompanied by wetting/drying processes and by the associated sample density changes.

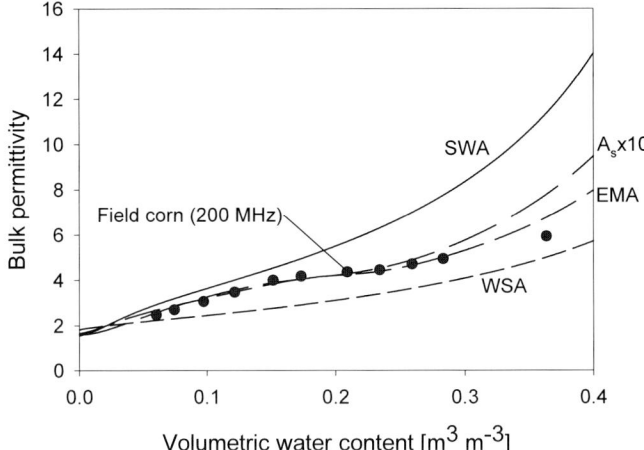

Fig. 4.10. Modeled permittivity of field corn using a sigmoidal increase in A_s (A_sx10) and using an effective medium approximation (EMA) of the two configurations, SWA and WSA with a sigmoidal increase in the WSA fraction with water content increase

4.4 Summary

Several factors influencing dielectric permittivity measurements in porous media have been identified using measurements and modeling approaches. Bound water associated with large surface area materials may lead to reduced dielectric, but other effects may also play a significant role in alteration of the permittivity-water content relationship. The thermodielectric effect observed in high surface area porous media might enhance or reduce the permittivity, depending on the water status and content and direction of temperature shift. Particle shape effects demonstrated in low-surface area media are a function of particle geometry and orientation with respect to the electrical field. In addition to bound water and geometrical effects, water-phase configuration may explain alteration of the bulk permittivity of wet porous media. Physical characteristics of porous media describe parameters critical for modeling permittivity, such as porosity and bulk density in addition to estimates of surface area. In general these can be broken down into representative estimates of solid, water and air fractions, with the volumetric water content being the variable of interest both for modeling purposes and for determination of water content from permittivity measurements. Identifying and separating bound water, particle shape and water-phase configuration effects in addition to other competing and perhaps confounding effects is an important step in understanding dielectric measurements in high surface area porous media for water content determination.

Acknowledgments

Partial funding for this work was provided by the Unites States-Israel Binational Agricultural Research and Development (BARD) Fund (Project IS-2839-97). We gratefully acknowledge research grants from Campbell Scientific and Harvest-Master (Logan, Utah) and Pioneer Hi-Bred International (Johnston, IA).

References

1. Topp GC, Davis JL, Annan AP (1980) Electromagnetic determination of soil water content: Measurements in coaxial transmission lines. Water Resour. Res. 16: 574-582
2. Sihvola A (1999) Electromagnetic mixing formulas and applications. Michael Faraday House, Stevenage, Herts, SG1 2AY, UK
3. Dasberg S, Dalton FN (1985) Time domain reflectometry field measurements of soil water content and electrical conductivity. Soil Sci. Soc. Am. J. 49: 293-297
4. Sihvola A, Lindell IV (1990) Polarizability and effective permittivity of layered and continuously inhomogeneous dielectric ellipsoids. J. Electromagn. Waves Appl. 4: 1-26
5. Jones SB, Friedman SP (2000) Particle shape effects on the effective permittivity of anisotropic or isotropic media consisting of aligned or randomly oriented ellipsoidal particles. Water Resour. Res. 36: 2821-2833
6. Kraszewski AW, Nelson SO (1996) In: Kraszewski A (ed) Microwave Aquametry. IEEE Press, NY, pp 177-203
7. Kelley JM, Stenoien JO, Isbell DE (1953) Wave-guide measurements in the microwave region on metal powders suspended in paraffin wax. J. Appl. Phys. 24: 258-262
8. Sillers RW (1936) The properties of a dielectric containing semi-conducting particles of various shapes. J. Inst. E.ect. Eng. 80: 378-394
9. Friedman SP, Robinson DA (2002) Particle shape characterization using angle of repose mesurements for predicting the effective permittivity and electrical conductivity of saturated granular media. Water Resour. Res. 38
10. Hillel D (1998) Environmental Soil Physics. Academic Press, San Diego
11. Brooker DB, Bakker-Arkema FW, Hall CW (1992) Drying and Storage of Grains and Oilseeds. Van Nostrand Reinhold, New York
12. Lamm G, Pack GR (1997) Local dielectric constants and Poisson-Boltzmann calculations of DNA counterion distributions. Int. J. Quant. Chem. 65: 1087-1093
13. Thorp JM (1959) The dielectric behaviour of vapours adsorbed on porous solids. Trans. Faraday Soc. 55: 442-454
14. Bockris JOM, Devanathan MAV, Muller K (1963) On the structure of charged interfaces. Proc. Roy. Soc. (London) A274: 55-79
15. Olhoeft GR (1981) In: Roy RF (ed) Physical Properties of Rocks and Minerals (CINDAS Data Series on Material Properties), vol II. McGraw-Hill, New York, pp 257-328
16. Nelson S, Lindroth D, Blake R (1989) Dielectric properties of selected and purified minerals at 1 to 22 GHz. J. Microwave Power Electromagnetic Energy. 24: 213-220
17. Booth F (1951) The dielectric constant of water and the saturation effect. J. Chem. Phys. 19: 391-394
18. Robinson DA, Jones SB, Wraith JM, Or D, Friedman SP (2003) A Review of Advances in Dielectric and Electrical Conductivity Measurement in Soils Using Time Domain Reflectometry. Vadose Zone J 2: 444-475

19. Oliver BM, Cage JM (1971) Electronic measurements and instrumentation. McGraw-Hill, New York
20. Weast RC (1986) CRC Handbook of Chemistry and Physics. CRC Press, Boca Raton, FL
21. Or D, Wraith JM (1999) Temperature effects on soil bulk dielectric permittivity measured by time domain reflectometry: A physical model. Water Resour. Res. 35: 371-383
22. Debye P (1929) Polar Molecules. Dover, Mineola, New York
23. Landau LD, Lifshitz EM (1960) Electrodynamics of Continuous Media. Permagon Press, New York
24. Schaap MG, deLange L, Heimovaara TJ (1996) TDR calibration of organic forest floor media. Soil Technology 11: 205-217
25. Jones SB, Or D (2002) Surface area, geometrical and configurational effects on permittivity of porous media. J. Non-Crystalline Solids. 305: 247-254
26. Dirksen C, Dasberg S (1993) Improved calibration of time domain reflectometry soil water content measurements. Soil Sci. Soc. Am. J. 57: 660-667
27. Ndife MK, Sumnu G, Bayindirli L (1998) Dielectric properties of six different species of starch at 2450 MHz. Food Research International. 31: 43-52
28. Jones RN, Bussey HE, Little WE, Mezker RF (1978). National Bureau of Standards, Boulder
29. Carter DL, Mortland MM, Kemper WE (1986) In: Klute A (ed) Methods of soil analysis, Part 1, Physical and mineralogical methods. ASA, Madison, WI, pp 413-423
30. Robinson DA, Cooper JD, Gardner CMK (2002) Modelling the relative permittivity of soils using soil hygroscopic water content. J. Hydrology. 255: 39-49
31. Leung HKH (1975) Capacity and force of water binding by carbohydrates and proteins as determined by nuclear magnetic resonance, Ph.D., University of Illinois at Urbana-Champaign
32. Leung HK, Steinberg MP, Wei LS, Nelson AI (1976) Water binding of macromolecules determined by pulsed NMR. J. Food Sci. 41: 297-300
33. Ryden BE (1992) In: Lund LJ (ed) Indirect methods for estimating the hydraulic properties of unsaturated soils. University of California Riverside, Riverside, CA, pp 693-706
34. Robinson RA, Stokes RH (1959) Electrolyte Solutions. Butterworths, London
35. Anderson DM, Low PF (1958) The density of water adsorbed by lithium-sodium- and potassium-bentonite. Soil Sci. Soc. Am. Proc. 22: 99-103
36. Low BW, Richards FM (1954) Measurements of the density, composition and related unit cell dimensions of some protein crystals. Am. Chem. Soc. J. 76: 2511-2518
37. Israelachvili JN (1992) Intermolecular and surface forces. Academic Press, San Diego
38. Gur-Arieh C, Nelson AI, Steinberg MP (1967) Studies on the density of water adsorbed on low-protein fraction of flour. J. Food Sci. 32: 442-445
39. Mackenzie RC (1958) Density of water sorbed on monmorillonite. Nature. 181: 334
40. Schoen M, Diestler DJ, Cushman JH (1987) Fluids in micropores. I. Structures of a simple classical fluid in a slit-pore. J. Chem. Phys. 87: 5464-5476
41. Agmon N (1996) Tetrahedral displacement: The molecular mechanism behind the Debye relaxation in water. J. Phys. Chem. 100: 1072-1080
42. Ratkovic S, Pissis P (1997) Water binding to biopolymers in different cereals and legumes: Proton NMR relaxation, dielectric and water imbibition studies. J. Material Sci. 32: 3061-3068
43. Konsta AA, Pissis P, Kanapitsas A, Ratkovic S (1996) Dielectric and conductivity studies of the hydration mechanisms in plant seeds. Biophysical Journal. 70: 1485-1493

44. Leschansky YI, Lebedeva GN, Shumilin VD (1971) Electrical parameters of sandy and loamy grounds in the range of centimeter, decimeter, and meter wavelength. Izv. Vyss. Ucheb. Zaved. Radiofiz. 14: 562-569

45. Thies-Weesie DME, Philipse AP, Kluijtmans SGIM (1995) Preparation of sterically stabilized silica-hematite ellipsoids: Sedimentation, permeation, and packing properties of prolate colloids. J. Colloid Interface Sci. 174: 211-223

46. Cumberland DJ, Crawford R (1987) In: Allen T (ed) Handbook of Powder Technology, vol 6. Elsevier, Amsterdam

47. Sokhansanj S, Lang W (1996) Prediction of kernel and bulk volume of wheat and canola during adsorption and desorption. J. Agric. Engng. Res. 63: 129-136

48. Milewski JV (1978) The combined packing of rods and spheres in reinforcing plastics. Ind. Eng. Chem. Prod. Res. Dev. 17

49. Mohsenin NN (1970) Physical properties of plant and animal materials: Structure, physical characteristics and mechanical properties. Gordon and Breach Science Publishers, New York

50. Sokhansanj S, Falacinski AA, Sosulski FW, Jayas DS, Tang J (1990) Resistance of bulk lentils to airflow. Trans. ASAE. 33: 1281-1285

51. Cherkaeva E, Golden KM (1998) Inverse bounds for microstructural parameters of composite media derived from complex permittivity measurements. Waves in Random Media 8: 437-450

52. Trabelsi S, Nelson SO (1998) Density-independent function for on-line microwave moisture meters: a general discussion. Meas. Sci. and Tech. 9: 570-578

53. Nelson SO (1979) RF and microwave dielectric properties of shelled, yellow-dent field corn. Trans. ASAE. 22: 1451-1457

54. Brusewitz GH (1975) Density of rewetted high moisture grains. Trans. ASAE 18: 935-938

55. Leopold AC (1983) Volumetric components of seed imbibition. Plant Physiol. 73: 677

56. Friedman SP (1998) A saturation degree-dependent composite spheres model for describing the effective dielectric constant of unsaturated porous media. Water Resour. Res. 34: 2949-2961

5 Model Systems for Materials with High Dielectric Losses in Aquametry

Ari Sihvola

Helsinki University of Technology, Electromagnetics Laboratory,
P.O. Box 3000, FIN–02015 HUT, Finland

5.1 Introduction

This article discusses dielectric properties of two-phase mixtures where one of the component phases is highly lossy, as often is the case in aquatic materials. A special enhancement effect is under study: a strong increase in the real part of the effective permittivity of the mixture for certain fractional volume and geometrical shape of the inclusion phase. For planar geometries, the enhancement effect is known as Maxwell–Wagner effect. Using classical homogenization rules for mixtures (like Maxwell Garnett, Bruggeman, Coherent potential), analytical results are derived and presented for the magnitude and critical parameters of the enhancement effect. Let us start with background and motivation to the study.

Modelling of the dielectric properties of materials in aquametry is a great challenge. Water as a material, and especially as an electromagnetically responding material, is very sensitive to the parameters of the environment and the dielectric excitation: the permittivity depends strongly on temperature and other parameters, like for example, the possible contamination by foreign components as it seldom appears in totally pure state; furthermore, it is dispersive, meaning that the temporal variation of the excitation has to be accounted for carefully, etc.

And this is, of course, not the whole problem. These uncomfortable properties apply to water as homogeneous substance which is already well studied, tabulated, and documented. When we are faced with applications where materials are moist and aqueous the difficulties in modelling increase manyfold. The boundaries between the liquid and solid phases, even if they can be uniquely defined, are often complicated in geometrical description. Boundary layer effects are created. Chemical interactions may take place between two liquid phases or between liquid and solid components of the mixture. Complexity lurks everywhere.

The message of the present article may sound strange, in particular in light of the above background. I am claiming that even if we neglect all the complicatedness of aquametry that follows from its molecular and dipolar microstructure, from chemical interactions and from complex geometry, basic dielectric homogenization of moist substances may still bring forth very unexpected observable behavior in the macrosopic dielectric response of this matter.

In other words: let us simplify the situation: forget the sensitivity to external parameters of the permittivity of water and play with the value at a given temperature, frequency, etc. In addition, neglect complex, possibly fractal, geometries and take a simple mixture: spheres, ellipsoids, discs, plates, laminates, and assume it to consist of two clearly separate phases where the other is plain homogeneous liquid water. Even if we assume all these simplifications to hold, interesting things may happen: using basic dielectric mixing rules and homogenization principles, the macroscopic effective response of the mixture may display unexpected behavior. And yet this behavior can be understood from physically appealing principles.

Furthermore, I would like to emphasize that this phenomenon, to be described in the following sections in more detail, where complexity grows out of simple elements, is a different effect than the phenomena of complexity, very much in fashion today, that arise in discrete systems from simple local cellular rules. Cellular automata can certainly bring forth extremely rich and emergent behavior using algorithms with astonishingly simple regulation, as Stephen Wolfram has shown in his recent book [1]. But the present, "aquatic" complexity is connected with the basic laws of dielectric response in materials and as such it is a consequence of electrostatic forces and their effects in polarizable matter, and not – at least very directly – with the way the state of a unit cell is affected by its history and the states of the neighboring cells.

The structure of this article is the following: starting with a reminder of the classical formulas to describe the effective permittivity of simple two-phase mixtures, these predictions are applied to aquatic mixtures where one of the phases is liquid water. Characteristic in its dielectric response is the high degree of losses. On the level of permittivity, this means that the quantity ε is a complex number with a large imaginary part. First, for planar mixtures – meaning that the inhomogeneity structure is one-dimensional – it is shown that, with certain assumptions, a strong enhanced polarization can follow. This gives an interpretation to the so-called Maxwell–Wagner effect. Then, a more detailed analysis follows on the conditions and characteristic properties for the dielectric enhancement effect. Analytical results are derived for the critical parameters for its appearance. A short discussion of the results concludes the article.

5.2 Classical Two-Phase Mixing Formulas

Mixing rules are developed for enumerating the effective permittivity ε_{eff} (or any other macroscopic material parameter) of a mixture, or of a heterogeneous medium that consists of medium components that are dielectrically different from each other. Another term, "homogenization," is often used to describe the operation of assigning a single number for the dielectric response of a sample of mixed material. The homogenization problem is indeed a classical problem in statistical physics [2] and a great amount of literature exists on the problems of homogenization, mixing rules, and theory of composites (see, for example, [3–5]).

Here, rather simple mixtures are analyzed. They are assumed to consist of two phases, of which one forms the background, or environment, and the other one is embedded as a guest, forming the inclusions phase. Let the dielectric relative permittivity of the host material (forming the background) be ϵ_e, and that of the guest ϵ_i. In the following, the important parameter is the volume fraction of the inclusions, f. Then the fractional volume occupied by the host is $1 - f$.

Note, once again, that in the following, the permittivities are *relative*, in other words dimensionless quantities. This is to avoid carrying everywhere the free-space permittivity along.

Maxwell Garnett formula. Perhaps the best-known classical mixing rule is the one that carries the name after Maxwell Garnett [6]. The prediction for the effective permittivity ϵ_{eff} is

$$\epsilon_{\text{eff}} = \epsilon_e + 3f\epsilon_e \frac{\epsilon_i - \epsilon_e}{\epsilon_i + 2\epsilon_e - f(\epsilon_i - \epsilon_e)} \qquad (5.1)$$

for a mixture where spherical inclusions with permittivity ϵ_i occupy a volume fraction f in a host material with permittivity ϵ_e.

If the Maxwell Garnett rule is turned "upside down", in other words the roles of the host and guest are reversed, we arrive at the so-called inverse Maxwell Garnett formula, which obviously reads

$$\epsilon_{\text{eff}} = \epsilon_i + 3(1 - f)\epsilon_i \frac{\epsilon_e - \epsilon_i}{\epsilon_e + 2\epsilon_i - (1 - f)(\epsilon_e - \epsilon_i)} \ . \qquad (5.2)$$

It may seem trivial to define this formula to be an inverse with respect to some given and a more primary, Maxwell Garnett formula. Is it not, anyway, quite arbitrary which phase we choose to be the environment in which the guest phase appears, especially if we are dealing with volume fractions of the order of 50% ? This type of objection can certainly be justified; however, in many connections, the differences in the character of the direct and inverse MG rules are analyzed with interesting conclusions (for example, [7]). Furthermore, the difference is also connected to the often-used concepts of a raisin-pudding mixture (inclusions are "heavier" than the background) and the Swiss-cheese mixture (hole-type inclusions in solid matrix). The contrast between the Maxwell Garnett and inverse Maxwell Garnett philosophies is schematized in Fig. 5.1.

Bruggeman formula. Certainly other rules, too, like the famous Bruggeman rule [8]

$$(1 - f)\frac{\epsilon_e - \epsilon_{\text{eff}}}{\epsilon_e + 2\epsilon_{\text{eff}}} + f\frac{\epsilon_i - \epsilon_{\text{eff}}}{\epsilon_i + 2\epsilon_{\text{eff}}} = 0 \qquad (5.3)$$

can be written for the effective permittivity ϵ_{eff} of this mixture. Here, again, the inclusions are assumed to be spherical.

The essence of the Bruggeman formalism is the equality between the phases in the mixture. There is no more host-versus-guest hierarchy. Instead, the homogenized medium itself is considered as the background against which polarizations

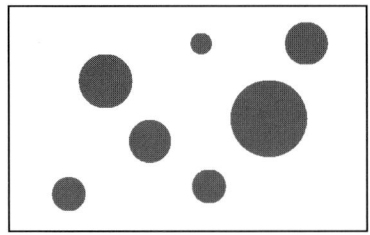
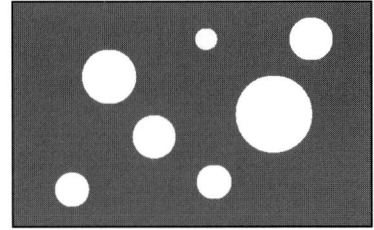

Maxwell Garnett   *Inverse Maxwell Garnett*

Fig. 5.1. Maxwell Garnett treats the phase ϵ_i as spherical (dark) inclusions in background (light) ϵ_e (raisin pudding -type mixture), whereas the inverse Maxwell Garnett considers ϵ_e as guest "inclusions" in the host ϵ_i (Swiss cheese type)

are measured. And now polarization with respect to the effective medium is to be accounted for as well from the environment as from the inclusions.[1] Equation (5.3) shows that we could not talk about the "inverse Bruggeman formula:" a mixture and its complement (which emerge through the transformation $\epsilon_i \rightarrow \epsilon_e$, $\epsilon_e \rightarrow \epsilon_i$, $f \rightarrow 1 - f$) have exactly the same effective permittivity.

Coherent potential formula. Another well-known formula which is relevant in the theoretical studies of wave propagation in random media is the so-called Coherent potential formula [9]. One of its forms (again, spherical inclusions) looks like

$$\epsilon_{\text{eff}} = \epsilon_e + f(\epsilon_i - \epsilon_e)\frac{3\epsilon_{\text{eff}}}{3\epsilon_{\text{eff}} + (1 - f)(\epsilon_i - \epsilon_e)} . \tag{5.4}$$

which is, like Eq. (5.3), implicit for the unknown effective permittivity ϵ_{eff}.

Unified mixing formula. The mixing approach presented in [10] collects all the previous aspects of dielectric mixing rules into one family. For the case of isotropic spherical inclusions ϵ_i in the isotropic environment ϵ_e, the formula looks like

$$\frac{\epsilon_{\text{eff}} - \epsilon_e}{\epsilon_{\text{eff}} + 2\epsilon_e + \nu(\epsilon_{\text{eff}} - \epsilon_e)} = f \frac{\epsilon_i - \epsilon_e}{\epsilon_i + 2\epsilon_e + \nu(\epsilon_{\text{eff}} - \epsilon_e)} . \tag{5.5}$$

This formula contains a dimensionless parameter ν. For different choices of ν, the previous mixing rules are recovered: $\nu = 0$ gives the Maxwell Garnett rule, $\nu = 2$ gives the Bruggeman formula, and $\nu = 3$ gives the Coherent potential approximation.

Mixing rules are often required to satisfy the basic "boundary conditions" for vanishing guest and vanishing host phases: $\epsilon_{\text{eff}} = \epsilon_e$ for $f = 0$, and $\epsilon_{\text{eff}} = \epsilon_i$ for $f = 1$. Both these conditions are easily seen to be valid for all the formulas above.

---

[1] Equation (5.5) can be interpreted as a balance condition for the polarizabilities due to the host and guest phases, both weighted with their own volume fractions.

In the following sections, these mixing rules are applied to aquatic mixtures. Here the inclusion phase was assumed to have spherical form. But the formulas can be written also for cases where the inclusions are general ellipsoids [4], and in the following applications, also such cases are treated.

5.3 Interfacial Polarization in Aquatic Mixtures

5.3.1 Interfacial Polarization and Dispersion

Perhaps the simplest geometry to illustrate how the dielectric mixing process may create surprising macroscopic effects is a planar structure, a purely one-dimensional inhomogeneity. Such a structure will be considered in this section: two dielectrically different materials are stacked in a laminar manner.

Geometries of this type may appear in several applications in aquametry. Especially this may happen in connection with bioelectrical measurements and applications, where one surely deals with complicated materials. Biological tissues may be seen as complex composites containing insulating sheets and also ionized liquids which often possess a high degree of conductivity. This kind of aggregate causes a challenge for the modelling of interface between electronic equipment and living matter.

Various dispersion mechanisms have been identified for the behavior of the tissue permittivity as a function of frequency [11]. One of these is the interfacial polarization in the kilohertz–megahertz-region which is due to the charge accumulation on the boundaries of conducting regions. The dissipation affiliated with such a mechanism carries the name Maxwell–Wagner losses [12, 13].

Indeed, the mechanism of interfacial polarization can be modelled by the mixing problem where the geometry is laminar. This certainly could be applicable for muscle sheaths or cells with membranes, or then in other aquametric applications. Let us therefore try to apply the principles of mixing theories for the dielectric permittivity of such systems. It turns out that the real part of the effective permittivity may become considerably enhanced. The advantage of the analytical treatment is that then also closed-form results can be derived for the geometrical conditions that correspond to maximum permittivity of the macroscopic structure and for other relevant parameters connected with this effect.

5.3.2 Effective Permittivity of a Layered Structure

Figure 5.2 shows a simple model for a planar structure that could model, for example, a biological cell. There, insulating walls of thickness d_i and relative permittivity ϵ are bounding a conducting volume with thickness d_c that contains an electrolyte or other medium with a very large imaginary part of the permittivity. Given its conductivity σ it is appropriate to normalize it:

$$s = \frac{\sigma}{\omega \epsilon_0} .\tag{5.6}$$

Here the conductivity is made dimensionless by using the angular frequency ω and the free-space permittivity ϵ_0. The complex relative permittivity[2] of the electrolyte is therefore $1 - js$.

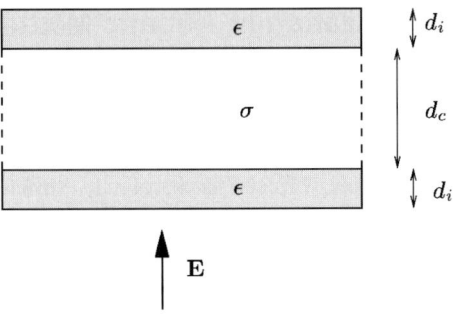

Fig. 5.2. A model for a laminated structure in aquametry. An example could be the cell with membrane and cytoplasm; conducting electrolyte between thin insulating sheets

Let the electric field point perpendicularly to the interfaces. Then the effective permittivity of such a cell structure can be calculated from the relation

$$\epsilon_{\text{eff}} = \frac{1}{f/\epsilon + (1-f)/(1-js)} = \frac{\epsilon(1-js)}{\epsilon - (\epsilon-1)f - jfs}\tag{5.7}$$

where the volume fractions are $f = 2d_i/(d_c + 2d_i)$ for the insulating phase and $1 - f = d_c/(d_c + 2d_i)$ for the conducting phase. This result in Eq. (5.7) follows from an application of a dielectric mixing rule [4] with the depolarization factor of 1. Alternatively, it can be seen as a volume-average of the inverses of the permittivities.[3] The real part of ϵ_{eff} in Eq. (5.7) ,

$$\epsilon'_{\text{eff}} = \epsilon \frac{\epsilon + (s^2 + 1 - \epsilon)f}{\epsilon^2 - 2\epsilon(\epsilon-1)f + [(\epsilon-1)^2 + s^2]f^2} ,\tag{5.8}$$

is a function varying from the value 1 at vanishing membrane thickness ($f = 0$) to ϵ at the other limit case ($f = 1$) where the conducting phase shrinks to zero and the membrane fills the whole space.

---

[2] The convention exp(jωt) is taken here for the time-harmonic dependence of the fields. Therefore for passive and dissipative media, the imaginary part of the permittivity is negative, and it is convenient to use the notation $\epsilon = \epsilon' - j\,\epsilon''$.

[3] In the earlier Maxwell Garnett formula Eq. (5.1), the depolarization factor was 1/3, which holds for spheres in three dimensions. The general form is $\epsilon_{\text{eff}} = \epsilon_e + f\epsilon_e\,(\epsilon_i - \epsilon_e)/[\epsilon_e + (1 - f)N(\epsilon_i - \epsilon_e)]$ for aligned ellipsoids with depolarization factor N.

However, ϵ'_{eff} can attain larger values than either of these limits if the conductivity s is large enough, and can become extremely large for good electrolytes. This enhancement phenomenon (the fact that the maximum for ϵ'_{eff} occurs somewhere inside the interval $0 < f < 1$) can be seen clearly in Fig. 5.3 where ϵ'_{eff} is shown in the (f, s)-plane for $\epsilon = 2$. The limiting values of ϵ'_{eff} are independent of s: for $f = 0$, $\epsilon'_{eff} = 1$, and for $f = 1$, $\epsilon'_{eff} = 2$. But when s increases sufficiently, a maximum in ϵ'_{eff} appears which is clearly larger than 2.

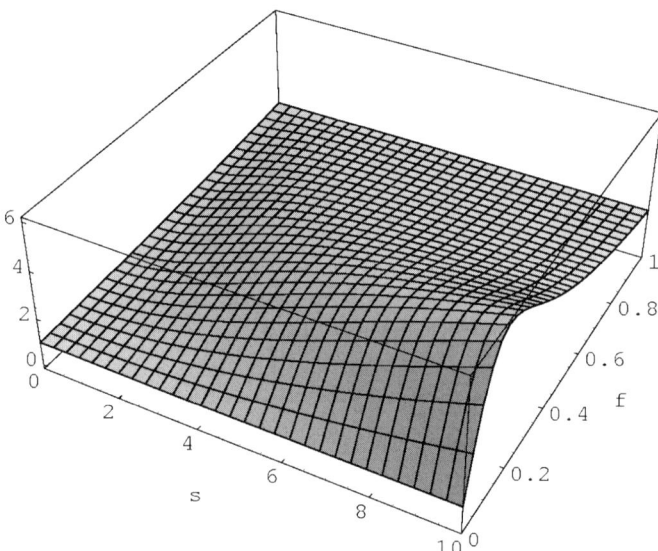

Fig. 5.3. The real part of the effective permittivity ϵ'_{eff} Eq. (8) as function of the volume fraction of the membrane $0 < f < 1$ and the relative conductivity $0 < s < 10$. The relative permittivity of the membrane is $\epsilon = 2$

Let us state the results more quantitatively. After not insignificant algebraic efforts, the following findings can be justified. The real part of the effective permittivity experiences a maximum (larger than 1 or ϵ) if

$$s > \sqrt{\epsilon - 1} \qquad (5.9)$$

and this happens for the volume fraction of the insulator phase f is

$$f_{max} = \epsilon \frac{(\epsilon - 1)^2 + s^2 - s\sqrt{1 + s^2}\sqrt{(\epsilon - 1)^2 + s^2}}{(\epsilon - 1)^3 - s^2[(\epsilon - 1)(\epsilon - 2) + s^2]}. \qquad (5.10)$$

The maximum value of ϵ_{eff} for given ϵ and s is then (using Eq. (5.10) in Eq. (5.8))

$$\epsilon'_{eff,max} = \frac{\epsilon s + \sqrt{1 + s^2}\sqrt{(\epsilon - 1)^2 + s^2}}{2s}. \qquad (5.11)$$

The volume fraction giving the maximum c'_{eff} is illustrated in Fig. 5.4 for some parameter values. There the tendency is visible of the maximum to become shifted to thinner membranes as the conductivity increases. At the same time, $c'_{eff, max}$ increases.

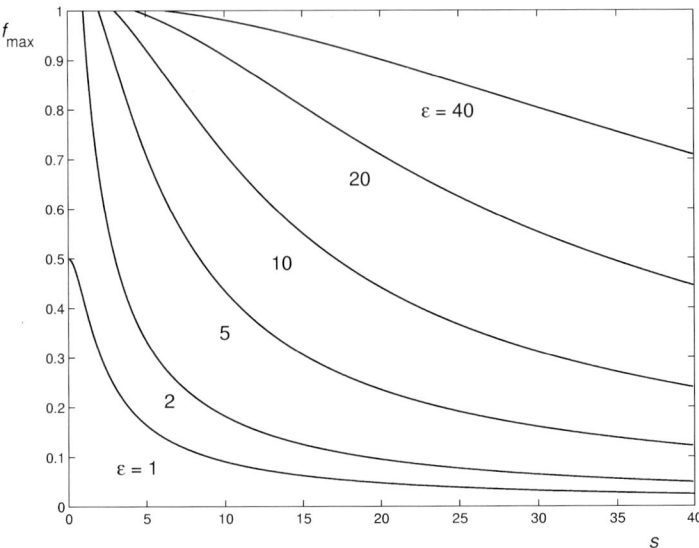

Fig. 5.4. The fraction of membrane of the whole volume (Eq. (5.10) for f_{max}) that gives the maximum to the real part of the effective permittivity c'_{eff}, given the membrane permittivity c and the relative conductivity of the electrolyte s.

5.3.3 Enhancement Effect

In light of these results, it is obvious that thin membrane layers can significantly increase the real part of the effective permittivity of the composite cell structure. Indeed, looking at the limiting cases when $s \gg c$, we can see that volume fraction f_{max} that gives the maximum c'_{eff}, tends to c/s, in other words to thin insulating layers. Furthermore, the maximum attainable value for c'_{eff} Eq. (5.11) becomes in this limit $(s + c)/2$. This number can well reach the level of millions because the imaginary part of the electrolyte permittivity, $s = \sigma/(\omega c_0)$, increases easily to such large values as the frequency decreases.

In the following sections, this enhancement effect and its parameters are studied from a more general point of view and adding more freedom into the geometrical structure of the mixture.

5.4 Analysis of the Enhancement Effect in High-Loss Mixtures

The previous section showed that even for a simple planar structure, classical mixing formulas predict that if very lossy dielectric particles are embedded in neutral background, the complex macrosopic permittivity of the mixture may be such that its real part is considerably larger than the real parts of the components that constitute the mixture. The parameters of this effect – the conditions for it to happen, the volume fraction, and the maximum attainable permittivity value – are certainly dependent on the shapes of the inclusions: whether the mixture consists of spherical, needle-shaped, disc-shaped, or ellipsoidal particles.

To estimate the effective and macroscopic behavior of inhomogeneous and random media, the simplest approach is again to use mixing formulas, the ones presented before, or generalized as in [4, 14]. A purely algebraic manipulation of these formulas in their complex form reveals properties of the critical parameters.

5.4.1 Peculiarities in High-Loss Mixing

When does the mixture display the enhancement effect? Of course for a mixture with two lossless materials whose permittivities are both real numbers, the effective permittivity is also real. Then also the effective permittivity is limited above by the larger of the two component permittivities. (Note that the effective permittivity discussed in the present article is a low-frequency concept, and scattering phenomena are neglected. The wavelength of the field has to be much larger than the inhomogeneities in the mixture.)

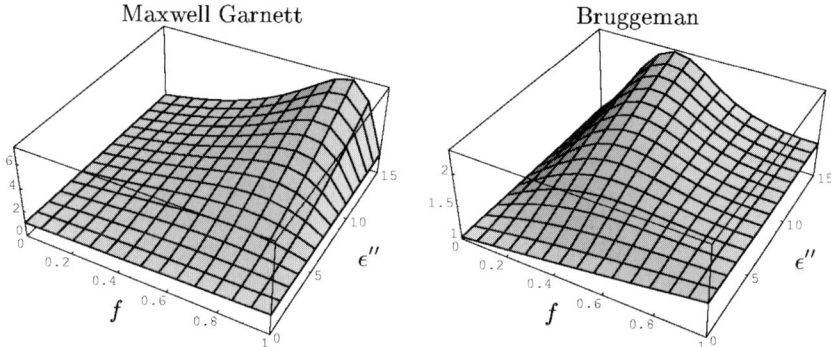

Fig. 5.5. The real part of the effective permittivity ϵ'_{eff} of a mixture according the Maxwell Garnett formula (left) and Bruggeman formula (right). Spheres with relative complex permittivity $\epsilon = 1.5 - j\epsilon''$ occupy a volume fraction f in vacuum. Note the difference in the amplitude of the maximum ϵ'_{eff} and the volume fraction where it takes place in these two surfaces

But now we know that for the case where one of the components is lossy to such an extent that the imaginary part of its permittivity is larger than the real part, it may happen that the real part of the effective permittivity behaves non-monotonically as a function of the volume fraction: a mixture may possess higher permittivity (in terms of its real part) than that of either of its components. Let us define that this is the point beyond which the enhancement effect takes place.

The enhancement effect, defined in this manner, is illustrated again in Fig. 5. There the real part of the effective permittivity of a mixture is shown. The mixture consists of lossy spheres with relative permittivity $1.5 - je''$ in vacuum, occupying a volume fraction f. The effective permittivity is calculated according to two much-used mixing rules (Maxwell Garnett (Eq. 5.1) and Bruggeman (Eq. 5.3)), and clearly, when the e'' increases sufficiently, a "hump" appears in the $e''_{\text{eff}}(f)$ curve.

So, can we find closed-form description for the salient properties of this hump phenomenon? This section answers in the affirmative using analysis based on the Maxwell Garnett formula [4, 6, 14–16].

5.4.2 Maxwell Garnett Formula for Lossy Mixtures

Consider a mixture where lossy, small, homogeneous particles with complex relative permittivity $\epsilon = \epsilon' - je''$ are randomly embedded in vacuum. Then the background permittivity is $\epsilon_e = 1$. In the long-wavelength limit where the scatterers are very small incomparison with the wavelength, the macroscopic permittivity of this mixture $\epsilon_{\text{eff}} = \epsilon'_{\text{eff}} - je''_{\text{eff}}$ can be calculated according to various mixing rules, as was noted above.

Spheres. If the inclusions are spheres, the relative effective permittivity reads (cf. Eq. (5.1))

$$\epsilon_{\text{eff}} = 1 + 3f \frac{\epsilon - 1}{\epsilon + 2 - f(\epsilon - 1)} \tag{5.12}$$

where f is the volume fraction of the inclusion phase with permittivity ϵ.

Counterintuitively, as was seen already, the behavior of the real part of the effective permittivity ϵ'_{eff} is not simple when we deal with complex permittivities. We saw earlier that if there are losses in the mixture, a maximum of ϵ'_{eff} can be attained with a certain volume fraction $0 < f < 1$, and this value is higher than the real part of the permittivity of either component. Let us now analyze the condition in more detail.

For small losses ϵ'_{eff} is a monotonic function of the volume f, increasing from the vacuum value $\epsilon'_{\text{eff}} = 1$ at $f = 0$ to $\epsilon'_{\text{eff}} = \epsilon'$ at $f = 1$. However, when the inclusion phase is lossy enough, such that its imaginary part ϵ'' surpasses the limit

$$\epsilon''_{\text{lim}} = \sqrt{(\epsilon' - 1)(\epsilon' + 2)} \tag{5.13}$$

there appears a maximum where $\epsilon'_{\text{eff}} > \epsilon'$.

After a certain amount of algebraic manipulations, one can solve the volume fraction which gives the maximum for ϵ'_{eff}. This is[4]

$$f_{\max} = \frac{|\epsilon - 1||\epsilon + 2|^2 - 3\epsilon''|\epsilon + 2|}{|\epsilon - 1| \left(|\epsilon|^2 + \epsilon' - 2\right)} \,. \tag{5.14}$$

For this volume fraction of the inclusion phase with permittivity ϵ, the real part of the effective permittivity is

$$\epsilon'_{eff,\max} = \frac{|\epsilon - 1||\epsilon + 2| - \epsilon''}{2\epsilon''} \,. \tag{5.15}$$

If we increase very much the imaginary part of the inclusion phase, the real part of the effective permittivity ϵ'_{eff} can be increased to a value approaching $\epsilon''/2$.

Aligned ellipsoids. Sphere is a special case of an ellipsoid. For an anisotropic mixture where dielectric ellipsoids of similar shape are aligned, the effective permittivity in the direction where the ellipsoids have a depolarization factor N reads

$$\epsilon_{eff} = 1 + f \frac{\epsilon - 1}{1 + (1 - f)N(\epsilon - 1)} \,. \tag{5.16}$$

The formulas corresponding to Eqs. (5.13–15) of the sphere case (where $N = 1/3$) are the following. The threshold value for the imaginary part of the permittivity of the inclusions phase is (generalization of Eq. (5.13))

$$\epsilon''_{\lim} = \sqrt{(\epsilon' - 1)(\epsilon' - 1 + 1/N)} \tag{5.17}$$

for a maximum $\epsilon'_{eff} > \epsilon'$ to appear. And also the results Eq. (5.14) and Eq. (5.15) read for the ellipsoidal case

$$f_{\max} = \frac{|\epsilon - 1||1 + N(\epsilon - 1)|^2 - \epsilon''|1 + N(\epsilon - 1)|}{N|\epsilon - 1| \left(N|\epsilon - 1|^2 + \epsilon' - 1\right)} \tag{5.18}$$

$$\epsilon'_{eff,\max} = \frac{|\epsilon - 1||1 + N(\epsilon - 1)| - \epsilon''(1 - 2N)}{2N\epsilon''} \tag{5.19}$$

As in the case of spheres, with increasing the imaginary part of the inclusion phase, the ϵ'_{eff} can be increased towards a value $\epsilon''/2$.

Needles with random orientation. The previous formulas were for an anisotropic mixture where ellipsoidal inclusions were aligned in orientation (although their positions could be random). However, if the inclusions are mixed such that their orientation is totally random the mixture becomes macroscopically isotropic. The effective permittivity for a mixture where the inclusions are randomly oriented needles[5] is

---

[4] In the formulas, the absolute value of a complex number $a = a' - ja''$ is $|a| = \sqrt{a'^2 + a''^2}$.
[5] For a needle with circular cross-section, the three depolarization factors are 0, 1/2, 1/2.

$$\epsilon_{\text{eff}} = 1 + f \frac{(\epsilon - 1)(\epsilon + 5)}{(3 - 2f)\epsilon + 3 + 2f} \ . \tag{5.20}$$

Once again, the real part of the effective permittivity, ϵ'_{eff}, can be made larger than ϵ' for a certain mixing ratio provided that the inclusions are lossy enough. Now the lower limit reads

$$\epsilon''_{\text{lim}} = \sqrt{\frac{(\epsilon'^2 - 1)(\epsilon' + 5)}{5 - \epsilon'}} \ . \tag{5.21}$$

From this it also follows that if $\epsilon'_{\text{eff}} \geq 5$, there is no hump effect.

The volume fraction for which the maximum is achieved is

$$f_{\text{max}} = \frac{3(\epsilon' + 5)|\epsilon - 1||\epsilon + 1|^2 - 6\epsilon''|\epsilon + 1||\epsilon + 5|}{2|\epsilon - 1|\{(\epsilon' + 5)(|\epsilon|^2 - 1) + 2\epsilon''^2\}} \tag{5.22}$$

and the maximum value for ϵ'_{eff}, using Eq. (5.22) in Eq. (5.20), given Eq. (5.21), is

$$\epsilon'_{\text{eff,max}} = \frac{|\epsilon^2 - 1||\epsilon + 5| - \epsilon''|\epsilon + 1|^2}{8\epsilon''} \ . \tag{5.23}$$

Unlike in the case of lossy spheres, ϵ'_{eff} cannot be made – through this maximum effect – arbitrarily large. When ϵ'' increases, ϵ'_{eff} saturates into the value

$$\epsilon'_{\text{eff,sat}} = \frac{\epsilon'^2 + 6\epsilon' + 25}{16} \ . \tag{5.24}$$

The saturation is achieved at the maximum volume fraction $3(\epsilon' + 3)/(2\epsilon' + 14)$. We have to remember, of course, the limitation from Eq. (5.21) which means that $\epsilon' \leq 5$. Therefore the "hump" maximum is also limited above by 5.

Disks with random orientation. For randomly oriented disks[6] (with relative permittivity ϵ) in vacuum, the effective permittivity is

$$\epsilon_{\text{eff}} = 1 + f \frac{(\epsilon - 1)(2\epsilon + 1)}{(3 - f)\epsilon + f} \ . \tag{5.25}$$

Again, the "hump effect" arises, in other words the real part of the effective permittivity becomes larger than ϵ', for a certain mixing ratio provided that the inclusions are lossy enough: now the cross-over point is

$$\epsilon''_{\text{lim}} = \sqrt{\frac{\epsilon'(2\epsilon' + 1)(\epsilon' - 1)}{3 - 2\epsilon'}} \ . \tag{5.26}$$

This means a still stricter limitation than for the mixture with needles: if $\epsilon' \geq 1.5$, no maximum appears below $f = 1$ in the $\epsilon'_{\text{eff}}(f)$ curve.

The volume fraction for which the maximum is achieved is

$$f_{\text{max}} = \frac{3(2\epsilon' + 1)|\epsilon|^2|\epsilon - 1| - 3\epsilon''|\epsilon||2\epsilon + 1|}{|\epsilon - 1|\{(2\epsilon' + 3)|\epsilon - 1|^2 + 3(\epsilon' - 1)\}} \tag{5.27}$$

and the maximum value for ϵ'_{eff}, using Eq. (5.27) in Eq. (5.25), given Eq. (5.26), is

---

[6] For a disk, the depolarization factors are 1, 0, 0.

$$\epsilon'_{\text{eff,max}} = \frac{|\epsilon - 1||\epsilon||2\epsilon + 1| - \epsilon''(2|\epsilon|^2 - 1)}{2\epsilon''} . \tag{5.28}$$

Like in the needle case, ϵ'_{eff} cannot be made arbitrarily large. When ϵ'' increases, ϵ'_{eff} saturates into the value

$$\epsilon'_{\text{eff,sat}} = \frac{4\epsilon'^2 - 4\epsilon' + 9}{8} . \tag{5.29}$$

The volume fraction of the inclusion phase for which the saturation maximum appears, is $3(2\epsilon' - 1)/(2\epsilon' + 3)$. Of course, the saturation maximum is subject to Eq. (5.26), meaning that $\epsilon' \leq 1.5$. Therefore the largest achievable hump maximum is 1.5 in the disk case.

5.4.3 Discussion on the Enhancement Effect

Figure 5.6 shows the real part of the effective permittivity ϵ'_{eff} as a function of the volume fraction of the inclusions for the different shapes: spheres, randomly oriented needles and disks. Four different lossy dielectric values for the inclusion phase are chosen: $\epsilon = 1.1 - j2$; $2 - j5$; $5 - j8$; and $10 - j40$. In all examples the imaginary part of the inclusion phase ϵ'' is chosen sufficiently large in order that the maximum effect in ϵ'_{eff} appears provided that the theory allows it to take place. It can be seen that for the smallest value, $\epsilon' = 1.1$, the hump appears with all inclusion shapes. If ϵ' is increased to 2, the disk mixture behaves monotonically. And when $\epsilon' > 5$, only the mixture with spherical inclusions displays the enhancement effect. Also, the results for the places and amplitudes of the maxima can be checked to be in agreement with the theoretical formulas that were presented in the above subsection.

Of course, the Maxwell Garnett formula to model effective properties of mixtures is an idealized one [14] and it is true that very often critique is voiced against the validity of the Maxwell Garnett predictions for mixtures where the volume fraction is high. Therefore objections could be raised against such a global use of the model as is taken here. However, I think that the above results have value as concise and explicit results describing the peculiar effect of the "hump phenomenon" with which a mixture can have higher permittivity than either of its components. The effect is certainly no artifact of the idealized Maxwell Garnett model. As was discussed in the earlier section 5.3, with many aquametric applications, the Maxwell–Wagner effect [12] is well known,[7] which is the interfacial polarization phenomenon where inhomogeneity causes dispersion in permittivity and in such systems, very high real parts for the macroscopic permittivity (up to hundreds of thousands) have been measured.

---

[7] As a nonscientific side note, Maxwell–Wagner effect is named after J. Clerk Maxwell and K.W. Wagner whereas the label for Maxwell Garnett formula is due to one man, confusingly with the name J.C. Maxwell Garnett but different than J. Clerk Maxwell [14].

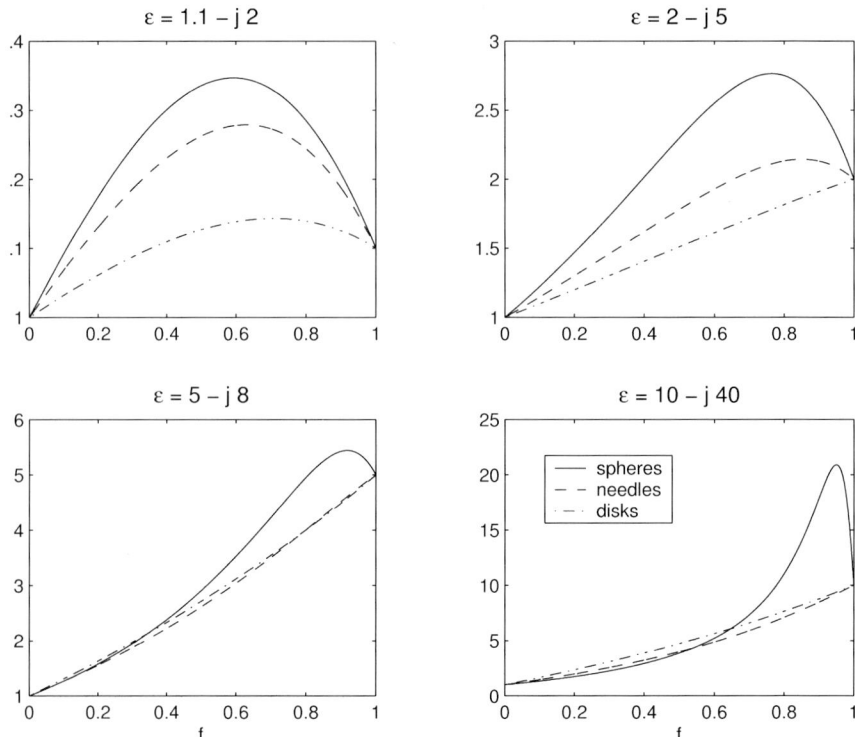

Fig. 5.6. The real part of the effective permittivity ϵ'_{eff} for various mixtures with lossy inclusions in vacuum as a function of the volume fraction f of the inclusions. The complex relative permittivity of the inclusions is ϵ, different in each of the four figures. The inclusions are spherical (solid line), needle-shaped (dashed line), or disk-shaped (dash-dotted line), all randomly oriented. Note that the "hump" effect vanishes for disks and needles as ϵ' increases. (The condition for the hump to appear is $\epsilon' < 5$ for needles and $\epsilon' < 1.5$ for disks.)

But since the Maxwell Garnett model is not the only mixing and homogenization principle that is used in the literature, let us expand the discussion in the next section to cover also other models.

5.5 Critical Parameters for Enhancement

Figure 5.5 showed that is is not only the Maxwell Garnett mixing model that predicts the "hump effect," in other words an enhanced real part for the effective permittivity when one of the components is lossy enough. The Bruggeman model is capable to that, too. But Fig. 5.5 shows, interestingly, that the enhancement phenomenon is not the same in the two models. The maximum permittivity value and conditions for it to happen depend on the model.

Indeed, in mixtures where one of the components is very lossy, the real part of the effective permittivity may behave in quite a nonlinear way.[8] In addition to the fact that this is bound to happen differently for different mixing rules, it also depends strongly on the shape of the microgeometry of the mixture; for different shapes of the inclusions in the mixture the behavior of the effective permittivity varies.

5.5.1 Mixing Rules for Varying Inclusion Shapes

To make quantitative observation on the enhancement phenomenon, let us first remind ourselves about the various predictions for ϵ_{eff} for different (isotropic) mixtures. In earlier chapters, the focus was mostly on spherical geometries.

The mixing formulas according to the various formulations are the following [4], with the different inclusions shapes (all randomly oriented to preserve macroscopic isotropy):

Maxwell Garnett. (Randomly oriented) spheres (cf. Eq. (5.1)):

$$\epsilon_{\text{eff}} = \epsilon_e + 3f\epsilon_e \frac{\epsilon_i - \epsilon_e}{\epsilon_i + 2\epsilon_e - f(\epsilon_i - \epsilon_e)} \tag{5.30}$$

Randomly oriented needles:

$$\epsilon_{\text{eff}} = \epsilon_e + f(\epsilon_i - \epsilon_e) \frac{\epsilon_i + 5\epsilon_e}{(3 - 2f)\epsilon_i + (3 + 2f)\epsilon_e} \tag{5.31}$$

Randomly oriented disks (the same as for Bruggeman mixing model, Eq. (5.35)):

$$\epsilon_{\text{eff}} = \epsilon_e + f(\epsilon_i - \epsilon_e) \frac{2\epsilon_i + \epsilon_e}{(3 - f)\epsilon_i + f\epsilon_e} \tag{5.32}$$

Bruggeman. (Randomly oriented) spheres; this is equivalent to Eq. (5.3):

$$\epsilon_{\text{eff}} = \frac{1}{4} \left(3f(\epsilon_i - \epsilon_e) - \epsilon_i + 2\epsilon_e + \sqrt{[3f(\epsilon_i - \epsilon_e) - \epsilon_i + 2\epsilon_e]^2 + 8\epsilon_i\epsilon_e} \right) \tag{5.33}$$

Randomly oriented needles:

$$\epsilon_{\text{eff}} = \frac{1}{6} \left((5f - 3)(\epsilon_i - \epsilon_e) \right.$$
$$\left. + \sqrt{(5f - 3)^2(\epsilon_i - \epsilon_e)^2 + 12\epsilon_i[3\epsilon_e + f(\epsilon_i - \epsilon_e)]} \right) \tag{5.34}$$

Randomly oriented disks (the same as for Maxwell Garnett, Eq. (5.32)):

$$\epsilon_{\text{eff}} = \epsilon_e + f(\epsilon_i - \epsilon_e) \frac{2\epsilon_i + \epsilon_e}{3\epsilon_i - f(\epsilon_i - \epsilon_e)} \tag{5.35}$$

---

[8] Here "nonlinearity" refers to a nonlinear function of a structural parameter, for example the volume fraction f, not to a nonlinearity in the constitutive $D(E)$ relation. The latter nonlinearity is present in many materials when very high-intensity fields are applied (cf. Kerr effect, Pockels effect).

Coherent Potential. (Randomly oriented) spheres; equivalent to Eq. (5.4):

$$\epsilon_{\text{eff}} = \frac{1}{6}\Bigg(4f(\epsilon_i - \epsilon_e) - \epsilon_i + 4\epsilon_e \tag{5.36}$$

$$+ \sqrt{[4f(\epsilon_i - \epsilon_e) - \epsilon_i + 4\epsilon_e]^2 + 12(1-f)\epsilon_e(\epsilon_i - \epsilon_e)}\Bigg)$$

Randomly oriented needles:

$$\epsilon_{\text{eff}} = \frac{1}{12}\Bigg(8f(\epsilon_i - \epsilon_e) - 3(\epsilon_i - 3\epsilon_e) \tag{5.37}$$

$$+ \sqrt{[8f(\epsilon_i - \epsilon_e) - 3(\epsilon_i - 3\epsilon_e)]^2 + 24[3\epsilon_e(\epsilon_i - \epsilon_e) + f(\epsilon_i^2 - 4\epsilon_i\epsilon_e + 3\epsilon_e^2)]}\Bigg)$$

Randomly oriented disks:

$$\epsilon_{\text{eff}} = \frac{1}{6}\Bigg(4f(\epsilon_i - \epsilon_e) - 3(\epsilon_i - 2\epsilon_e) \tag{5.38}$$

$$+ \sqrt{9\epsilon_i^2 + 12f\epsilon_e(\epsilon_i - \epsilon_e) + 16f^2(\epsilon_i - \epsilon_e)^2}\Bigg)$$

5.5.2 Example of the Critical Case

The lesson learnt from the earlier sections is that an enhancement effect may appear if there are enough losses in at least one of the components. But what kind of hump can appear? How high can the peak become?

From numerical tests it can be observed (as is known in percolation studies [17]) that for spherical inclusions the maximum enhancement happens at the percolation point, in other words for the volume fraction $f_{\text{max}} = 1$ for Maxwell Garnett ("generalized percolation threshold"), $f = 0.333$ for the Bruggeman mixing rule, and $f = 0.25$ for Coherent Potential. Very large values can be attained for ϵ'_{eff}; in fact, there is no limit how large it can become, as long as the loss factor of the inclusion phase is increased. But this is not the case if the inclusions are not spherically shaped. Then it happens that the enhancement effect is suppressed if ϵ'_i is large enough. In such cases, no matter how large losses are attributed to the inclusion phase (by increasing ϵ''_i), the hump in ϵ'_{eff} does not appear.[9]

Consider mixtures where the environment is vacuum ($\epsilon_e = 1$), and the inclusions (having various shapes) are lossy ($\epsilon_i = \epsilon_i - j\epsilon''_i$). Then we are ready to study the conditions for the enhancement effect. In particular, in the following Table 5.1, the results for the critical parameters are listed for the nine models Eqs. (5.30–38). In particular, the following two parameters are derived:
- The volume fraction f_{limit} for which the maximum attainable enhancement in ϵ'_{eff} takes place.
- The limiting value for the real part of the inclusion phase $\epsilon'_{i,\text{max}}$, defined by the condition that if ϵ'_i be larger than that, no hump is observed in ϵ'_{eff}.

---

[9] This phenomenon was shortly discussed already in Section 5.4.

By the "maximum effect" it is meant here that $c'_{\text{eff}} \geq \max\{c'_e, c'_i\}$ for $0 < f < 1$. Figure 5.7 shows the limiting situation, here calculated, as an example case, according to the Bruggeman mixing rule and with (very) lossy needles in vacuum (and in Fig. 5.8, another look at the same effect is taken). The needles are assumed randomly oriented in the mixture, meaning that Eq. (5.34) is applied. As is seen, the limiting value for c'_i for the mixture "going critical", is the divine proportion, *sectio aurea*, (≈ 1.618034), for the real part of the needle permittivity. This is in agreement with the corresponding critical parameter listed in Table 5.1.

Table 5.1. Characteristic parameters for the maximum effect (enhancement of c'_{eff} for the case of lossy inclusions in vacuum): the maximum real part for the inclusion permittivity $c'_{i,\text{max}}$ for which the enhancement effect in c'_{eff} still occurs, and the volume fraction f_{limit} to which the maximum tends when losses (c''_i) are increased

model	inclusion shape	$\epsilon'_{i,\text{max}}$	f_{limit}
Maxwell Garnett	spheres	no limit	1
	needles	5	1
	disks	1.5	1
Bruggeman	spheres	no limit	1/3
	needles	$\dfrac{1+\sqrt{5}}{2} \approx 1.61803399$	3/5
	disks	$3/2 = 1.5$	1
Coherent Potential	spheres	no limit	1/4
	needles	$4/3 \approx 1.3333$	3/8
	disks	$\dfrac{3+\sqrt{2}}{4} \approx 1.10355339$	3/4

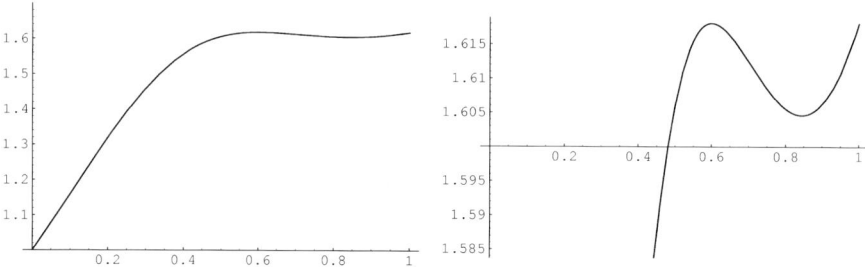

Fig. 5.7. Example of a mixture on the verge of going critical; starting to display the maximum effect (left: global behavior; right: blow-up in c'_{eff}). The real part of the effective permittivity c'_{eff} for lossy needles, randomly oriented in vacuum, with permittivity $\epsilon'_i = (1 + \sqrt{5})/2 - j\ 1000000$, according to the Bruggeman mixing rule. The horizontal axis is the volume fraction of the inclusions, showing the maximum effect at $f = 0.6$. (Compare with the theoretical values in Table 5.1.)

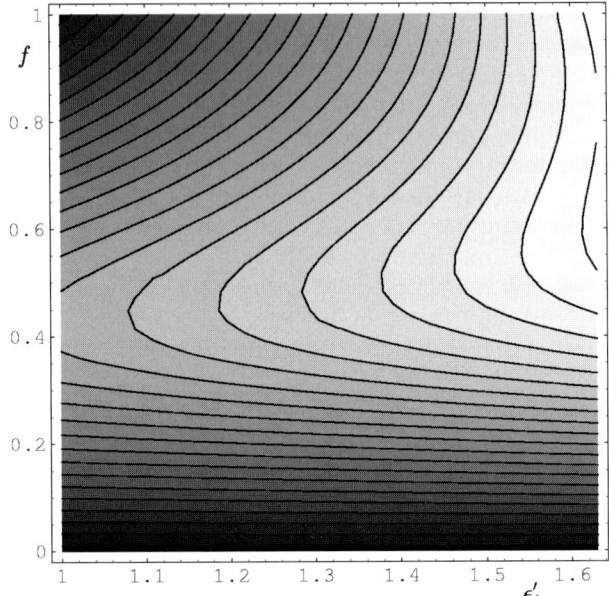

Fig. 5.8. Illustrating the same effect as in Fig. 5.7 for ϵ'_{eff}, as a contour plot, with one more variable parameter (ϵ'_i). The global maximum in the middle values for f vanishes, and becomes as large as ϵ'_i (the value of ϵ'_{eff} for $f = 1$), when ϵ'_i increases to the value of *sectio aurea*, $(1 + \sqrt{5})/2$

5.6 Discussion

Mixed phases can certainly display very interesting and unexpected character in their macroscopic behavior. Like in composite materials, this may be visible in mechanical properties, it can be an exceptional thermal property, or like has been discussed in the present article, it may appear in the dielectric and electromagnetic characteristics. Here the new effect was the enhanced polarization for mixtures where one of the phases was highly lossy. The real part of the effective permittivity of a mixture could become much larger than that of either its components. Such a behavior could be thought of breaking natural bounds that the laws of physics force on mixtures; see [18] for discussion on the strictness of such limits.

Dielectric mixtures in general, and aquatic mixtures in particular give therefore confirmation to the old expression that "the whole is more than the sum of parts." Qualitatively new properties are "created" in the mixing process. Ice cream is an everyday example [19]. It consists of crystalline solids, liquid, and gas, which, offered separately, might provide the same nutritional value as the original dessert, but the pleasure of consumption would be lost. Ice cream and other aquatic materials surely acquire emergent properties [20] in the process of mixing. And in this respect one needs to bear in mind that in this analysis we only have played with isotropic materials![10]

The results that have been presented in the previous sections can be used in the analysis of the composition of samples that contain lossy dielectric phases. Microwave measurements in aquametry often provide us with average permittivity values that have to be inverted to get quantitative information of moisture and other internal parameters of the measured samples. The critical parameters for the enhancement effect may be of help, but it is important to keep in mind the main message of the present article: aquatic mixtures certainly belong to the class of complex materials. One should carefully check the measurements and modelling calculations before claiming that the macroscopic dielectric response is understood.

References

1. Wolfram S (2002) A new kind of science, Wolfram Media, Inc., Champaign, IL
2. Bergman DJ (1978) The dielectric constant of a composite material — a problem in classical physics. Phys Rep C 43:377–407
3. Milton GW (2002) The theory of composites, Cambridge University Press
4. Sihvola A (1999) Electromagnetic mixing formulas and applications, IEE Publishing, Electromagnetic Wave Series 47, London
5. Neelakanta PS (1995) Handbook of electromagnetic materials, CRC, Boca Raton, Florida
6. Maxwell Garnett JC (1904) Colours in metal glasses and in metal films. Trans Royal Society (London) CCIII:385–42
7. Hashin Z, Shtrikman S (1962) A variational approach to the theory of the effective magnetic permeability of multiphase materials, J Appl Physics 33(10):3125–3131
8. Bruggeman DAG (1935) Berechnung verschiedener physikalischen Konstanten von heterogenen Substanzen, I. Dielektrizitätskonstanten und Leitfähigkeiten der Mischkörper aus isotropen Substanzen. Annalen der Physik (Ser. 5) 24:636–679
9. Elliott RJ, Krumhansl JA, Leath PL (1974) The theory and properties of randomly disordered crystals and related physical systems. Revs Modern Physics, 46(3):465–543
10. Sihvola A (1989) Self-consistency aspects of dielectric mixing theories. IEEE Trans. Geoscience Remote Sensing 27(4):403–415
11. Foster KR, Schwan HP (1989) Dielectric properties of tissues and biological materials: a critical review. CRC Critical Revs in Biomedical Engineering 17(1):25–104
12. Wagner KW (1914) Erklärung der dielektrischen Nachwirkungsvorgänge auf Grund Maxwellscher Vorstellungen. Arch Elektrotechnik, 2(9):371–387
13. Nyfors E, Vainikainen P (1989) Industrial microwave sensors, Artech House, Norwood, Mass.
14. Landauer R (1978) Electrical conductivity in inhomogeneous media. In: Garland JC, Tanner DB (Eds.) Electrical transport and optical properties of inhomogeneous media, American Institute of Physics, Conference Proc., 40:2–45.

---

[10] Although complex in the microscopic character, these media are simple in the sense that there is no anisotropy, nor magnetoelectric coupling in their response.

15. Taylor LS (1965) Dielectric properties of mixtures. IEEE Trans. Antennas Propagation 13(6):943–947
16. Tinga WR, Voss WAG, Blossey DF (1973) Generalized approach to multiphase dielectric mixture theory. J Appl Physics 44(9):3897–3902
17. Sihvola A, Saastamoinen S, Heiska K (1994) Mixing rules and percolation. Remote Sensing Revs 9:39–50
18. Sihvola, A (2002) How strict are theoretical bounds for dielectric properties of mixtures? IEEE Trans Geoscience Remote Sensing 40(4):880–886
19. Banhart J, Weaire D (2002) On the road again: metal foams find favor, Physics Today, July 2002, 37–42
20. Sihvola, A (2002) Electromagnetic emergence in metamaterials. Deconstruction of terminology of complex media. In: Zouhdi S, Sihvola A, Arsalane M (Eds.) Advances in electromagnetics of complex media and metamaterials, Kluwer, Dordrecht, 3–17

6 Moisture Measurement in Multi-Layer Systems

Kailash Prasad Thakur

Electromagnetic Sensing, Imaging and Sensing Team,
Industrial Research Limited, 24 Balfour Road, Parnell, Auckland, New Zealand

6.1 Introduction

In principle, one should be able to measure the moisture content of a material in any form. However, the layered form of moist material offers an extra benefit of the application of a variety of electromagnetic phenomena for which the basic principles of interaction of electromagnetic waves with materials are well understood. The present chapter describes a few different ways to measure the electrical property and moisture content of layered materials.

6.2 Waveguide Techniques for Layered Dielectrics

Measurement techniques for the dielectric constant and moisture content of a material completely filling a waveguide cell are well developed [1, 2]. However, there exists considerable interest in the measurement of the dielectric constant of inhomogeneous materials, and materials which only exist as a mixture of two or more components, such as layers of fat and meat in animal products, layers of asphalt on the pavement, ceramic layers, laminated veneer lumber, layered mineral deposits and also unevenly shaped materials that do not completely fill the waveguide cell leaving an air void. The usual practice in such cases is to measure the permittivity of the mixture and use a linear mixture equation to deduce the dielectric constant of the material [3], namely

$$\varepsilon_{eff} = v\varepsilon_d + (1-v)\varepsilon_0 \qquad (6.1)$$

where v is the volume fraction of the dielectric material, ε_{eff} is the effective dielectric constant of the mixture and ε_d and ε_0 are the dielectric constants of the material and air, respectively.

The validity of Eq. (6.1) is limited to few simple systems only. Let us investigate the validity of Eq. (6.1) for the layered dielectrics in a waveguide cell. In particular, consider the case shown in Fig. 6.1 where a material is in the form of a number of longitudinal slayer interspersed with air gaps, and to determine the upper value of the electrical width of the material layer for which Eq. (6.1) is valid to within some specified accuracy. This investigation is undertaken by computing the cut-off frequency of the waveguide filled with dielectric layers using the transverse resonance method.

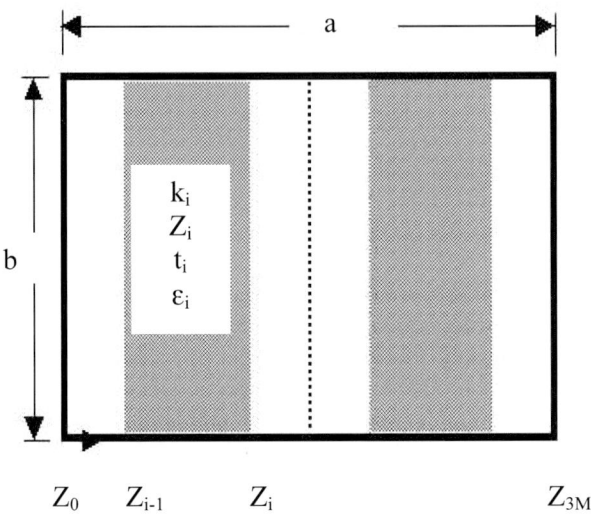

Z_0 Z_{i-1} Z_i Z_{3M}

Fig. 6.1. Multiple layers of dielectric in Waveguide cell for $M = 2$

6.2.1 Theoretical Considerations

Consider a metal walled rectangular waveguide of dimensions a and b in the x and y directions, respectively. If the cut-off frequency of the waveguide when filled with a homogeneous dielectric, or a mixture of dielectrics, is $f_c(\varepsilon_{eff})$ and that of an air-filled waveguide is $f_c(\varepsilon_0)$, then

$$\frac{\varepsilon_{eff}}{\varepsilon_0} = \left(\frac{f_c(\varepsilon_0)}{f_c(\varepsilon_{eff})} \right)^2 \qquad (6.2)$$

Consequently the measurement, or calculation, of the cut-off frequency of a waveguide filled with a mixture of dielectrics permits the effective dielectric constant to be determined by Eq. (6.2). Consider now the situation of a slab-loaded waveguide where the volume fraction of the dielectric in the waveguide cell is v. In particular, we consider the case where the dielectric is present as M parallel layers aligned in the longitudinal direction, each layer of dielectric (of thickness $t_d = av/M$) surrounded on both sides by layers of air of thickness $a(1 - v)/2M$, with this pattern (of air-dielectric-air) being repeated M times across the width of the waveguide. The case for $M = 2$ is shown in Fig. 6.1. This arrangement provides symmetric loading across the cell, and as M increases the electrical thickness of each layer decreases but the ratio of dielectric to air in the cell remains constant.

This waveguide filled with dielectric layers can be considered as $3M$ regions across the waveguide width where the dielectric constant, ε_i, and thickness, t_i, of the ith layer are given by the following for $p = 1, 2 .. M$:

$$
\begin{aligned}
&\varepsilon_i = \varepsilon_d \text{ for } i = (3p - 1), \text{ and} \\
&\varepsilon_i = \varepsilon_0 \text{ for } i = 3p \text{ and } i = (3p - 2); \text{ and} \\
&t_i = t_d = av/M \text{ for } i = (3p - 1), \text{ and} \\
&t_i = a(1 - v)/(2M) \text{ for } i = 3p \text{ and } i = (3p - 2).
\end{aligned}
\tag{6.3}
$$

We now employ the transverse resonance technique to find the cut-off frequency of the multiple-slab loaded waveguide and adopt the following notation: z_i is the impedance at the interface plane between the $(i + 1)$th and ith layer, $k_i (= (k_c^2 - \beta^2)^{1/2})$ is the wave number in the ith layer, $Z_i (= b\omega\mu_0/k_i)$ is the characteristic impedance of a unit length of the ith layer section while the cutoff wave number is k_c, the free-space wave number is k_0, and the longitudinal propagation constant is β. The transmission (chain) matrix [4, 5, 6] that relates voltages and currents at the ends of the ith layer may be represented by

$$
B^i = \begin{bmatrix} \cos(k_i t_i) & j.Z_i \sin(k_i t_i) \\ \dfrac{j.\sin(k_i t_i)}{Z_i} & \cos(k_i t_i) \end{bmatrix} = \begin{bmatrix} b^i_{11} & j.b^i_{12} \\ j.b^i_{21} & b^i_{22} \end{bmatrix}
\tag{6.4}
$$

Taking into account the total number of slabs ($i = 1$ to $i = 3M$), a cumulative transmission matrix can be defined for the whole structure by successively multiplying the respective B-matrices defined by Eq. (6.4) to obtain:

$$
C = \prod_{i=1}^{3M} B^i = \prod_{i=1}^{3M} \begin{bmatrix} b^i_{11} & jb^i_{12} \\ jb^i_{21} & b^i_{22} \end{bmatrix} = \begin{bmatrix} c_{11} & jc_{12} \\ jc_{21} & c_{22} \end{bmatrix}
\tag{6.5}
$$

which relates voltages and currents at the two terminating impedances z_0 (at $x = 0$) and z_{3M} (at $x = a$) respectively. The resonant condition of the network is easily found to be

$$
z_{3M} = -\frac{z_0 c_{11} + c_{12}}{z_0 c_{21} + c_{22}}
\tag{6.6}
$$

For a rectangular waveguide $z_0 = 0$ and $z_{3M} = 0$, and hence we obtain the condition

$$
c_{12} = 0
\tag{6.7}
$$

The problem here is to solve for the values of transverse propagation constants, k_i, so that Eq. (6.7) is satisfied when $\beta = 0$ hence determining the cut-off frequency f_c. The computational technique employed in this study involved evaluating the function c_{12} for $\beta = 0$ and searching for the lowest value of f_c that satisfied Eq. (6.7).

6.3 Mixture Equation for Multi-Layered Dielectrics

As a typical example, consider a rectangular waveguide with dimensions $a = 72.136$ mm and $b = 34.036$ mm within which there are M layers of a dielectric with $\varepsilon_d = 11$. Figure 6.2 presents the computed values of ε_{eff} from Eq. (6.2) as a function of M for different values of v. In Fig. 6.2, ε_{eff} converges to a constant value generated by Eq. (6.1) as the value of M increases. Figure 6.3 shows the surface of relative error in the effective dielectric constant computed by the method outlined in the previous section to that predicted by Eq. (6.1), as a function of the volume fraction v and M.

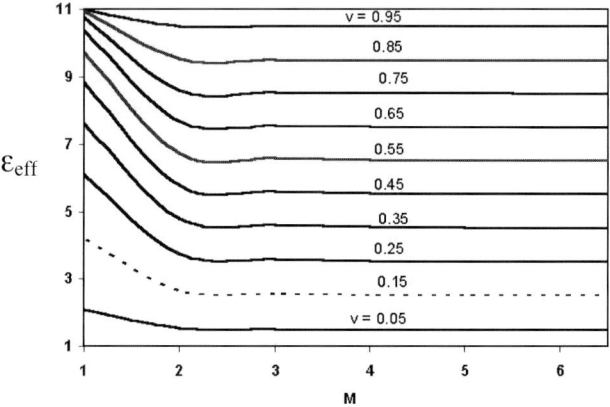

Fig. 6.2. Computed values of effective dielectric constant as a function of number of dielectric layers for different values of v

As expected Eq. (6.1) is almost correct in the two extremes $v = 0$ and $v = 1$, but note that Eq. (6.1) is most in error when v is the range 0.2 to 0.3, approximately. Calculations have shown that the situation is similar for other values of ε_d. Note also that the value of ε_{eff} predicted by Eq. (6.1) is always smaller than that determined from the theory presented here and, the difference between the results decreases as the value of M increases (Figs. 6.2 and 6.3), that is as the ratio of thickness of the dielectric to the wavelength in the dielectric decreases.

The surface in Fig. 6.3 is convex with respect to v-axis. Values of ε_{eff} obtained from Eq. (6.2) for lowest and highest values of M, (ie, $M = 1$ and $M = 20$) are shown in Fig. 6.4, which indicates that for large values of M, ε_{eff} is almost linear with v and hence close to the results of Eq. (6.1). The convex behaviour of effective dielectric constant of the mixture of two components with respect to v in Figs. 6.3 and 6.4 can be explained by the mixture equation [7]

$$e^{\alpha \varepsilon_{eff}} = v_1 e^{\alpha \varepsilon_1} + v_2 e^{\alpha \varepsilon_2} \tag{6.8}$$

Unlike Eq. (6.1), Eq. (6.8) has capability to reproduce the computed values of ε_{eff} with reasonable error using $\alpha = 0.3$ for $M = 1$, $\alpha = 0.03$ for $M = 2$, and $\alpha = 0.00908$ for $M = 3$. Equation (6.8) reduces to Eq. (6.1) for α tending to zero. However, incorporating an additional empirical error term $d\varepsilon_{eff} = q \sin(\pi v^w)$ in Eq. (6.1), i. e.,

$$\varepsilon_{eff} = v\varepsilon_d + (1 - v)\varepsilon_0 + q\sin(\pi v^w) \tag{6.9}$$

can also reproduce the computed values of ε_{eff} using $q = 3.4$ and $w = 0.93$ for $M = 1$. These results are also shown in Fig. 6.4, which compare well with those obtained from

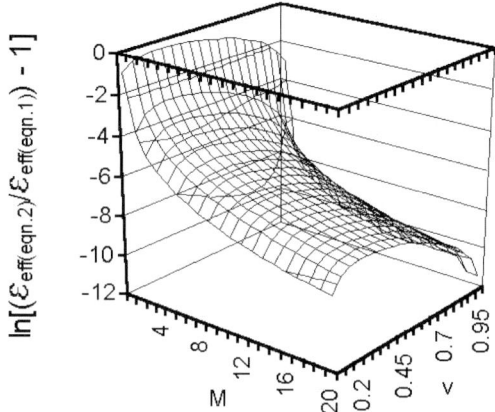

Fig. 6.3. Relative error in computed values of effective dielectric constant with respect to that obtained from Eq. 6.1 for the waveguide filled with multiple layers as a function of volume fraction of material of layer, v, for various M and $\varepsilon_r = 11$

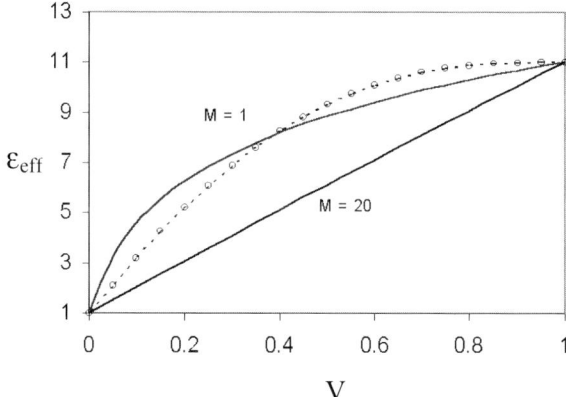

Fig. 6.4. Values of effective dielectric constant of mixture, ε_{eff} as a function of volume fraction of material of dielectric layer, v for $M = 1$, and $M = 20$. Open circles are values obtained from Eq. (6.2) for $M = 1$, solid curve is from Eq. (6.8) for $M = 1$ and dotted curve is from Eq. (6.9) also for $M = 1$; solid line is from Eq. (6.2) for $M = 20$

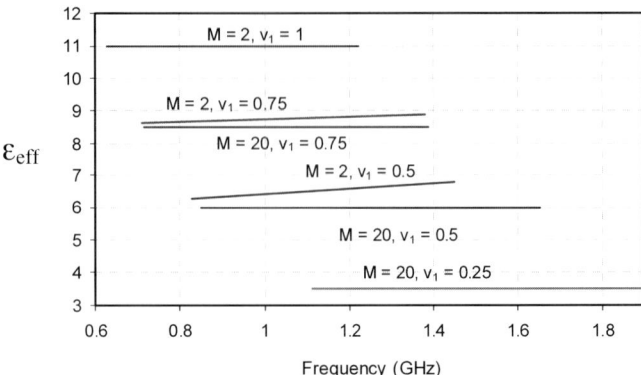

Fig. 6.5. Variation of ε_{eff} as a function of frequency for different values of v and M

Eq. (6.2). For the same size waveguide, one can obtain the propagation constant [4] and hence the permittivity at elevated frequencies. Figure 6.5 shows the variation of ε_{eff} as a function of frequency for selected values of M and v. For the homogeneous mixture, M = 20, Eq. (6.1) holds good. At high frequency, Eq. (6.8) with constant value of α cannot work for small values of M unless α is taken as a function of frequency.

At the cut-off frequency, for dielectric constants up to about 11 it would be necessary for the layer thickness to be less than 5% of the wavelength in the dielectric in order for the linear mixture equation (6.1) to predict the effective dielectric constant with an error of less than 1%. To achieve an error of less than 0.1% the thickness would have to be reduced to less than 1.4% of the wavelength in the dielectric.

Measurement of the cut-off frequency permits us to obtain the effective dielectric constant of multi-layered dielectrics, which can be used to obtain the dielectric constant of an individual layer using the mixture equation. However, this technique is not very practical because layers will have to be inserted in the waveguide cell with a high degree of accuracy in their alignment within the cell. The next section describes an easy to use method for the measurement of the electrical property and moisture content of a multi-layered system.

6.4 Free-Space Measurement

The non-destructive estimation of electrical properties of materials has been extensively studied using a variety of microwave sensors, which employ the axiomatic characteristics of interaction of electromagnetic waves with materials. Measurements and computation of the reflection coefficients of plane electromagnetic waves from the layered dielectrics is of fundamental importance since the Fresnel's equations became available [8–12]. There exists an enormous application of this fundamental phenomenon of electromagnetics in the determination of material properties of dielectrics from the measured reflection coefficients. Such applications are very important in the development of sensors where there are no other non-destructive and non-contact techniques to estimate the thickness and electrical property of layers. Such applications include evaluation of thickness and properties of buried objects, determination of moisture contents in layered materials, detection of disbounding and delamination in layered-dielectric-slabs, estimation of thickness of asphalt and bitumen on the road, and thickness of ash deposited by volcanic eruption. Non-contact non-destructive evaluation of disbonds, delaminations and minute thickness variations in layered materials generally backed by a metal plate is of great importance in several industrial applications. The behaviour of multiple dielectric slabs in waveguide has been thoroughly investigated [9]. However, in practice, the multiple dielectric layers might also exist in free space or in contact with a very thick dielectric, which might not be metal [10, 11]. In the present section, the reflection of plane electromagnetic waves from multiple layers of plane dielectrics at an arbitrary angle of incident has been studied to carry out an inversion process to generate the thickness and permittivity of one of the dielectric layers.

6.4.1 Theoretical

A generalised treatment of reflection from multiple layers of dielectrics has been presented elsewhere [10–11] in which the angle of incident as well as the number of layers can vary. A brief description of the technique is given here for ready reference. The Fresnel's equations as mentioned in the textbooks for perpendicular polarisation are given by [12]

$$R_\perp = \frac{\eta_2 \cos\theta_1 - \eta_1 \cos\theta_2}{\eta_2 \cos\theta_1 + \eta_1 \cos\theta_2} \tag{6.10}$$

$$T_\perp = \frac{2\eta_2 \cos\theta_1}{\eta_2 \cos\theta_1 + \eta_1 \cos\theta_2} \tag{6.11}$$

$$\eta_i = \sqrt{\frac{\mu_i}{\varepsilon_i}} \tag{6.12}$$

where η_1 and η_2 are the intrinsic impedances of the media and θ_1 and θ_2 are the propagation angles in the two media, having wave vectors k_1 and k_2 which are related to each other through the Snell's law

$$k_1 \sin\theta_1 = k_2 \sin\theta_2 \tag{6.13}$$

For a multi-layer (inhomogeneous) system, the reflection and transmission coefficients can be obtained using the chain-matrix-rule. For multiple media, we have [12].

$$\begin{bmatrix} c_1 \\ b_1 \end{bmatrix} = \begin{bmatrix} \dfrac{e^{j\phi_1}}{T_1} & \dfrac{R_1 e^{-j\phi_1}}{T_1} \\ \dfrac{R_1 e^{j\phi_1}}{T_1} & \dfrac{e^{-j\phi_1}}{T_1} \end{bmatrix} \begin{bmatrix} \dfrac{e^{j\phi_2}}{T_2} & \dfrac{R_2 e^{-j\phi_2}}{T_2} \\ \dfrac{R_2 e^{j\phi_2}}{T_2} & \dfrac{e^{-j\phi_2}}{T_2} \end{bmatrix} \cdots \begin{bmatrix} \dfrac{e^{j\phi_n}}{T_n} & \dfrac{R_n e^{-j\phi_n}}{T_n} \\ \dfrac{R_n e^{j\phi_n}}{T_n} & \dfrac{e^{-j\phi_n}}{T_n} \end{bmatrix} \begin{bmatrix} c_n \\ b_n \end{bmatrix} \qquad (6.14)$$

Equations (6.10) to (6.14) allow us to obtain the reflection and transmission for any number of layers.

6.4.2 Calibration of the Focussed System and Measurement

For the measurement of reflection coefficients, two plano-convex lenses were used as a focussing system. The focussing system is widely used in a variety of microwave sensing applications [13–17]. At the beam waist of the Gaussian beam produced by the focussing system, one finds a good approximation to a plane elec-tromagnetic wave where the measurements must be carried out. For the calibration of the focussing system the technique proposed earlier [17] is adopted which has been described extensively in [10].

6.4.3 Inverse Techniques

In order to obtain the thickness and permittivity of a dielectric sheet from the measured values of reflection coefficients, an inverse technique is used. In order to investi-gate reflection from a single sheet backed by a metal reflector one can adopt sev-eral well-known techniques including the exact analytical solution. However, these techniques work only when the measured data is almost free from noise.

6.4.4 Dielectric on Metal Reflector

If we have a single lossless dielectric layer backed by a perfect reflector, it is sim-ple and straightforward to obtain the thickness and permittivity from the measured reflection coefficients. In this case, for normal incidence, the reflection coefficient becomes

$$R_\perp = \frac{j \tan(kd_1 \sqrt{\varepsilon_r}) - \sqrt{\varepsilon_r}}{j \tan(kd_1 \sqrt{\varepsilon_r}) + \sqrt{\varepsilon_r}} \qquad (6.15)$$

where d_1 is the thickness of layer and ε_r its relative permittivity. Using $R_\perp = u + jv$ and defining a reduced parameter Γ as

$$\Gamma = \frac{1 + R_\perp}{1 - R_\perp} = \frac{j}{\sqrt{\varepsilon_r}} \tan kd \sqrt{\varepsilon_r} \qquad (6.16)$$

we have

$$\frac{1}{\sqrt{\varepsilon_r}}\tan kd\sqrt{\varepsilon_r} = \left[\left(\frac{u}{v}\right) \pm \sqrt{\left(\frac{u}{v}\right)^2 + 1}\right] \qquad (6.17)$$

If the reflection coefficients are measured over a wide range of frequencies, the thickness of the dielectric layer and its permittivity can be explicitly determined [18]. Since the dielectric layer is lossless, the reflection coefficients must lie on the unit circle and we obtain from Eq. (6.17) when $v \to 0$,

$$d_e = d\sqrt{\varepsilon_r} = \frac{(n+1)}{2} \cdot \frac{c}{f_0} \qquad \text{(for } \mu < 0 \text{)} \qquad (6.18)$$

or

$$d_e = d\sqrt{\varepsilon_r} = \frac{(2n+1)}{4} \cdot \frac{c}{f_0} \qquad \text{(for } \mu > 0 \text{)} \qquad (6.19)$$

where n is an integer. Thus the electrical length d_e is accurately determined, by simply calculating the frequency f_0 at which $v \to 0$. To determine the permittivity of the dielectric layer, we consider the case when $\mu \to 0$, then

$$\frac{1}{\sqrt{\varepsilon_r}}\tan kd\sqrt{\varepsilon_r} = 1 \qquad (6.20)$$

or

$$\varepsilon_r = \tan^2\left[\frac{2\pi f_1 \cdot d_e}{c}\right] \qquad (6.21)$$

where f_i is the frequency at which $\mu \to 0$. Various other techniques also exist for the measurement of thickness of dielectric backed by a metal reflector [19–21]. However, they work well if the phase of the reflection coefficient is changing smoothly with frequency, ie, for a noise-free data, and fail if the noise level is too high.

6.4.5 Inhomogeneous Dielectrics

The technique described in [18] fails for a multi-layered system and one needs to develop an inverse technique to estimate the electrical properties of any dielectric layer within the stack of layers of dielectrics. In this case the reflection coefficient can be calculated using Eqs. (6.10–14). The reflection coefficient is then transformed to a reduced variable Γ_c according to Eq. (6.16), which is compared with the same measured parameter, Γ_m. For this purpose, we define the error parameters as

$$E_1 = Re(RMS((\Gamma_m - \Gamma_c)/\Gamma_m)) \qquad (6.22a)$$

$$E_2 = Re(RMS(\Gamma_m - \Gamma_c)) \qquad (6.22b)$$

$$E_3 = RMS(Re(R_m) - Re(R_c)) \qquad (6.22c)$$

$$E_4 = RMS(Im(R_m) - Im(R_c)) \qquad (6.22d)$$

Here *RMS* stands for root of mean of squared values, and R_m and R_c are the measured and computed values of reflection coefficients.

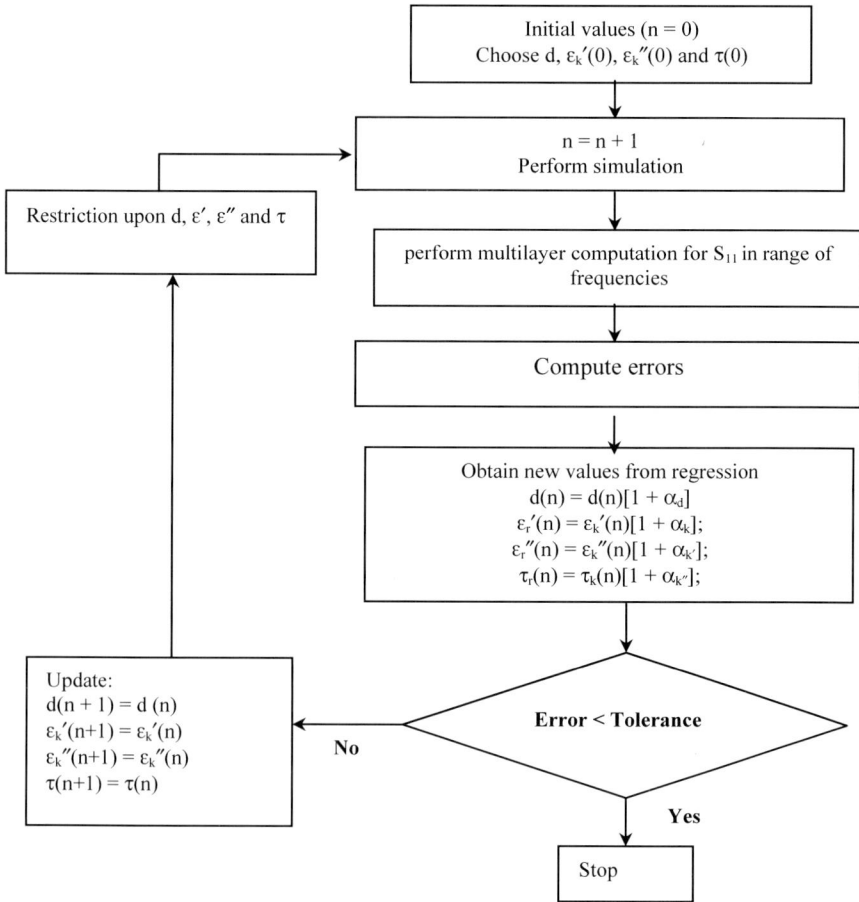

Fig. 6.6. Procedure for coupling the computation of reflection coefficients from multilayer algorithm and non-linear least-squares regression technique

The purpose here is to minimise the values of errors for specific values of dielectric thickness and permittivity. In order to obtain the thickness and dielectric constant, we start with an initial guess for thickness, $d = d_0$ and $\varepsilon'_r = \varepsilon'_{r0}$. Let us consider that the reduced reflection coefficient, Γ is a function of thickness, d, and dielectric constant ε'_r ie, $\Gamma = \Gamma(d, \varepsilon_r)$, and expand it in Taylor series around $\Gamma_0 = \Gamma(d_0, \varepsilon'_{r0})$ and neglect the higher-order terms

$$\Gamma = \Gamma_0 + \left(\frac{\partial \Gamma}{\partial (d)}\right)_{d_0} (d - d_0) + \left(\frac{\partial \Gamma}{\partial (\varepsilon'_r)}\right)_{\varepsilon'_{r0}} (\varepsilon'_r - \varepsilon'_{r0}) \qquad (6.23)$$

In Eq. (6.23), Γ is the measured value and Γ_0 is the computed value of reduced reflection coefficient obtained from Eqs. (6.10–14, 16) for thickness, $d = d_0$ and $\varepsilon'_r = \varepsilon'_{r0}$. The derivatives with respect to thickness, d and dielectric constant ε'_r are also computed from Eqs. (6.14) and (6.16). The method of obtaining derivatives is discussed in Appendix A. Now the error between Γ and Γ_0 in Eq. (6.23) is minimised by the least square technique to obtain thickness, d and dielectric constant ε'_r. Since the system is highly non-linear, we do not expect to get the correct answer for thickness, d and dielectric constant ε'_r straightway. Therefore, we repeat the process with the new values of thickness, d, and dielectric constant, ε'_r, as the initial values of thickness, d_0, and dielectric constant, ε'_{r0}. This process is repeated until the measure of errors E_1, E_2, E_3 and E_4 are minimised, that is, until the errors are lower than a specified tolerance. It must be remarked here that all these errors cannot be minimised simultaneously. At the same time, minimising only one measure of error sometimes leads to some spurious solution. Hence we have considered minimising the weighted sum of errors

$$E_m = w_1 {}^* E_1 + w_2 {}^* E_2 + w_3 {}^* E_3 + w_4 {}^* E_4; \qquad (6.24)$$

where $w_1 + w_2 + w_3 + w_4 = 1$.

It must be remarked that the presence of trigonometric functions in Eqs. (6.10–14) generates multiple solutions and is hence are not acceptable. To get rid of this problem, constraints are placed upon the values of thickness, d, and dielectric constant, ε'_r, so that they are within the predefined lower and upper limited values. Moreover, the measurement of the reflection spectrum at wide bandwidth permits one to get rid of spurious solutions. Measurement of the reflection coefficient of plane electromagnetic waves in the X-band frequencies shows a non-linear dependence upon the thickness of the layer. The computational steps involved in the process of minimisation of errors are shown in Fig. 6.6.

6.4.6 Moist Dielectrics

In the case of moist and lossy dielectrics, the magnitude of reflection coefficient is much less than unity as compared with a loss free material. Due to the effect of dispersion the magnitude of reflection coefficient becomes a function of frequency which is prominent at frequencies that are close to the absorption frequency. Figure 6.7 shows the plot of the magnitude of reflection coefficient as a function of frequency for acrylic, dry Medium Density Fibreboard (MDF) and wet MDF backed by a metal reflector. For acrylic and dry MDF, which are low loss materials, the magnitude of reflection coefficient remains almost constant in the entire frequency range of 8 to 12 GHz. However, for the moist MDF, the magnitude of reflection coefficient decreases with frequency at X-band. This is because of the presence of water in the material. For pure water, the dispersion relationship is given by the Debye or the Debye-Drude equations [22]. The slope of the dispersion curve for pure water is very steep at the X-band. A change in frequency from 8 to 10 GHz, reduces the dielectric constant of pure water by 9.4%, and increases the loss factor and loss tangent by 11.9% and 19.5%, respectively. In order to investigate the permittivity of moist materials within the X-band frequency spectrum using the inverse technique, the dispersion relationship for the moist material must be incorporated in the model. Since the complete dispersion relationship for the moist materials is not known, our objective is limited to the variation of permittivity of moist material at X-band (particularly between 8 and 12 GHz). We have incorporated the slope of the Debye dispersion curve that converts dielectric constant and loss factor of water at 8 GHz to those at elevated frequencies as follows:

$$\varepsilon'(f) = \varepsilon'_m \left[\frac{1}{\varepsilon'_w} \left(\varepsilon_\infty + \frac{\varepsilon_s - \varepsilon_\infty}{1 + (2\pi f \tau)^2} \right) \right] = \varepsilon'_m \, \phi(f, \tau) \tag{6.25}$$

$$\varepsilon''(f) = \varepsilon''_m \frac{1}{\varepsilon''_w} \left(\frac{(\varepsilon_s - \varepsilon_\infty) 2\pi f . \tau}{1 + (2\pi f \tau)^2} \right) = \varepsilon''_m \, \psi(f, \tau) \tag{6.26}$$

where ε_s (= 80), and ε_∞ (= 4.23) are static and high frequency dielectric constants of pure water, ε'_w and ε''_w are dielectric constant and loss factor of pure water at frequency, f = 8 GHz, and ε'_m, dielectric constant of the moist sample, and ε''_m, the loss factor of the moist sample at frequency, f = 8 GHz. Equations (6.25) and (6.26) represent an exact dispersion curve for water. Assuming the slope of the dispersion curve for moist material to be the same as that for pure water, these equations can convert dielectric constant, ε'_m and loss factor, ε''_m of a moist sample at f = 8 GHz to those at any frequency f around f = 8 GHz. Equations (6.25) and (6.26) use three unknown variables for the moist sample (viz. dielectric constant, ε'_m, loss factor, ε''_m at 8 GHz, and relaxation time τ) which have been used to convert permittivity values of moist material from 8 GHz to different frequencies within X-band. It is highly unlikely that the slope of dispersion curve for moist material will be governed exactly by Eqs. (6.25) and (6.26) and hence we have incorporated some room for variation of slope by adjusting the relaxation time for moist material. Hence the transfer functions in Eqs. (6.25) and (6.26) are functions of relaxation time of material and frequency. Now our objective in the inverse procedure is to estimate three unknowns for each moist sample, dielectric constant, ε'_m, loss factor, ε''_m at 8 GHz and relaxation time τ.

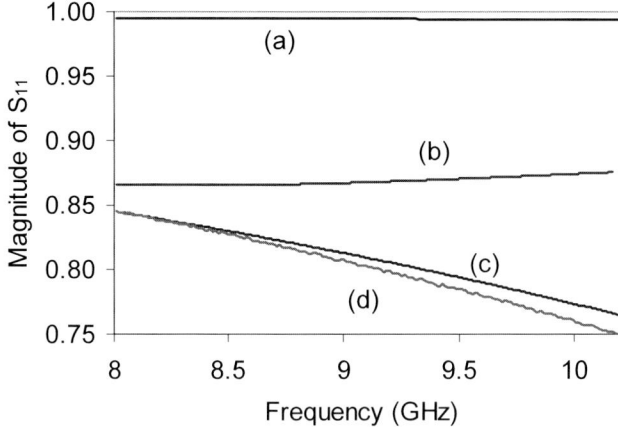

Fig. 6.7. Amplitude of reflected radiation as a function of frequency for (a) acrylic sample, (b) dry MDF (Q3C2), (c) wet MDF (12 % moisture, sample QW35r) and (d) wet MDF (13.4 % moisture, QW44C1)

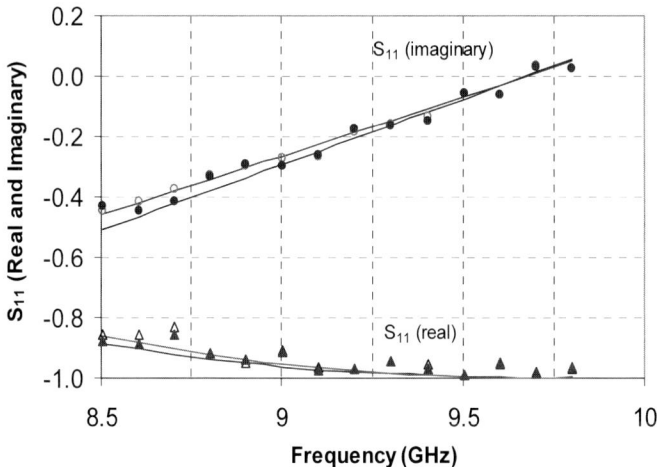

Fig. 6.8. Two sets of measured and simulated results of the reflection coefficients for 9.5mm acrylic sheet placed upon metal sheet. The simulated thickness is 9.4558 mm and 9.499 mm in two separate measurements. The simulated permittivities are $2.67 - j\,0.000279$ and $2.64 - j\,0.000306$

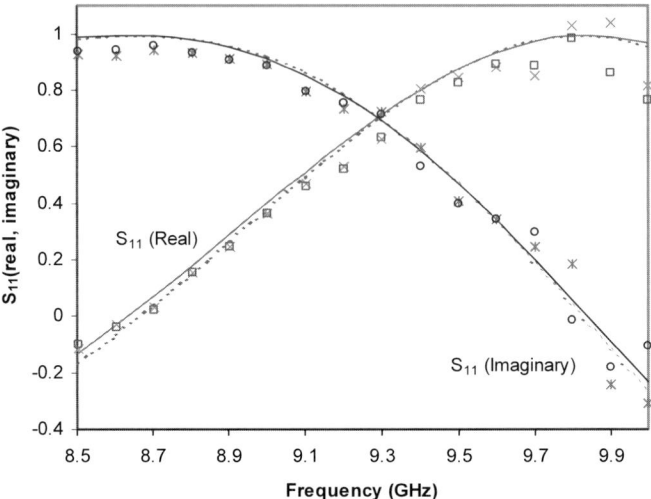

Fig. 6.9. Measured and simulated results for 14.1mm acrylic sheet placed upon metal sheet. The simulated thickness is 14.19 mm and 14.87 mm according to two separate measurements. The simulated permittivity are $2.60 - j\,0.00029$ and $2.36 - j\,0.00033$

6.4.7 Results and Discussion

In order to substantiate the present model let us consider a simplified system of a single dielectric backed by a metal reflector. Several measurements for acrylic sheets of different thickness backed by a metal reflector have been carried out [10]. The results for two acrylic sheets of thickness 9.5 mm and 14.1 mm having permittivity $\varepsilon_r = 2.6 - j\,0.0003$ are shown in Figs. 6.8 and 6.9, respectively. For acrylic sheets, the dispersion equations (6.25) and (6.26) have not been used. Hence the simulation procedure determines only three unknowns for the sample, viz. dielectric constant, loss factor and the thickness.

For the moist materials, medium density fibreboard (MDF) is used [11], which is soaked in water or dried in air to change its moisture content. For the moist samples, which show strong dispersion effects in the X-band, the dispersion relations (6.25) and (6.26) have been used. Hence the task for the simulation process in this case is increased to determine a fourth parameter, the relaxation time τ apart from dielectric constant, the loss factor and thickness.

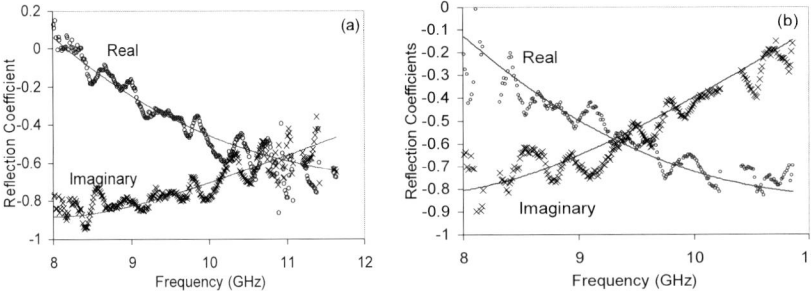

Fig. 6.10. Measured and simulated values of Reflection coefficient for (**a**) dry MDF (sample Q1C1_Aug) and (**b**) wet MDF (Mc 3.6%) W00C2_Aug

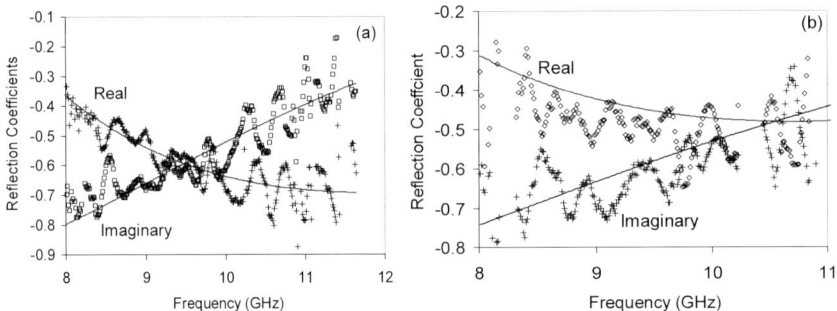

Fig. 6.11. Measured and simulated values of Reflection coefficient for wet MDF (**a**) Mc 6% (QW10C1_Moist) and (**b**) Mc 10.5%, (QW24C2-2Debye)

Figures 6.10 to 6.12 show the presence of noise in the measured data whereas the simulated results are smooth. Readjusting the focussing system and shifting the plane of measurement can reduce the noise in the measurement. The variation of errors during the simulation process is shown in Fig. 6.14a. Figure 6.14b presents the convergence of estimated values of thickness and permittivity.

Table 6.1 summarises the results for some of our measurements for MDF using this system for different values of moisture ranging from 0 to 24%. Sometimes the noise in experimental data for reflection coefficients is too high for the model to exactly determine the correct answer for the thickness and permittivity values for the dielectric. Although the simulated values of thickness and permittivity are a little bit far from the actual values, the product of thickness with square root of dielectric constant is very close to the actual value.

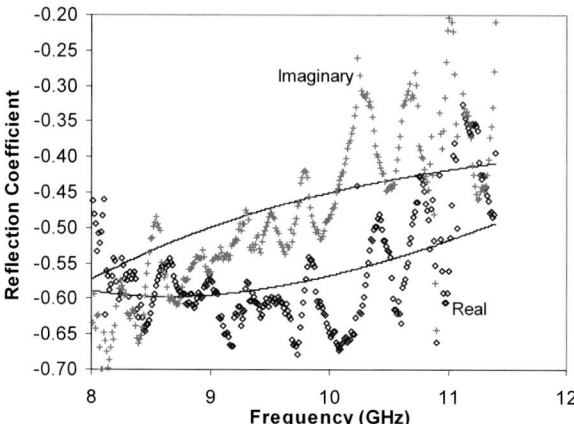

Fig. 6.12. Measured and simulated values of Reflection coefficient for wet MDF Mc 12%, QW35c1_2Debye

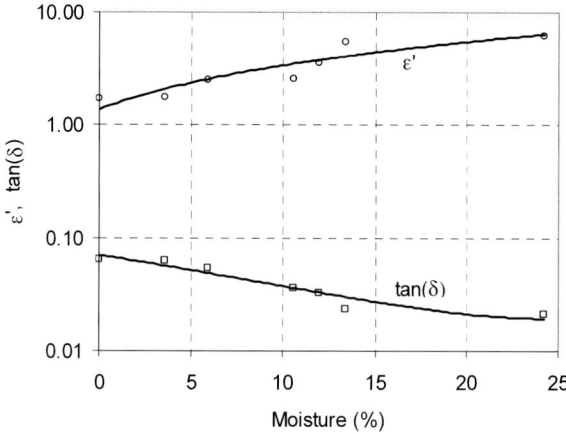

Fig. 6.13. Variation of dielectric constant and loss factor for moist material as a function of moisture content

This is because of the noise present in the measured reflection coefficients. In principle there is no limit on the measurement of thickness as long as electromagnetic energy can travel through the reflection and transmission path in the dielectric with little attenuation. However, some practical limitations exist since our focussing system cannot generate the exact plane wave that we would like to use for the measurement. The plane wave exists only around the beam waist and hence the restriction of thickness measurement will be of the order of a wavelength of an electromagnetic wave. Nevertheless, in this study we have measured electrical thickness of up to $1.7\lambda_0$ with reasonable accuracy.

Table 6.1. Simulated thickness, permittivity at 8 GHz and relaxation time for moist MDF

Sample	Moisture (%)	d (mm)	ε'	ε''	τ (ps)	$\varepsilon''/\varepsilon'$
Q	0.00	9.17	1.70	0.109	7.617	0.064
Q0	3.59	10.29	1.78	0.112	8.494	0.063
Q1	5.97	8.53	2.53	0.136	9.032	0.054
Q2	10.56	10.72	2.57	0.092	13.060	0.036
Q3	11.94	10.49	3.55	0.116	15.595	0.033
Q4	13.38	10.79	5.42	0.128	19.339	0.024
Q5	24.22	9.02	6.23	0.131	18.179	0.021

Assigning the values to the weights, w_i, is not straightforward. The non-linear least square error minimisation technique itself has its own error to minimise, which does not always converge correctly. This is because of the trigonometric and complex nature of the function. Hence the iteration gets trapped into the secondary minimum. Therefore, further measures are introduced to ensure that the solution does not converge in the secondary minimum. This is carried out by an additional set of errors all of which must be minimised simultaneously at the location of the principal minimum. As we can see in Fig. 6.8, there exist some secondary minima for each error type which do not always occur simultaneously. In the example, all weights were initially set equal to 0.25. This gave good results for some test samples but not for all test samples. The final best values of the weights are, $w_1 = 0.05$, $w_2 = 0.05$, and $w_3 = 0.2$, and $w_4 = 0.7$, which generated good answers for all samples investigated in this study.

Error measures reported for the acrylic sample of 14.1 mm thickness show [10] very large values of E_1 and E_2. This could be due to presence of a singularity in the value of Γ at some frequency calculated from Eq. (6.16). Giving higher values to weights to w_1 and w_2 will never produce the correct answer in a noisy set of data as shown in [10]. However, it must be remarked here that there is no need for the additional error checks if the noise level in the measured reflection coefficients is negligibly small. For low-noise data, the non-linear least square error minimisation technique will safely generate correct results.

There are many different sources of error in the measured reflection data. The most important one is the mismatch of plane of the object with the plane of the incident wave front, i.e. the angle of incident becomes different from zero. This is important not only during the measurements but also during the calibration of the focussing system. Some multiple reflections between the lens and the object under study exists, the amount of which varies with the electrical property of the sample. There exists some noise from the electronic components as well. Although attempts have been made to keep the noise level to a minimum, it cannot be eliminated completely.

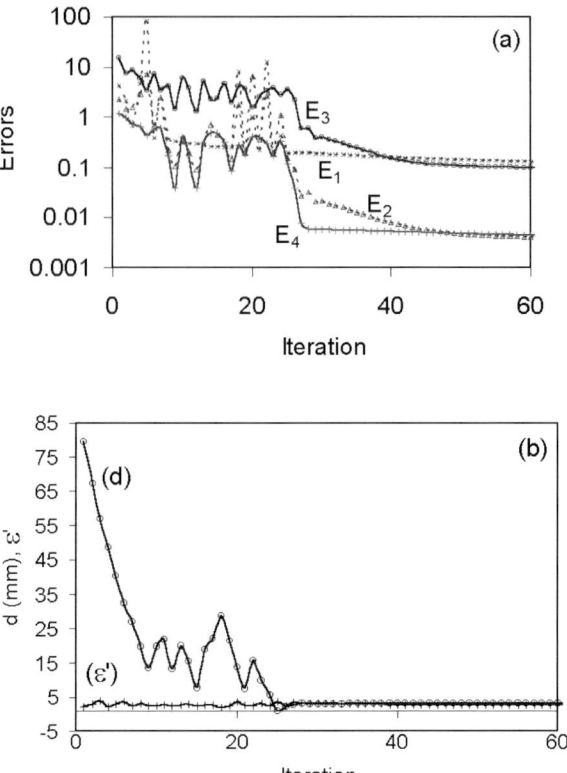

Fig. 6.14. Convergence of (**a**) errors and (**b**) thickness and permittivity during simulation process for high noise data

6.5 Conclusion and Future Research

Measurements of reflection coefficient of plane electromagnetic wave in the X-band frequency shows a non-linear dependence upon the thickness of the layer [10]. Once the reflection spectrum for a multi-layered dielectrics is known, it is possible to estimate the thickness and permittivity of one of the dielectric layers of the multi-layer system. For the moist materials, the value of relaxation time can also be obtained to generate an approximate dispersion relationship.

For the iterative non-linear least squares minimisation technique described, the convergence is reached within one second on a Pentium (600 MHz) computer. In principle, the present technique can be extended to estimate several unknown parameters at the cost of computer time, and measurements must be done at wide bandwidth to avoid spurious solutions.

Although, the numerical and experimental results are presented for the simplified case of a single layer dielectric backed by metal surface, the present method is very generalised and can be used to determine layer properties in multilayered dielectrics as well.

Even if a considerable amount of noise exists in the reflection measurement, the present technique works well because the model tries to obtain the best fit of experimental data with the theoretical model.

The non-linear least square error minimisation technique presented here can be used equally well in a variety of electromagnetic problems for obtaining the inverse of the direct problems [23]. An alternative procedure for obtaining the inverse could be to carry out a computer simulation of reflection data obtained for multiple layers of dielectrics to generate three unknowns, namely, dielectric constant, loss factor and thickness of one or more than one layer, using a different error minimisation technique [24].

References

1. Thakur KP (1999) Modelling the dielectric constant of rice grains. In: Proceedings of third workshop on electromagnetic wave interaction with water and moist substances, USDA, ARG, Athens, April 1999, pp 239–243
2. Thakur KP, Holmes WS (1999) Dielectric constant of Lentil and dry Peas. In: Proceedings of 34th annual microwave symposium, International Microwave Power Institute, Washington, July 1999, pp 13–16
3. Van Beek LKH (1967) Dielectric behaviour of heterogeneous systems. In: Progress in dielectrics, Vol 7. Heywood, London, pp 69–114
4. Sorrentino R (1989) Transverse resonance technique. In: Itoh T (ed) Numerical techniques for microwave and millimeter-wave passive structures, John Wiley & Sons, New York, pp 637–696
5. Collin RE (1970) Field theory of guided waves. IEEE, New York, pp 612–35
6. Frederick JT (1970) H-Guide with Laminated Dielectric Slab. IEEE Trans. Microwave Theory Tech 18: 9–15
7. Thakur KP, Cresswell KJ, Bogosanovich M, Holmes WS (1999) Non-interactive and distributive property of dielectrics in mixture. IEE Electronics Letters 35: 1143–1144
8. Thakur KP, Holmes WS (2001) Reflection of plane wave from multi-layered dielectrics. In: Proceedings of IEEE Microwave Conference, APMC 2001, Asia-Pacific, Taipei, Vol 2, pp 910–913.
9. Thakur KP, Williamson AG (2001) Multiple dielectric slabs in waveguide cell. IEEE Microwave Wireless Comp Lett 11: 121–123
10. Thakur KP, Chan KL, Holmes WS, Carter G (2002) An inverse technique to evaluate thickness and permittivity using reflection of plane wave from inhomogeneous dielectrics. In: Proceedings of 59th automatic RF techniques group conference, Seatle, pp. 77–83
11. Thakur KP, Holmes WS (2004) Noncontact measurement of moisture in layered dielectrics from microwave reflection spectroscopy using an inverse technique. IEEE Trans Microwave Theory Tech 52(1): 76–82

12. Collin RE (1991) Field theory of guided waves. IEEE Press, New York, pp 181–3
13. Nelson SO, Trabelsi S, Kraszewski AW (2001) RF sensing of grain and seed moisture content. IEEE Sensor J. 1: 119–126
14. Kraszewski AW, Trabelsi S, Nelson SO (2002) Broadband microwave wheat permittivity measurements in free space. J Microwave Power Elect Energy 37: 41–54
15. Afsar MN, Xiaohui L, Hua C (1990) An automated 60 GHz Resonator system for precision dielectric measurement of loss tangent and permittivity. IEEE MTT-S Int. Microwave Symp Dig 3: 1125-1128
16. Afsar MN, Ding H (2001) A novel open-resonator system for precise measurement of permittivity and loss-tangent. IEEE Trans. Instrum Meas 50: 402–405
17. Gagnon DR (1991) Highly sensitive measurements with a lens-focussed refractometer. IEEE Trans Microwave Theory Tech S Digest, pp 1017–1018
18. Judah SR, Holmes WS (1998) Prototype non-contact electromagnetic sensor. Industrial Research Limited Report, Auckland, #8810800-1-98
19. Zoughi R, Bakhtiari S (1990) Microwave nondestructive detection and evaluation of disbonding and delamination in layered-dielectric-slabs. IEEE Microwave Theory Tech., vol 39 (6), pp 1059–63
20. Bakhtiari S, Qaddoumi N, Ganchev SI, Zoughi R (1994) Microwave noncontact examination of disbond and thickness variation in stratified composite media. IEEE Microwave Theory Tech. 42: 389–95
21. Bramanti M (1992) A nondistructive diagnostic method based on swept-frequency ultrasound transmission-reflection measurements. IEEE Trans Instrum Meas 41: 490–494
22. Hasted JB (1973) Aqueous dielectrics. Chapman and Hall, London, pp 19–47
23. Thakur KP (2003) Imaging embedded objects from 3D inverse scattering using finite element method. In: Proceedings of inverse problems in Engineering symposium, Tuscaloosa, June 9–10 (http:///www.me.ua.edu/ipes2003/Abstracts.htm)
24. Thakur KP, Holmes WS (2001) An inverse technique to evaluate permittivity of material in a cavity. IEEE Trans Microwave Theory Tech 49: 1129–1132

Appendix A

The amplitudes of input and output is related according to Eq. (6.14)

$$\begin{bmatrix} c_1 \\ b_1 \end{bmatrix} = \begin{bmatrix} a_{00} & a_{01} \\ a_{10} & a_{11} \end{bmatrix} \begin{bmatrix} c_n \\ b_n \end{bmatrix} \tag{A1}$$

where

$$\begin{bmatrix} a_{00} & a_{01} \\ a_{10} & a_{11} \end{bmatrix} = \begin{bmatrix} \dfrac{e^{j\phi_1}}{T_1} & \dfrac{R_1 e^{-j\phi_1}}{T_1} \\ \dfrac{R_1 e^{j\phi_1}}{T_1} & \dfrac{e^{-j\phi_1}}{T_1} \end{bmatrix} \begin{bmatrix} \dfrac{e^{j\phi_2}}{T_2} & \dfrac{R_2 e^{-j\phi_2}}{T_2} \\ \dfrac{R_2 e^{j\phi_2}}{T_2} & \dfrac{e^{-j\phi_2}}{T_2} \end{bmatrix} \cdots \begin{bmatrix} \dfrac{e^{j\phi_n}}{T_n} & \dfrac{R_n e^{-j\phi_n}}{T_n} \\ \dfrac{R_n e^{j\phi_n}}{T_n} & \dfrac{e^{-j\phi_n}}{T_n} \end{bmatrix} \begin{bmatrix} c_n \\ b_n \end{bmatrix} \tag{A2}$$

and

$$\phi_i = k_0 d_i \sqrt{\varepsilon_{ri} - \sin^2(\theta_i)} \tag{A3}$$

The reflection coefficient is obtained from

$$R_\perp = \left(\frac{a_{10}}{a_{00}} \right)_{b_n = 0} \tag{A4}$$

and

$$\Gamma = \frac{1 + R_\perp}{1 - R_\perp} \tag{A5}$$

The derivative of Γ with respect to thickness of ith layer d_i is obtained from

$$\frac{d\Gamma}{d(d_i)} = \frac{2.}{(1 - R_\perp)^2} \frac{dR_\perp}{d(d_i)} \tag{A6}$$

where

$$\frac{dR_\perp}{d(d_i)} = \frac{a_{00} a'_{10} - a_{10} a'_{00}}{a_{00}^2} \tag{A7}$$

$$\begin{bmatrix} a'_{00} a'_{01} \\ a'_{10} a'_{11} \end{bmatrix} = \begin{bmatrix} \dfrac{e^{j\phi_1}}{T_1} & \dfrac{R_1 e^{-j\phi_1}}{T_1} \\ \dfrac{R_1 e^{j\phi_1}}{T_1} & \dfrac{e^{-j\phi_1}}{T_1} \end{bmatrix} \cdots \dfrac{j\phi_i}{d_i} \begin{bmatrix} \dfrac{e^{j\phi_i}}{T_i} & -\dfrac{R_i e^{-j\phi_i}}{T_i} \\ \dfrac{R_i e^{j\phi_i}}{T_i} & -\dfrac{e^{-j\phi_i}}{T_i} \end{bmatrix} \cdots \begin{bmatrix} \dfrac{e^{j\phi_n}}{T_n} & \dfrac{R_n e^{-j\phi_n}}{T_n} \\ \dfrac{R_n e^{j\phi_n}}{T_n} & \dfrac{e^{-j\phi_n}}{T_n} \end{bmatrix} \begin{bmatrix} c_n \\ b_n \end{bmatrix} \quad (A8)$$

Similarly, derivative of Γ with respect to ε_{ri} can be obtained with caution since ε_{ri} is involved in R_i, T_i, and also in R_{i+1}, T_{i+1}. One can also use a numerical technique to obtain the derivatives, to the first approximation, by computing the function $\Gamma(x)$ and $\Gamma(x + \delta x)$

$$\frac{d\Gamma}{dx} = \frac{\Gamma(x + \delta x) - \Gamma(x)}{\delta x} \qquad (A9)$$

Measurement Methods
and Sensors
in Frequency Domain

7 Methods of Density-Independent Moisture Measurement

Klaus Kupfer

Materialforschungs- und –prüfanstalt an der Bauhaus-Universität Weimar
Amalienstraße 13, D-99423 Weimar, Germany

7.1 Introduction

Accurate determination of the moisture content or mass of material is required in many different fields of industry, including high speed production processes, structural analysis in buildings and quality certification on the basis of ISO 9000 standards. Applications of radiometric density sensors, which radioactively irradiate and ionize the material, are unsuitable or prohibited for food or agricultural products.

Moisture measurements using dielectric methods in the radio frequency (RF) or microwave range are indirect measuring methods. During the calibration the measured values of attenuation, phase shift, or frequency shift, etc. containing the complex value of permittivity will be assigned to the moisture content determined by the oven dry method. The measured complex value of the relative permittivity will be influenced by the moisture content MC, but also by density variations $\Delta\rho$, ionic conductivity σ, temparature T, and scatterings at coarse corn sizes.

$$\underline{\varepsilon}_r = \varepsilon_r' - j\varepsilon_r'' \propto F\,(MC,\ \Delta\rho,\ \sigma,\ T\,) \tag{7.1}$$

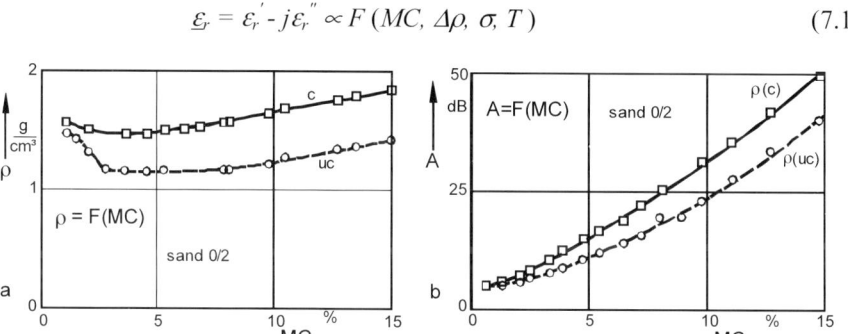

Fig. 7.1. Density variations $\Delta\rho$ for uncompacted (uc) and compacted (c) material versus moisture content MC (**a**) and their effects on the calibration curves of attenuation [1] (**b**)

In order to increase the accuracy of measurement, different ways exist to reduce the number of disturbances. Bulk-density variations cause compressions of capillary tubes and measuring errors in nearly all indirect moisture measuring methods. These measurement errors are in many cases bigger than those caused by the influence of salt content in the RF range. Different investigations were carried out in order to reduce the influence of density variations on the measuring results or to determine the density. During the application of one-parameter methods, density variations have to be kept small, constant, or their effects have to be compensated for. Figure 7.1 shows possible density variations and their influence on the parameter attenuation of measurements in the microwave field.

7.2 Density Variations in One-Parameter-Methods

Microwave absorption or loss tangent as moisture dependent parameters were used during the development of the first microwave setups for moisture determination. In laboratory setups the density has been kept constant by using a probe with a certain mass in a container with a constant volume. It was calibrated for different materials, e.g. by   Aquatrace   from the firm Rotronic Switzerland. Sensors must be installed in the flowing material stream on conveyor belts or at aggregate bins. The bulk-density must be kept constant during the measurement with different probe former systems [2, 3, 4]:

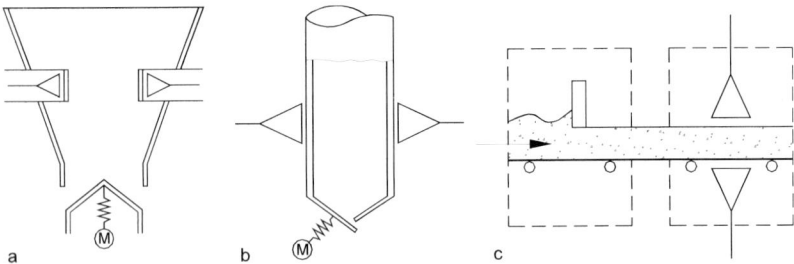

Fig. 7.2. a, b Mechanical means for stabilising the layer thickness of grain and its density during on-line measurements [4]; **c** Stripper plate on a conveyor belt

Figure 7.2 shows examples for probe former systems. Horn antennas were installed inside of a metal chute (Fig. 7.2a), and outside a dielectric pipe used for grain transport (Fig. 7.2b). M denotes a motor vibrating the bottom flow limiter to minimize density fluctuations. In Fig. 7.2c a stripper belt is installed on a conveyor belt to realize a material stream with a smooth surface and a constant thickness.

In the concrete industry, microwave moisture sensors operate in most cases as open ring resonator systems at a single frequency of 433 MHz [5]. The electromagnetic field of the inductor penetrates the ion conductive material to be measured and induces eddy currents. The system shows a higher penetration depth compared with an electric field sensor. The one-parameter sensors have to be installed at locations with small density variations e.g. in the flowing material stream of aggregates (Fig. 7.3). By installation of microwave moisture sensors below a baffle plate or at a chute below the discharge of an aggregate bin, a homogenizing of the material stream but also the determination of mass variations with a pressure gauge is possible. Density variations are also decreased during the homogenizing of fresh concrete in the mixer. A microwave sensor at the bottom of a mixer can measure the moisture content at different mixing phases and can detect the status of homogenizing of fresh concrete. This can reduce the mixing time up to 30 % [6].

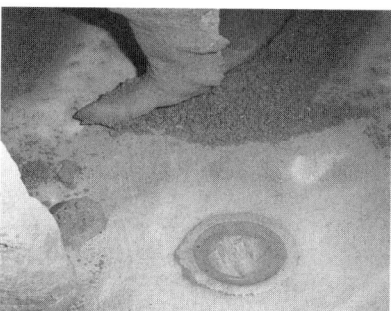

Fig. 7.3. Reducing of density variations in the flowing material stream

7.3 Density Determination by Using Gamma Rays

The density dependence of microwave absorption during the moisture measurement (Fig. 7.1) can be reduced by a ratio of attenuation and density. The density can be determined by gamma rays. The first device Accusense operating on this measuring method was developed by the firm Kay-Ray at the beginning of the eighties (Fig. 7.4) [7]. It was firstly applied at a chute, which was installed in the product stream. Later the density independence was realized by using the ratio of microwave phase-shift and density, which was measured simultaneously by gamma rays. It could be established that the phase shift was less influenced by ionic conductivity, temperature changes and scattering as the attenuation [8]. The measurement set-up operating as a homodyne network analyser at 3 GHz was produced and first successfully applied in the coal industry by the firm Berthold (1989) [9]. The microwave measuring value is related to the mass per unit area and the density-independent measuring value can be calculated with the following equation:

$$MC_{\Phi} = a\frac{\Phi_M}{m_A} + c \tag{7.2}$$

A combination of attenuation and phase shift gave the following result for the moisture content determination of inhomogeneous materials:

$$MC_{A\Phi} = a\frac{\Phi_M}{m_A} + b\frac{A_M}{m_A} + c \tag{7.3}$$

where m_A is mass per unit area, Φ_M, A_M are measuring values of phase shift and attenuation, a, b, c are coefficients.

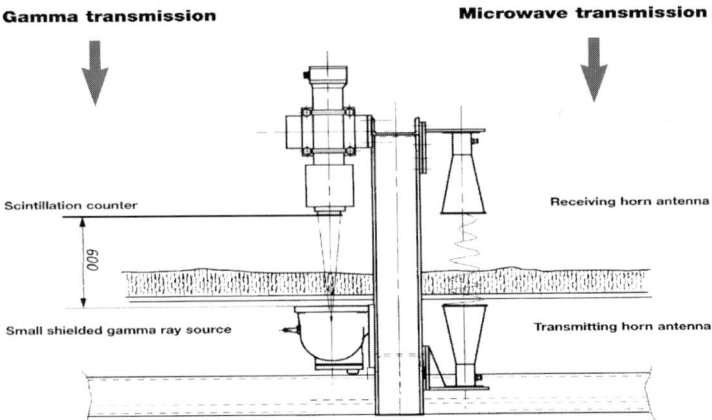

Fig. 7.4. Microwave and gamma transmission measurements for density compensation during the moisture determination [10]

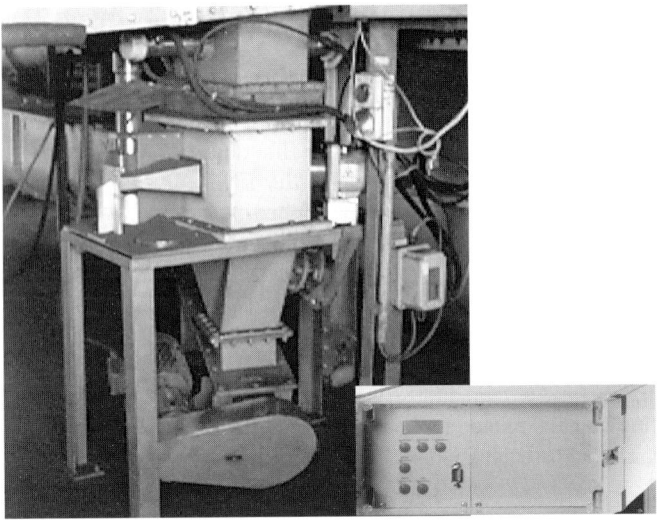

Fig. 7.5. Microwave transmission and gamma density measurements for moisture content determination in a lignite stream [10]

Many different applications exist e. g. in lime-sand-brick-stone production, brick production, coal industry, production of wood chipboards (Fig. 7.5), processing of agricultural products, etc.

7.4 Density Compensation using Model Equations in RF and Microwave Range

In the RF and microwave range the most successful method that is used in developments and applied all over the world is the multi-parameter measurement with different algorithms and modifications of instruments. Two basic electrical methods exist: simultaneous measurement of two electrical parameters of the substance at one single frequency and the simultaneous measurement of one or two electrical parameters at different frequencies. These methods can be applied using a frequency sweep. The resulting changes of attenuation and phase shift show a bigger accuracy than with those ones measured at a single frequency.

Microwave vector parameters of a transmission measurement, attenuation A and phase shift Φ or the scalar parameters of a resonator curve, change of bandwidth ΔBW and frequency shift Δf are functions of the relative complex permittivity $\underline{\varepsilon}$ which depends on the moisture content MC, the material thickness t, the density ρ, and the temperature T:

$$A; \Phi \propto F\left[\underline{\varepsilon}\left(MC;t;\rho;T\right)\right] \tag{7.4}$$

$$\Delta BW; \Delta f_r \propto F\left[\underline{\varepsilon}\left(MC;t;\rho;T\right)\right] \tag{7.5}$$

$$\Phi\left(f_1\right); \Phi\left(f_2\right) \propto F\left[\underline{\varepsilon}\left(f_1\right);\underline{\varepsilon}\left(f_2\right)\right] \tag{7.6}$$

7.4.1 Multiparameter-Multifrequency Method According to Chope

The first method and apparatus for measuring multiple properties of material by applying electric fields at multiple frequencies and combining detection signals was filed from Chope in July 1960 and patented in November 1964 [11].

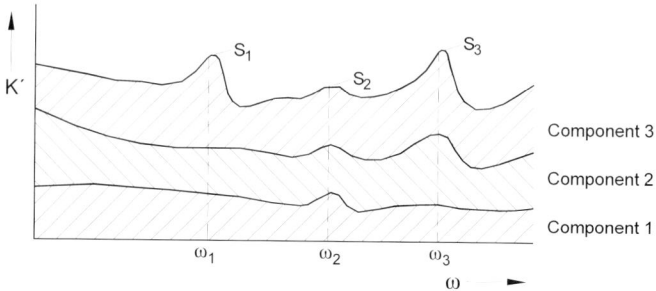

Fig. 7.6. Relative capacity of a composite material [11]

The primary object was to provide a method and system for measurement and control of material in a continuous industrial process by means of simultaneous analysis of that material with a plurality of different frequency signals and deriving from the detected signals the influence of the composite effect of the individual components and other properties of the material. By utilizing the derived signals the calculation of the quantitative values of the individual components of the material can be obtained.   Figure 7.6 shows the schematic representation of the effect of the three material components 1, 2, 3 on the dielectric constant of a mixture. K' is a plot of the relative capacity of the composite material as a function of frequency. At the points ω_1, ω_2, ω_3, the component contributions to K' generate the proportional signals S_1, S_2 and S_3. A change in the relative quantity of component 1, 2, or 3 will cause corresponding but contrasting changes in all three signals. Measurements of K' or K'' in the microwave region require other techniques such as the determination of wave guide impedance as modified by the presence of material or by the determination of reflection, transmission and absorption of electromagnetic radiation.

It was assumed that the relative concentration of the individual components of the material could be considered to have a linear relation to the signal produced in an associated electric circuit when the material is examined by an electromagnetic field. A solution of a general set of independent linear equations relating n material unknowns $Y_1....Y_n$ to the measured signal quantities $S_1...S_n$ has the following form:

$$Y_1 = A_{11}S_1 + A_{12}S_2 + ... + A_{1n}S_n$$
$$Y_2 = A_{12}S_1 + A_{22}S_2 + ... + A_{2n}S_n$$
$$.\qquad .\qquad\quad .\qquad .\qquad .$$
$$.\qquad .\qquad\qquad .\qquad\quad .$$
$$Y_n = A_{n1}S_1 + A_{n2}S_2 + ... + A_{nn}S_n$$

$$(7.7)$$

where S_1, S_2 ... S_n are measured signals at distinct frequencies f_1, $f_2 ... f_n$; Y_1, $Y_2...Y_n$ are respective characteristics such as the relative concentration of constituents of the mixture; A_{11}, $A_{12} ... A_{1n} ...A_{nn}$ are coefficients of the linear equations. With the coefficients A_{ij} of these equations known and with the $S_1...S_n$ measured at the distinct frequencies, the n equations are completed and can be solved for the $Y_1 ... Y_n$ variables.

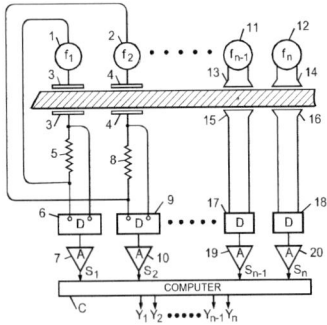

Fig. 7.7. Setup for the simultaneous material measurement in the RF and microwave range [11]

A specific embodiment of the invention is shown for use with frequencies in the radio and microwave spectrum (Fig. 7.7). The oscillators 1 and 2 operate in the radio frequency range. The material properties will be determined between plates

3 and 4 of the relating capacitors. The microwave measurements will be carried out at the frequencies f_n and f_{n-1} using horn antennas. To calibrate the computer the coefficients of the terms in the set of n equations must be determined using standard-samples of the composite material in which known deviations of a particular component from the specification value for that component are present.

A set of n standard calibration samples each having a known deviation in one of the respective components Y_1, Y_2 ... Y_n is applied to the measuring system and the resulting set of n equations are solved for the coefficients A_{ij} of the Y_n where the Y_n being known for the standard sample under test.

7.4.2 Two-Parameter Method According to Kraszewski and Stuchly

Kraszewski and Stuchly were the first to create a new concept of the microwave technique for the measurement of both moisture and bulk-density. The idea was based on a simultaneous measurement of amplitude and phase of an electromagnetic wave passing through a layer of a substance under test [12, 13]. The parameters are related to water content and the mass of dry material thus the moisture content of the material under test. It was subsequently developed into a device that measured both phase (Φ) and attenuation (A) and was calibrated with simultaneous linear equations dependent on both of the material variables.

$$A = t\left(\frac{m_w}{V}a_1 + \frac{m_d}{V}a_2\right) \tag{7.8}$$

$$\Phi = t\left(\frac{m_w}{V}a_3 + \frac{m_d}{V}a_4\right) \tag{7.9}$$

where m_w is the weight of water present, m_d is the weight of dry material, V is the volume of the wet material, t its thickness, and a_i are constants found by calibration. The variables are expressed as weights and not as percentages. The bulk-density ρ was given by,

$$\rho = \frac{m_w + m_d}{V} \tag{7.10}$$

The moisture content $MC = m_w/(m_w + m_d)$ can be expressed as;

$$MC = \frac{Aa_4 - \Phi a_2}{\Phi(a_1 - a_2) - A(a_3 - a_4)} \tag{7.11}$$

The material can be located in a wave guide, resonator or in a free space, and the microwave frequency of operation should be appropriate for the structure and properties of the material. The microwave measuring setup has been proposed for direct determination of the moisture content of the material. It can be applied to the automatic control of drying or moistening of the material in industrial processes. Using a transmission setup (Fig. 7.8), Eq. (7.11) was tested for moisture content determination of different materials [13].

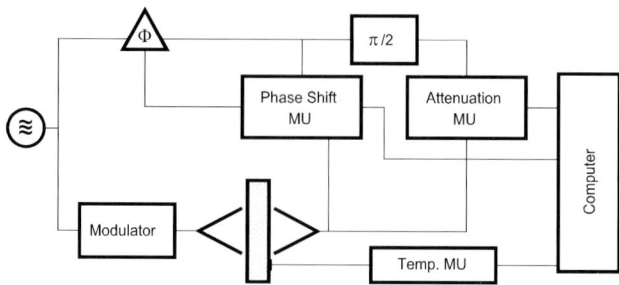

Fig. 7.8. Combined measurement of attenuation and phase shift in a transmission setup [13]

The coefficients of sand were found by regression analysis at a measurement frequency of 9.3 GHz: $a_4 = 4.21$; $a_2 = 0.014$; $a_1 - a_2 = 3.01$; $a_3 - a_4 = 77.56$. The correctness of Eq. (7.11) was proved later by microwave measurements of rye (standard deviation $S = 0.2154\%$ MC), barley ($S = 0.2187\%$ MC) and wheat/rye ($S = 0.2489\%$ MC) [14].

Kraszewski and Stuchly used a linear approximation of the relationships between attenuation, phase shift, and water content. The assumed linearity of the calibration equations is applicable only over a narrow range of water content for many materials, but on the other hand in a majority of industrial processes moisture content does not vary from zero to the saturation of the material.

This approach was far ahead of its time. If the computing power and speed commonplace now had been available at that time it would have been implemented widely. Unfortunately that limitation plus the cumbersome electromechanically operated microwave variable components militated against proper commercialisation of the technique.

An attenuation measurement method at three different frequencies, which allows to determine the values of moisture content, density, and temperature, was introduced by Stuchly [15].

7.4.3 Two-Parameter Method According to Meyer and Schilz

7.4.3.1 Transmission Measurement

Meyer and Schilz developed a density-independent model and a measurement system with different applicators for a variety of materials [16, 17, 18]. Using a heterodyne network analyser operating at 9 GHz (Fig. 7.9), Eq. (7.12) was tested with different materials e.g. tobacco [19, 20]. It was revealed a density-independent function (Fig. 7.10).

$$\frac{\Phi}{A} = \frac{\varepsilon_r' - 1}{\varepsilon_r''} \frac{2\sqrt{\varepsilon_r'}}{1 + \sqrt{\varepsilon_r'}} \qquad (7.12)$$

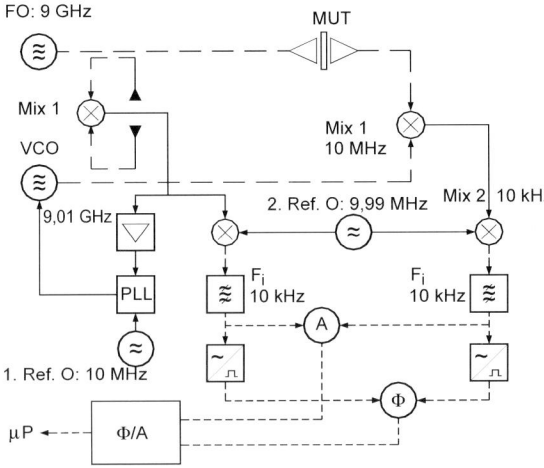

Fig. 7.9. Heterodyne NWA with a transmission line for density-independent moisture measurement [19]; broken lines: microwave connections (9 GHz); straight lines: hf-connections (10 MHz); dotted lines: lf-connections (10 kHz)

The microwave part consists of the measuring path with the applicator, the measuring device under test (MUT) and two Gunnoscillators, which are kept at a constant frequency distance of 10 MHz by a PLL (Reference Oscillator 1). From this first IF by a second down conversion (Ref. Osc. 2) the final processing-frequency of 10 kHz is gained where phase and amplitude detection are accomplished with digital techniques.

The overall error for the microwave transmission coefficient was 0.1 dB in amplitude (A) and ± 1 deg in phase shift (Φ) measurement.

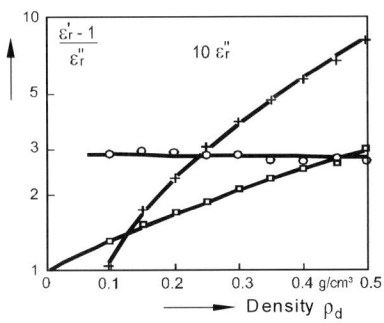

Fig. 7.10. Density dependence of ε_r', ε_r'' and $(\varepsilon_r' - 1)/\varepsilon_r''$ for tobacco at 12.5 GHz

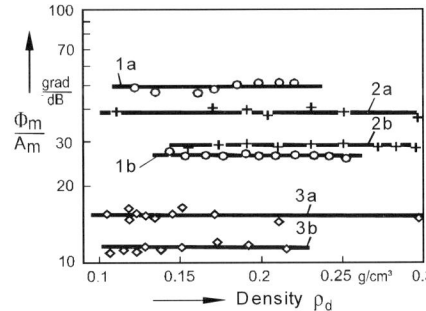

Fig. 7.11. Ratio Φ/A versus density using different applicators

Figure 7.11 shows the ratio Φ/A versus density using different applicators:

1. Horn applicator
   a)   tobacco 12 % MC, 9 GHz;
   b)   tobacco 18 % MC, 9 GHz;
2. Microstrip line 50 Ω
   a)   on a teflon substrate, tobacco 12 % MC, 9 GHz;
   b)   on a Al_2O_3-substrate, tobacco 18% MC, 9 GHz;
3. Inverted image guide
   a)   tobacco 12 % MC, 12.4 GHz;
   b)   tobacco 18 % MC, 12.4 GHz

It has been stated that a complete compensation of sample thickness and density could not be reached. The remaining density dependence expressed by the factor $2(\varepsilon_r')^{1/2}/[1+(\varepsilon_r')^{1/2}]$ in Eq. (7.12) ranging from $1(\varepsilon_r' = 1)$ to 2 $(\varepsilon_r' \to \infty)$ and was negligible for small $\varepsilon_r' \cong 1$ [17]. The ratio Φ/A measured for plane wave propagation with horn antennas was close to the ratio $R = (\varepsilon_r' - 1)/\varepsilon_r''$ and insensitive to the sample density [21].

The relations of permittivity dependent on density at different frequencies from coal and wheat were shown by Nelson [22]. Equation (7.12) was also used by Vainikainen in measuring moisture content of moving wood veneer sheets with a stripline resonator [21]. The two-variable-technique was investigated by Kent and Kress-Rogers in materials such as fish meal, wheat, coffee, milk powder and coal [24, 25] (see Sect. 7.4.5.2).

7.4.3.2 Resonator Measurement

Moisture measurement could also be realised with resonant cavities by using scalar measuring values of the resonance curve. The real and imaginary parts of permittivity given by the resonant frequencies and quality factors could be correlated with the moisture content:

$$R(MC) = \frac{\varepsilon_r' - 1}{\varepsilon_e''} = 2\frac{\dfrac{f_0 - f_1}{f_1}}{\dfrac{1}{Q_1} - \dfrac{1}{Q_0}} \tag{7.13}$$

where f_0 and f_1 are the resonant frequencies, Q_0 and Q_1 are the Q-factors before and after insertion of the sample [26, 27]. The central part was a cylindrical resonant cavity (Fig. 7.12). The sample was placed in the centre of cavity and consisted, for example, of a bunch of cotton threads. Two Gunn oscillators, arranged above and below the probe, were coupled to the cavity by means of "cut-off" technique. One oscillator excites the E_{012} mode, the other excites the H_{011} mode of the cavity. The frequency of the two modes differ only slightly around 11.5 GHz.

Figure 7.13 shows the power loss and frequency shift as a function of relative density and Fig. 7.14 as function of moisture content of a bunch of cotton threads. Both quantities are density-dependent on ρ_r and therefore not suitable separately to measure the moisture content of the sample. The ratio $\Delta f/\Delta P$ is independent of the density and can be used as a quantity of the moisture content (Figs. 7.13, 7.15).

Fig. 7.12. Sensor head of the cavity-type moisture meter

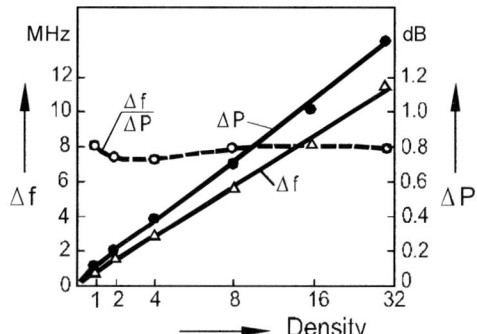

Fig. 7.13. Power loss ΔP and frequency shift Δf versus relative density of cotton threads

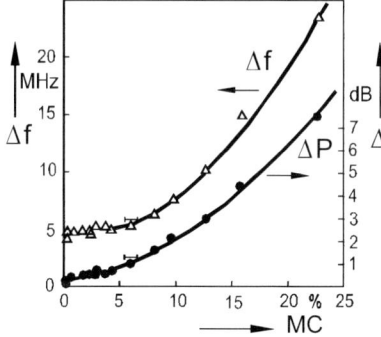

Fig. 7.14. Power loss ΔP and frequency shift Δf versus moisture content

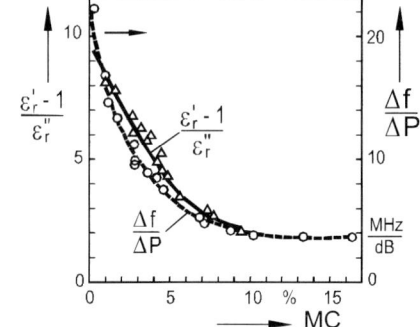

Fig. 7.15. Density-independent functions $(\varepsilon_r'-1)/\varepsilon_r''$ and $\Delta f/\Delta P$ versus moisture content

The ratio $\Delta f/\Delta P$ remains relatively constant within (7.5 ± 0.2) MHz/dB at density variations up to 32 g/cm^{-3} and a corresponding range of moisture content of $(4.5 \pm 0.1)\%$ MC. The plots of density-independent functions $(\varepsilon_r'-1)/\varepsilon_r''$ and $\Delta f/\Delta P$ versus moisture content are represented in Fig. 7.15. The change of Δf and ΔP correspond the frequency shift and power loss of oscillations in the E_{012} mode after inserting the sample. $R(MC)$ can be determined by measuring the quality factor and the frequency shift in one of the possible modes before and after insertion of the sample using Eq. (7.13). The system indicated the feasibility of the method below

10 % MC. Above 10 %, the sensitivity of the Eq. (7.12) decreased to unacceptable values (Fig. 7.15).

A resonator measurement system based on Eq. (7.13) for different industrial applications was developed and is produced by the firm Tews (see Chap. 9) [28]

7.4.4 Methods in the RF Range

Microwave systems for permittivity and moisture content measurement are generally more expensive than their lower frequency counterparts. Thus, systems were developed to realize the density-independent measurement in the lower frequency range.

In 1993 Zoerb developed an on-line dielectric moisture meter for monitoring moisture content [29]. He applied both coaxial and parallel plate sample holders, using them as the timing capacitor of a multivibrator. The grain moisture content was correlated to the multivibrator's period of oscillation. The standard error of the predicted moisture content ranged from 0.07 % to 1.18 % for barley; the correlation coefficient was 0.96. An advanced density-independent function was derived by Powell et. al. for grain moisture determination at microwave frequencies [30]. A homodyne system for simultaneous measurement of attenuation and phase shift was developed at 10.53 GHz in order to predict the moisture content of grain by sensing dielectric properties under flowing conditions. He proved that the relationship Eq. (7.12) of Meyer and Schilz was valid only over a small range of densities. Based on permittivity and density relationships published previously by Nelson [31], the following function was derived:

$$R(MC) = \frac{\sqrt[3]{\varepsilon_r'} - 1}{\sqrt[2]{\varepsilon_r''}} \qquad (7.14)$$

This function could be verified in the moisture range between 5 and 25 % MC, was density-independent over a wider range of densities and showed less deviations than Eq. (7.12).

7.4.4.1 Parallel Plate Capacitor

The relation Eq. (7.14) was also applied from Lawrence and Nelson, in order to determine the moisture content of hard red winter wheat at a frequency of 1 MHz. Under static conditions the conductance and susceptance of the shielded parallel plate capacitor with and without grain sample have been measured (Figs. 7.16, 7.17).

The range of moisture content was approximately 11 to 22 % MC, and that of bulk densities 660 to 855 kg/m³ [32, 33]. Standard error of calibration SEC ≈ 0.6 % and standard error of prediction SEP ≈ 0.5 % has been obtained.

The two-frequency method has been applied at frequencies of 1 MHz and 10 MHz to measure the moisture content of wheat by Lawrence [33]. For hard red winter wheat the following equation (Eq. (7.15)) was used to estimate moisture content in the called ranges of moisture and density.

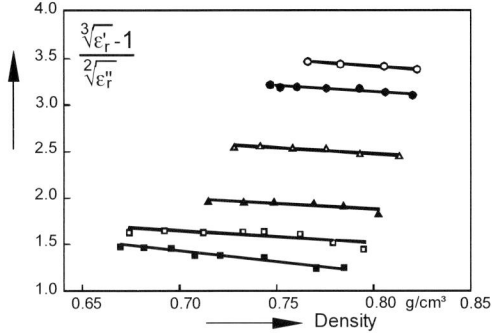

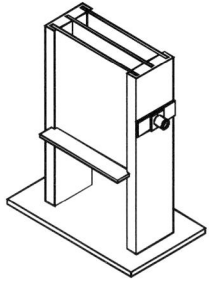

Fig. 7.16. Values of Eq. (7.14) of hard red winter wheat of different moisture content versus density at 1 MHz

Fig. 7.17. Shielded parallel plate capacitor

$$MC = 28.04 - 2.01\,\varepsilon'_{10\,MHz} + 4.83\,ln\,\varepsilon'_{1\,MHz} + 4.4\cdot10^{-4}\left[\left(\frac{\varepsilon'_r - 1}{\varepsilon''_r}\right)_{1\,MHz} - \left(\frac{\varepsilon'_r - 1}{\varepsilon''_r}\right)_{10\,MHz}\right]^2$$

(7.15)

7.4.4.2 Coaxial Sample Container

Equation (7.15) is well suited for high moisture content values > 14 % MC. The extension of the frequency range up to 110 MHz and swept frequency admittance measurements enables sensing moisture content independently of density. Advanced multivariate analysis tools were used to select optimum measurement frequencies, which can be applied in a commercial instrument. Standard errors were reduced to 0.4 % . By using measurement frequencies at 2.3, 24, and 84 MHz, the Eq. (7.16) was derived [34]. With this equation a SEC-value of 0.34 % MC, a correlation coefficient of $R^2 = 0.99$, and, for validation, a SEP value of 0.39 % MC could be realized.

$$MC = 29.96 + 0.62\left[\frac{(C_m - C_a)}{(G_m - G_a)}\right]_{2.3} - 19.81\left[\frac{(C_m - C_a)}{(G_m - G_a)}\right]_{24} - 18.23\left[\frac{(C_m - C_a)}{(G_m - G_a)}\right]_{83}$$

(7.16)

Examinations showed that this equation was not entirely density independent; however, it is much less sensitive to density fluctuations than measurements based on the permittivity alone. It could be established that a four frequency model indicates only small improvements compared to the three frequency method.

An analysis of density-independent equations for determination of moisture content of wheat with a shielded coaxial sample holder operating between 500 kHz and 5 MHz was published by Berbert and Stenning [35, 36]. By using non-linear regression analysis on wheat varieties Mercia and Hereward, the following density-independent relation for moisture content ranging from 11 to 22.1% MC was obtained at 1 MHz:

$$MC = 1.54\ln(\zeta) + 14.61 \qquad \text{where} \qquad \zeta = \frac{13.44}{\left(\dfrac{\varepsilon_r' - 1}{\varepsilon_r''}\right)_{1\,\text{MHz}} - 2.86} - 1 \qquad (7.17)$$

Its limitation is the fact that it only could be used in the range of $2.9 \ll [(\varepsilon_r'{-}1)/\varepsilon_r''] < 16.3$. The applicable range of moisture content was reduced to $11\,\% \le MC \le 19\,\%$. During the measurements of 97 wheat samples and comparison of different equations they determined that Eq. (7.17) at 1 MHz gave the best results with a coefficient of determination of 0.99, a standard error of calibration of 0.3 % MC, and a standard error of prediction of 0.4 % MC. Moisture measuring systems for grain in the RF-range were published in [37].

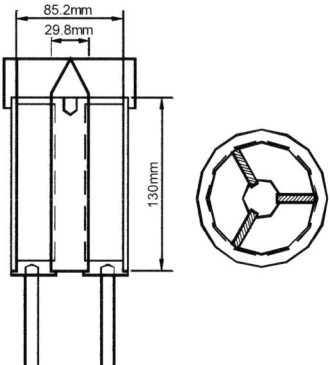

Fig. 7.18.
Coaxial sample container for dielectric measurements on wheat [36]

7.4.4.3 Stray-Field Capacitor

The general validity of the Eq. (7.12) was proved with a capacitive measurement system. It operates using a stray-field capacitor at a measurement frequency of 5 MHz.

Fig. 7.19. Stray-field capacitor for density-independent moisture measurement in tobacco [38]

The resonance frequency and attenuation of a free oscillating resonance circuit will be influenced by the dielectric properties of the measuring capacitor where the oscillation versus time is described as follows:

$$U(t) = U_0 \cdot e^{-\gamma t} \cdot \cos(\omega \cdot t + \varphi) \tag{7.18}$$

Resonance frequency
without probe
$$\omega_0 = \frac{1}{\sqrt{LC_m}} \tag{7.19}$$

Resonance frequency
with probe
$$\omega = \sqrt{\left(\omega_0^2 - \gamma^2\right)} \tag{7.20}$$

Attenuation constant
$$\alpha = \frac{\omega}{2} \cdot \tan \delta \tag{7.21}$$

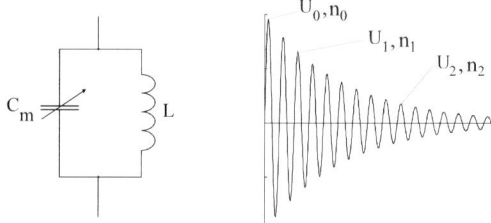

Fig. 7.20.
Free oscillating circuit at 5 MHz
for determination of ε_r' and $\tan \delta$

The ratio of oscillating amplitudes U_1 and U_2 after n_1 and n_2-periods delivers the following voltage ratio

$$\frac{U_2}{U_1} = e^{-\frac{\alpha 2\pi(n_2 - n_1)}{\omega}} \tag{7.22}$$

and the loss factor:
$$\tan \delta = \frac{1}{\pi(n_2 - n_1)} \ln \frac{U_1}{U_2} \tag{7.23}$$

The measuring capacity C_m is influenced by the parallel capacity C_p:

$$C_m + C_p = \frac{1}{4\pi^2 L f^2} \tag{7.24}$$

Using the measuring capacity of the empty capacitor C_{m0} and the corresponding frequency f_L Eq. (7.24) delivers the following result:

$$\frac{C_m - C_{m0}}{C_{m0} + C_p} = \frac{f_L^2}{f^2} - 1 = \frac{C_{m0}}{C_{m0} + C_p}(\varepsilon_r' - 1) \tag{7.25}$$

With some mathematical transformations using Eq. (7.23 and 7.25) there is the following result for a density-independent model equation:

$$R = \frac{\tan \delta}{\varepsilon_r' - 1} = \frac{1}{\pi(n_2 - n_1)} \cdot \frac{\ln \dfrac{U_1}{U_2} - \ln \dfrac{U_{1L}}{U_{2L}}}{\dfrac{f_L^2}{f^2} - 1} = a_1 \cdot (1 + \alpha\vartheta) + a_2\psi + a_3\psi \tag{7.26}$$

where U_{1L}, U_{2L}, f_L are measuring values of the empty resonator, ψ is MC, ϑ is temperature. In Eq. (7.26) the increment of loss factor has to be considered in the mathematical model. It was possible to realize a density-independent moisture measurement with a free oscillating resonance circuit [38] (precondition: high quality factor and loose coupling of the material under test).

Using the measuring system in Fig. (7.21) the results of measurements are presented in the Figs. (7.22 and 7.23); the ratio R shows the remaining density-dependence.

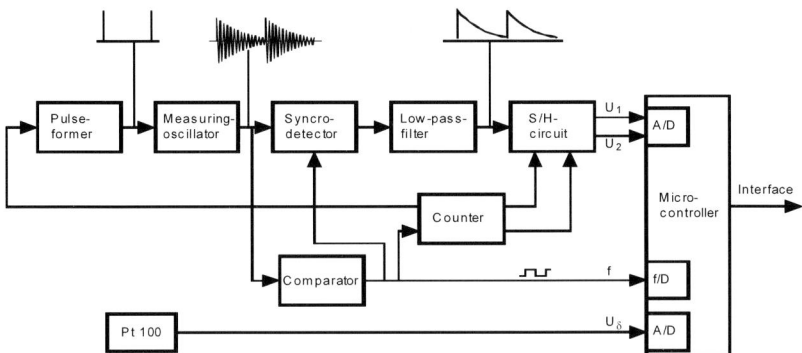

Fig. 7.21. Block diagram of the measuring system

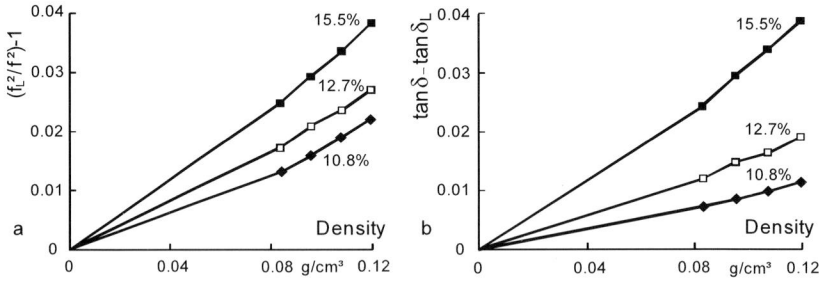

Fig. 7.22. a Relative permittivity $(\varepsilon_r'-1)$ versus density and MC as parameter for tobacco; **b** loss factor tan δ versus density and *MC* as parameter

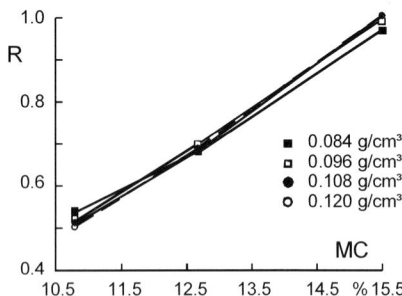

Fig. 7.23. Ratio R versus *MC* and density as parameter

7.4.5 The A-Φ-Diagram

7.4.5.1 Transmission Measurements of Anorganic Materials

An electromagnetic plane wave with frequency ω, which transmitted a material of thickness t and density ρ in a limited volume received an attenuation A and a phase shift Φ:

$$A = \frac{\omega}{c}\rho \cdot t \cdot \sqrt{\varepsilon_r'}\,\frac{\varepsilon_r''}{\varepsilon_r'} \tag{7.27}$$

$$\Phi = \frac{\omega}{c}\rho \cdot t \cdot \sqrt{\varepsilon_r'} + 2\pi n \tag{7.28}$$

The values A and Φ depend on frequency, density, thickness, and on the real and imaginary part of permittivity. Klein established that the values A_S and Φ_S referred to the mass per area (product of density and thickness) are independent of the density [8].

$$A_s = \frac{A}{\rho \cdot t} = F_1(MC) \qquad \text{and} \qquad \Phi_S = \frac{\Phi}{\rho \cdot t} = F_2(MC) \tag{7.29}$$

The ratio R of both is a density-independent value if the complex value of permittivity is relatively constant. This can be assumed in a range of small moisture variations in monofrequency investigations

$$R = \frac{A/\rho \cdot t}{\Phi/\rho \cdot t} = F_3(MC) \tag{7.30}$$

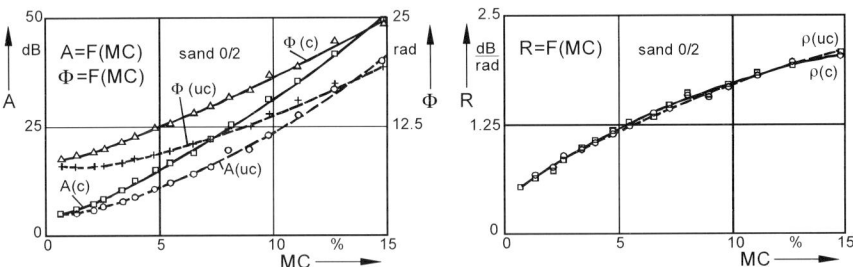

Fig. 7.24. Differences of attenuation and phase shift versus *MC* at variation of densities for sand 0/2

Fig. 7.25. The ratio of attenuation and phase shift versus *MC* at variation of densities

Density variations between uncompacted and compacted materials (see Fig.7.1) cause differences of attenuation and phase shift versus *MC* during the measurements, e.g. at 9.3 GHz (Fig. 7.24). The resulting ratio of both values $R = A/\Phi$ versus *MC* shows hardly any differences between the calibration curves (Fig. 7.25). In this case it was possible to realize a density-independent measurement in the range from 1 to 15 % MC with an error $\Delta MC \leq \pm 0.5$ %. In most cases the calibration curves are parabolic functions versus moisture content.

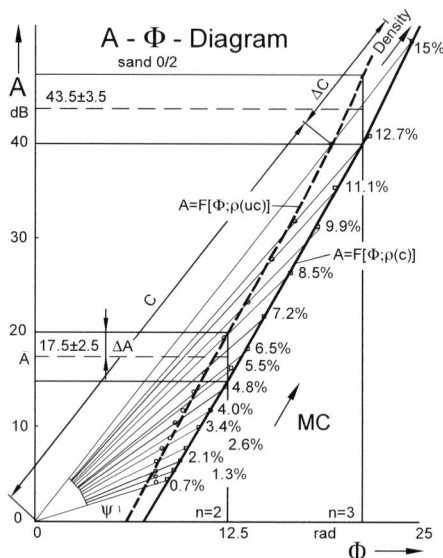

Fig. 7.26. Calibration curves versus *MC* in the A-Φ-diagram; the angle ψ shows a density-independent value of *MC* [1, 39]

Plotting the attenuation calibration curves of uncompacted and compacted material versus phase shift, linearity is demonstrated (Fig. 7.26). Straight lines through equal points of *MC* on both calibrating curves $A=F[\Phi;\rho(uc)]$ and $A=F[\Phi;\rho(c)]$ lead under an angle ψ to the origin of the A-Φ-diagram. Note that tan ψ shows independence of bulk-density and layer thickness, but only depends on the moisture content of the material. It is defined as follows:

$$\tan \psi = \frac{A(uc)}{\Phi(uc)} = \frac{A(c)}{\Phi(c)} = F(MC) \tag{7.31}$$

At a linear dependence the length of straight lines is a function of density:

$$C(uc) = \sqrt{A^2(uc) + \Phi^2(uc)} = F\left[\rho(uc)\right] \tag{7.32}$$

The density variation *ΔC* between the uncompacted and compacted values of calibration curves can be described with the following equation:

$$\Delta C = C(c) - C(uc) = \sqrt{\Delta A^2 + \Delta \Phi^2} = F[\Delta\rho] \tag{7.33}$$

Equation (7.33) is a function of density variations.

The A-Φ-diagram is shown as a function of gradual increasing bulk-density (Fig. 7.27). All values of density lie on a straight line between the limiting calibration curves $\rho(uc)$ and $\rho(c)$. The angle ψ of the straight line is independent of density and thickness and is only a function of *MC*.

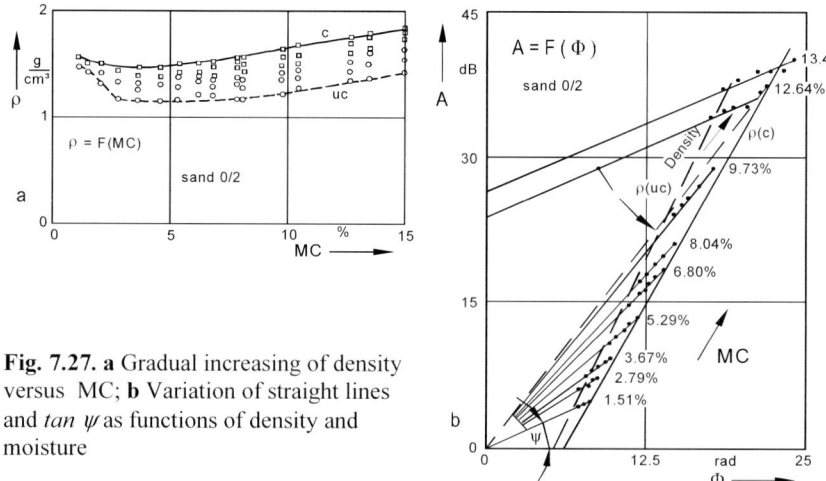

Fig. 7.27. a Gradual increasing of density versus MC; **b** Variation of straight lines and *tan ψ* as functions of density and moisture

Randomly scattered measured values are intensified in the moisture saturation range. The calibration curves of the ratio R of uncompressed and compressed materials are in most cases divergent (Fig. 7.25) and a density-independent measurement is incorrect. At high values of MC the straight line is displaced along the attenuation coordinate A (Fig. 7.27b). The possibility for a density-independent moisture measurement in the saturation range is demonstrated in this figure with the straight line $MC = 12.64$ %. Plotting all values of density proportional values along the straight line, it is displaced to higher A-values. Taking only the highest density values into account, it is possible to lay the straight line through the origin of the coordinate system. It means a precompaction has to be carried out for materials with high moisture content values to realize a density-independent measurement in the saturation range. That is also a reason why the repeatability of a calibration curve of medium or high compacted materials recorded with a higher accuracy than found for uncompacted materials.

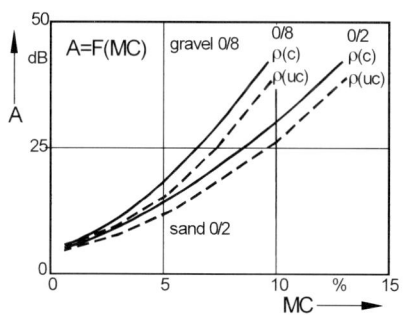

Fig. 7.28. Attenuation versus *MC* measured in sand with different grain sizes

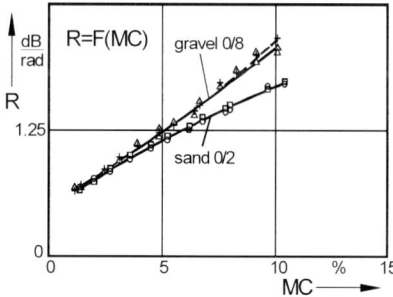

Fig. 7.29. Density-independent function *R* versus *MC* measured in sand 0/2 and gravel 0/8

The model will be disturbed by influences which affect the attenuation or loss tangent, e.g. one-sided temperature changes, inhomogeneities like coarse grain sizes, and salt content. Thus, the slope of the calibration curve of attenuation is increased for materials with coarse grain sizes compared to materials with fine corn sizes (Fig. 7.28). Figure 7.29 shows the same properties of density-independent calibration curves of sand 0/2 and gravel 0/8.

The moisture range of a density-independent measurement is limited (Fig. 7.29). Because of a smaller surface area the saturation range of a coarser-grain material is shifted to lower moisture contents. The calibration curves of uncompacted and compacted materials diverge above this point. Using precompression it was possible to realize a density-independent measurement for sand 0/2 in the moisture range from 1 to 10.6 % with an error of $\pm$ 0.25% MC. This range was reduced at gravel sand 0/8 until 1.5 to 9.2 % MC with an error of $\pm$ 0.5% MC.

Materials with coarse corn sizes are homogenized by precompression and the influence of scattering is reduced. This density-independent model was developed on the basis of aggregates for the concrete industry using a microwave bridge with a transmission measurement set-up [1, 39].

7.4.5.2 Transmission Measurements of Organic Materials

Kent and Kress-Rogers established with a microwave bridge operating in the X-band for measure-ments of fish meal, wheat, coffee, and milk powder that the ratio A/Φ was suitable for materials with low moisture content. The ratio shows an increasing density dependence at high moisture values. It was found that this ratio was slightly less sensitive to density variations than the ratio $\varepsilon_r''/(\varepsilon_r' - 1)$. The small residual density-dependence of the ratio A/Φ could be eliminated and the bulk-density determined in-line with an iteration procedure but with the phase shift instead of the permittivity ε_r'. The temperature dependence of the ratio $\varepsilon_r''/(\varepsilon_r' - 1)$ for coffee and milk powder was smaller than for ε_r'' alone [24, 25].

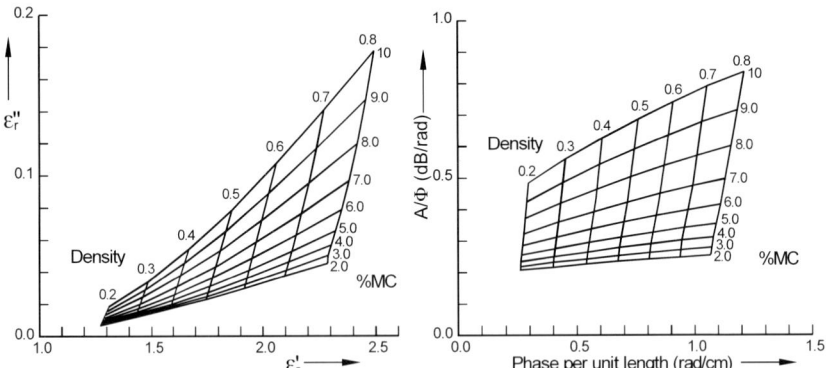

Fig. 7.30. Dielectric properties of milk Powder with varyinng density and moisture content

Fig. 7.31. Plot of the ratio A/Φ versus phase shift for the datas of Fig. 7.30

Fig. 7.30 shows the dielectric properties of milk powder with varying density and moisture content measured at 10 GHz. In most cases the complex permittivity could not be measured directly. It was measured attenuation and phase shift containing real and imaginary part of permittivity. Kent and Kress-Rogers combined the relatively density-independent ratio $R = A/\Phi$ with the phase measurement. The values for density and moisture content could be obtained from the diagram of the measured quantities (Fig. 7.31). They stated that the two-variable technique is useful for the determination of one compositional variable in the presence of another, uncontrolled variable e.g. for moisture and density. This technique was extented for moisture and fat, and concentration and temperature of aqueous solutions [40, 41].

7.4.5.3 The A-Φ-Diagram Transformed onto a Resonator System

The A-Φ-diagram was also transformed onto a ΔBW-Δf_r-diagram of the scalar measuring values of an open resonator system operating at 150 MHz (Fig. 7.32) [5, 42].

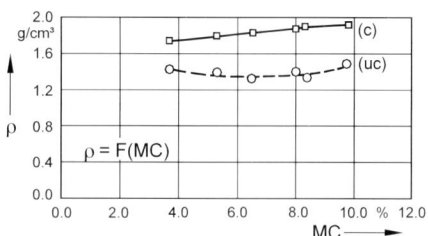

Fig. 7.32. Open resonator system

Fig. 7.33. Density of a compacted and uncompacted material versus moisture content

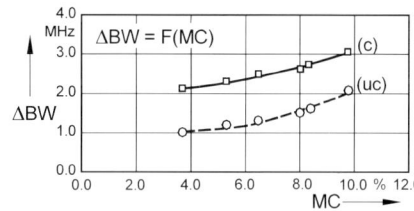

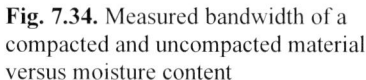

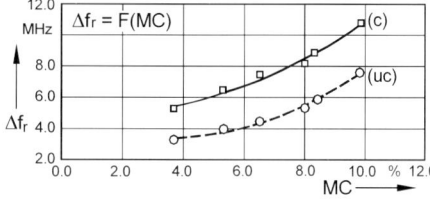

Fig. 7.34. Measured bandwidth of a compacted and uncompacted material versus moisture content

Fig. 7.35. Measured frequency shift of a compacted and uncompacted material versus moisture content

Figure 7.33 shows the values of density-variations versus *MC* which were transferred to the calibration curves $\Delta BW = F(MC)$ (Fig. 7.34) and $\Delta f_r = F(MC)$ (Fig. 7.35). The model of a density-independent measurement system is plotted in Fig. 7.36.

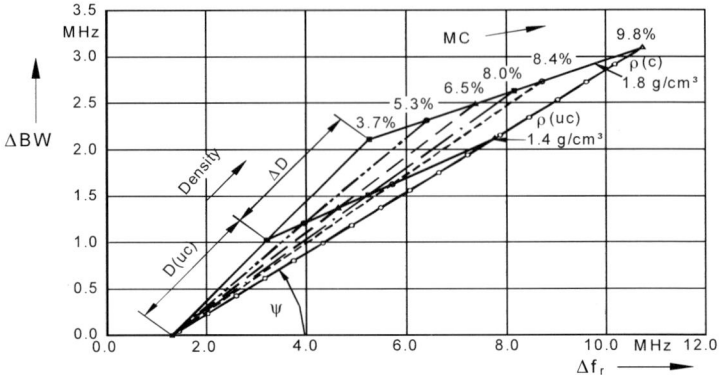

Fig. 7.36. Model of density-independent measurement using an open resonator system

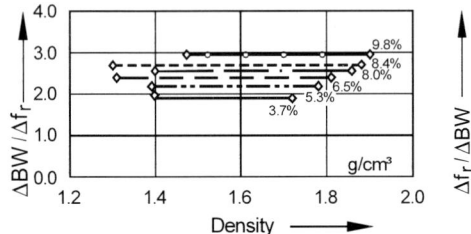

Fig. 7.37. The ratio $\Delta BW/\Delta f_r$ is constant versus density at different MC

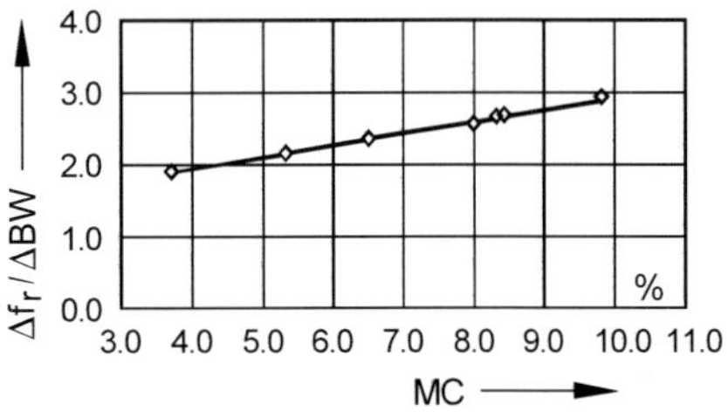

Fig. 7.38. The density-independent ratio $\Delta f_r/\Delta BW$ versus MC

The straight lines start on the Δf_r–axis at a value of the empty resonator and lead under the angle ψ to the values $\Delta BW\,[f_r;\rho\,(uc)]$ and $\Delta BW\,[f_r;\rho\,(c)]$ on the calibration curve. The angle ψ is a function of moisture:

$$\cot\psi = \frac{\Delta f_r(uc)}{BW(uc)} = \frac{\Delta f_r(c)}{BW(c)} = F(MC) \tag{7.34}$$

The ratio $\Delta BW/\Delta f_r$ is constant versus density at constant MC (Fig. 7.37). The inverse ratio $\Delta f_r/\Delta BW$ is a function of moisture content with a positive slope and is equivalent to the cotangens ψ (Fig. 7.38). The length of the straight line depends on the density:

$$D(uc) = \sqrt{\Delta BW^2(uc) + \Delta f_r^2(uc)} = F[\rho(uc)] \tag{7.35}$$

The density variation ΔC between the uncompacted and compacted values of calibration curves can be described with the following term:

$$\Delta D = D(c) - D(uc) = \sqrt{\Delta BW^2 + \Delta f_r^2} = F[\Delta\rho] \tag{7.36}$$

At a constant value of moisture content a density determination can be realised with the length of straight line using a suitable calibration.

7.4.6 Density-Independent Measurement with an Open Dielectric Resonator Pair

A further system was developed by the firm NUTech at a frequency of 2.45 GHz (Fig. 7.39). The resonance curves of two dielectric resonators are tuned symmetrically to one operation frequency ω_{op} (Fig. 7.40)[43]. The density-independent measurement can be realised by application of differences of frequency shift and quality factors (Eq. (7.37)). This system is based on the operation mode of a discriminator.

Fig. 7.39. Moisture measurement system with a symmetric resonator pair [44]

The need of the symmetric resonator pair and their required equal characteristics are disadvantages of the system. Furthermore moisture gradients between the different measuring spots cause essential measuring errors. The following equations are valid in the neighbourhood of resonance frequencies $\Delta\omega << \omega_{01} \; ; \; \omega_{02}$. The amount of the amplitudes u_i at the frequency $\omega = \omega_{op}$ can be recorded by a peak detector.

$$\Delta\varepsilon_r''(\omega_{op}) \propto \left.\frac{u_{01} - u_{02} - (u_1 + u_2)}{u_{01} + u_{02}}\right|_{\omega=\omega_{op}} \qquad \Delta\varepsilon_r'(\omega_{op}) \propto \left.\frac{u_1 - u_2}{u_{01} + u_{02}}\right|_{\omega=\omega_{op}} \qquad (7.37)$$

Fig. 7.40. Two resonance curves symmetrically to the operating frequency ω_{op}; the influence of $\Delta\varepsilon_r'$ causes frequency shifts $\Delta\omega$, the increase of the dielectric losses $\Delta\varepsilon_r''$ reduces the quality-factors and increases the band-width.

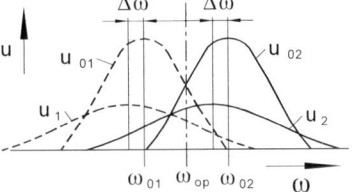

For the accurate determination of the real and imaginary part of the permittivity the following limitations of validity have to be taken into account:

-    Negligible asymmetry of resonance curves U_{01} and U_{02}
-    Application of loose coupling between resonator and material under test
-    Other resonance modes of resonators should be in sufficiently far frequency distance
-    The temperature coefficients (TC) of resonators have to be equal

These demands require the limitation of the measuring range and/or the measuring sensitivity.

7.4.7 Density-Independent Measurement Using Frequency Swept-Transmission

The Meyer-Schilz equation Eq. (7.12) was investigated for a mono-frequency density-independent measurement for different materials from Menke and Knöchel; its limitations were shown.

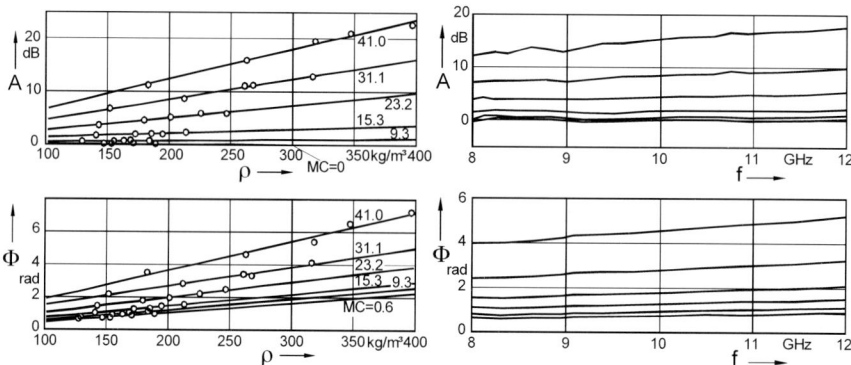

Fig. 7.41. Attenuation A and phase shift Φ versus density

Fig. 7.42. Attenuation A and phase shift Φ versus frequency

They used broadband frequency swept-transmission measurements to realize differences of the frequency characteristics of ε_r'' and $\varepsilon_r' - 1$ or A and Φ in order to increase sensitivity versus MC. Inspections of the curves A and Φ indicated that density and layer thickness influence A and Φ in a similar manner like frequency changes would do (Figs. 7.41, 7.42).

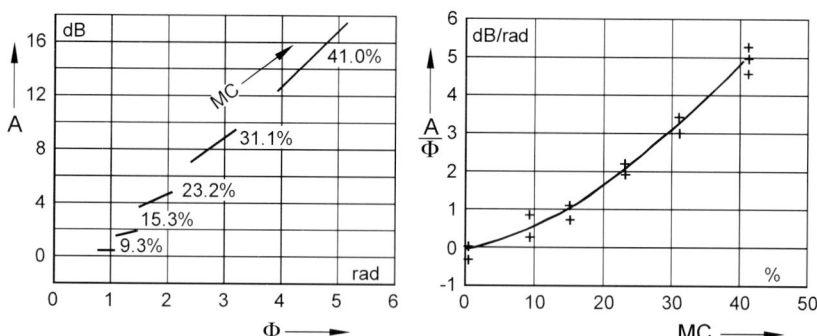

Fig. 7.43. Attenuation versus phase shift from tobacco; parameter: $\psi\,(MC)$

Fig. 7.44. Gradient of A-Φ-straight lines of tabacco versus moisture content

Using the ratio A/Φ versus frequency a density-independent measurement method could be developed over the frequency range of the X-band [45, 46].

$$X_f = \frac{\Delta A(\omega)/\Delta\omega}{\Delta\Phi(\omega)/\Delta\omega} \tag{7.38}$$

Figures 7.41 and 7.42 show that attenuation A and phase shift Φ depend obviously linearly on density and frequency. The slope of straight lines versus frequency increases with growing moisture content (Fig. 7.43). The slope of the calibration curve is smoothed and a saturation at high moisture values does not exist (Fig. 7.44). The measurement accuracy could be improved, especially at high moisture content values. The model was demonstrated for different materials with a frequency-swept microwave transmission technique in X-band. Compared to Eq. (7.12) the moisture range could be expanded. The calibrations have been performed not only for one material but also for material groups. The following formula uses the frequency dependence of the ratio A/Φ for evaluation. It gives the extension of density-independent measurement in the range of high moisture contents which is suitable for on-line calibration:

$$X_{\partial f} = \omega\frac{d}{d\omega}\left(\frac{A}{\Phi}\right) = \omega\frac{A}{\Phi}\left(\frac{1}{A}\frac{dA}{d\omega} - \frac{1}{\Phi}\frac{d\Phi}{d\omega}\right) \tag{7.39}$$

Using these measuring methods the phase measurement is greatly simplified because of phase tracking versus frequency. In contrast the integral phase through a material layer can only be measured in the range $\pm\pi$.

7.4.8 Microwave Free-Space Technique at Two Frequencies

A method using microwave free-space technique for density-independent moisture measurement at two frequencies was proposed by Zhang and Okamura and verified for timber and green tea [47, 48]. The ratio of phase shifts at two frequencies can be expressed as,

$$\frac{\Delta\Phi_2}{\Delta\Phi_1} = \frac{\sqrt{\varepsilon'_{r2}}-1}{\sqrt{\varepsilon'_{r1}}-1}\cdot\frac{\lambda_1}{\lambda_2} \tag{7.40}$$

where $\Delta\Phi_1$, $\Delta\Phi_2$ the phase shifts, ε'_{r1}, ε'_{r2} the dielectric constants, and λ_1, λ_2 the free-space wavelengths. The ratio λ_2/λ_1 is a constant. A function of dielectric properties at two frequencies η can be obtained:

$$\eta = \frac{\sqrt{\varepsilon'_{r2}(12\,GHz)}-1}{\sqrt{\varepsilon'_{r1}(9\,GHz)}-1} \tag{7.41}$$

The function dependent on density and moisture content was investigated at 9 and 12 GHz in a waveguide using samples of green tea (Fig. 7.45). Compared to the Meyer-Schilz function (Eq. 7.12) the function η is able to re-present a wider moisture range than the function $R = \Phi/A$, but the function η is less sensitive to the moisture content than the function R (Fig. 7.46).

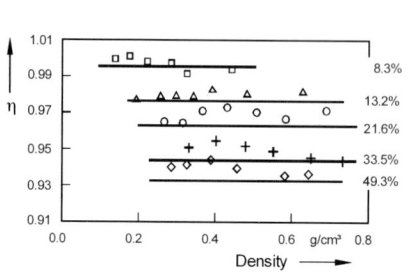 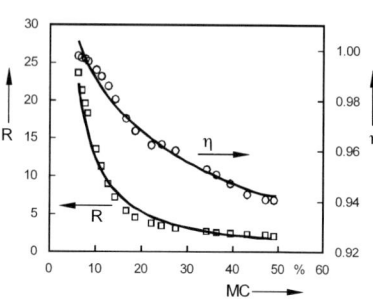

Fig. 7.45. The function η via dry density ρ of the sample

Fig. 7.46. Comparison of the function η to the function $R = \Phi/A$ (Eq. 7.12)

7.4.9 Methods of Density-Independent Measurement for Grain in the Microwave Range

The density-independent moisture measurement of soft red winter wheat at a frequency of 9 GHz was investigated by Kraszewski and Nelson [49]. The measuring values of attenuation and phase-shift have been normalised in the first case to the product of thickness t by density ρ. The calibration equation related to the moisture content may be obtained by solving the normalised values of attenuation and phase shift $(A/t\cdot\rho;\ \Phi/t\cdot\rho)$ (Eq. 7.42):

$$MC = \frac{109.9}{\left(\dfrac{\Phi}{A}\right)^{0.49}} \qquad (7.42) \qquad MC = \frac{440}{\dfrac{\Phi}{A}} + 7.4 \qquad (7.43)$$

Another way of presenting the experimental results is to correlate the ratio Φ/A with the given moisture content for every density (Eq. 7.43). In the first case (Eq. 7.42) a standard error of prediction SEP ≈ 0.26 % MC could be achieved, in the second case (Eq. 7.43) was SEP ≈ 0.30 % MC. The relationship between the moisture content determined by the standard oven method and the moisture content of the grain sample of arbitrary density determined from the microwave measurements could be expressed in the form:

$$MC_{oven} = 1.012MC - 0.028 \quad r = 0.9949 \qquad (7.42a)$$

$$MC_{oven} = 0.996MC + 0.032 \quad r = 0.9932 \qquad (7.43a)$$

Using the mass per unit area related values A/m_A, Φ/m_A, and the ratio of A/Φ the density-independent measurement at 3 and 9.3 GHz has been applied successfully also during the analysis of the processing technology of calcium silicate brick production by Kupfer [50].

7.4.9.1 Artificial Neural Network

An artificial neural network (ANN) was used for determination of moisture content of hard red winter wheat by Bartley, Nelson et. al. [51]. The ANN was trained to recognize moisture content (wet basis) in the range 10.6 to 19.2 % MC from transmission measurements of wheat samples of varying bulk densities (0.72 to 0.88 g/cm^3) at 8 frequencies in the range 10 to 18 GHz. The ANN has three layers: the input layer, one hidden layer and the output layer. Each layer consists of nodes or neurons (Fig. 7.47). The number of neurons in the input layer was determined by the number of inputs. Each of the transmission coefficients contains two parts – an amplitude and a phase shift. This results in 16 inputs. The number of neurons in the output layer was one, the moisture content. The number of neurons associated with this hidden layer was determined by training the network for varying numbers of neurons. It was established that 15 hidden nodes delivered the best prediction with a mean absolute error of 0.135.

The 179 available samples were divided into a training set of 70 measurements, a test set of 48 measurements and a production set of 61 measurements. The training set was used to train the network. The test set was used in the feed-forward mode only during training to determine the optimal point of training stop in order to avoid overfitting the training set. The trained ANN was evaluated with the production set. The effectiveness of applying the trained network to the production set of measurements is shown in Table 7.1.

Table 7.1. Statistical results of processing the production set by trained ANN

	Transmission coefficient	Amplitude transmission coefficient	Permittivity
R^2	0.993	0.982	0.992
Mean square error	0.028	0.073	0.032
Mean absolute error	0.135	0.219	0.142

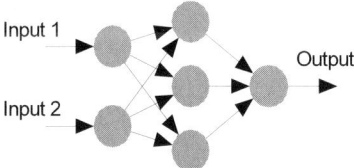

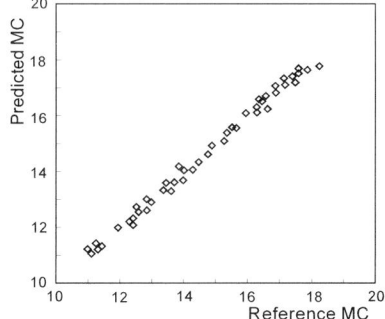

Fig. 7.47. Symbol of the ANN

Fig. 7.48. Predicted MC by trained ANN compared to reference MC of the oven-dry method

There is a strong correlation between the measurements and the moisture content provided by the trained ANN (Correlation coefficient $R^2 = 0.99$). Figure 7.48 shows the result obtained when all the data (training, test, and production sets) were applied to the trained ANN. The resultant mean absolute error from transmission coefficient measurement was 0.135 % MC compared to the oven dry method. The mean absolute error increased to 0.219 % MC when the amplitude of the transmission coefficient measurements were used as inputs of the ANN. In this case the need for phase measurement and the connected hardware effort could be eliminated. The same level of accuracy (0.142 % MC) could be obtained with the calculated values of permittivity. These values are independent of measurement technique and sample holders. It could be established that by using an ANN a further way exists for density-independent moisture measurements.

7.4.9.2 Argand-Diagram According to Trabelsi, Kraszewski and Nelson

Recently Trabelsi, Kraszewski and Nelson have described a new density-independent function (see Chap. 18) [52]. In the Cole-Cole-Diagram the values of $\varepsilon_r^{''}$ and $\varepsilon_r^{'}$ have been normalized to the density. At increasing moisture content and temperature the locus or the values of imaginary part versus real part of permittivity at different frequencies show linear dependence. The measuring values for increasing moisture content have the same slope as the temperature (Fig. 7.49). This will be declared for a higher mobility of water molecules at increasing temperature and results in larger values $\varepsilon_r^{''}$ and $\varepsilon_r^{'}$. The linear relationship between the normalized components of the relative permittivity can be expressed as:

$$\frac{\varepsilon_r^{''}}{\rho} = a_f \left(\frac{\varepsilon_r^{'}}{\rho} - k \right) \qquad (7.44)$$

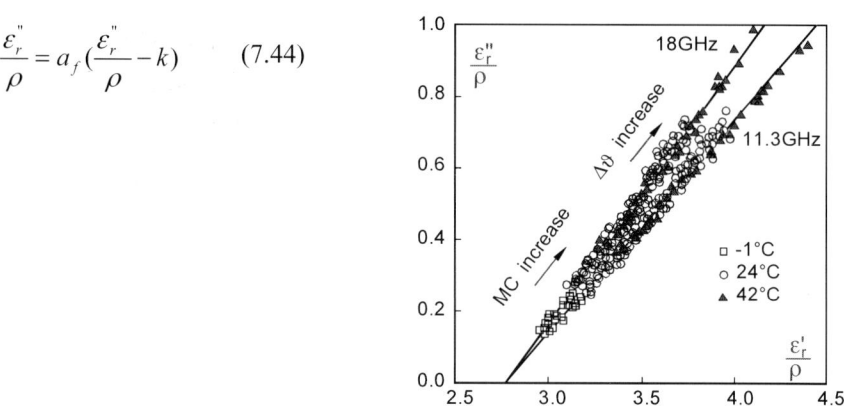

Fig. 7.49. Argand-Diagram of the relative complex permittivities normalized to the bulk-density for hard red winter wheat at 11.3 and 18 GHz [52]

The loss tangent varies with bulk-density (Eq. 7.45 and Fig. 7.50), where a_f (slope) is a linear function of frequency, and $k \cdot a_f$ is a constant for a given material and a given frequency f.

It was developed the following density-independent function (Eq. 7.46 and Fig. 7.51):

$$\frac{\tan \delta}{\rho} = k \cdot a_f \left[\frac{\varepsilon_r^{''}}{\varepsilon_r^{'} (a_f \varepsilon_r^{'} - \varepsilon_r^{''})} \right] \tag{7.45}$$

$$\xi = \frac{1}{a_f \varepsilon_r^{'} - \varepsilon_r^{''}} \left(\frac{\varepsilon_r^{''}}{\varepsilon_r^{'}} \right) \tag{7.46}$$

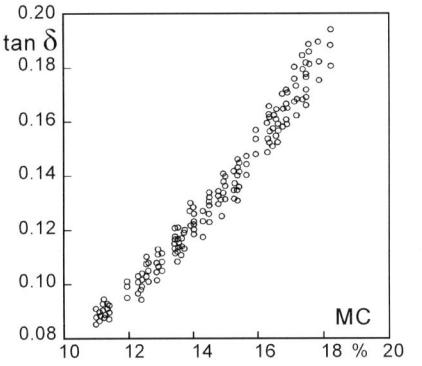
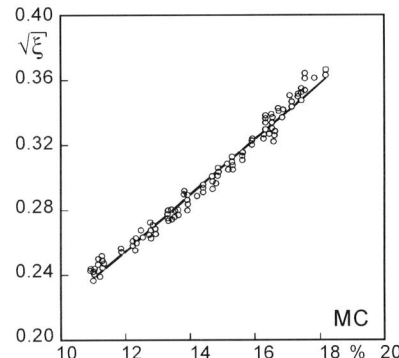

Fig. 7.50. Variation of *tan δ* versus moisture content at 14.2 GHz and different densities

Fig. 7.51. Variation of the square root of *ξ* versus moisture content at 14.2 GHz and different densities [52]

7.4.9.3 Comparison of Density-Independent Functions

Nelson compared the density-independent functions Eqs. (7.12, 7.13 and 7.46) in Fig. 7.52 [53].

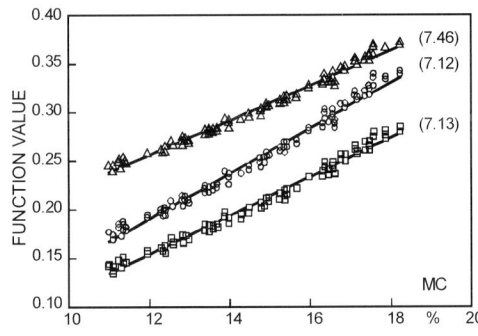

Fig. 7.52. Moisture dependence of the density independent functions at 14.2 GHz for hard red winter wheat [53]

The results showed that all three functions evaluated with microwave measurements on hard red winter wheat over frequency ranges from 11 to 18 GHz, moisture content 11 to 19 %, and bulk-densities 0.72 to 0.88 g/cm³, were equivalent for density-independent moisture measurements with standard errors of performance of about 0.3 % MC.

Comparison of density-independent expressions for moisture content in wheat at microwave frequencies were published by Kraszewski [54] and Trabelsi [55].

A density-independent measurement system for grain, wood flakes and different bulk goods was developed and is produced by King [56]. The application of the density-independent Meyer-Schilz function could be realised by Volgyi during the monitoring of particleboard production [57].

7.5 Summary

During moisture determination in the RF and microwave ranges the measured complex value of permittivity will be mainly influenced by the moisture content, by density variations, and salt content. Different investigations were carried out in order to reduce the influence of density variations on the measuring results. The application of one-parameter methods required small or constant density variations or the compensation of their effects e.g. the arrangement of moisture sensor in the flowing material stream. A constant material thickness can be realized by using a stripper plate, which is installed in the material stream of a conveyor belt. The microwave measuring value, influenced by density variations, can be compensated by a parallel density measurement with gamma-rays. Basic investigations were performed by Klein with the microwave set-up "MicroMoist". The microwave values of attenuation and phase shift will be compensated with the mass per unit area determined by a parallel gamma-density gauge.

In the RF and microwave range two basic electrical methods exist: simultaneous measurement of two electrical parameters of the substance at one single frequency or the simultaneous measure-ment of one or two electrical parameters at different frequencies. The moisture measurement can be carried out with a scalar measurement by recording the parameter of a resonance curve frequency shift and bandwidth or with vector measurement methods by using the transmission values of attenuation and phase shift. The first density-independent equation was developed by Kraszewski using a vector measuring unit. In the eighties further basic progress made by Meyer and Schilz. The density-independent equations were derived from the perturbation theorie. The measuring parameters attenuation and phase shift, which included the values of complex permittivity, were recorded with different transmission applicators and a vector network analyser in the X-band. Different cavity resonators were also applied for scalar measurements. The scalar measuring parameters frequency shift and bandwidth served for the calculation of the density-independent measuring value at a single frequency.

Different models were developed by Kent, Kupfer and Trabelsi to draw the locus of imaginary and real part of permittivity or the measuring values attenuation and phase shift or change of bandwidth and frequency shift. By using these diagrams it was possible to depict the calibration curves, the density-independent values and the values for density determination. Many different investigations for a density-independent moisture measurement of grain were made by Nelson, Kraszewski, Lawrence and Trabelsi in the RF and microwave ranges. Lawrence has applied the multifrequency method for density-independent measurements of grain in the RF range. Menke and Knöchel developed a density-independent model using the ratio A/Φ as a function of frequency. The model was demonstrated at different materials with a frequency swept technique in X-band. Compared to Eq. (7.12) the moisture range could be expanded. A further method enabled the extension of density-independent measurement in the range of high moisture contents. It could be applied for on-line calibration. Different models were also presented by using newer mixing rules. The calibration for density-independent functions could be extended by using artificial neural netwoks. The variety of methods results from the multitude of developments for different applications. Each material requires connected with the relating measuring device the evidence of density independent moisture measurement. It does not exist a unique method for universal applications. Some points of view for optimising criterions are: application of a nondestructive measurement method e. g. in high-speed production processes or in structural analysis of building physics, transmission measurement or sensing from one side, penetration depth, moisture and density ranges, accuracy, repeatability, expenses.

References

1. Kupfer K (1990) Feuchtemessung an Zuschlagstoffen für die Betonherstellung unter Verwendung der Mikrowellenmeßtechnik. Hochschule für Architektur und Bauwesen Weimar / Baustoffverfahrenstechnik, Diss. A
2. Kalinski J (1979) "Einige Probleme der industriellen Feuchtemessung mit Mikrowellen". *TIZ (Fachberichte)* 103 pp. 145-153
3. Kalinski J, Rakowski J (1984) "On-Line Measurements of Material Quality by Microwaves".-*Proc. of the Int. Symp. on Metrology for Quality Control in Prod.* pp 94-99, Tokyo
4. Kraszewski A (1988) Microwave Monitoring of Moisture Content in Grain – Further Considerations. Journ. of Microwave Power and Electromagnetic Energy, 23 (4), p 236
5. Kupfer K (2000) "Radiofrequency and Microwave Moisture Sensing of Building Materials". *Sensors Update Vol. 7, RF&Microwave Sensing of Moist Materials, Food and other Dielectric,* Wiley-VCH Verlag Weinheim / New York, Guest Editors: Kupfer, Kraszewski, Knöchel
6. Kupfer K et. al. (1997) "Genauigkeitsanforderungen an Feuchtemeßsysteme bei der Betonherstellung". *Technisches Messen* 64, pp 433-439
7. Kay-Ray (1982) "Accu-sense" on-line noncontacting moisture measurement system. Data sheet USA
8. Klein A (1981) "Comparison of Attenuation and Phase Measurement". *Journ. of Microwave Power* 16, pp 289-304

9. Berthold (1989) "MICROMOIST LB 354" Manual Bad Wildbad

10. Kupfer K (2000) "Microwave Moisture Measurement Systems and their Applications". *Sensors Update Vol. 7, RF&Microwave Sensing of Moist Materials, Food and other Dielectric,* Wiley-VCH Verlag Weinheim / New York, Guest Editors: Kupfer, Kraszewski, Knöchel

11. Chope RH (1960) Method and apparatus for measuring multiple properties of material by applying electric fields at multiple frequencies and combining detection signals. US Patent 3,155,898, 3 November 1964

12. Stuchly S, Kraszewski A (1965) "Method for the determination of water content in solids, liquids and gases by means of microwaves and arrangement for application of this method", *Polish Patent,* 51,731

13. Kraszewski A, Kulinski S (1976) "An improved Microwave Method of Moisture Content Measurement and Control". *IEEE Trans. on Ind. Electr. and Contr. Instr.* IECI 23, pp 364-370

14. Mlodzka-Stybel A (1990) "Practical Verification of the Microwave Two-Parameter Method of Moisture Monitoring in Grain in Harvest Time". *Proc. of the 20th Europ. Microwave Conference,* pp 1679-1682 Budapest

15. Stuchly S, Hamid M (1972) "State of the art in microwave sensors for measuring non-electrical quantities". *Int. Journ. Electronics* 33, pp 617-633

16. Meyer W, Schilz W (1980) "A microwave method for density independent determination of the moisture content of solids". *J. Phys. D: Appl. Phys.,* 13, pp 1823-1830

17. Meyer W, Schilz W (1981) Feasibility study of density-independent moisture measurements with microwaves. IEEE Trans. On MTT 29; 7 pp 732-739

18. Meyer W, Schilz W (1979) Verfahren zur Messung der relativen Feuchte eines Messgutes mit Hilfe von Mikrowellen im GHz-Bereich. DE 2928487 Anmeldetag 14. 7. 79

19. Jacobsen R, Meyer W, Schrage B (1980) "Density independent moisture meter at X-band". *Proc. of the 10th EuMc,* pp 216-220, Warschau

20. Meyer W, Schilz W (1982) "High Frequency Dielectric Data on Selected Moist Materials". *Journ. of Microwave Power* 17, pp 67-77

21. Kent M, Meyer W (1982) A density-independent microwave moisture meter for heterogeneous foodstuffs. J. Food Eng. 1, pp 31-42

22. Nelson SO (1983) "Observations on the Density Dependence of Dielectric Properties of Particulate Materials". *Journ. of Microwave Power* 18, pp 143-152

23. Vainikainen PV, Nyfors EG, Fischer MT Radiowave sensor for measuring the dielectric properties of dielectric sheets: Application to veneer moisture content and mass per unit area measurement. *IEEE Trans.Instr.Meas.* IM-36(4), pp 1036-1039

24. Kent M, Kress-Rogers E (1986) "Microwave moisture and density measurements in particulate solids". *Trans. Inst. MC* 8, pp 161-168

25. Kress-Rogers E, Kent M (1987) Microwave Measurement of Powder Moisture and Density. Journ. of Food Eng. 6, pp. 345-376

26. Meyer W, Schilz W (1982) "Microwave measurement of moisture content in process material *Philips techn. Review* 40, pp 112-119

27. Hoppe W, Meyer W, Schilz W (1981)Vorrichtung zur Feuchtemessung mit Hilfe von Mikrowellen. DE 2942971 A1 Anmeldung 24. 10. 81

28. Herrmann R (1997) "Mikrowellen-Feuchtemessung mit Resonatoren und ihre Anwendungen". *Materialfeuchtemessung,* Renningen-Malmsheim expert-Verlag, Editor K. Kupfer

29. Zoerb GC, Moore GA, Burrow RP: Continuous measurement of grain moisture content during harvest. *Trans. Of the ASAE* 36(1), pp 5-9

30. Powell SD et.al. (1988) "Use of a Density-Independent Function and Microwave Measurement System for Grain Moisture Measurement". *Trans. of the ASAE* 31, pp 1875-1881
31. Nelson SO (1984) "Density dependence of the dielectric properties of wheat and whole-wheat flour". *Journal of Microwave Power* 19, pp 55-64
32. Lawrence KC, Nelson SO (1993) "Radio-frequency density independent moisture determination in wheat". *Trans. Of the ASAE* 36, pp 477-483
33. Lawrence KC (1997) "Density-independent multiple-frequency technique for measuring moisture content in grains with a radio-frequency permittivity sensor". *Ph.D. Dissertation*, University of Georgia, Athens, Georgia
34. Lawrence KC, Windham WR, Nelson SO: "Wheat Moisture Determination By 1- to 110 MHz Sweptfrequency Admittance Measurements". *Trans. of the ASAE* 41, pp 135-142
35. Berbert PA, Stenning BC (1996) Analysis of Density-independent Equations for Determination of Moisture Content of Wheat in the Radiofrequency Range. J. agric. Engng. Res., Vol. 65, pp 275-286
36. Berbert PA, Stenning BC (1996) On-line Moisture Content Measurement of Wheat. . J. agric. Engng. Res., Vol. 65, pp 287-296
37. Lawrence KC, Nelson SO (2000) Radifrequency sensing of moisture content in cereal grains. *Sensors Update Vol. 7, RF&Microwave Sensing of Moist Materials, Food and other Dielectric,* Wiley-VCH Verlag Weinheim / New York, Guest Editors: Kupfer, Kraszewski, Knöchel pp 377-390
38. Heck B, Hohenstein N, Schröder D (1994) "Verfahren zur dichteunabhängigen kapazitiven On-line-Messung des Wassergehaltes fester Stoffe". *Technisches Messen* 61 S 421-428
39. Kupfer K (1996) "Possibilities and Limitations of Density-Independent Moisture Measurement with Microwaves" Chapter 21 pp 313-327. *Microwave Aquametry*"; New York, IEEE Press Book-Series, Editor A. Kraszewski
40. Kent M (1989) Application of two-variable microwave techniques to composition analysis problems. Trans Inst MC Vol. 11 No. 2; April-June, pp 58-62
41. Kent M (2000) Simulteaneous determination of Composition and Other Materials by Using Microwave Moisture Sensors. *Sensors Update Vol. 7, RF&Microwave Sensing of Moist Materials, Food and other Dielectric,* Wiley-VCH Verlag Weinheim / New York, Guest Editors: Kupfer, Kraszewski, Knöchel
42. Kupfer K (1999) Methods and Devices for Density-independent Moisture Measurements. Proc. on 3. Workshop on Electromagnetic Wave Interaction with Water and Moist Substances; Athens GA April, pp 11-19
43. Stang G Verfahren und Vorrichtung zur Messung der Dielektrizitätskonstante von Probenmaterialien. Patentschrift DE 43 42 505 C1
44. Datasheets Microwave resonator sensor. Keller GmbH Ibbenbühren-Langenbeck
45. Menke F, Knöchel R (1996) New Density-IndependentMoisture Measurement Methods using Frequency swept Microwave Transmission. IEEE MTT-S Digest 1996 Vol. 3, pp 1415-1418
46. Menke F (1998) Zerstörungsfreie Feuchtemeßverfahren mit Mikrowellen Fortschrittsberichte VDI Reihe 8 Meß-, Steuerungs- und Regelungstechnik Nr. 690; VDI Verlag Düsseldorf
47. Zhang Y, Okamura S (1999) "New Density-independent Moisture Measurement Using Microwave Phase Shifts at Two Frequencies", IEEE Transactions on Instrumentation and Measurement, vol. 48, (6), pp1208-1211.
48. Zhang Y, Okamura S (2000) "Moisture content measurement for green tea using phase shifts at two microwave frequencies", Subsurface Sensing technologies and Applications, vol. 1, (4), pp129-136.

49. Kraszewski AW, Nelson SO (1991) Density-independent moisture determination in wheat by microwave measurement. *Trans. of the ASAE* Vol. 34, pp 1776-1783
50. Kupfer K, Klein A (1992) Experiments on the Suitability of Microwave Measuring Techniques for Moisture Measurement in Calcium Silicate Brick Production. Mineral processing 33 (4), pp. 213-221
51. Bartley Ph, Nelson SO, McClendon RW, Trabelsi S (1998) "Determining Moisture Content of Wheat with an Artificial Neural Network from Microwave Transmission Measurements". *Trans. on Instr. and Meas.* Vol. 47, pp 123-126
52. Trabelsi S, Kraszewski AW, Nelson SO (1998) "A Microwave Method for On-line Determination of Bulk Density and Moisture Content of Particulate Materials. *Trans. on Instr. and Meas.* Vol. 47, pp 127-132
53. Nelsson SO, Trabelsi S, Kraszewski AW (1998) "Advances in Sensing Grain Moisture Content by Microwave Measurements".*Trans. of the ASAE* Vol. 41, pp 483-487
54. Kraszewski AW, Trabelsi S, Nelson SO (1998) "Comparison of Density-independent Expressions for Moisture Content Determination in Wheat at Microwave Frequencies". *J. agric. Engng Res.* 71, pp 227-237
55. Trabelsi S, Nelson SO (1998) "Density-independent functions for on-line microwave moist meters: a general discussion" *Meas. Sci. Technol.* 9, pp 570-578
56. King R (2000) On-line industrial Applications of Microwave Moisture Sensors. *Sensors Update Vol. 7, RF&Microwave Sensing of Moist Materials, Food and other Dielectric,* Wiley-VCH Verlag Weinheim / New York, Guest Editors: Kupfer, Kraszewski, Knöchel pp 109-170
57. Volgyi F (2000) Monitoring of Particleboard Production using Microwave Sensors. *Sensors Update Vol. 7, RF&Microwave Sensing of Moist Materials, Food and other Dielectric,* Wiley-VCH Verlag Weinheim / New York, Guest Editors: Kupfer, Kraszewski, Knöchel pp 249-274

8 Microwave and RF Resonator-Based Aquametry

Alexander S. Sovlukov

Institute of Control Sciences, 65, Profsoyuznaya str., Moscow, 117997, Russia

8.1 Introduction

Among known measurement methods used for determination of various nonelectrical quantities, certain advantages are inherent to radiofrequency (RF) and microwave techniques [1–14]. RF and microwave methods are most widely used for measurement of water content in various liquid, gaseous, particulate and solid substances. Such methods-based moisture and humidity measuring devices are very high sensitive to water content in the appropriate frequency bands and provide possibilities of accurate metering, contactless and averaged through measuring volume measurements. Theory and applications of known RF and microwave moisture/humidity measuring devices are contained in numerous publications (monographs [1–7], papers, patents). Certain RF/microwave resonator-based moisture/humidity meters are proven to be very effective devices. Because physical processes in moist materials are described in said publications in detail, here will we considered principles of design and some applications of known and new RF and microwave resonator-based sensors and measuring devices for water content determination in various substances.

8.2 Traditional and New Types of Resonator Sensors and some of their Applications for Microwave and RF Aquametry

8.2.1 Basic Types of Resonator-Based Moisture/Humidity Sensors

Microwave measurement methods imply the use of various transmission lines, resonators, and radiating elements as sensors. They provide accurate contact and distant measurements. A variety of RF and microwave resonator sensors exist. Multiple constructions of traditional RF/microwave resonator sensors are suggested for aquametry. They are built from basic RF/microwave components or modified and used as sensors. The latter are specially constructed for the solution of definite measuring tasks. The oscillation characteristics (resonance frequency f, quality factor Q) of a resonator sensor are changed depending upon the moisture/humidity of a substance in the electromagnetic field of such a resonator.

A moist substance can be placed within a RF or microwave (coaxial, two-wire, stripline, cavity, etc.) resonator, outside it in the fringing field or in some other way in order to provide interaction with the electromagnetic field and thus change the value of a sensor parameter (f, Q). Such sensors for resonator aquametry are described in the literature (for instance, in monographs [1–7]).

8.2.2 New Types of Resonator Sensors and some of their Applications in Aquametry

On-line determination of water content in various liquid, solid, gas and particulate substances is needed in many industrial applications. Microwave sensors are considered effective for contactless moisture measurement of substances in stationary conditions, transported on conveyor belts, or in pipelines [1–7]. However their sensitivity to moisture and/or measurement accuracy are not enough in some applications. Besides, if non-homogeneous moist gas, liquid, particulate substances are monitored, in many cases known sensors don't provide independent measurement results for the random distribution of these non-homogeneities. In tasks where the position of a monitored moist object is not stable, which is often true for production of sheet-like (paper, etc.) materials, then the output characteristics of known microwave sensors used for such purposes are dependent on this disturbing factor, resulting in a decrease of measurement accuracy. New principles for microwave sensor design are considered here, which allow us to realize sensors without the above drawbacks.

8.2.2.1 Waveguide-Based Resonators with Separate Interaction of Oppositely-Directed Waves with Moist Substances

Waveguide resonators are used in particular as sensing elements in traditional microwave resonator aquametry. A monitored object is located in such a sensor in the standing wave field disturbing initial electromagnetic field distribution. However in some applications principally achievable measurement accuracy or/and sensor sensitivity are non-satisfactory. These drawbacks result from nonuniform field distribution along the resonator length: the position of a monitored object in the resonator field is not often exactly known or may change during monitoring process.

Possible methods for the design of microwave sensors are considered here, as well as measuring devices based on them that give ability to solve the listed problems. They have higher sensitivity or/and accuracy compared with existent ones. Design of the considered microwave resonator sensors can be realized using passive microwave components for controlling the propagation characteristics of guided electromagnetic waves (direction of propagation, amplitude, phase, polarization) and, as a result, of sensor output characteristics. Advantages of such sensors are: improved measurement accuracy; increased sensitivity to measured moisture/humidity; invariance of measuring results to the influence of different disturbances. For the control of sensor output characteristics, various functional microwave components (circulators, directional couplers, phase shifters, etc.) can be applied. These microwave components, as well as waveguides, are designed and manufactured for the use in certain, rather broad frequency bands.

With the proposed microwave techniques, significant improvement of functional characteristics (measurement accuracy, sensor sensitivity) as compared with existent analogous ones can be realized. The improvements are achievable due to the design of these sensors on the base of waveguide-based resonators where *separate* interaction of oppositely-directed travelling electromagnetic waves with monitored moist substances takes place.

In such sensors are performed the following functions: interaction of only one-directed travelling wave with a monitored object; said interaction of both such waves separately; different degree of propagation characteristics change for oppositely directed waves under said interaction. Phase shifts of waves in such sensors determine values of informative parameters that can be much more sensitive to measured quantities, in particular to moisture/humidity. In order to provide said interaction of waves with an object at least one non-reciprocal component (circulator, directional coupler) is introduced into the sensor construction. Similar schemes may also be listed for separate interaction of both oppositely-directed waves. For contactless (on-line) measurements, the proposed waveguide/resonator sensors contain aerial antennae. Some of the schemes listed above are considered here in detail and are described for measurement of moisture. Examples of schemes for certain applications based on similar approaches are presented in [15].

The appropriate interaction schemes are considered here using initially linear waveguide resonators (Fig. 8.1). A monitored object is sounded separately at least by one of the oppositely-directed travelling waves.

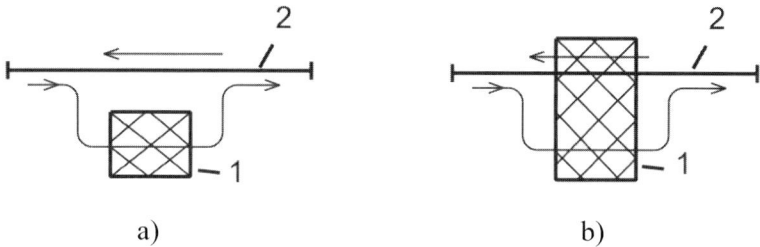

Fig. 8.1. Schemes for separate one-direction (**a**) and two-direction (**b**) interaction of waves in waveguide resonator with a monitored object (examples)
1 – monitored object, *2* – initial waveguide resonator

For the provision of wave–object interaction at least one of the oppositely-directed waves is lead out from the resonator and returned there again after the completion of such an interaction; Fig. 8.1a shows such a scheme for a single path interaction. Object 1 is a dielectric one, possibly with losses; multiple interactions of waves with objects in such schemes are also possible; there can be various modifications of these interaction schemes depending on the tasks to be solved.

Also, separate interaction of both oppositely directed waves with an object 1 is possible (Fig. 8.1b).

In general, using microwave scheme components allows the behavior of each of the oppositely-directed waves to be controlled. In particular, both the waves that are initially oppositely-directed may be guided in one direction, providing nevertheless their separate interaction with a monitored object.

Resonator devices with controlling passive components for guided waves may be designed according to the wave–object interaction schemes in Fig. 8.1. They are applicable for on-line moisture determination. Here we consider only some examples of such devices. Figure 8.2a shows the scheme of the device that corresponds to the interaction scheme in Fig. 8.1a. A moist substance 1 is sounded by travelling, one-direction waves excited in the waveguide resonator. Three-port circulators 3 are used for the provision of such a sounding. They are installed into the waveguide 2 along its length. One of them (left) is used for leading one of the oppositely-directed waves out of the resonator while the other one (right) is used for input of the wave again into the resonator after interaction with the substance. In particular, radiating (left) and receiving (right) antennas 4 are used as sensitive elements. The resonator is formed here by the plurality of microwave components: resonator, circulators, the other parts of wave trajectories containing the sensitive elements. Oscillator 5 is connected to the resonator; electronic unit 6 is used for measurement of resonance frequency. Waves reflected from surfaces of a substance 1 and propagating at the opposite direction, don't really influence measurement results. They circulate within the ring circuit restricted by circulators. Such a scheme provides measurement results independent of position and/or sizes of a substance relative to antennas.

Microwave resonator devices with controlling passive components for guided waves may be designed according to the wave–object interaction schemes in Fig. 8.1. They are useful for on-line moisture determination. Here we consider only some examples of such devices.

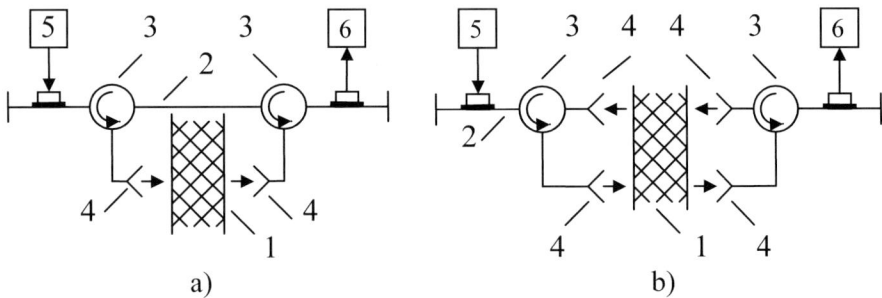

a)   b)

Fig. 8.2. Schemes of microwave measuring devices with separate one-direction (**a**) and two-direction (**b**) wave–object interaction (examples)
1 – monitored object, *2* – waveguide, *3* – circulator, *4* - antenna, *5* – oscillator, *6* – measuring unit

The scheme of the microwave device corresponding to the interaction scheme in Fig. 8.1b is shown in Fig. 8.2b. A moist substance 1 is probed separately by both oppositely-directed waves in the resonator. Wave–object interaction is realized with the application of sensitive elements that are two pairs of transmitting and receiving antennas 4. Also other types of sensitive elements may be applied. The direction of the wave propagation is shown by arrows. Two circulators 3 serve for provision of separate interaction of travelling waves with an object 1. Like the scheme in Fig. 8.2a, here also waves reflected from the surfaces of a moist substance 1 really don't influence the measurement results. These waves circulate in the opposite direction within the circuit restricted by circulators. Such a scheme, being similar to the scheme in Fig. 8.2a, is characterized by twice sensitivity to a measured quantity. The position of an object between antennas in the direction of sounding doesn't influence the measurement results.

The natural (resonance) frequencies of the synthesized resonator are determined by phase relationships according to trajectories of waves, as well as by other characteristics of guided waves, and by the dielectric/geometrical parameters of moist substances.

The resonator structures considered above are neither more linear nor more ring-shaped than previously examined structures. Instead of them there exist now more complicated resonator structures. They may be called linear-ring resonators.

The described schemes of wave–object interaction in a waveguide resonator allow broader application areas for microwave sensors because of the possibility of contactless measurements, and the increase of sensitivity using multiple soundings of an object separately by one and both of the oppositely-directed waves. By varying scheme parameters and choosing an informative parameter, these schemes may be optimized taking into account the specifics of a solved measurement problem and the needed sensitivity. In particular the natural (resonance) frequency of oscillations may be chosen as the informative parameter of a microwave sensor.

Resonance Frequencies Versus Parameters of Monitored Objects

The resonance frequencies of resonator sensors can be used as informative parameters. Their values can be estimated using a simplified approach assuming TEM-approximation – free-space propagation of transverse electromagnetic (TEM) waves and position of a monitored object in designed resonators.

The natural (resonance) frequencies of the synthesized resonator are determined by phase relationships according to the trajectories of wave propagation, by other characteristics of guided waves, and by dielectric/geometrical parameters of moist substances.

The scheme in Fig. 8.1a is considered as an example (other schemes can be described in a similar way). For this case the following relationship for resonance conditions can be written:

$$\beta_0 (2l - \Delta l + l_s - l_m) + \beta\, l_m = 2\pi n \qquad n = 1,\, 2,... \qquad (8.1)$$

where β_0 is the propagation factor for waves in the initial resonator 2 and in a path (free-space, waveguides) outside it; for simplification it is assumed that its value is the same in the designed resonator: for a free-space path $\beta_0 = 2\pi f_r (\varepsilon_w)^{1/2}/c$, for a waveguide path $\beta_0 = 2\pi[\varepsilon_w - (f_c/f_r)^2]^{1/2}/c$, where f_r and f_c are resonance frequency and cut-off frequency in the waveguides, respectively, ε_w is relative dielectric permittivity of a substance in a wave pathway (free-space, waveguides), c is velocity of light; $\beta = 2\pi f_r [\mathrm{Re}(\varepsilon_m(W)]^{1/2}/c$ is propagation factor in a monitored substance 1 with moisture W; l is the length of the initial resonator 2; Δl is the length of the part of the resonator 2 between points of leading waves out of this resonator and introducing them into it again; l_s is sum length of waveguide pathway outside of the initial resonator; l_m is pathway length of waves in a substance 1; ε_m is relative dielectric permittivity of a substance 1.

For lossy dielectrics

$$\varepsilon_m = \varepsilon_m' + j(\varepsilon_m'' + \frac{\sigma_m}{2\pi f \varepsilon_0}) \tag{8.2}$$

where ε_m' and ε_m'' are the real and imaginary parts of ε_m. The value of ε_m'' corresponds to dielectric losses due to displacement currents, σ_m is conductivity of a substance; f is frequency; ε_0 is permittivity of vacuum (8.85×10^{-12} F/m).

From Eq. (8.1) follows the expression for natural (resonance) frequency f_r, assuming free-space wave propagation ($f_c = 0$):

$$f_r = \frac{nc}{(2l - \Delta l + l_s - l_m)\sqrt{\varepsilon_w} + l_m(\mathrm{Re}\langle\sqrt{\varepsilon_m}\rangle)}, \quad n = 1, 2, ... \tag{8.3}$$

By measuring f_r, the value of ε_m and W may be determined. From Eq. (8.3) it follows the formula for the change of frequency f_r relative to its initial value f_{r0} follows:

$$\frac{\Delta f_r(W)}{f_{r0}} = \frac{f_r(W) - f_{r0}}{f_{r0}} = -\frac{l_m(\mathrm{Re}\langle\sqrt{\varepsilon_m}\rangle)}{nc} f_r(W) \tag{8.4}$$

here f_{r0} is initial (at $l_m = 0$ or $\varepsilon_m = \varepsilon_w$) value of $f_r(W)$.

The following data can be used in Eq. (8.3) as a numerical example of the scheme considered: $l = 20$ cm, $\Delta l = 10$ cm, $l_s = 20$ cm, $l_m = 10$ cm, $\varepsilon_w = 1$, $\varepsilon_m = 2$. Then it follows: $f_r = 0.6n$ GHz. So if $n = 5$ then $f_r = 3.0$ GHz. For the change of ε_m to the value $\varepsilon_m = 2,5$ due to water content, the corresponding relative change of f_r is $\Delta f_r/f_r = 1.63$ (%). Such a change of ε_m corresponds to the range $W = 0 - 8.3$ % if water content in oil or oil products (with $\varepsilon_m = 2$ at $W = 0$) is determined.

8.2.2.2 *RF TEM Line-Based Resonator Sensors*

Devices based on RF sensors are rather simple and cost-effective to implement [1]. They are effectively used for the measurement of various technological parameters. In particular RF sensors are applicable in the building industry [16, 17]. RF resonator sensors for contactless characterization of various materials can be synthesized on the base of specially designed sections of two-wire lines, striplines, etc. Specifically, elements of technological installations may be effectively used as RF sensors.

Lattices of metal rollers are used in some installations for conveying manufactured items. Such a lattice is shown in Fig. 8.3. Two neighboring rollers 1 and 2 isolated from others may be considered as a two-wire TEM transmission line. Being short-circuited on both ends such a construction represents a RF resonator that is a half-wave TEM-line section.

Water content in a material on the lattice can be determined by exciting electromagnetic oscillations in this line section and measuring the value of an informative parameter (resonance frequency, quality factor). Consider the following lattice for transportation of clay mass in the production of bricks: distance between rollers 50 mm, length of a roller 400 mm, total length of the table with rollers 600 mm, its height 800 mm. Contactless coupling elements may be placed at the short-circuited ends of the line section in order to excite electromagnetic oscillations and receive information for further processing.

If resonance frequency f_r is used as informative parameter for determination of water content W, then

$$f_r = \frac{cn}{2l\sqrt{\varepsilon_{eff_0}}} \tag{8.5}$$

where c is speed of light, l is length of a roller, $n = 1,2, \ldots$ is number of harmonics excited in the resonator, $\varepsilon_{eff\,0}$ is effective dielectric permittivity of two-layered (air and coatings on wires) substance. If no coatings are on the wires then $\varepsilon_{eff} = 1$.

For a half-wave TEM-line section $n = 1$ and $f_r = 375$ MHz. This value of f_r also provides the ability to receive averaged moisture data. However, if a substance with rather high water content is monitored then quality factor Q of such a resonator may be inadequate for registration of the informative parameter. There are a few methods to increase Q: covering both the conductors of such two-wire resonator with dielectric coatings; providing a gap between a monitored substance and the rollers; or both methods.

Several consequently connected rollers of the lattice isolated from the others may also be used for the synthesis of a TEM-line section. A conductor with zigzag form represents one of two conductors of the TEM-line; a nearby metal plate may serve as the other conductor. Both ends of this line section may be short-circuited. The resonance frequency of this resonator is thus decreased, and the depth of electromagnetic wave penetration into a moist substance and degree of moisture data averaging are appropriately increased.

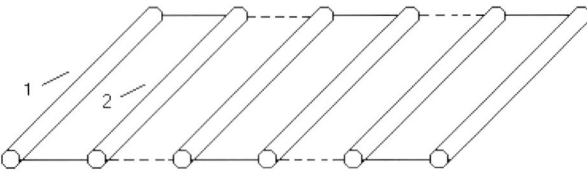

Fig. 8.3. Lattice made from metal rollers
1 and *2* – rollers as conductors of two-wire TEM transmission line

For contactless RF measurements several ways of placing a RF TEM transmission line-based sensor and a monitored material are possible. Figures 8.4a and 8.4b show the one-sided and two-sided contactless position of conductors 1 and 2 of two-wire RF-sensors relative to a monitored object 3. A monitored object is placed so that it provides effective influence on the electromagnetic field of such a sensor. As shown here, it is often valuable to cover conductors with dielectric coatings. It allows us to increase the quality factor of TEM line resonator sensors [1]. Various types of RF resonator sensors may be applied; their choice depends on the measuring problem to be solved.

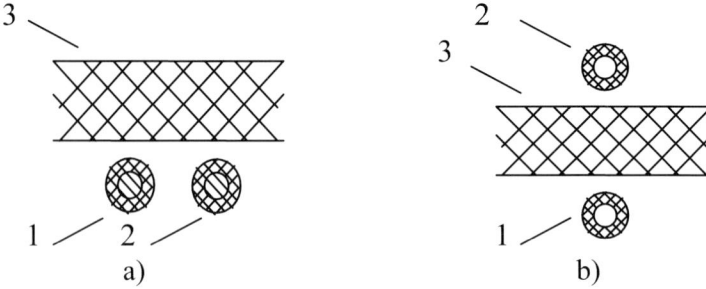

Fig. 8.4. One-sided (**a**) and two-sided (**b**) position of two-wire line conductors relative to a monitored object; *1* and *2* – conductors of two-wire transmission line, *3* – monitored object

Figure 8.5 presents one possible scheme for contactless on-line RF materials characterization. Here a two-wire RF transmission line section short-circuited at both ends is used as sensitive element. Characterized material 1 is placed within the area between the conductors of resonator 2. RF-oscillator 5 and measuring unit 6 (providing determination of resonance frequency f) are connected to resonator 1 via coaxial cables through coupling elements 3 and 4 that are inductive loops (in this simplified scheme we don't show matching elements for connecting coaxial cables with the two-wire line). Such a line section is half-wave resonator. The initial (without a material) resonant frequency f_0, used as the informative parameter, is expressed by Eq. (8.5).

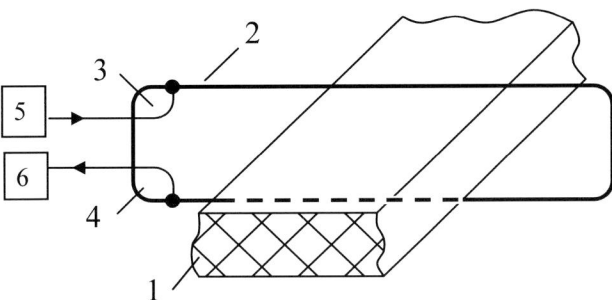

Fig. 8.5. RF-device containing two-wire resonator sensor
1 – monitored material, *2* – two-wire resonator, *3* and *4* – coupling elements,
5 – RF-oscillator, *6* – measuring unit

If a material is present between the conductors then for determination of the appropriate value of the resonance frequency f, the value of ε_{eff0} in Eq. (8.5) should be replaced by ε_{eff}. Air is substituted by the real layered substance that is composed of wire coatings, a monitored material and air.

For example, if production of building materials (bricks) is considered and water content W in clay mass is determined [16] then we get the following data for a moist clay mass between the conductors of resonator 2. If f_0 = 300 MHz (l = 0.5 m), air gaps are 10 mm, then f = 0.251f_0 = 75.3 MHz. If density of the dry substance (brick) is 1.2 g/cm^3 at temperature t = 20 – 60°C, brick dimensions are 24×12×6 cm^3 (clay volume is 1602.4 cm^3), produced bricks are porous, number of pores is 19, then for W = 25%, f/f_0 = 0.223, that is f = 67 MHz.

Figure 8.6 shows the calculated dependence of f/f_0 versus moisture W = 22 – 28 % (real values) at f_0 = 300 MHz. Here lines 1 and 2 are related to flows of continuous clay mass (without gaps between line conductors and this clay mass) and of porous clay mass (also without said gaps); line 3 is received for porous mass flow if gaps of 10 mm are present on both sides of this clay mass. From these data it follows the rather high sensitivity of the RF resonator moisture sensor.

For characterization of flowing materials in pipelines special TEM line-based resonator sensors are used [1, 18]. Their conductors are placed on the surface of the inner dielectric layer of a pipeline within the measuring section. No sensor parts are inside the pipeline. The electromagnetic field of such a resonator sensor with sufficient sensor sensitivity provides reliable determination of characterized material parameters. Distribution of the electromagnetic field intensity is uniform through the cross-section of the measuring section of a pipeline. It follows that various parameters (concentration of two-component substances, including in particular continuity of liquid-gas flows, etc.) of non-uniform flows can be measured with high accuracy. Measuring results are independent from distribution of flow components within the measuring section.

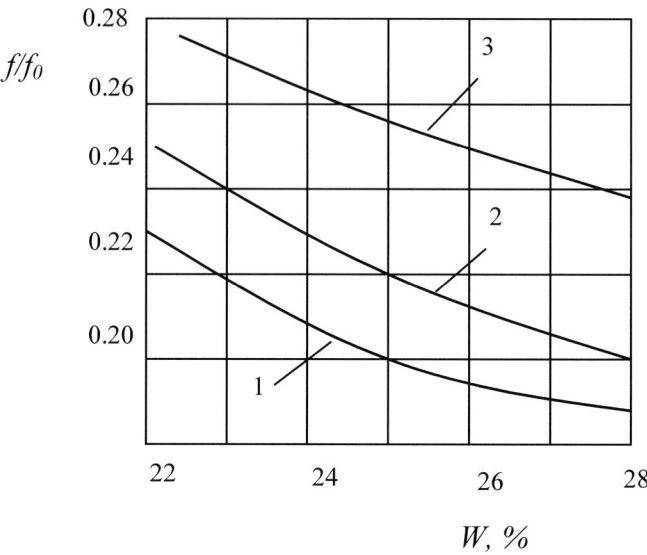

Fig. 8.6. Dependence of resonance frequency on clay moisture

8.2.2.3 Multiple-Probe Moisture Sensors for Measurements in Pipelines

Microwave moisture measurements in pipelines are often made using the sounding flow of a substance in the direction transverse to the flow [2, 3]. Transmitting and receiving antennas are connected to the pipeline. A drawback of such an approach is low measurement accuracy caused by a small layer of a monitored substance through which the microwave beam is propagated. In pipelines of large diameter (tens of centimeters), sounding of only a part of flow cross-section results in appropriate error. This is typical problem for flows of oil, oil products and chemicals. This error is significant especially for monitoring of non-uniform flows, in particular gas-liquid flows where distribution of gas inclusions is random. Use of the metal pipeline itself as a waveguide sensor [2, 5] doesn't allow for increased measurement accuracy because of non-uniform field distribution in the sensor's cross-section.

The use of microwave sensors with multiple probes can significantly increase measurement accuracy and sensitivity to measured variables, in particular to moisture of a substance. Simultaneous multiple probing of a flowing substance results in a significant increase of wave–substance interaction that is characterized by growth of phase shift, and power absorption of waves propagated through a substance.

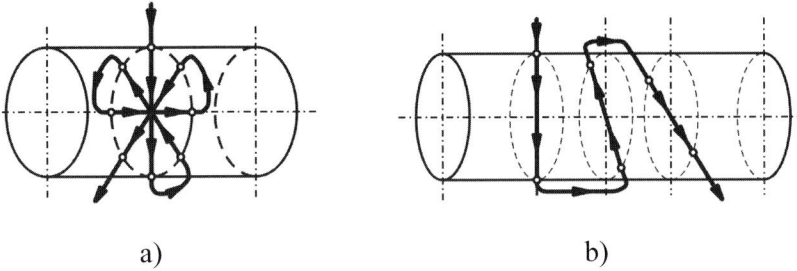

a)   b)

Fig. 8.7. Schemes of multiple soundings of a flowing substance in a pipeline through one
(**a**) and various (**b**) cross-sections

A scheme of multiple soundings of a flowing substance in a pipeline through its
cross-section plane using waves at different angles is shown in Fig. 8.7a. In particular such k soundings ($k = 2,3, \ldots$) may be at equal angles $360°/k$.

Multiple sounding of a flowing substance means increasing the path length by
k-times for waves propagated through the monitored substance. Now path length
is $d_k = kd$ (d is inner diameter of a pipeline, k is the number of soundings) in the
cross-section of this pipeline. The value of a used informative parameter is also
increased by k.

Significant decrease of non-monitored areas in the cross-section of a pipeline
results in the averaging of moisture data. An advantage of this scheme in comparison with the single sounding scheme is that it gives measurements in narrow substance layer of the pipeline cross-section. The thickness of such a layer is determined by directivity diagrams of the antennas. In order to avoid interaction of
electromagnetic waves under different angles (if it takes place), sounding of flow
at different angles may be done in various sections along pipeline, transversely to
flow direction in each such section (Fig. 8.7b).

A device with multiple soundings of a moist substance in a pipeline can be realized using k pairs ($k = 2,3, \ldots$) of transmitting antennas 2a, 2b, ..., 2k and corresponding receiving antennas 3a, 3b, ..., 3k that are connected to the pipeline. Each
preceding antenna is connected to the next one by an appropriate waveguide. Excitation of waves is done by an oscillator. Waves received after multiple propagation
through the cross-section of the pipeline are registered. It contains appropriate units
depending on chosen informative parameter (amplitude, phase shift, etc.).

A sensitivity increase follows also from analytical consideration. If the
informative parameter is the power or the amplitude of received waves then the
loss factor N (in decibels) is

$$N = 8.686\alpha_w W\rho\gamma d + /R/ \qquad (8.6)$$

where $/R/$ is module of reflection coefficient from interface "antenna-substance",
d is the thickness of a substance layer, that is the inner diameter of a pipeline in

this case, γ is the empirical constant taking into account structure of a substance, ρ is the density of the moist substance, α_w is the loss factor for water [7].

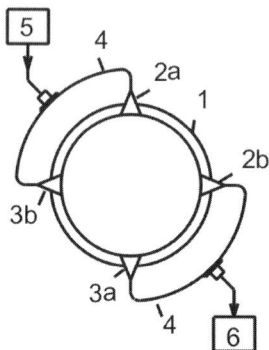

Fig. 8.8. Microwave device with multiple soundings of a moist substance in a pipeline (example)

It is seen that N is proportional to the thickness of the substance layer. Wave path length for $k = 2,3, \ldots$ soundings is kd; appropriately is increased the value of N.

Based on the above-described schemes, it is possible to design appropriate devices with resonator sensors. For this purpose the output of the last receiving antenna 3k is connected with input of antenna 2a. The resonator oscillations are excited in the cross-section of pipeline 1. Figure 8.8 shows an example of such a resonator moisture-measuring device. Antennas 2a, 2b and 3a, 3b here are transmitting-receiving. Pairs 2a, 3b and 3a, 2b are connected by waveguides 4 and are used for flow sounding in the direction transverse to the flow. Oscillator 5 and register 6, used to measure the resonance frequency of the synthesized resonator are also shown.

The natural (resonance) frequency f_n of electromagnetic oscillations of such a resonator can be expressed as:

$$f_n = \frac{n/k}{(d\sqrt{\varepsilon_m(W)})/c + l/v_{ph}} \qquad (8.7)$$

ere $\varepsilon_m(W)$ is the dielectric permittivity of a substance with moisture W; l is the length of each waveguide 4 (it is admitted here equal for all the waveguides but this condition is not principal); v_{ph} is the phase velocity of waves in the waveguides 4; $n = 1.2,\ldots$; $k = 2.3,\ldots$. So, if $d = 0.3$ m; $l = 0.1$ m; $k = 2$ and $\varepsilon_m = 2$, then $f_n = 0.3n$ GHz (n is the number of the type of oscillations). It follows from Eq. (8.7) that the frequency change Δf_n of frequency f_n due to the change $\Delta \varepsilon_m(W)$ can be written as:

$$\Delta f_n(W) = \frac{f_n \Delta \varepsilon_m(W)}{2n^2 ck\varepsilon_m^{3/2}} \qquad (8.8)$$

For instance it follows from this relationship that for $\Delta\varepsilon_m(W) = 1$, $n = 4$, $f_n = 1.2$ GHz we receive $\Delta f_n = 40$ MHz (~ 3.3 % of f_n); such a frequency change is practically enough for its registration. These data correspond in particular to the change of moisture in oil or oil products with $\varepsilon_m = 2$ within the range $W = 0 - 16.7$ %.

The considered approach is applicable for moisture measurements in pipelines of different diameters and those made from various materials (metal, dielectric, layered material, etc.). Areas of possible application may cover transportation of oil, oil products, chemicals, etc.

8.2.2.4 Split-Cavity Moisture Sensors for Sheet-Like Materials

The use of proposed split-cavity sensors provides both contactless monitoring of needed parameters of sheet-like materials and also independence of measurement results on a sheet (web) position within the split in a direction that is transverse to the direction of movement [17].

For this purpose uniform transverse distribution of electromagnetic energy must occur in this area (Fig. 8.9a). In the proposed resonator on the base of a rectangular waveguide 1, this approximately uniform distribution occurs (Fig. 8.9a shows the distribution of electric field amplitude E). Dielectric slabs 2 and 3 with thickness d and permittivity ε_s are placed on the opposite broad sides of the rectangular waveguide cross-section along its length. A TEM-mode field exists within the free central part of the resonator cross-section of this waveguide with initial mode H_{10} [19, 20].

The electromagnetic wave is transversal (TEM) if the following condition is met for the length λ of the electromagnetic wave for plane wave operation:

$$\lambda = 4d\sqrt{\varepsilon_s - 1} \tag{8.9}$$

The nearly constant distribution of electromagnetic energy in the waveguide free space may be considered as the one present in the same substance within the split area.

Such an approach appeared effective in microwave heating applicators, which are used particularly in biology and medicine [21, 22]. Various dielectrics may be used as slabs: plexiglass ($\varepsilon_s = 2.59$), stycast ($\varepsilon_s = 7$), alumina ($\varepsilon_s = 10.07$), etc. So, for a waveguide with cross-section dimensions ≈ 7 by 3.5 cm and slabs with permittivity $\varepsilon_s = 7$ and thickness $d = 1.3$ cm, the needed operation conditions with TEM-field in the central area were provided at frequency 2450 MHz ($\lambda = 12.45$ cm) [21]. Uniform field distribution in this area is expected to be approximately the same if λ is not significantly changed due to a change of oscillator frequency or insertion of a dielectric object into the free space slightly perturbing the electric field.

A cavity resonator on the base of such a waveguide may serve as a sensor for measurements of some parameters, in particular those of moving sheet-like materials. Uniform energy distribution in such a waveguide 1 and in the resonator sensor based on it results in independence of measurement results on a sheet position in transverse direction in the central part.

Figure 8.9b shows a split-cavity sensor containing a moving dielectric material 4 within the split. Dielectric slabs 2 and 3 are in both parts of the cavity.

The resonator is supplied by cut-off waveguides formed by metal sheets 5 and 6 at the free edge on each half of the split-cavity.

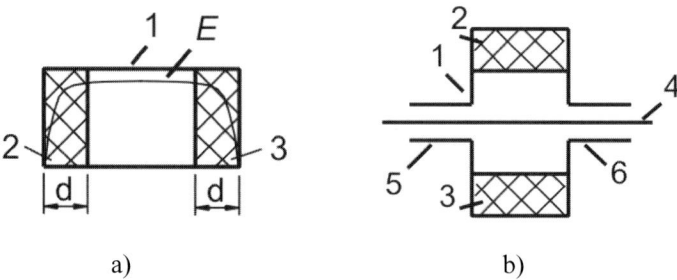

a)   b)

Fig. 8.9. Waveguide with TEM-mode field in its central part (**a**) and split-cavity sensor with a sheet within the split (**b**)
1 – waveguide, *2* and *3* – dielectric slabs, *4* - sheet-like material, *5* and *6* – metal sheets

Location of a sheet material within the split along the electric field component results in maximum sensor sensitivity. The same uniformity of energy distribution remains when the sheet thickness as well as its moisture W and density are changed, provided the sheet thickness a is small as compared with the resonator height l ($a \ll l$).

It can be shown that the dependence of resonance frequency f_r on sheet thickness a, permittivity ε_m is expressed as:

$$\frac{f_r}{f_{r0}} \approx 1 - \frac{a(\varepsilon_m(W) - 1)}{l} \tag{8.10}$$

here f_{r0} is initial value of f_r (in the absence of the sheet), l is length of the broad side of the split.

The value of ε_m depends on the density and the moisture of a material. Each of these parameters can be measured if sheet thickness a is constant. Measurements of a and/or W and also mass per unit area M may be done by measuring both resonance frequency f_r and amplitude A of the resonance pulse, or two (or more) resonant frequencies of various resonator modes, and by their subsequent functional transformation.

These measurements are needed in particular in pulp and paper industry. For example, a moving sheet may have width 4300 mm and have drift within ± 10 mm transverse to movement direction. When several resonator sensors are disposed across the monitored sheet profile of a, W and/or M distribution can be determined. This determination can be made simultaneously if more than one informational parameter is measured. In practice 10–20 split-cavity resonators can be installed in papermaking machines for manufacturing broad paper sheets (4300 mm). It may be done by measuring both resonance frequency f and amplitude A of resonance pulse and by their subsequent functional transformation. Thus current values of *a, W* and *M* may be found.

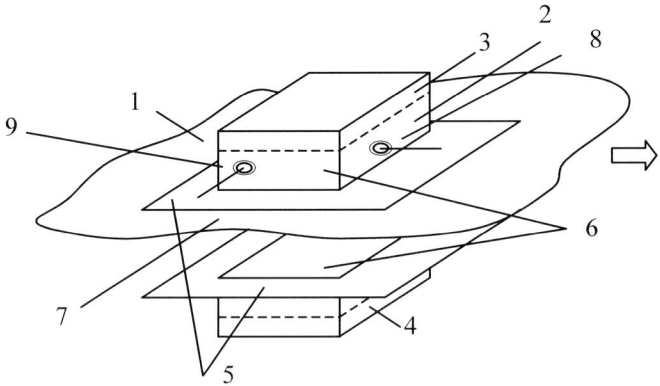

Fig. 8.10. Split-cavity sensor with a sheet within the split
1 – sheet-like material, *2* - resonator, *3* and *4* – dielectric slabs, *5* - metal sheet,
6 - open surface, *7* – split, *8* and *9* – coupling elements

Such an approach is also useful for invariant measurements when independence of measurement results of each of these parameters from other disturbing variables is required. Similar results are also obtained by measuring two (or more) resonant frequencies of various resonator modes. This approach is also applicable when independence of measurement results of each of these parameters from other disturbing variables is needed.

Figure 8.10 shows a split-cavity sensor containing a moving dielectric material 1 within the split of the resonator 2. Dielectric slabs 3 and 4 are contained in both parts of the split cavity. For on-line contactless measurements of moving sheet materials 1 this resonator is supplied by cut-off waveguides. They are formed by metal sheets 5 at the free edge on each half of the split-cavity resonator 2. Thus electromagnetic energy radiation from the cavity is prevented. Location of a monitored sheet material between open-ended surfaces 6 of the cavity 2 within the split 7 along the electric field line directions results in maximum sensitivity of the resonator-based sensor. The same energy distribution uniformity remains when sheet thickness, moisture and density are changed, if the measured sheet thickness *a* is small as compared with the resonator height *l* ($a \ll l$). Coupling elements 8 and 9 serve for excitation of electromagnetic oscillations in the cavity and its connection with an electronic unit measuring the value of the informative parameter (resonance frequency *f*).

8.2.2.5 *Waveguide Resonator Moisture Sensors for Flowing Substances*

A similar approach based on the use of dielectrically-loaded waveguides can be used for the design of resonator sensors for characterization of non-homogeneous dielectric substances in pipelines. In this case rectangular (Fig. 8.11a) or elliptical waveguide cavities can be synthesized. Waveguide 1 of resonator 5 contains dielectric slabs 3 and 4 at its broad walls and dielectric tube 2 with flowing material along its length. Thus independence of resonance frequency from distribution of inclusions (air bubbles, solid particles, etc.) in substances is provided. Tube 2 may have an exterior metal coating, which acts as cut-off waveguide relative to the waveguide cavity; sections of the metal pipelines 8 of the same diameter may also serve as such cut-off waveguides (Fig. 8.11b). Longitudinal twisting (90° or more as it is shown in Fig. 8.11c) of the resonator can be used for further accuracy improvement. This occurs due to spatial averaging of measurement results because of the angular change of the field pattern. Coupling elements 6 and 7 are used for connection of the sensor with the oscillator and electronic unit.

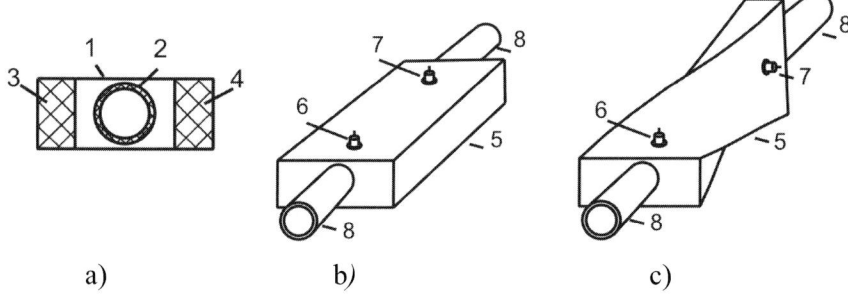

a)   b)   c)

Fig. 8.11. Rectangular waveguide (cross-section) with dielectric slabs and tube (**a**); waveguide resonator sensor (**b**); twisted waveguide resonator sensor (**c**)
1 – waveguide, *2* – dielectric tube, *3* and *4* – dielectric slabs, *5* – waveguide resonator, *6* and *7* – coupling elements, *8* – pipeline section (cut-off waveguide)

8.3 Dielectric Permittivity/Density-Independent Resonator-Based Aquametry

During the determination of water content in various substances, the accuracy of measurement results may be significantly decreased by the influence of non-measurable parameters such as density, temperature, type, etc. Therefore it is very important to provide moisture measurements that are independent of the influence of non-measurables. Within microwave moisture measurement techniques some effective density-independent methods and means are known applicable for many actual areas [1–3, 10, 23–26].

Here we consider the microwave method that provides moisture measurement results independent of the dielectric permittivity of a monitored substance [27]. In turn this permittivity is functionally dependent on the type of a monitored substance or on various parameters (density, type, quality, etc.) of a certain substance.

Measurement Principles

This method is based on the fact that electrophysical parameters of water are frequency-dependent in the microwave frequency range. In particular, the dielectric permittivity ε_w of water is decreased with the increase of frequency f. Dependence $\varepsilon_w(f)$ is shown in Fig. 8.12. So, if $f = 16$ GHz then $\varepsilon_w = 42.5$; if $f = 37.5$ GHz, then $\varepsilon_w = 21.8$ (for the temperature 20°C). Therefore the permittivity ε of a moist substance is dependent on the value of ε_w for various microwave frequencies. Conversely, frequency dispersion absence is known for dielectric permittivity ε_s values of many solids and non-polar liquids (with the uncertainty of $2 \cdot 10^{-4}$) at the frequency range $10^{-2} - 10^{11}$ Hz [28].

This behavior of ε_w is the basis for water content W determination independently of the value of ε_s. In turn we assume the independence of non-measured parameters of a substance functionally connected with ε_s.

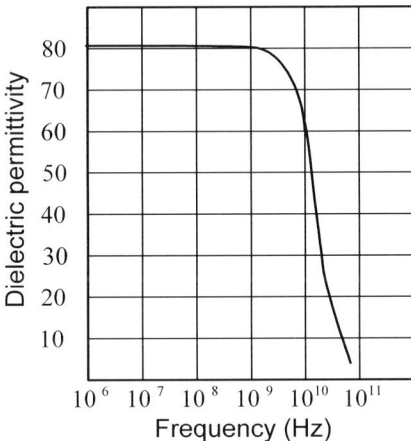

Fig. 8.12. Dielectric permittivity of water versus frequency

According to the proposed method, measurements of water content W are required at two frequencies f_1 and f_2. These frequencies are in ranges where ε_w has different corresponding values ε_{w1} and ε_{w2}. Values $\varepsilon_1(\varepsilon_s, \varepsilon_{w1}, W)$ and $\varepsilon_2(\varepsilon_s, \varepsilon_{w2}, W)$ of the permittivity of a moist substance are measured by a microwave measuring method (known or to be designed) according to certain measurement conditions. The frequencies f_1 and f_2 may be chosen in the following ways: 1) frequency f_1 is chosen on the part of the curve $\varepsilon_w(f)$ where the considered dispersion is absent, in particular within the frequency range less than 100 MHz; frequency f_2 is chosen on the dispersion curve part being in microwave range (Fig. 8.13a); 2) both frequencies f_1 and f_2 correspond to the dispersion part of the curve $\varepsilon_w(f)$ (Fig. 8.13b). For some substances, like non-polar dielectrics, ε_s is non-changeable at frequencies f_1 and f_2. Joint processing of $\varepsilon_1(\varepsilon_s, \varepsilon_{w1}, W)$ and $\varepsilon_2(\varepsilon_s, \varepsilon_{w2}, W)$ and allows for the determination of water content W independently of ε_s.

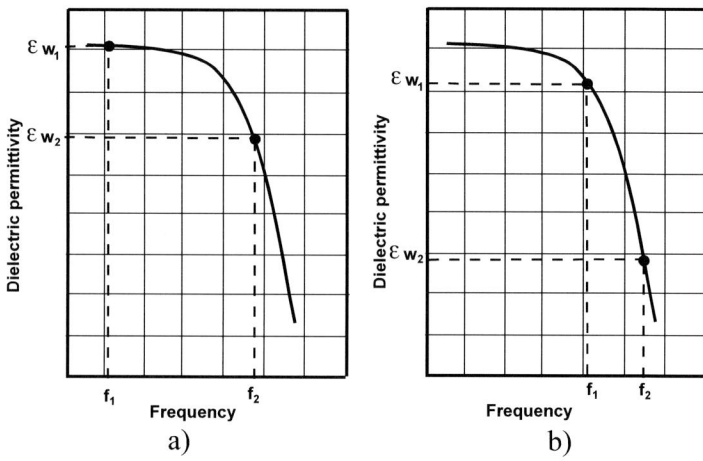

Fig. 8.13. Choice of two frequencies at the curve $\varepsilon_w(f)$

There is a need to know analytical dependence, $\varepsilon(\varepsilon_s, \varepsilon_w, W)$. If the dependence is not known for some substances it can be determined empirically and then with its approximation by an appropriate analytical expression.

Generally, the method is principally applicable for moist substances with various electrophysical parameters. Here application of the method is considered for some non-polar moist dielectrics, in particular water-in-oil emulsions. Their dielectric permittivity can be described analytically. As noted above, in the general case there is no requirement that the substance be a non-polar dielectric.

So, for some non-polar liquids (oil, oil products, etc.) at $W \leq 0.1$ it can be written [6]:

$$\varepsilon = \varepsilon_s \left(1 + \frac{3W}{D - W} \right) \tag{8.11}$$

where $D = \dfrac{\varepsilon_w + 2\varepsilon}{\varepsilon_w - \varepsilon_s}$.

Such an approach provides measurements of water content in a substance independently of ε_s. Taking into consideration expression Eq. (8.11), that is written for the frequencies f_1 and f_2, the system of these two equations is as follows:

$$\varepsilon(f_1) = \varepsilon_s \left(1 + \frac{3W}{D(f_1) - W}\right),$$ (8.12)

$$\varepsilon(f_2) = \varepsilon_s \left(1 + \frac{3W}{D(f_2) - W}\right)$$ (8.13)

here $D(f_1) = \dfrac{\varepsilon_w(f_1) + 2\varepsilon_s}{\varepsilon_w(f_1) - \varepsilon_s}$, $D(f_2) = \dfrac{\varepsilon_w(f_2) + 2\varepsilon_s}{\varepsilon_w(f_2) - \varepsilon_s}$.

The values $D(f_1)$ and $D(f_2)$ can be considered as constants for frequencies f_1 and f_2.

So, if measurements are undertaken at frequencies $f_1 = 10$ GHz and $f_2 = 37.5$ GHz, then corresponding values of D at $\varepsilon_s = 2$ are the following ones: $D(f_1) = 1.095$, $D(f_2) = 1.383$.

Equations (8.12) and (8.13) in the above system and its solution can be simplified under these conditions: 1) $D(f_1) - W \approx D(f_1)$, $D(f_2) - W \approx D(f_2)$; 2) these values don't depend on ε_s. These simplifications are permissible under small values of water content (up to $\cong 5\%$) and real relatively small limits of ε_s change.

Then after solution of the system of Eqs. (8.12) and (8.13) relative to W with exclusion of ε_s from the result, it can be found that:

$$W = \frac{1}{3} \frac{\varepsilon(f_1) - \varepsilon(f_2)}{\dfrac{\varepsilon(f_2)}{D(f_1)} - \dfrac{\varepsilon(f_1)}{D(f_2)}}$$ (8.14)

Various practical measurement methods are known providing determination of $\varepsilon(\varepsilon_s, \varepsilon_w, W)$ at two frequencies [1, 2]. Among them are methods for determination of phase shift, of reflection coefficient values, etc.

Values of $D(f_1)$ and $D(f_2)$ are constant at a fixed (measured) temperature for corresponding frequencies f_1 and f_2. It is so because the values of permittivities ε_s and ε_w are constant in formulae for $D(f_1)$ and $D(f_2)$. The value of ε_s is constant in a broad range of frequency f change. The value of ε_w is constant under changing f within the non-dispersive part of the curve $\varepsilon_w(f)$ and is considered constant on the dispersive part of this curve. This holds in the following cases:

– under measurements at fixed frequencies;
– under measurements of small values of water content using devices with changing frequencies, a situation often met in practice (in such cases f is changed non-significantly and, as a result the change of ε_w is very small and doesn't result in the error increase of water content determination over admittable value).

Change of temperature can be taken into account in Eqs. (8.12), (8.13) and (8.14), if the value of ε_w at a current (measured) temperature is used. Temperature change of ε_s doesn't significantly influence the values of $D(f_1)$ and $D(f_2)$.

It can be shown in practice that coefficients $3/[D(f_1) - W]$ and $3/[D(f_2) - W]$ at W in Eqs. (8.12) and (8.13) don't depend (with some admittable error) on ε_s. So, for a 10% change of ε_s compared with the initial value $\varepsilon_s = 2$, that is until the value

$\varepsilon_s = 2.2$, we obtain at $f_1 = 10$ GHz and $f_2 = 37.5$ GHz: $D(f_1) = 1.095$, $D(f_2) = 1.303$ for $\varepsilon_s = 2$; $D(f_1) = 1.105$, $D(f_2) = 1.337$ for $\varepsilon_s = 2.2$. It follows that the relative change of $D(f_1)$ is near 0.9%, and the relative change of $D(f_2)$ is near 2.6%.

It follows from these data that the relative change of both $3/[D(f_1) - W]$ and $3/[D(f_2) - W]$ is near 0.9% that is nearly ten times less than the relative change of ε_s. It means that real changes of ε_s don't practically influence on the above coefficients at W.

The considered method can also be realized using two resonator sensors (Fig. 8.14). Their informative parameters are resonant frequencies f_{r1} and f_{r2} that change depending on water content in a substance. Note that the dielectric permittivity of water $\varepsilon_w(f)$ is thus also changed on the dispersive part of this curve. Nevertheless such changes are known (because the current frequency of each resonator is known), as well as temperature-dependent change of ε_w (at a measured temperature). They can be taken into account in the computer unit of the resonator device.

For resonator measurements of water content independent of substance permittivity, the system of equations can be written:

$$\varepsilon_s\left(1 + \frac{3W}{D(f_{r1}) - W}\right) = \frac{f_{r1_0}^{\,2}}{f_{r1}^{\,2}}, \tag{8.15}$$

$$\varepsilon_s\left(1 + \frac{3W}{D(f_{r2}) - W}\right) = \frac{f_{r2_0}^{\,2}}{f_{r2}^{\,2}} \tag{8.16}$$

here f_{r1o} and f_{r2o} are initial (at $\varepsilon_s = 1$, $W = 0$) values of the resonant frequencies f_{r1} and f_{r2}, accordingly. Coefficients at W in Eqs. (8.15) and (8.16) for its small values are changed non-significantly with the change of frequency. Such changes in the general case are known and can be taken into account during computation of water content.

After solving this system of equations relative to W, we receive

$$W = \frac{1}{3} \cdot \frac{\dfrac{f_{r1_0}^{\,2}}{f_{r1}^{\,2}} - \dfrac{f_{r2_0}^{\,2}}{f_{r2}^{\,2}}}{\dfrac{1}{D(f_{r1})}\dfrac{f_{r2_0}^{\,2}}{f_{r2}^{\,2}} - \dfrac{1}{D(f_{r2})}\dfrac{f_{r1_0}^{\,2}}{f_{r1}^{\,2}}} \tag{8.17}$$

The value of W in Eq. (8.17) is independent of dielectric permittivity ε_s.

In the scheme for resonator measurements of water content in pipelines (Fig. 8.14) a monitored substance can flow through the basic pipe (shown) or through additional pipelines (bypasses).

Measuring resonators can be realized as constructions that allow free flowing through of monitored substances at both RF and microwave frequency ranges. Also they can have no parts inside the pipeline disturbing the flow. For example, such resonators can be open-ended with pipe parts at its both ends as cut-off waveguides; many other constructions can be used [2]. As it is shown in Fig. 8.14, resonators can be used as frequency-determining elements of corresponding self-excited oscillators. Output signals with frequencies f_1 and f_2 of these oscillators come to the computer unit for determination of water content independent of a substance permittivity.

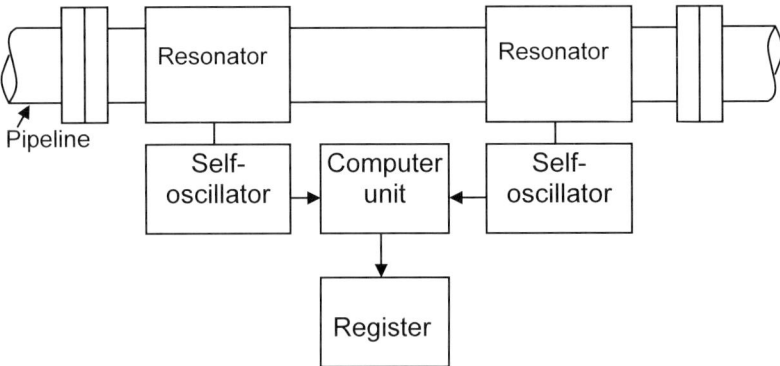

Fig. 8.14. Scheme for water content determination based on measurement of resonance frequencies of two resonators

8.4 Conclusion

The new design principles for RF and microwave resonator sensors of moisture/humidity considered here and the application of known RF/microwave resonator sensors in aquametry show that these techniques are applicable for highly accurate determination of water content in various liquid, solid, particulate, and gaseous substances. Many other technological parameters may also be effectively measured on the basis of such approaches.

References

1. Viktorov VA, Lunkin BV, Sovlukov AS (1980) High frequency method for measurement of non-electrical quantities, Nauka Publ., Moscow (in Russian)
2. Viktorov VA, Lunkin BV, Sovlukov AS (1989) Microwave measurements of technological processes parameters, Energoatomizdat Publ., Moscow (in Russian)
3. Nyfors E, Vainikainen P (1989) Industrial microwave sensors, Artech House, Norwood, MA
4. Zoughi R (2000) Microwave non-destructive testing and evaluation, Kluwer Academic Publ., the Netherlands
5. Benzar VK (1974) Techniques for microwave moisture metering, Vysheyshaya shkola Publ., Minsk (in Russian)

6. Krichevskiy ES, Benzar VK, Venediktov MV, et. al. (1980) Theory and practice of express moisture monitoring in solid and liquid materials. Krichevskiy ES (ed) Energiya Publ., Moscow (in Russian)
7. Krichevskiy ES, Volchenko AG, Galushkin SS (1987) Monitoring of moisture of solid and particulate materials. Krichevskiy ES (ed) Energoatomizdat Publ., Moscow (in Russian)
8. Jain RC, Voss WAG (1994) Dielectric measurement methods for industrial, scientific and medical applications in microwave frequency range. IETE Technical Review 11(5-6): 297–311
9. Nyfors E Industrial microwave sensors – a review (2000) Subsurface Sensing Technologies and Applications 1(1): 23–43
10. Kraszewski A (2001) Microwave aquametry: an effective tool for non-destructive moisture sensing. Subsurface Sensing Technologies and Applications 2(4): 347–362
11. Trabelsi S, Kraszewski AW, Nelson SO (2001) New calibration technique for microwave moisture sensors. IEEE Trans. on Instrumentation and Measurement 50(4): 877–881
12. Kupfer K (1999) RF & microwave instrumentation for moisture measurement in process and civil engineering. Subsurface Sensors and Applications, Nguen C (ed) Denver, Colorado, USA. Proceedings of SPIE, 1999, Vol. 3752, pp 39–46
13. Daschner F, Knoechel R, Kupfer K (2001) Resonator based microwave moisture meter with digital phase signal processing. In: Kupfer K, Huebner C (eds) Proceedings of the 4th international conference on electromagnetic wave interaction with water and moist substances, Weimar, 2001, pp 125–131
14. Larsi T, Glay D, Mamouni A, Leroy Y (1999) Microwave sensors for nondestructive testing of materials. Subsurface Sensors and Applications, Nguen C (ed) Denver, Colorado, USA. Proceedings of SPIE, 1999, Vol. 3752, pp 29–38
15. Sovlukov AS (1989) Designing principles for highly sensitive microwave sensors to be applied in metallurgy. Preprints of the 6th symposium IFAC on automation in mining, mineral and metal processing, Buenos-Aires, 1989, Vol. 2, pp 267–271
16. Sovlukov AS (1997) Radiofrequency methods design for on-line measurement of moisture content in produced building materials. Proceedings of the conference "Non-Destructive Testing in Civil Engineering", Bungey JH (ed) Liverpool, 1997, Vol. 2, pp 783–793
17. Sovlukov AS (2002) RF and microwave resonator sensors for contactless materials characterization. Proceedings of the conference "11.Feuchtetag 2002", Weimar, 2002, pp 247–256
18. Lunkin BV, Sovlukov AS, Ivanov AV (1981) Radiowave measurements of two-phase flow continuity in pipelines. Preprints of the international symposium on flow: its measurement and control in science and industry, St. Luis, USA, 1981, Vol.2, pp 209–217
19. Heeren RG, Baird JR (1971) An inhomogeneously filled rectangular waveguide capable of supporting TEM propagation. IEEE Trans. on Microwave Theory and Techniques 19(11): 884–885
20. Bernhard JT, Joines WT (1995) Electric field distribution in TEM waveguides versus frequency. Journal of Microwave Power and Electromagnetic Energy 30(3): 109–116
21. Cheung AY, Dao T, Robinson JE (1977) Dual-beam TEM applicator for direct-contact heating of dielectrically encapsulated malignant mouse tumor. Radio Science 12(6s): 81–85
22. Kumar A (1984) Dielectric-loaded rectangular waveguide applicator. International Journal of Electronics 57(2): 299–303

23. Kupfer K (1999) Methods and devices for density-independent moisture measurement. Third workshop on electromagnetic wave interaction with water and moist substances. Collection of papers, Athens, Georgia, USA, 1999, pp 11–19
24. Kupfer K (2000) Advances for density-independent moisture measurement: a review. Subsurface Sensing Technologies and Applications II, Nguen C (ed) San Diego, USA. Proceedings of SPIE, 2000, Vol. 4129, pp 68–81
25. Trabelsi S, Kraszewski AW, Nelson SO (1999) Density-independent permittivity functions for moisture sensing in foods and agricultural products. Proceedings of the 34th microwave power symposium, Arlington, USA, 1999, pp 9–12
26. Okamura S, Zhang Y (2000) New method for moisture content measurement using phase shifts at two microwave frequencies. Journal of Microwave Power and Electromagnetic Energy 35(3): 175–178
27. Sovlukov AS (2001) Microwave method for determination of water content in a substance independent of its dielectric permittivity. In: Kupfer K, Huebner C (eds) Proceedings of the 4th international conference on electromagnetic wave interaction with water and moist substances, Weimar, 2001, pp 446–453
28. Gudkov OI (1986) Estimation of dispersion mechanisms for relative dielectric permittivity at low and microwave frequencies. Izmeritelnaya tekhnika. N 2, pp 44–45 (in Russian)

9 Density and Moisture Measurements Using Microwave Resonators

Thorsten Hauschild

TEWS Elektronik, Hamburg, Germany

9.1 Introduction

Resonators using microwave resonance frequencies are very well suited for the determination of moisture and density. These resonators can be partially filled with a moist material, which can be seen as a dielectric medium with the complex dielectric constant (DC) $\underline{\varepsilon} = \varepsilon_r' - j\varepsilon_r''$. The use of resonators with resonance frequencies in the microwave region has several advantages. Since the penetration depth into a product is roughly due to the absorption condition caused by water molecules, the wavelength of the electromagnetic wave used for the measurement, (approx. 10 cm wavelength), can penetrate deep into the material. Thus not only the surface moisture but also the moisture in the core of the material under test is measured. In addition, the microwave measurement technology is relatively independent from salt and mineral concentrations, which can vary to a large extent since the ionic conductivity of salts dissolved in water does not have any influence in this frequency range.

In the electrical field of the resonator the material produces a resonance frequency shift and a broadening of the resonance curve (bandwidth) compared to the empty state. From this measurement the DC or direct moisture of the material under test can be determined. Therefore, small amounts of a test material inside the resonator cause clear effects because the dielectric material is in constant reaction with the energy inside the resonant structure, which changes continuously between the electrical and magnetic state.

From that, resonant applicators with resonant frequencies in the microwave region show high sensitivity to test materials with different moisture content. In a typical application, a test material that causes only a small deviation of the resonance frequency is inserted into the resonant structure. The resonant frequency is shifted and the width of the resonant curve increases. The deviation in resonance frequency shift and bandwidth can be measured by a simple scalar system. So it is possible to measure the moisture of a material, being density independent, with a two-parameter scalar measurement. Density fluctuations of the material can be eliminated due to the assumption that the resonant structure, which is used for the moisture measurement, reacts only with the dielectric material under test. Changes of resonance frequency and bandwidth should not be caused by wall losses in the resonator or due to radiating fields. On the other hand it is possible to measure the density of material, being moisture independent, with the same scalar measurement. In this case moisture variations of the material can be eliminated. This is possible due to the same assumption for the resonant structure.

To fulfill the above assumptions it is necessary to design resonant structures that are not overloaded by a moist material. An essential demand on the construction is that through the load of the material under test the field pattern does not change. The mode of a resonant circuit must remain the same to secure the calculation base, which is the empty resonator. Also the frequency-distance between the wanted mode and sturgeon modes must be so wide that sturgeon modes cannot withdraw energy from the wanted mode. The resonance curve would become unsymmetrical in this case and the exact measurement of resonance frequency and bandwidth is impossible. So it is necessary to secure this requirement to ensure the accuracy of the moisture measurement. This accuracy should be, depending on the application, down to fractions of a percent or even per thousand. Not only internal sturgeon modes can cause problems, but also external reflections through connected, non-reflection-free incoming lines.

Errors in the determination of the moisture can be caused by radiation losses as well. These effects would enlarge the bandwidth and destroy the density-independent measurement. Since every resonant applicator requires apertures for the supply of the material under test, restrictions on size and construction demands of the material supply exist. One has to look not only for radiation losses of an empty resonator but for the filled structure as well.

In the following chapter practical resonant structures will be described. These structures are designed in order to fulfill the constraints. After that the analysis of the resonant frequency behavior for determining the moisture and/or density will be described in detail. The calibration will be discussed and several successful industrial application examples will be given.

9.2 Resonators

The type of resonator being used for measuring the moisture and density of a material under test depends, to a large extent, on practical requirements for the handling of the material. Shielded and open resonators are to be distinguished.

Shielded resonators are closed on all sides so that apertures into the cavity are necessary to fill the resonator. A sample tube made of plastic, glass, or ceramic is inserted through the aperture. The size of the resonator results from the chosen resonance mode, the demand of the measurement, and from the chosen operating frequency and of course from the practical construction of the resonator. The last point often results in the use of circular cylindrical resonators. A minimum or maximum size of the resonator may perhaps be necessary due to the structure of the material under test and the desired measurement. For example, the desirable design of a resonator for measuring the moisture and density of single cigarettes depends on whether one wants to measure the parameter from the whole cigarette or be location-dependent. The size and location of the aperture depends on the desired resonance mode. Radiation through the aperture has to be prevented, especially when the resonator is loaded with moist material, which means a high dielectric constant. The aperture behaves like a wave-guide, which has to be used below the cut-off frequency.

Open resonators are much more difficult to design due to their open structure. Prevention of radiation is not as easy as with cavity resonators. They offer advantages compared to the shielded resonators because material can be more easily brought into contact with the resonance structure. The limits in practical application of cavity resonators are imposed due to the fact that the product to be measured must be put into the sample tube of the resonator. In process measurements this is only possible in the case of products with good flowing or pouring behavior. This can be done in a bypass by transferring product from the main stream with the aid of conveying elements, filling it into the measurement tube and then conveying it back into the main stream. Using open resonators, the resonator can be brought directly into the main stream of the product flow without interfering with the process [1, 2].

Several examples for resonators, which are in use in industrial processes, are given in the following sections of the article.

9.2.1 Cavity Resonators

9.2.1.1 E-Type Resonators

Cavity resonators using resonance modes with only an electrical field into a cylindrical-axis are useful for light and dense materials. Three examples will be given in this section. In Fig. 9.1 a simulation of a typical circular cylindrical resonator is shown as a case of a specific resonance mode (the fundamental mode of a circular cylindrical resonator, E_{010}). For a better view only a half of the resonator is shown. In the case of a resonance in a given spatial region, a spatially stable distribution of field lines can be set up which interact in a controlled way with the moist material under test. To accomplish this, use an appropriate plastic container for feeding the product through the measurement field region or insert a plastic tube into the resonator to feed the product directly through the resonator in an industrial process. This type of resonator is very sensitive to moist materials due to the high electrical field in the zone where the material is. Hence materials with high water content may attenuate the field too much. Simulations have shown that the resonator can be filled with nearly every moist material without radiating through the aperture. As a rule of thumb, the critical dielectric constant ε_r^{crit} depends for the basic mode only on the diameters of the resonator d_{res} and aperture d_{ap}:

$$\varepsilon_r^{crit} = \left(\frac{d_{res}}{d_{ap}} \right)^2 \tag{9.1}$$

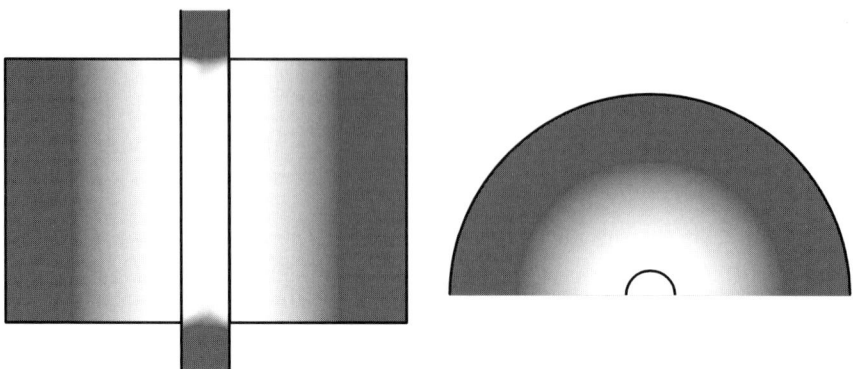

Fig. 9.1. Circular cylindrical resonator, E_{010} mode, resonance frequency 2.7 GHz, view of a half of the resonator, chimney, left: magnitude of E-type, right: magnitude in a middle slice of the resonator, white high magnitude, black low magnitude

For the resonator example given in Fig. 9.1 the critical DC is approximately $\varepsilon_r^{crit} \approx 70$.

In Fig. 9.2 another example for an E-type resonator is given. In this case the measurement field should be very constant over the aperture in the radial direction and extremely short in the z direction. This task was achieved using a resonator filled with a low loss dielectric material concentrating the field in the resonator but not in the aperture. A resonator like this can be used to measure, for example, cigarettes, location-dependent. The measurement zone in the z-direction is as low as 3 mm [3, 4].

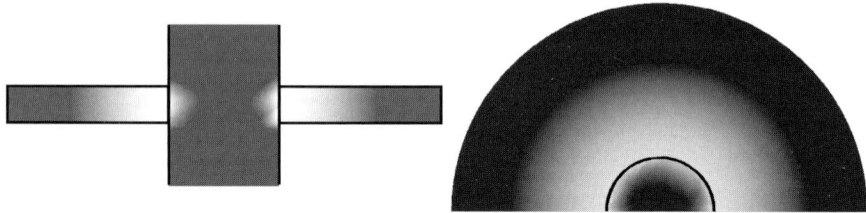

Fig. 9.2. Circular cylindrical resonator, E_{010} mode, resonance frequency 2.5 GHz, resonator is filled with dielectric material except in the aperture, view of halve of the resonator, chimney, left: magnitude of E-type in a cross section, right: magnitude in a middle slice of the resonator, white high magnitude, black low magnitude

As already mentioned the E-type resonators are very sensitive to moisture. This is an advantage, but often also a problem. If you want to measure, for example, cigars over a wide moisture range you would normally use various resonators. A resonator has been designed that can measure over the wanted range [5]. Using an aperture not placed in the middle of the resonator and the resonance mode E_{110} the resonator can be used in two ways. In position A the aperture is in the lower section of the resonator. Hence the cigar is laying in the stronger part of the field

(Fig. 9.3). Moisture contents from 6 to 18% can be measured. If the moisture content becomes too high the resonator is rotated into position B. Now the cigar is laying in the weak part of the field and moisture contents from 16 to 26% can be measured. This can be done by using one calibration due to the density-independent procedure.

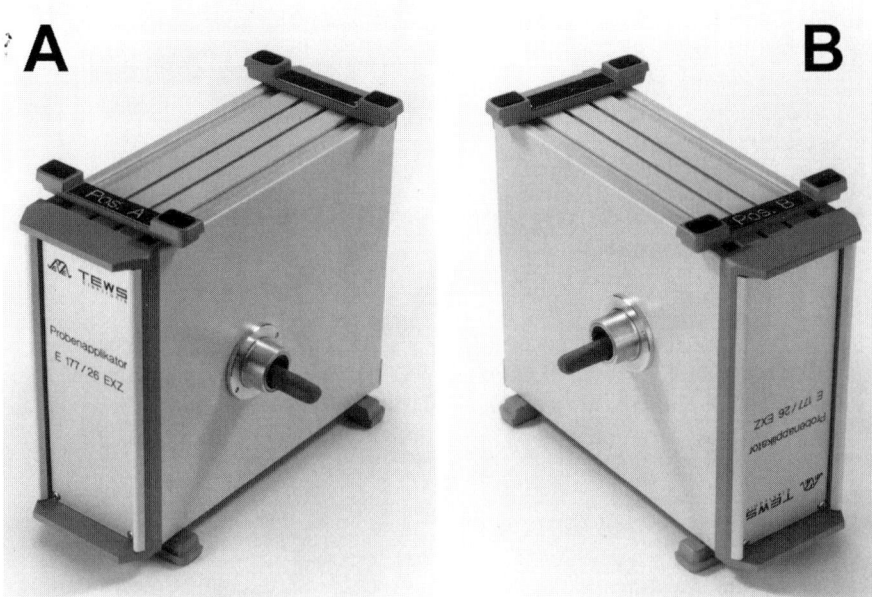

Fig. 9.3. Circular cylindrical resonator, E_{010} mode, resonance frequency 2.5 GHz, aperture is not in the middle of the resonator, position **A** for low moisture content, position **B** for high moisture content

9.2.1.2 H-Type Resonators

The main advantage of cavity resonators using resonance modes with only a magnetic field into cylindrical-axis is the possible size of the aperture between 40 up to 120 mm. Due to the radial electrical field components only, the size of the aperture can be larger than with E-type resonators. Another advantage of H-type resonators is the easy possibility of calculating the dielectric constant from the resonant behavior.

A special feature of H-type resonators is the nonlinear behavior of frequency shift and the broadening of the resonant curve with moisture, but it is possible to linearize this effect [6].

In Fig. 9.4 a typical H-type resonator is shown which can be used in a wide variety of applications for moisture measurement due to the large aperture. The chosen mode ensures a small electrical field in the middle of the resonator.

For density or mass measurements this type of resonator can not be applied because of the inhomogeneous electric field.

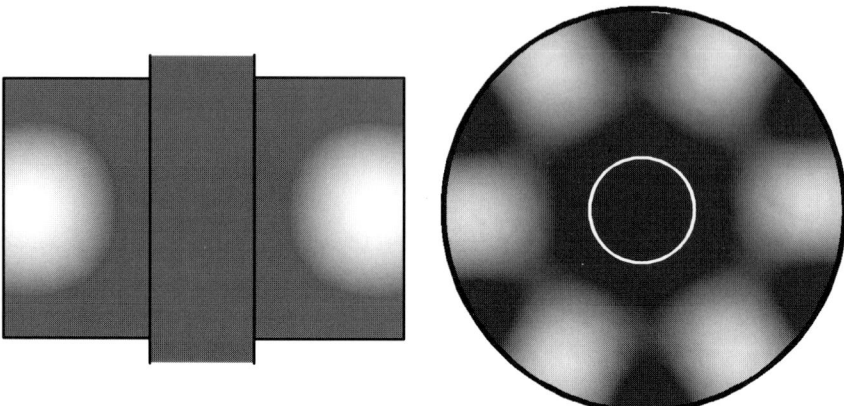

Fig. 9.4. Circular cylindrical resonator, H_{311} mode, resonance frequency 2.8 GHz, view of
half of the resonator, chimney, left: magnitude of E-type in a cross section, right: magnitude
in a middle slice of the resonator, white high magnitude, black low magnitude

9.2.2 Planar Resonant Structures

Planar resonant structures are desirable for process measurements where they can
be directly incorporated into the main stream of product flow without interfering
with the process. Using simple open cavities, the microwave can react into the free
space. In doing so, an antenna-like radiation of microwave energy takes place
which, as already mentioned, effects the shifting of the resonant frequency and the
widening of the resonant curve and thus corrupts the measurement result.

Planar resonators have been developed which provide a leakage microwave
field in the space in front of the resonators without the undesirable antenna effect.
This is achieved by the microwave leakage field sensor which is of rotationally
symmetrical design with an alternating field of standing waves in it. The spatial
period of this field is less than the vacuum wavelength at the frequency of the alternating field. As a result of this difference in the wavelengths, the microwaves
penetrate a certain distance into the free space above the planar sensor, but decrease to a very great extent with distance from the sensor. So there is no microwave radiated. This is based on the fact that negative interference occurs in the far
field and only the near field remains [1, 2].

Using this development, planar sensors can be designed which are microwave
resonators. An electromagnetic field is generated above the planar sensor surface,
which penetrates into the product to be measured and has no radiation losses as
long as the dielectric constant of the product lies below a critical limit value.

Under these conditions, the losses measured by the planar resonator technique arise not due to radiation effects but due to the heat conversion within the product.

These planar sensors can be used for measurements where only one-sided access is possible, for example, wooden boards, masonry slabs, etc. The measurement is also possible by incorporation in container walls, silos or in the moving product stream within process installations, etc.

9.2.3 Overview Resonators

In this section only a short overview on the great variety of resonators is given. In Table 9.1 the previously described E- and H-type cavity resonators and the E planar resonator is given in an overview. Possible areas of application and the volume needed for measurement are given.

Table 9.1. Overview resonators

Resonator Type	Sampler	Areas of Application
H normal Laboratory version and control process version	plastic tube with internal diameter of 40 or 46 mm (sample volume approx. 300 to 400 ml)	sprinklable and pourable products in all areas of industry, which process chemicals, pharmaceuticals, and natural products. Stuffable products (tobacco, hops, etc.)
H cup Laboratory version	in plastic cup with internal diameter of 33, 40 or 50 mm (sample volume of 35 to 150 ml).	pastes and fluid products for fast and precise measurement (pastes, oils, marzipan, syrups, etc.)
H large Version for measurement in the laboratory and control process	plastic cup or tube with internal diameter of 96, 128 or 197 mm (sample volume approx. 2 to 5 liters)	products with course granularity or difficult flow behavior (rapeseed grist at high temperature and moisture, tobacco leaves and cut tobacco, diced beets, cookies, etc.)
E Laboratory version and control process version	plastic cup or tube with internal diameter of 10 mm (optionally 26 mm) (sample volume 1 to 27 ml) or flat bed made of plastic for foil or paper.	powders in small quantities, cigarettes and filter sticks, plastic foils, paper or sheeted products
Open resonator for process use	Planar structure, product flow over the resonator	nearly every product possible, depends just on dielectric constant of the product

Fig. 9.5. Overview of various cavity resonators, from left to right: H-type with 96 mm tube and sample volume 1800 ml, H-type with 46 mm tube and sample volume 400 ml, E-type with 26 mm tube and sample volume 27 ml, H-type with 40 mm aperture for cup and sample volume 150 ml, E-type with 10 mm aperture for cup and sample volume 5 ml

Fig. 9.6. Two possible planar resonators, left resonator useful for light and relatively dry products, right for dense and relatively moist products

9.3 Density Independent Moisture Determination

Using the resonator method, one can directly measure the resonance frequency and the bandwidth of the resonator. These parameters depend both on the material moisture and the packing density (see Fig. 9.7). The directly measured quantities are first transformed into the linearized quantities A and B.

Figure 9.8 shows, using the so-called *A–B representation*, how to separate the influence of moisture and density. A linear relation between the two indirect measurement quantities A and B can be seen due to variation of the sample density at a constant moisture. So the separation of the respective influence of moisture and density is easy.

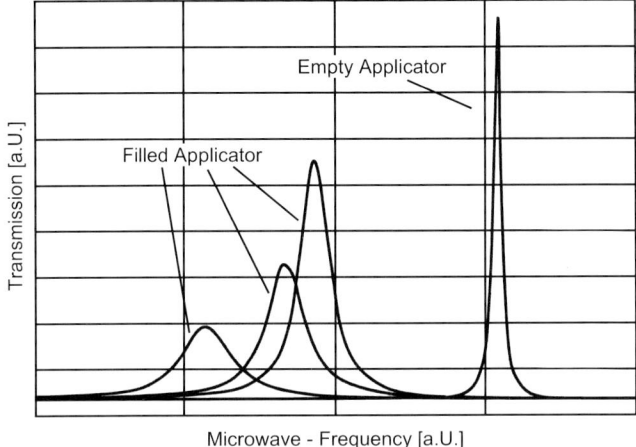

Fig. 9.7. Changes in resonance behavior of a resonator due to the interaction between microwave, moisture content and density

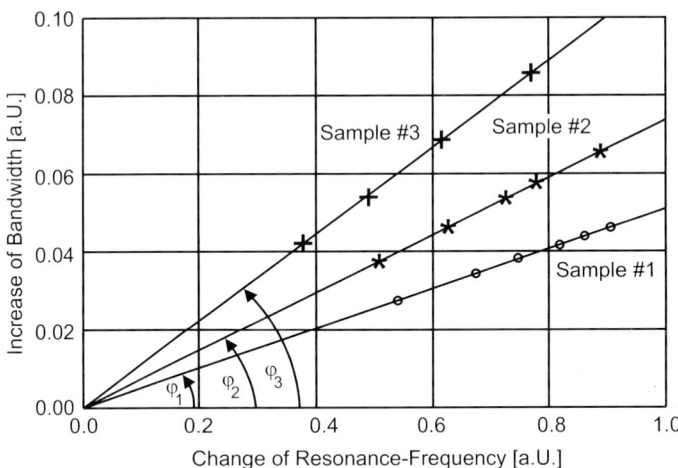

Fig. 9.8. Change of resonance frequency (*A*) versus increase of bandwidth (*B*) for 3 samples with varying moisture, density varies along the slopes

In fact this linearity between *A* and *B* is maintained even in the case of stronger coupling between the resonator electric field and the dielectric sample material, a situation that often occurs at higher material moistures and densities. Of course the connection between the indirectly measured quantities *A* and *B* and the direct quantities (resonance frequency and bandwidth) become highly non-linear. In spite of this, a transformation is possible so that the linear relationship shown in Fig. 9.8 continues to hold [6].

Based on the linearity between A and B, several patents [7, 8] where claimed. Using these patents the method for separate determination of moisture and density will work as follows:

Let ψ be the material moisture (expressed in percent relative to the wet mass of the product) and ρ the material density, then quantities A and B, which are displayed in Fig. 9.8, are to be computed from:

$$A = A_F - A_E = \left[A_D(1-\psi) + A_W(\psi)\psi\right]f(\rho) \tag{9.2}$$

$$B = B_F - B_E = \left[B_D(1-\psi) + B_W(\psi)\psi\right]f(\rho) \tag{9.3}$$

The index F designates the measured parameter for the filled resonator, E the empty resonator. D stands for the influence of the dry material of which the moisture should be determined, W for the influence of water. The possibility of separating moisture from density is based decisively on the fact that both B and A depend on the sample packing density in the same way. This is represented in the empirically established formulas above by the appearance of the common factor $f(\rho)$.

The effect of this feature can be seen in Fig. 9.8. Compressing a sample, keeping moisture constant, the A and B values increase along a straight line in the A-B coordinate plane. This line runs through the A and B values of the empty resonator. The slope of this line depends on the moisture but not on the density. Hence, after a calibration, the moisture can be determined from the microwave values without knowing the density. An easy way to do this is to measure the slope of a moisture line in Fig. 9.8, which is called the microwave value in the following sections.

$$\varphi(\psi) = \arctan\left(\frac{A}{B}\right) \tag{9.4}$$

9.4 Moisture Independent Density Determination

In the previous chapter the density independent behavior of the moisture determination was explained. Using the same measurement setup one can measure the density of the sample as well. In fact both moisture and density can be determined with the same measurement using an appropriate calibration. First one has to determine the slope of the moisture line (see Fig. 9.8) to get the moisture of the material under test, then the A value depends on the known moisture and the unknown density, hence the density can be calculated.

The density measurement based on microwaves has become more and more important over the last few years. Especially natural materials, where mixtures of various original and constantly changing moistures are used, can be processed to a target density using this technology without problems.

9.5 Measurement Devices

A great variety of moisture measuring devices are available. Depending on the application one can distinguish between three groups, laboratory devices, process devices, and handheld devices. Every device is manufactured using the same basic concept. A generator (GEN) is controlled by a microprocessor (CPU) which sends the microwave through a resonator (see Fig. 9.9). The transmitted microwave power is sampled by a detector (DET). The microprocessor calculates from these sampled values the moisture and density. Several interfaces to the user and/or process-control computer complete the setup.

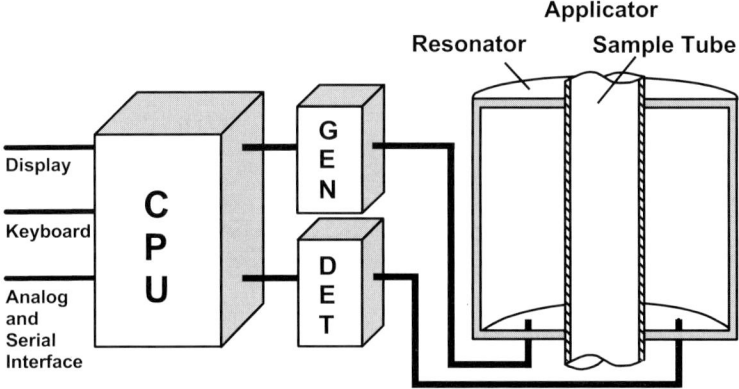

Fig. 9.9. System setup of moisture measurement devices

The MW 3150 measuring system (see Fig. 9.10) has been designed to be easy to use and provide fast and accurate moisture analysis, especially by laboratory technicians for quality control. The moisture results can be obtained, usually with no sample preparation, in less than a millisecond. The measurement procedure using the MW3150 is very easy. The empty state of the resonator is determined automatically. After filling of the resonator, the moisture is displayed within fractions of a second. Finally one has to empty the resonator. The system can use different sensors depending on the application. (i.e. granule, powders, liquids, pastes, or slurries). Using an appropriate resonator the systems are used in a great variety of industries (see Table 9.2); (i.e. food, chemical, pharmaceutical, paper, plastic and agriculture). The accuracy of the moisture determination is typically better than 1% relative to the maximum measurement range (e.g. +0.1% at 10% moisture measurement range). Other laboratory devices use larger displays, external keyboards or touchpanels, but the main measurement principle is the same. They mainly distinguish differences in the software, especially in the man/machine interface, which allows the user to calibrate the device and save the measurement values and the storage capacity.

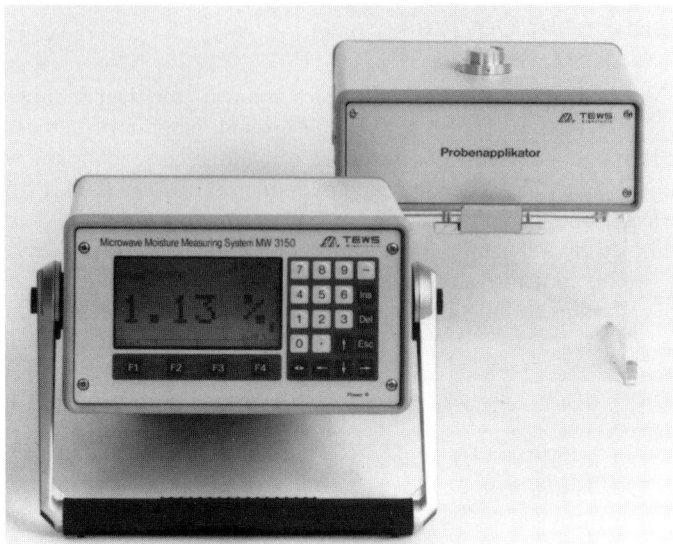

Fig. 9.10. Microwave measuring system MW3150 (shown with cavity resonator). The system is used mainly in quality control of laboratories

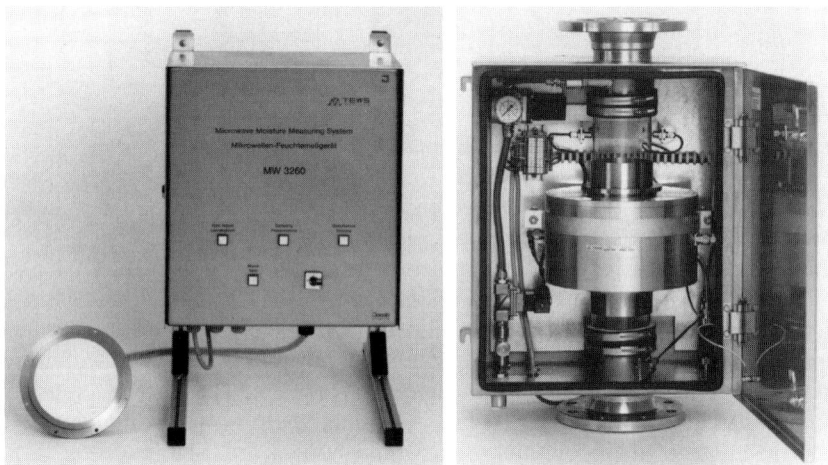

Fig. 9.11. Microwave measuring system MW3260 (shown on the left with planar resonator, on the right a typical cavity resonator for process applications is shown). The system is used for online quality control in process applications

Figure 9.11 shows the MW3260, a typical moisture measuring setup for online process quality control. The monitoring system is encased in a watertight housing and manages the sampling, measuring, and output of the measured values. Here again several resonators can be used. In the left part of Fig. 9.11 the MW3260 is shown with a planar resonator, which is used in a process (i.e. when product flows from one conveyor belt to another or in silos). In the right part of Fig. 9.11 a cavity resonator is shown, which is used with the MW3260 as well. Samples of a main product flow via an assemblage of conveyer devices (compressed air, screws, pushers, valves etc.) into the cavity. The moisture and/or density can be determined then the cavity is emptied and filled again. The entire measurement procedure can be controlled by the user from a macro-program, which can be arbitrarily tailored to the specifications of the measuring station. Control signals from the production line can be accommodated in addition to the normal measurement specific monitoring. A temperature sensor provides information about the product temperature; an infrared full/empty gauge monitors filling of the sample tube.

The MW3230 and the MW3220 (Fig. 9.12) are tailored for measurement of cigarettes to control the weight and moisture of each single cigarette in a quality laboratory. Using the sensor from Fig. 9.2 the measurement field is very narrow, so that the density of a small part of the cigarette can be measured. Knowing the velocity of the cigarette, the weight can be calculated. An example using the MW3220 will be given in one of the following sections.

Fig. 9.12. Microwave measuring system MW3230 and MW3220. These stand alone systems are used for quality control in cigarette production. Density and moisture can be determined integral or position dependent to get the profile along the cigarette

The MW1000 (Fig. 9.13) is a handheld device. It is battery powered, lightweight and performs some basic measurement tasks. The accuracy of this device is reduced compared to the other systems at 5% relative to the maximum measurement range, but this is still acceptable for many applications. The system uses a built-in planar resonator, a simple man/machine interface and very reduced calibration schemes. Using this system one has to wait approximately 5 seconds for a single measurement.

Fig. 9.13. Handheld microwave measuring system MW1000

9.6 Temperature Compensation

When using microwave resonators for measuring the moisture of a material under test, the measuring device has to be calibrated first. This has to be done in two steps. Before calibrating the moisture with a reference method, the measuring device itself must be initially calibrated. For H-type resonators one has to calibrate for the correct behavior with moisture [6].

Another effect in the resonance-based measuring technique is the temperature behavior of the system. Due to temperature, the size of the resonator varies. So the resonance frequency of the empty resonator varies as well. Due to the evaluation process, where the resonance frequency and the bandwidth are compared to the empty resonator, it is extremely important to know the parameters for the empty resonator. They can be measured directly before using the system, but this is often not possible in process systems. In this case the empty resonator will be measured once and the measured parameters will be used for hours or often days or months. The resonance frequency of the empty resonator can change with temperature, an effect that cannot be measured easily. There are several solutions to this problem. Three possibilities are introduced here. They can be used in combination or alone, depending on the temperature range, which should be covered.

First of all it is possible to build a resonator with a material where the temperature variation is very low (as it is well known from resonator based filter design techniques, where Invar (Fe64/Ni36) has been used for decades). This can be done but often there is not only one material used to build the resonator. Tubes inside the cavity are necessary for the product stream, ceramic shielding is used in the case of planar resonators to protect the resonator from abrasion. Another point against the use of temperature compensated material is often the price of this material.

Another solution is heating of the resonator. In this case the resonator is heated to a temperature above the highest possible product temperature. Then the resonance parameters of the empty resonator can be measured once at this temperature and then they are valid for the following measurements, assuming that the temperature of the resonator does not change. The heating has to be controlled very accurately. Another positive effect of the heating is that water from the moist, often warm material under test cannot condense on the metal surface of the colder resonator.

The last solution is a precalibration of the temperature behavior. In this case the resonator is heated over the whole temperature range. The resonant frequency and the bandwidth are stored for every temperature and the temperature dependency can be calculated. Afterwards, using the resonator in the laboratory or process, the resonance parameters are measured as often as possible together with the temperature. Then the system can be used for a long period without emptying the resonator. Only the temperature is measured and the empty parameters can be corrected in relation to temperature variation. The measurement of the empty resonator is still necessary because, due to dirt or abrasion, the empty resonator parameters can vary slightly. Using this precalibration it is assumed that the resonance behavior of the empty and filled resonator acts like it is operating at the same temperature. This assumption has been well proven during the last few years.

Using the mentioned possibilities the system can measure a sample accurately over temperature if the corrected microwave measured value is connected with the laboratory reference value for the same sample. This connection is actually the calibration of the system, which is explained in the following chapter.

9.7 Calibration

Moisture measurement with microwaves is an indirect measurement technique, which has to be calibrated with a reference method (e.g. Karl-Fischer titration). It is possible to use two samples, a moist one and a dry one, to calibrate the system but a large number of moist samples will improve the calibration. Every sample is measured with the measurement system and the reference value is determined afterwards. Depending on the number of samples the calibration can be linear or non-linear. Using this type of calibration scheme the moisture can be measured easily afterwards.

Due to the mobility of the water molecules in the material under test the resonant parameters change with water content. But the amount of water inside the material is not only responsible for this change. The mobility also varies with temperature. So the calibration has to be performed with samples of the same temperature. If the material under test will change the temperature with respect to the calibration temperature, then the calibration has to be performed over this temperature range. Figure 9.14 shows how the effect of temperature on the measurement signal can be entirely compensated by the user simply by taking a sufficient number of calibration points in the temperature and moisture range of interest. Using linear or non-linear regression it is then possible to compute a calibration surface in the space spanned by temperature, moisture and the microwave axis. By simultaneously measuring the microwave value and the product temperature, the moisture can be determined from this surface without an error induced by product temperature variation.

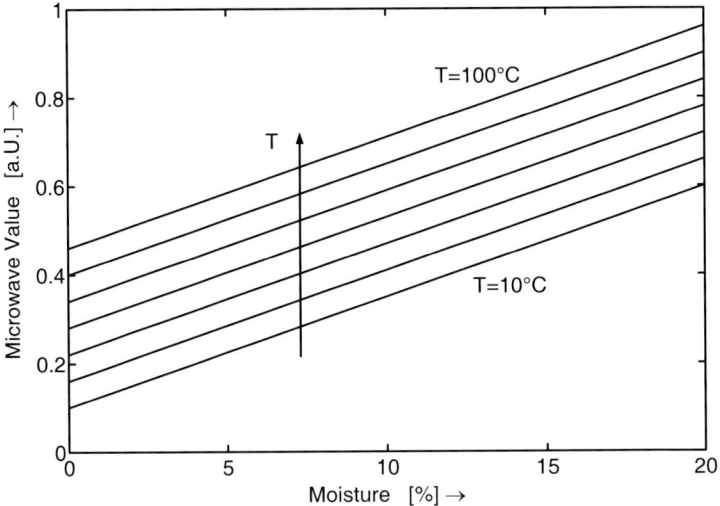

Fig. 9.14. Typical calibration for a product, temperature varies from 10 to 100°C

For example, tobacco samples provided by the research institute of a major tobacco company were measured with microwave moisture meters. They represent a wide variety of plant types (Virginia, Burley etc.) as well as place(s) of origin (Thailand, USA, France, Malawi, etc.). Nevertheless the calibration curve is independent of the type of host material over a wide moisture range. In this way arbitrary mixtures of products, even if they have a variable constitution, can be measured. By contrast with the microwave method, all other moisture-measuring technologies such as NIR, which are sensitive to surface effects, lead to a large scatter of the measurement points as the type of material is varied and/or the place of origin and/or the quality of the tobacco is changed. In the realm of natural products (such as tobacco, coffee, chocolate, sugar, cotton), where mixtures of various origin are constantly changing, independence of the calibration from type, origin, salt content, grain size of the particles etc. as well as the density independence are of great advantage for a moisture-measuring device.

9.8 Industrial Applications

In this chapter one example for each moisture and density measurement will be given out of the various successful applications (see Table 9.2), where moisture meters are in use.

Table 9.2. Examples for successful applications, where moisture meters have been used to determine moisture and/or density

Product	Moisture Range	Error
Cocoa, Whole Bean	2.0–6.0%	0.05%
Caramel	8.0–11.0%	0.03%
Carob Seeds	6.0–18.0%	0.05%
Alumniaoxid	0.5–1.0%	0.01%
Coffee Beans	3.0–7.0%	0.03%
Cracker	0.6–1.2%	0.03%
Dry Soup	2.5–8.0%	0.03%
Dye Suspension	9.0–30.0%	0.4%
Duroplast	0.9–3.0%	0.02%
Fertilizer	0.18–0.64%	0.004%
Gelatine Capsules	10.0–15.0%	0.05–0.1%
Herbicide	0.02–3.0%	0.01%
Hops	7.0–16.0%	0.2%
Cork	5.0–10.0%	0.03%
Lime	0.1–8.0%	0.01–0.2%
Milk Powder	1.0–4.0%	0.03%
Noodles	6.0–14.0%	0.16%
Polystyrol	0.01–0.11%	0.005%
PVC	0.12–5.0%	0.02%
Silica Gel	2.0–45.0%	0.01–0.09%
Soya, Rape Seed	8.0–15.0%	0.08%
Starch, Potato	6.0–18.0%	0.09%
Starch, Corn	2.4–10.9%	0.1%
Sulfate	0.1–0.5%	0.01%
Sugar	0.01–0.4%	0.01%
Tablet Powder	0.5–6.0%	0.01%
Tobacco	2.0–35.0%	0.1%
Vitamin A	3.6–4.7%	0.02%
Waffles	1.0–7.0%	0.1%
Zeolite	2.0–18.0%	0.04%

9.8.1 Moisture Measurement in Coffee Beans

Coffee beans have a extremely low moisture content after roasting. Normally water is added after roasting to cool down the coffee beans, to prevent them from burning. The beans are really extinguished. The coffee producer must respect certain regulations and prevent the coffee from getting moldy as a result of being soaked by the extinguishing water. The upper limit for coffee is in the region between 4 and 5%. Figure 9.15 shows the calibration for coffee, using a planar sensor and the MW3260 process moisture meter. Several probes were used in the interesting moisture level from 3 up to 4%, where the producer of this grinded coffee has set his target moisture depending on the final product. Higher roasted dark coffee (mocha) has normally a slightly lower moisture level than mild coffee. Figure 9.16 shows the typical behavior of the process for this two sample products.

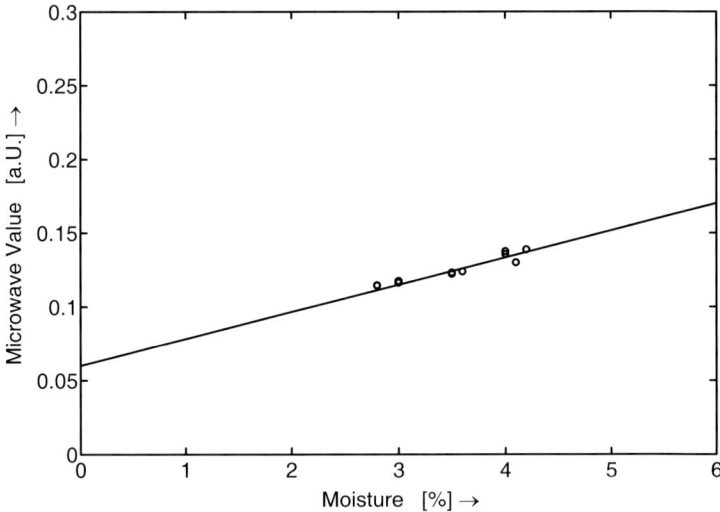

Fig. 9.15. Calibration for grinded coffee beans

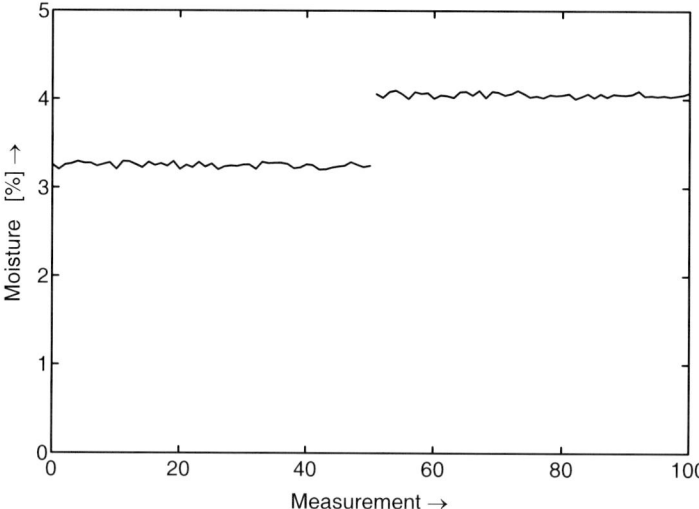

Fig. 9.16. Typical measurement of grinded coffee in a process before vacuum sealing the coffee, using planar sensor and MW3260, first 50 measurements were taken on a dark roasted coffee (mocha), next 50 measurement value show typical behavior for a mild coffee, which normally has a slightly higher moisture

9.8.2 On Site Moisture Measurement of Concrete

The MW1000 is designed to determine moisture in all kinds of material. The measuring field penetrates the product up to 3 cm deep. Here as an example the MW1000 is used to determine the moisture of concrete. In a laboratory three samples of concrete have been prepared. Two samples with nearly the same moisture content (5.2% determined by oven-method afterwards) but various thickness (4 cm compared to 5 cm) were stored laying in a plastic bag before measured by the MW1000. The MW1000 was able to determine the effect of the gravity of water. The moisture content was much higher at the bottom of the sample than on the top. In Fig. 9.17 one can see that the determined values for sample 1 (measurement 1 to 6) and sample 2 (measurement 7 to 12) or nearly the same. The measurements on the upper surface (1, 2, 3 and 7, 8, 9) show a lower moisture value than the measurements on the bottom surface (4, 5,6 and 10, 11, 12). Afterwards a dry sample was measured (measurements 13 to 18) with moisture value of 0%. This measurement shows the possibility of using the MW1000 for a quick on-site determination of moisture content in concrete walls or ceilings, which is necessary prior to plastering.

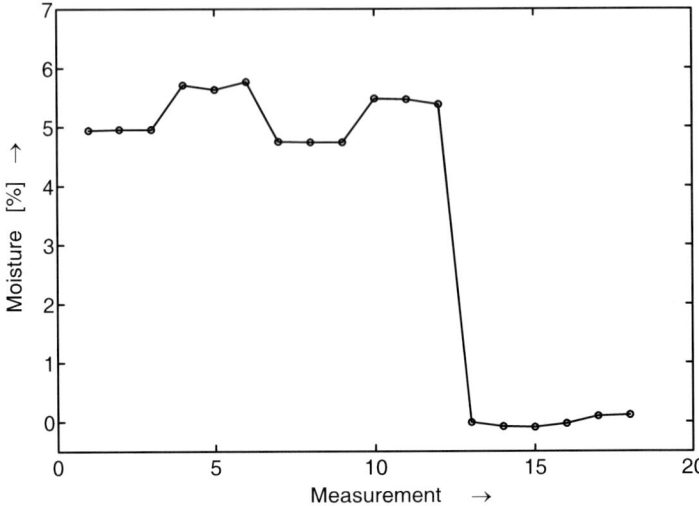

Fig. 9.17. Measurements of 3 concrete samples. Each sample has been measured six times, three times on the top surface, three times on the bottom surface. Moisture content sample 1 and 2: 5.2%, moisture content sample 3: 0%.

9.8.3 Density Measurement in Cigarettes

As already mentioned in the previous chapters a great variety of measurement instruments is offered, which can be used during the production of tobacco and cigarettes. Here an example for a density measurement will be given. Cigarettes are produced on so-called makers, where the tobacco is pressed into a rod and wrapped with the cigarette paper. Afterwards the rod is cut into pieces and a filter is set on the piece to complete the cigarette. The density of the rod has to be controlled in quality for two main reasons. First the weight of the cigarette must reach a target value with an accuracy of +/- 1 mg. Typical target values for a cigarette are in the region of 900 to 1000 mg. Another reason for the quality control is the filling of the cigarette, which is not constant. The density should be lower in the middle of the cigarette and higher in front and at the filter end. So the maker presses the tobacco in this way into the rod. It is now necessary to cut the rod at the appropriate positions to ensure the correct filling.

This can be done after production in the laboratory with the MW3220 and MW3230. The MW3230 can measure the moisture and weight of the complete cigarette integral. The MW3220 can perform profile measurements and is used to control the correct cutting of a cigarette to ensure the lowest density in the middle of the cigarette. To do this up to 100 cigarettes can be measured together in one batch. Each cigarette moves slowly through the resonator and is measured every millimeter. Now the density and moisture can be displayed. In Fig. 9.18 the result for a batch of 10 cigarettes is shown. The resolution for the location is, due to the use of a resonator with a measurement zone of approximately 3 mm, in the range of a few millimeters. So, at the beginning of the cigarette, when the cigarette moves into the resonator, and at the end, when the filter starts, the density and

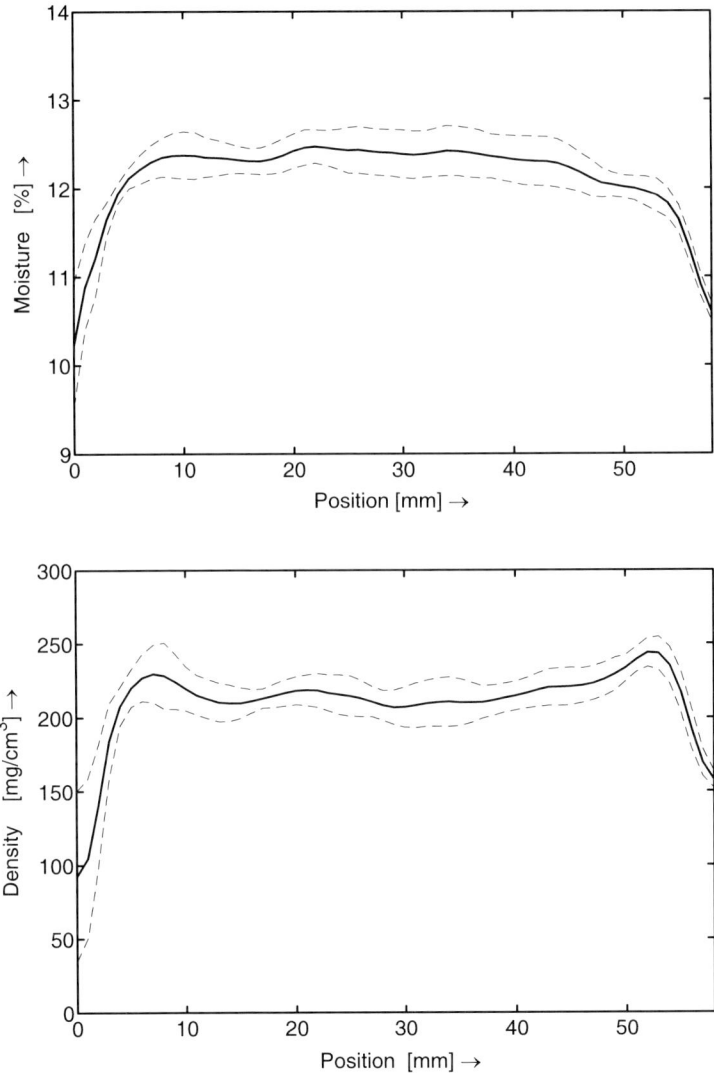

Fig. 9.18. Density and moisture of cigarettes measured with the MW3230. A batch of 10 cigarettes, mean value (solid line) and standard deviation (dashed) are shown, filter is on the right side, beginning at 58 mm.

moisture measurement values are a little bit smeared. But it can be clearly seen from the results that the cutting position for this cigarette batch was correct.

9.9 Conclusions

Microwave resonator-based moisture and density measurement techniques are well suited for use in industrial applications. Using this method, the density is determined independent of the product moisture and the moisture is determined independent of the packing density and grain size of the particles. Minerals, salts and various mixture-ratios of products of different types have no influence over a wide range of fluctuations. This is of great importance for the measurement accuracy of the device. Not only the moisture on the surface of a product is measured but also the core moisture inside the product. The systems can be used for residual moisture detection (e.g., moisture and density in plastics in the range from 0 to 0.2%) as well as in regions of much higher moisture (sewage sludge with up to 60%, cellulose ether up to 70%). Using this solution, only a proper selection of the resonator appropriate to the application being considered must be done.

The moist material is measured, depending on the selected measurement system, within milliseconds, and the moisture and density are displayed. Temperature effects are taken into account in respect to the system itself and the material behavior. Based on this technique a measuring instrument is available that determines rapidly with high accuracy the material moisture. The result of this measurement is independent of the type of the product, the surface of the product, the color and, in the case of natural products of the area of cultivation.

Acknowledgment

I wish to acknowledge the help and the valuable discussions I had with Dipl.-Phys. R. Herrmann, Dr. E. Pilz, Dr. U. Schlemm, Dr. J. Sikora (all from TEWS-Elektronik) and J.A. Carpenter (from Instrument Distributors International Inc., Gaithersburg, MD, USA, North American distributor of TEWS Elektronik).

References

1. Herrmann R, Zaage S (1998) Mikrowellen-Streufeldsensor zur Feuchte- und/oder Dichtemessung. Europäisches Patent, Publication Number EP 0 908 718 A1
2. Herrmann R, Zaage S (2001) Microwave leakage field sensor for measuring moisture and/or density. US Patent 6,316,946B2
3. Herrmann R (1998) Feuchte- und Dichtesensor. Europäisches Patent, Publication Number EP 0 889 321 A1
4. Herrmann R (1999) Moisture and density sensor. US Patent 5,977,780

5. TEWS Elektronik (1999) Cigar Applicator. Product leaflet
6. Herrmann R, Sikora J (1997) Mikrowellen-Feuchtemessung mit Resonatoren und ihre Anwendungen. In Kupfer, K (editor) Materialfeuchtemessung, expert verlag, Renningen-Malmsheim, pp 291-310
7. Tews M, Sikora J, Herrmann R (1993) Verfahren zur Messung der Feuchte eines Meßgutes mit Hilfe von Mikrowellen und Vorrichtung zur Durchführung des Verfahrens. Patentschrift Bundesrepublik Deutschland, Patent Number DE 4004119C2
8. Tews M, Sikora J, Herrmann R (1995) Method for measuring the material moisture content of a material under test using microwaves. US Patent 5,397,993

10 Microwave Semisectorial and Other Resonator Sensors for Measuring Materials under Flow

Ebbe Nyfors

Roxar Flow Measurement AS, P.O.Box 112, N-4065 Stavanger, Norway

10.1 Introduction

In industry there is often a need to measure the properties of materials flowing in pipes, and sometimes also the flow rates. Many of these needs can be met by using microwave resonator sensors [1]. The material under test (MUT) may be a mixture of liquids, liquids and gases, or solids in pneumatic or liquid assisted transportation. Regardless of the MUT, a general requirement is that the sensor structure must be open enough to let the MUT pass through it. The required degree of openness may vary depending on whether the flow contains solids or not, or whether the flow speed and viscosity tend to create a differential pressure over the sensor. An open structure is inherently prone to radiate energy, i.e. to let the energy escape into the pipe, but for a resonator sensor to work properly this radiation must be limited to a minimum by the choice of the design.

The main advantages with microwave resonator sensors, when measuring in pipes, are:

- The possibility to measure the whole flow, not just close to the pipe wall.
- The resonant frequency of the empty sensor usually depends on the mechanical dimensions, and is therefore less prone to drift than in the case of capacitive sensors.
- A resonator is not affected by reflections in the pipe in the same way as transmission sensors [1].

A disadvantage is that the losses must generally be low enough not to kill the resonance, i.e. $\varepsilon'' \ll \varepsilon'$. In cases where this is a problem, it may be overcome by using fringing field sensors, where most of the energy is outside the MUT.

This chapter gives an account of various methods of implementing microwave resonator sensors in pipes for measuring materials under flow. Because many practical solutions are based on various sectorial or semisectorial structures, a brief description of the corresponding waveguide modes is given. A few commercial sensors (Roxar), which are largely based on such structures, are described together with their applications for measuring mixtures of oil, water, and gas.

10.2 Implementing a Resonator Sensor in a Pipe

There are several different requirements for a resonator sensor in a pipe. Many of these are specific to the application, but there are two general requirements: The structure must confine the energy to the resonator, and it must be open enough for the MUT to pass practically unhindered. There are four different principles that can be used in realizing such a sensor [2]:

1) Resonant frequency below cut-off
No electromagnetic energy can propagate in a pipe at a frequency lower than the cut-off frequency of the lowest mode, which is the TE_{11} mode in a circularly cylindrical waveguide. By designing the resonator such that the resonant frequency is lower than this, the energy cannot escape from the resonator into the pipe [3]. The cylindrical fin resonator (CFR) sensor, and the V-Cone sensor that are described later in this chapter are examples of this principle. Even completely non-intrusive structures can be realized for example by creating a cavity resonator from an enlargement of the diameter of the pipe. The section with the larger diameter can be fitted with an internal dielectric pipe, with the same inner diameter as the main pipe, so that the flow is not disturbed [3].

2) End grids
A section of the pipe can be isolated by so-called end grids. The space between the end grids is then a cavity resonator. The end grids are structures with holes through which the MUT can pass, but not the microwaves. For the microwaves the holes are waveguides, with a cut-off frequency that is higher than the resonant frequency of the sensor.

3) Fringing field sensors
The main part of the energy may also be confined to a cavity outside the pipe. Only a fringing field penetrating from, for example, a slot, or from an open end of a coaxial structure, is then in contact with the MUT. A disadvantage is that the whole flow is not measured, and an advantage is that higher-loss materials can be measured than with the other sensors.

4) Non-coupling modes
In some cases it may be enough to use a resonance mode, with a field pattern that does not couple to the pipe modes, even though the pipe modes may have a lower cut-off frequency. An example is the TE_{011} mode, which has a circularly symmetrical tangential electric field that does not couple to the TE_{11} mode in the pipe. The end grids, which are needed as shorts for the resonance mode, may then have larger holes, without the constriction of the cut-off frequency. This principle is best suited for measuring homogeneous materials, like in the case of measuring the humidity in gases [2]. Inhomogeneities in the flow may distort the field pattern and cause radiation, which is falsely interpreted as dielectric losses.

A common feature of all these principles is that even though the energy does not propagate into the pipe, there is a fringing field that penetrates some distance into the pipe. Even in the case of a frequency below the cut-off of the pipe, or the hole in an end grid, some energy may escape coupled by the fringing field if the section with the high cut-off frequency is short. This puts some requirements on the thickness of the end grids, or the length of pipe with constant diameter outside the below-cut-off resonator. Generally, the smaller the holes in an end grid, the thinner the grid may be, because the decay of the power (P) in the fringing field as a function of the distance (z) depends on the ratio between the cut-off frequency (f_c) and the resonant frequency (f_r):

$$P(z) = P_0 e^{-\left(\frac{4\pi f}{c}\sqrt{\left(\frac{f_c}{f_r}\right)^2 - 1}\right)\cdot z}$$

(10.1)

In the practical work of developing sensors for measuring oil, water, and gas in pipes, the author has in many cases ended up with structures based more or less on sectorial or semisectorial structures, both in the case of below-cut-off sensors and end grids. The waveguide modes in such structures are not described in the textbooks. Lin and Omar [4] have described some properties of such waveguides, but no comprehensive study seems to be available in the literature. A summary of the most important results of a study performed by the author will therefore be given below [2].

10.3 Sectorial and Semisectorial Waveguides

The cross section of a sectorial waveguide is shaped like the slice of a pie, while that of a semisectorial waveguide lacks the centre part, i.e. a semisectorial waveguide is bounded between two concentric cylinders and two radial planes, as shown in Fig. 10.1. The sector angle is φ_0, the outer (larger) radius is a, and the inner (smaller) radius is b. Metal walls are where $\rho = a$ (for all φ), $\rho = b$ (for all φ), $\varphi = 0$ (for all ρ), and $\varphi = \varphi_0$ (for all ρ). Because the sectorial structure is a special case of the semisectorial case ($b = 0$), the modes are here derived for the latter.

The derivation follows the same pattern as presented for ordinary circularly cylindrical waveguides in the textbooks, e.g., [5], and [6], but with other boundary conditions.

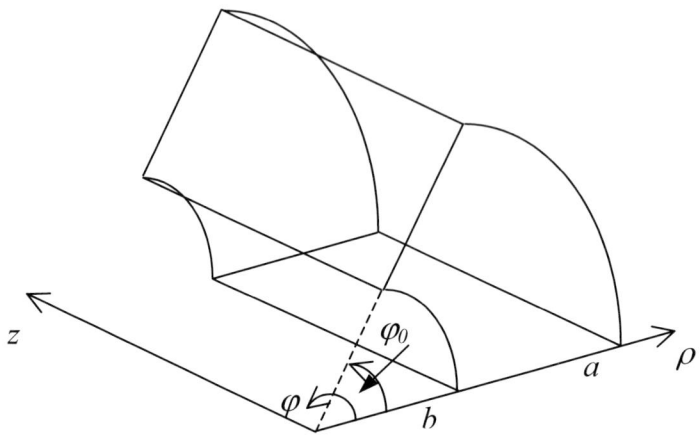

Fig. 10.1. A semisectorial waveguide in cylindrical coordinates

10.3.1 TM Modes in Semisectorial Waveguides

For transverse magnetic (TM) modes the axial magnetic field is zero ($H_z = 0$). The waveguide modes can be derived from the transversal Helmoltz's equation for the axial electric field E_z:

$$\nabla_T^2 E_z(\rho,\varphi) + k_c^2 E_z(\rho,\varphi) = 0 \qquad (10.2)$$

where ∇_T is the transversal Laplacean operator and k_c is the cut-off wave number. Eq. (10.2) can be solved by separating

$$E_z(\rho,\varphi) = P(\rho)\,\Phi(\varphi) \qquad (10.3)$$

which leads to two differential equations:

$$\frac{1}{P(\rho)}\left\{\rho[\rho P'(\rho)]' + k_c^2 \rho^2 P(\rho)\right\} = -\frac{\Phi''(\varphi)}{\Phi'(\varphi)} = v^2 \qquad (10.4)$$

The solutions to Eq. (10.4) are known to be

$$\Phi(\varphi) = A\cos v\varphi + B\sin v\varphi$$
$$P(\rho) = C\,J_v(k_c\rho) + D\,Y_v(k_c\rho) \qquad (10.5)$$

where J_ν and Y_ν are Bessel functions of the first and second kind and order ν. This is the standard solution for a circularly cylindrical waveguide, see for example p.108 in [5]. The boundary condition for the semisectorial waveguide is that $E_z = 0$ on all metal walls. Applied on the flat walls it gives that

$$E_z(\varphi = 0) = 0 \quad \Rightarrow A = 0$$

$$E_z(\varphi = \varphi_0) = 0 \quad \Rightarrow \sin \nu \varphi_0 = 0 \quad \Rightarrow \nu \varphi_0 = n\pi \quad \Rightarrow \nu = \frac{n\pi}{\varphi_0} \tag{10.6}$$

where n is an integer: $n = 1, 2, 3, \dots$. Also $n = 0$ fulfils the boundary condition, but leads to zero fields and is therefore a non-physical solution. For semisectorial waveguides ν is not necessarily an integer, as for circularly cylindrical waveguides. On the curved walls the boundary condition is

$$E_z(\rho = a, b) = 0 \tag{10.7}$$

which leads to

$$CJ_\nu(k_c a) + D Y_\nu(k_c a) = 0$$

$$CJ_\nu(k_c b) + D Y_\nu(k_c b) = 0 \tag{10.8}$$

This system of equations can not be solved in closed form, except in the special case, when $\nu = \frac{1}{2}$. By slight manipulation Eq. (10.8) turns into

$$\frac{J_\nu(k_c a)}{J_\nu(k_c b)} \cdot \frac{Y_\nu(k_c b)}{Y_\nu(k_c a)} - 1 = 0 \tag{10.9}$$

which can be solved numerically. There is an infinite number of solutions $p_{\nu m} = k_c a$ for any ratio of the radii $r = b/a$. Here $p_{\nu m}$ denotes the mth solution to Eq. (10.9) and is generally not equal to a zero of any specific Bessel function. When the solution $p_{\nu m}$ has been found, $s = D/C$ is found by solving Eq. (10.8):

$$s = -\frac{J_\nu(p_{\nu m})}{Y_\nu(p_{\nu m})} \tag{10.10}$$

The cut-off frequency is

$$f_{c,\nu m} = \frac{c p_{\nu m}}{2\pi a} \tag{10.11}$$

The field equations for the $TM_{\nu m}$ modes in semisectorial waveguide are now

$$E_z = E_0 \left[J_v \left(\frac{p_{vm}\rho}{a} \right) + sY_v \left(\frac{p_{vm}\rho}{a} \right) \right] \cdot \sin v\varphi \cdot \exp\{- j\beta_{vm}z\}$$

$$E_\rho = \frac{- j\beta_{vm} p_{vm}}{ak_{c,vm}^2} E_0 \left[J_v' \left(\frac{p_{vm}\rho}{a} \right) + sY_v' \left(\frac{p_{vm}\rho}{a} \right) \right] \cdot \sin v\varphi \cdot \exp\{- j\beta_{vm}z\}$$

$$E_\varphi = \frac{- jv\beta_{vm}}{\rho k_{c,vm}^2} E_0 \left[J_v \left(\frac{p_{vm}\rho}{a} \right) + sY_v \left(\frac{p_{vm}\rho}{a} \right) \right] \cdot \cos v\varphi \cdot \exp\{- j\beta_{vm}z\} \qquad (10.12)$$

$$H_z = 0$$

$$H_\rho = -\frac{E_\varphi}{Z_{e,vm}}$$

$$H_\varphi = \frac{E_\rho}{Z_{e,vm}}$$

where

$$\beta_{vm} = \sqrt{k^2 - \left(\frac{p_{vm}}{a} \right)^2}$$

$$k = \frac{2\pi f}{c} \qquad (10.13)$$

$$Z_{e,vm} = \frac{\beta_{vm}}{k} Z_w$$

$$Z_w = \sqrt{\frac{\mu_r}{\varepsilon_r}}$$

10.3.2 TE Modes in Semisectorial Waveguide

For transverse electric (TE) modes the axial electric field is zero ($E_z = 0$). The waveguide modes can be derived from the transversal Helmoltz's equation for the axial magnetic field H_z:

$$\nabla_T^2 H_z(\rho,\varphi) + k_c^2 H_z(\rho,\varphi) = 0 \qquad (10.14)$$

This is the same differential equation for H_z as Eq. (10.2) for E_z. The solutions are therefore also the same:

$$H_z = P(\rho)\Phi(\varphi) \qquad (10.15)$$

where $P(\rho)$ and $\Phi(\varphi)$ are given by Eq. (10.5). The boundary condition is that the tangential electric field is zero on the metal walls. Before it can be applied, the transverse electric field must be calculated:

$$E_T = \hat{\mathbf{u}}_z \times \nabla_T H_z(\rho, \varphi) \tag{10.16}$$

where

$$\nabla_T H_z(\rho, \varphi) = \hat{\mathbf{u}}_\rho \frac{\partial H_z}{\partial \rho} + \hat{\mathbf{u}}_\varphi \frac{1}{\rho} \frac{\partial H_z}{\partial \varphi}$$

$$\frac{\partial H_z}{\partial \rho} = \Phi(\varphi)\left[C k_c J_v'(k_c \rho) + D k_c Y_v'(k_c \rho) \right] \tag{10.17}$$

$$\frac{\partial H_z}{\partial \varphi} = P(\rho)\left(- A v \sin v\varphi + B v \cos v\varphi \right)$$

The transverse electric field is now

$$E_T = -\hat{\mathbf{u}}_\varphi \frac{\partial H_z}{\partial \rho} + \hat{\mathbf{u}}_\rho \frac{1}{\rho} \frac{\partial H_z}{\partial \varphi} \tag{10.18}$$

The boundary condition on the flat walls is $E_\rho(\varphi = 0, \varphi = \varphi_0) = 0$, which gives

$$E_\rho(\varphi = 0) = 0 \quad \Rightarrow B = 0$$

$$E_\rho(\varphi = \varphi_0) = 0 \quad \Rightarrow \sin v\varphi_0 = 0 \quad \Rightarrow v\varphi_0 = n\pi \quad \Rightarrow v = \frac{n\pi}{\varphi_0} \tag{10.19}$$

where n is an integer: $n = 0, 1, 2, 3, \ldots$. This time $n = 0$ is also a solution, corresponding to modes with fields without dependence on φ. Because $v = 0$ always when $n = 0$, independent of φ_0, the wave mode solution, including cut-off frequency, is also independent of the sector angle φ_0.

On the curved walls the boundary condition is

$$E_\varphi(\rho = a, b) = 0 \tag{10.20}$$

which applied to Eq. (10.18) gives

$$C J_v'(k_c a) + D Y_v'(k_c a) = 0$$
$$C J_v'(k_c b) + D Y_v'(k_c b) = 0 \tag{10.21}$$

This system of equations can not be solved in closed form. By slight manipulation Eq. (10.21) turns into

$$\frac{J'_v(k_c a)}{J'_v(k_c b)} \cdot \frac{Y'_v(k_c b)}{Y'_v(k_c a)} - 1 = 0 \qquad (10.22)$$

which can be solved numerically. Again there is an infinite number of solutions $p'_{vm} = k_c a$ for any ratio of the radii $r = b/a$. This time p'_{vm} denotes the mth solution to Eq. (10.22) and is generally not equal to a zero of a derivative of any specific Bessel function. When the solution p'_{vm} has been found, $s = D/C$ is found by solving Eq. (10.21):

$$s = -\frac{J'_v(p'_{vm})}{Y'_v(p'_{vm})} \qquad (10.23)$$

The cut-off frequency is

$$f_c = \frac{c\,p'_{vm}}{2\pi a} \qquad (10.24)$$

The field equations for the TE_{vm} modes in semisectorial waveguide are now:

$$H_z = H_0 \left[J_v\left(\frac{p'_{vm}\rho}{a}\right) + sY_v\left(\frac{p'_{vm}\rho}{a}\right) \right] \cdot \cos v\varphi \cdot \exp\{-j\beta_{vm}z\}$$

$$H_\rho = \frac{-j\beta_{vm}p'_{vm}}{ak_{c,vm}^2} H_0 \left[J'_v\left(\frac{p'_{vm}\rho}{a}\right) + sY'_v\left(\frac{p'_{vm}\rho}{a}\right) \right] \cdot \cos v\varphi \cdot \exp\{-j\beta_{vm}z\}$$

$$H_\varphi = \frac{jv\beta_{vm}}{\rho k_{c,vm}^2} H_0 \left[J_v\left(\frac{p'_{vm}\rho}{a}\right) + sY_v\left(\frac{p'_{vm}\rho}{a}\right) \right] \cdot \sin v\varphi \cdot \exp\{-j\beta_{vm}z\} \qquad (10.25)$$

$$E_z = 0$$

$$E_\rho = Z_{h,vm} H_\varphi$$

$$E_\varphi = -Z_{h,vm} H_\rho$$

where k and Z_w are given by Eq. (10.13), and β_{vm} and $Z_{h,vm}$ are given by

$$\beta_{vm} = \sqrt{k^2 - \left(\frac{p'_{vm}}{a}\right)^2}$$

$$Z_{h,vm} = \frac{k}{\beta_{vm}} Z_w \qquad (10.26)$$

10.3.3 Values for p_{vm} and p'_{vm} for Waveguide Modes in Semisectorial Waveguide

The p_{vm} and p'_{vm} values depend on the sector angle φ_0 for all $n \neq 0$, and on r. No comprehensive table of values can therefore be given, although we can give some examples. Below a table is given (Table 10.1) as well as a set of diagrams (Fig. 10.2) with values solved from Eqs. (10.9) and (10.22) for different values of r. Each diagram is for one value of v. For $n = 0$ (only TE modes) the values are valid for any φ_0. For $n = 1$, the v values correspond to cases, where the sector angle is an even fraction of 2π (i.e. the cylinder has been divided into an even number of sectors: 1, 2, 4, 6, or 8). For modes with $n \geq 2$, v can be calculated from Eq. (10.19), and p_{vm} and p'_{vm} looked up in the table, or approximately from the graphs in Fig. 10.2. Alternatively Eqs. (10.9) or (10.22) are solved numerically on a computer, as was done in this work.

For some special cases of $v = l + \frac{1}{2}$, where l is an integer, there may be exact solutions to Eq. (10.9). For example from Chap. 11 in [7]:

$$J_{l+\frac{1}{2}}(x) = \sqrt{\frac{2x}{\pi}}\, j_l(x)$$

$$Y_{l+\frac{1}{2}}(x) = \sqrt{\frac{2x}{\pi}}\, y_l(x)$$

(10.27)

where j and y are so called spherical Bessel functions of the first and second kind. For integer order they have exact representations given by trigonometric functions. For example for $l = 0$ [7]:

$$j_0(x) = \frac{\sin x}{x}$$

$$y_0(x) = -\frac{\cos x}{x}$$

(10.28)

If Eq. (10.27) and Eq. (10.28) are substituted into Eq. (10.9), the result is

$$\frac{\tan k_c a}{\tan k_c b} - 1 = 0$$

(10.29)

The solutions to this equation are

$$p_{\frac{1}{2}m} = k_c a = \frac{m\pi a}{a-b} = \frac{m\pi}{1-r}$$

(10.30)

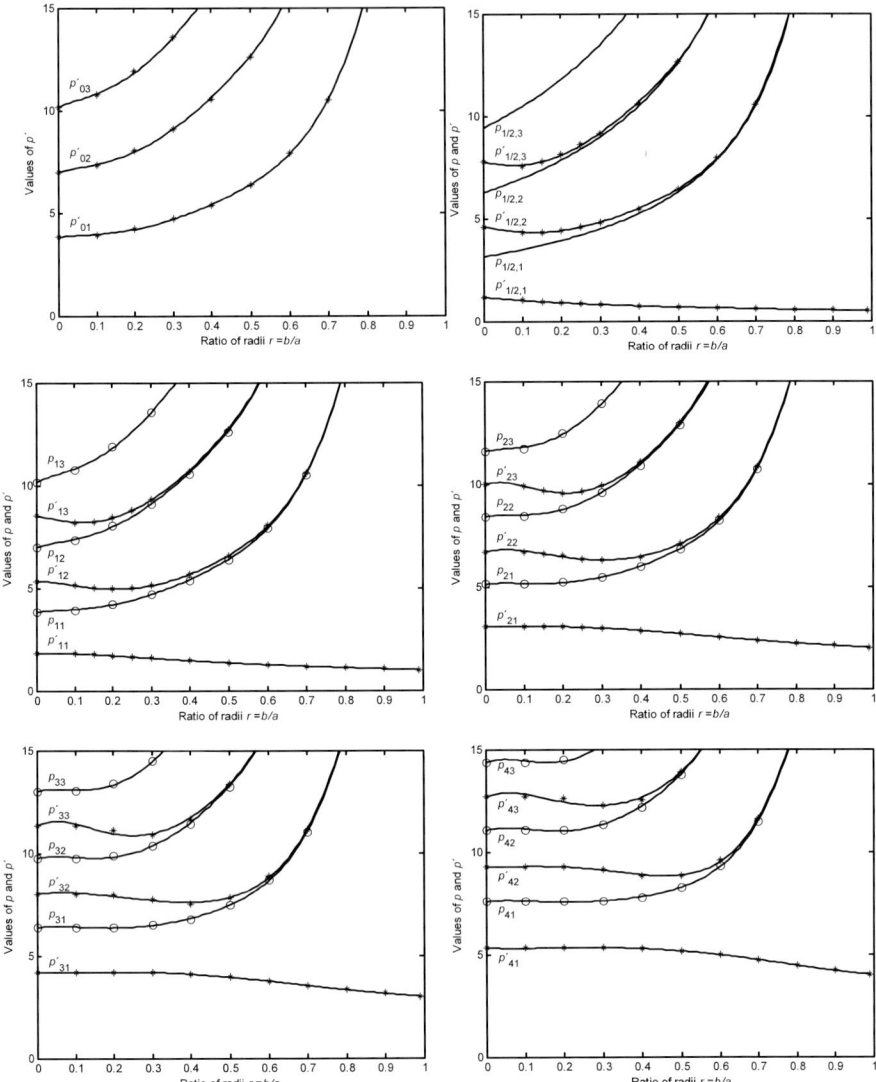

Fig. 10.2. Graphs for p and p' for TM and TE modes in semisectorial waveguides. The asterisks and rings denote solved values. The curves are fitted polynomials, except for the curves for $p_{\frac{1}{2}m}$, which were calculated using Eq. (30). Some of the other curves falsely indicate a hump between the solved values for small r and large v and m

For j and y for higher values of l, see e.g. [7]. Exact solutions to Eq. (10.9) may be found also for these special cases, which correspond to $v =$ integer $+ \frac{1}{2}$.

10.3.4 Discussion on Semisectorial and Sectorial Waveguide Modes

A sectorial waveguide is a special case of a semisectorial waveguide, with $r = 0$. In the other end, when $r \to 1$, the semisectorial waveguide starts to resemble a low rectangular waveguide that is bent. When the radius of curvature is large compared to the height of the waveguide, one can expect that the curvature has no influence on the cut-off frequency. Indeed, a study of the semisectorial modes shows that the cut-off frequency approaches that of the corresponding mode in the rectangular waveguide, when r approaches 1.

Especially in narrow sectors (or for small values of r) one would intuitively expect the fields to occupy the wider space closer to the periphery and escape from the tip. This is also the case for most of the modes, as can be seen by considering that the narrower the sector becomes, the larger v becomes. It is also large for large values of n, which correspond to cases where the basic field pattern (of a wave mode with $n = 1$) repeats n times over the sector, so that one pattern occupies only a narrow sector. From the equations above it is seen that the field distribution along ρ is then described by Bessel functions of high order or their derivatives. The higher the order of a Bessel function, the slower the function rises in the beginning. This also applies to its derivative, which means that the larger v becomes, the more the fields are concentrated close to the periphery and the less in the centre tip of the sector. Because the wave modes have little energy in the tip of a narrow sector, it is no surprise that the p values are practically independent of r as long as r is small, as is seen from the diagrams. There is one exception to the discussion above, the TE_{0m} modes. These modes have a field structure that is independent of φ, i.e. they have circular symmetry. They also exist in circular waveguides.

A semisectorial mode transforms continuously as the sector angle or ratio of radii changes, and must therefore be regarded as basically the same mode. For different sector angles the first index (v) will, however, be different according to Eq. (10.6) and Eq. (10.19). Therefore modes with the same n and m values should be regarded as the same mode. Figure 10.3 shows an example of the same mode in three different sectorial waveguides.

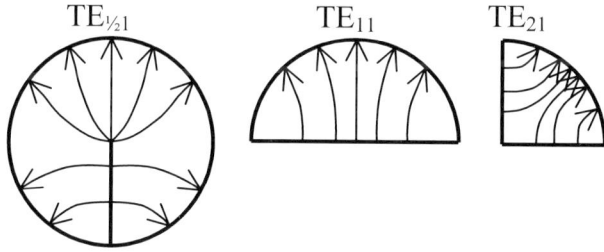

$$TE_{\frac{1}{2}1} \qquad TE_{11} \qquad TE_{21}$$

Fig. 10.3. A qualitative representation of the electric field of three special cases of the same mode in sectorial waveguide p'_{v1}, $n = 1$ for $\varphi_0 = 2\pi$, π, and $\pi/2$

Table 10.1. Some numerically solved p-values for semisectorial waveguide modes

v	r	TM$_{vm}$			TE$_{vm}$		
		p_{v1}	p_{v2}	p_{v3}	p'_{v1}	p'_{v2}	p'_{v3}
0	0.00	-	-	-	3.832	7.016	10.174
0	0.10	-	-	-	3.941	7.331	10.748
0	0.20	-	-	-	4.236	8.055	11.926
0	0.30	-	-	-	4.706	9.104	13.553
0	0.40	-	-	-	5.391	10.558	15.766
0	0.50	-	-	-	6.393	12.625	18.889
0	0.60	-	-	-	7.930	15.747	23.588
0	0.70	-	-	-	10.522	20.969	31.433
0	0.80	-	-	-	15.737	31.431	47.134
½	0.00	π	2π	3π	1.1656	4.604	7.790
½	0.10	3.491	6.981	10.472	1.0140	4.316	7.557
½	0.20	3.927	7.854	11.781	0.8973	4.429	8.149
½	0.30	4.488	8.976	13.464	0.8074	4.816	9.155
½	0.40	5.236	10.472	15.708	0.7366	5.459	10.589
½	0.50	6.283	12.566	18.850	0.6792	6.436	12.645
½	0.60	7.854	15.708	23.562	0.6316	7.958	15.761
½	0.70	10.472	20.944	31.416	0.5913	10.539	20.978
½	0.80	15.708	31.416	47.124	0.5567	15.748	31.436
½	0.90				0.5266		
½	0.99				0.5025		
1	0.00	3.832	7.016	10.174	1.8412	5.331	8.536
1	0.10	3.941	7.331	10.748	1.8035	5.137	8.199
1	0.20	4.236	8.055	11.926	1.7051	4.961	8.433
1	0.30	4.706	9.104	13.553	1.5821	5.137	9.308
1	0.40	5.391	10.558	15.766	1.4618	5.659	10.683
1	0.50	6.393	12.625	18.889	1.3547	6.565	12.706
1	0.60	7.930	15.747	23.588	1.2621	8.041	15.801
1	0.70	10.522	20.969	31.433	1.1824	10.592	21.004
1	0.80	15.737	31.431	47.134	1.1134	15.778	31.451
1	0.90				1.0531		
1	0.99				1.0050		
2	0.00	5.135	8.417	11.620	3.054	6.706	9.970
2	0.10	5.142	8.457	11.738	3.053	6.687	9.887
2	0.20	5.222	8.804	12.494	3.035	6.495	9.549
2	0.30	5.470	9.600	13.905	2.968	6.274	9.918
2	0.40	5.966	10.894	15.999	2.842	6.416	11.056
2	0.50	6.814	12.856	19.046	2.681	7.063	12.949
2	0.60	8.227	15.904	23.694	2.516	8.367	15.961
2	0.70	10.720	21.071	31.501	2.363	10.799	21.106
2	0.80	15.855	31.490	47.174	2.226	15.898	31.510
2	0.90				2.106		
2	0.99				2.010		
3	0.00	6.380	9.761	13.015	4.201	8.015	11.346
3	0.10	6.380	9.764	13.030	4.201	8.014	11.338
3	0.20	6.394	9.874	13.381	4.199	7.964	11.106

3	0.30	6.494	10.371	14.477	4.180	7.721	10.920
3	0.40	6.780	11.435	16.380	4.108	7.535	11.666
3	0.50	7.458	13.232	19.304	3.958	7.840	13.347
3	0.60	8.699	16.161	23.868	3.754	8.889	16.226
3	0.70	11.041	21.239	31.614	3.540	11.136	21.276
3	0.80	16.050	31.589	47.240	3.339	16.096	31.610
3	0.90				3.159		
3	0.99				3.015		
4	0.00	7.588	11.065	14.372	5.318	9.282	12.682
4	0.10	7.588	11.065	14.374	5.317	9.282	12.681
4	0.20	7.590	11.091	14.497	5.317	9.273	12.610
4	0.30	7.623	11.348	15.245	5.313	9.152	12.241
4	0.40	7.790	12.152	16.901	5.282	8.852	12.501
4	0.50	8.267	13.742	19.662	5.175	8.836	13.892
4	0.60	9.317	16.515	24.110	4.970	9.582	16.590
4	0.70	11.476	21.472	31.771	4.711	11.594	21.512
4	0.80	16.318	31.727	47.332	4.451	16.369	31.748
4	0.90				4.212		
4	0.99				4.020		

10.4 Cylindrical Fin Resonator (CFR) Sensor

Looking at the values in Table 10.1 reveals that for $r = 0$ and $\varphi_0 = 2\pi$, which is a cylinder with an axial fin extending to the centre of the pipe, as in the left picture in Fig. 10.3, $p'_{1/2.1} = 1.1656$. This value is only 63.3% of the smallest value for an ordinary cylindrical waveguide ($p'_{11} = 1.8412$). Because the cut-off frequency, according to Eq. (10.24), is directly proportional to this value, a waveguide with such a fin has a cut-off frequency which is 63.3% of that of the cylindrical waveguide without the fin. This provides a simple means to make a resonator sensor in a pipe. A resonator with open ends (contrary to shorted) can support TE_{vml} resonance modes with $l = 0$, which lack structure in the length direction. The resonant frequency of such a mode is equal to the cut-off frequency. Therefore, a fin of a suitable length attached to the pipe wall becomes a resonator, with a resonant frequency (TE_{v210}) that is 63.3% of the cut-off frequency of the pipe outside the section with the fin. Therefore the energy cannot escape even though the structure is very open in the pipe.

Roxar Flow Measurement has designed a watercut meter, i.e. a meter for measuring the water content of oil, according to this principle. A schematic picture of the sensor design is shown in Fig. 10.4. An advantage is that the sensor is simple to make. Only two probes for coupling of the energy must be added. Otherwise the sensor consists of a flanged piece of pipe with a fin welded to the wall. The length of the fin has been chosen to be equal to one inner diameter of the pipe. This provides an ideal compromise between robustness and distance to the higher resonance modes that do depend on the length of the fin. One of the probes is located opposite to, and centred on the fin. This eliminates the coupling to the

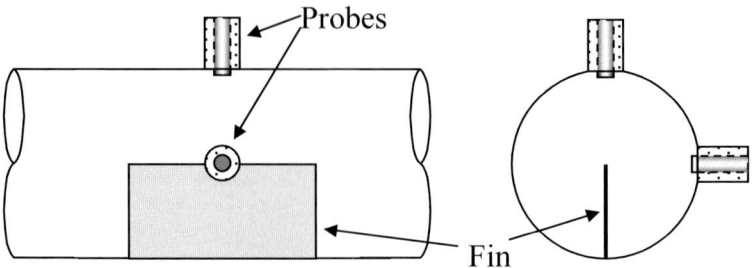

Fig. 10.4. The principal design of the CFR sensor of the Roxar WaterCut Meter

next two modes (TE_{110}, and $TE_{\frac{1}{2}11}$), which further eliminates the risk of confusion between peaks in a dynamic measurement situation.

Another advantage with the CFR sensor design is the minimal obstruction to the flow that the fin creates. It is also possible to make the fin with a slanting end, which should further reduce the risk of clogging by solids.

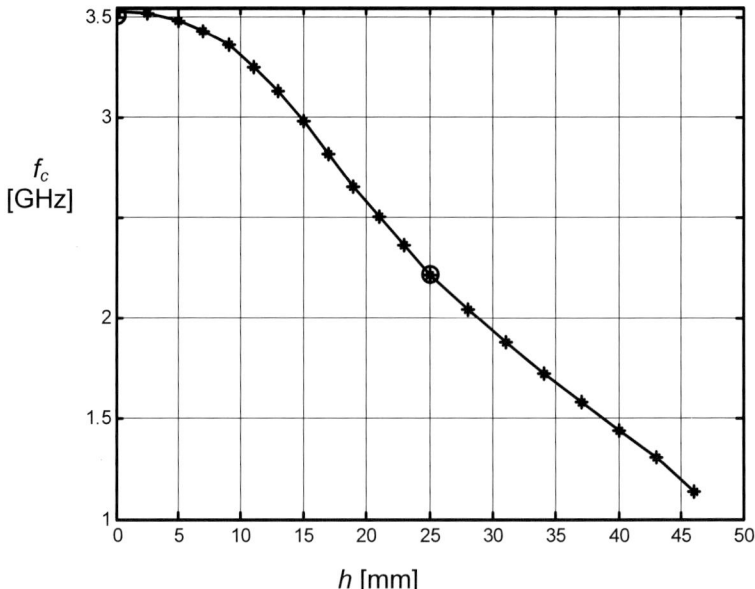

Fig. 10.5. The resonant frequency of a CFR sensor as a function of the height of the fin. The diameter of the pipe is 50 mm. The circles are exactly calculated results (pipe without fin, and sectorial waveguide), and the asterisks are simulations.

A slight disadvantage is that the sensor is longer than in the case of end grid resonator sensors (see below). This is because a length of pipe equal to one inner diameter of the pipe must be added to each end of the sensor to eliminate radiation, when the sensor is tested on the bench during manufacturing. Together with the length of the section with the fin, a sensor is therefore three pipe diameters long. In addition a thermo well for a temperature sensor is often added, which further adds to the length. In most cases this is, however, not a problem.

The Roxar CFR watercut sensor has been produced in sizes from 2" to 30" in diameter. Also a 1" version for fast loops (bypass on larger pipes) has been made. This is a semisectorial sensor, with a cylinder in the centre in addition to the fin. As can be seen from Table 10.1, increasing the value of r reduces the cut-off (and resonant) frequency. The semisectorial design was chosen to reduce the resonant frequency to be the same as in a 2" sensor.

Some properties of CFR sensors cannot be calculated based on ideal sectorial wave modes. To study the sensor further the author therefore used the Hewlett Packard high frequency structure simulator (HFSS) software. Figure 10.5 shows one of the results, i.e. how the resonant frequency depends on the height of the fin. It is seen that there is a continuous change as a function of the height. The basic version, with a fin extending to the centre of the pipe, was chosen for the Roxar sensor, because this design gives the most even field distribution. The results of the simulations have been reported in more detail in [2].

10.5 End Grid Resonator Sensor

A CFR sensor may sometimes be inconveniently long, especially for large diameters of the pipe. In such cases an end grid sensor may be a better choice [2]. The Roxar WaterCut Meter can be delivered with either version.

The Roxar end grid resonator sensor (EGR) uses the TM_{010} resonance mode, which lacks structure in the axial direction of the pipe due to the third index being "0". For the same reason the resonant frequency is equal to the cut-off frequency of that mode. When the end grids were designed, various structures were studied [2]. The one that was found to be the best consists of eight radial plates with a ring in the middle, as shown in Fig. 10.6. The radial plates provide a good short to the TM_{010} resonance mode, which due to the tangential magnetic field excites radial currents in the end structures. The main purpose of the ring in the centre is to make the radial plates shorter. Otherwise radial quasi-TEM modes may resonate in the plates on a too low frequency, and possibly cause leakage.

In accordance with Eq. (10.1) the end grids may be shorter the larger is the distance between the resonant frequency and the lowest cut-off frequency of the grids. The relation between the cut-off frequency in the cylindrical hole in the centre, in the semisectorial holes in the outer region, and the resonant frequency of the quasi-TEM mode in the radial plates was studied for various numbers of radial

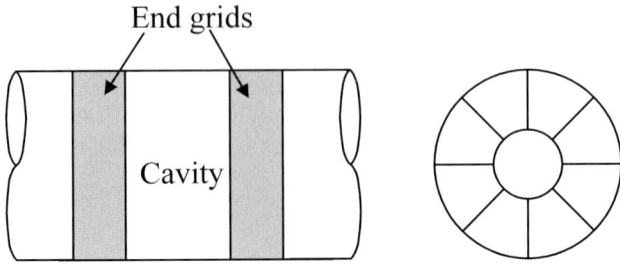

Fig. 10. 6. The principal design of the end grid resonator for the Roxar WaterCut Meter

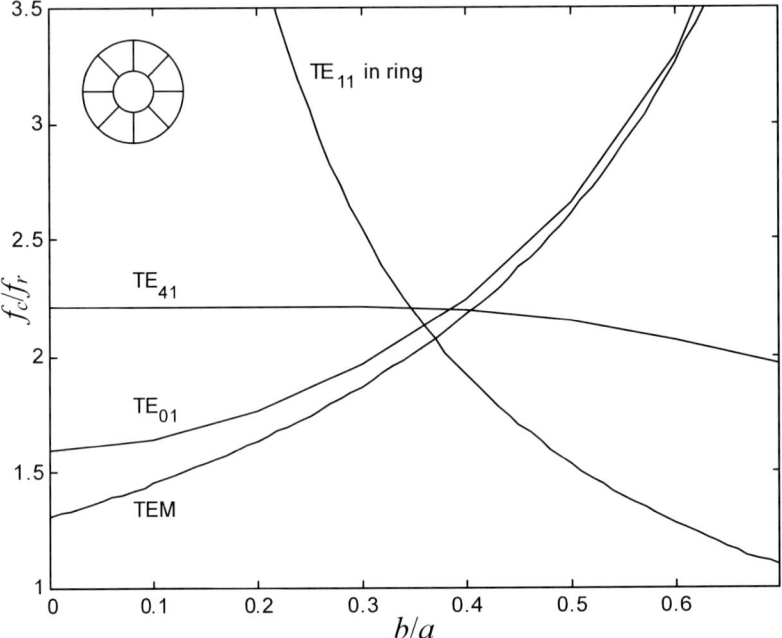

Fig. 10.7. The cut-off frequencies in the structure of an end grid compared to the resonant frequency of the sensor

plates and for various values of the ratio of radii r. The lowest mode in the ring is TE_{11}, and in the semisectors TE_{01} and, e.g., TE_{41} for the case of 8 plates. Figure 10.7 shows a diagram of the various frequencies as a function of r in the case of 8 plates. For all values of r, the cut-off frequency of TE_{41} is higher than either that of TE_{11} in the cylinder, or TE_{01} in the semisectors. Because the cut-off frequency of TE_{01} is independent of φ_0, i.e. of the number of plates, no further improvement

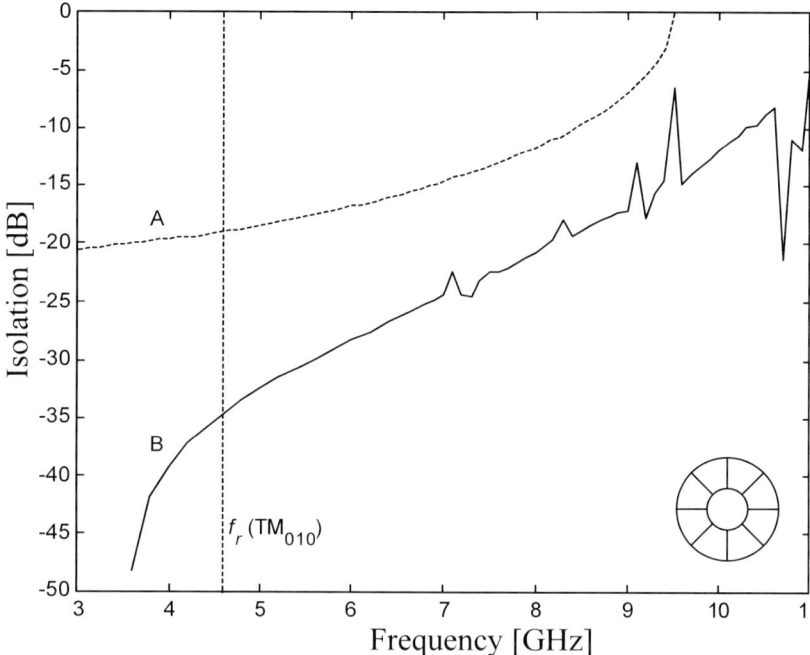

Fig. 10.8. The calculated (A) and simulated (B) isolation of the end grid. The inner diameter of the pipe was assumed to be 50 mm

can be achieved by increasing the number of plates. A reduction of the number of plates would cause the cut-off frequency of the mode, which corresponds to TE_{41} for 8 plates, to be the lowest in the middle range of the diagram. The optimal result is therefore achieved with 8 plates and $r \approx 0.37$. The performance of such grids was both simulated using HFSS and measured.

A resonator prototype was built with longer grids than judged to be necessary. The performance was measured while slices of the structure was cut off in small steps. The leakage did not affect the resonant frequency and Q-factor significantly until the grids were shorter than 0.25 times the inner diameter of the pipe. In the simulation a model of that length and with 8 plates (thickness 2 mm) and $r = 0.37$ was used. This grid was mounted in a waveguide, and the transmission through the grid was studied. Figure 10.8 shows the result together with the calculated isolation. The calculation was based on the cut-off frequencies, ignoring any coupling mismatch. The coupling mismatch is, however, significant because the TM_{010} mode has circular symmetry, while the lowest mode in the pipe (TE_{11}) does not. Therefore the simulated result is significantly better than the calculated. The spikes in the simulated isolation are probably caused by inaccuracies, and are not real.

10.6 Wet Gas Meter Based on a V-Cone Resonator

Oil wells usually produce a mixture of oil, water, and gas. The relative fractions vary during the lifetime of the well, and to be able to optimise the recovery, and the use of the production facility, the composition and the flow rate of the produced fluid must be measured. The conventional technology is to have a test separator, through which each well can be routed one at a time for testing. Since 1995 so-called inline multiphase meters have replaced the test separators in many places [8]. Future field developments will to an increasing degree include wet gas wells. These are wells that have a very high gas void fraction (GVF). In the case of wet gas wells a test separator is usually not a possible option because of the lenghty time constant, and multiphase meters are not accurate and sensitive enough. Therefore inline wet gas meters (WGM) must be used for well testing, continuous reservoir monitoring and production optimisation, allocation, and flow assurance.

A wet gas meter provides the flow rates of the constituents, and the volume fractions in the pipe at meter conditions. The gas and condensate rates are important because of the economical value of the production, and the water rate is important for the control and inhibition of corrosion and hydrate formation, particularly in long subsea pipelines.

Roxar has developed a WGM for the above-mentioned purposes. The first units were delivered in 2002 to Statoil for subsea installation on the wells at the Mikkel field in the North Sea. The meter concept is based on a combination of a microwave permittivity measurement, a differential pressure flow measurement, and pressure, volume, temperature (PVT) calculations. The microwave measurement is sensitive mostly to the water volume fraction (WVF), while the differential pressure measurement gives the impulse rate (i.e. *density×velocity²*), and the PVT calculations give the split between gas and condensate. For the PVT calculations the chemical composition of the hydrocarbons must be known. In addition the temperature and pressure are measured. By combining these measurements the flow rates (in kg/h) of gas, condensate, and water, and the volume fractions and the density of the flow can be calculated.

The microwave sensor of the WGM is a cavity resonator. Various versions of the WGM have been designed because of various demands by the customers, but the V-Cone based design is the most compact and will be described here.

The V-Cone was originally developed by others as a mixer a number of years ago, and has subsequently also been used for differential pressure measurements. In Roxar it was first considered for the WGM as a mixer and differential pressure device in combination with a CFR microwave sensor. It was, however, soon realized that the V-Cone itself could also be used as microwave resonator to create a highly compact WGM [9]. Only microwave probes had to be added for the coupling of the microwave energy (see Fig. 10.9), and various methods of attachment of the cone were considered both from a microwave and a mechanical point of view.

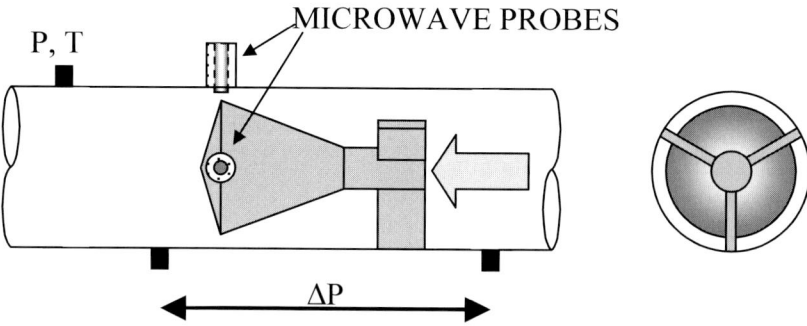

Fig. 10.9. The principal design of the Roxar V-Cone WGM

A conventional V-Cone is attached to the pipe wall by one support in the narrow end of the cone. The first idea was to extend the support to the whole length of the cone to create a semisectorial resonator with a varying diameter of the inner cylinder. A semisectorial waveguide has a cut-off frequency, which is lower than in the plane pipe, and a semisectorial structure with open ends will support a TE resonance mode, with a resonant frequency equal to the cut-off frequency. Therefore the energy will not escape. This structure was simulated using the HFSS software, and found to have a good frequency response. A disadvantage was the fact that the electric field distribution around the circumference of the pipe was uneven, because of the shorting effect of the support. The conventional design was therefore also considered as a possibility. Here the structure resonates more like a λ/4 resonator, shorted to the wall by the support in the narrow end and open in the other end. A simulation showed that this also had a good frequency response, and a somewhat lower resonant frequency. The oil industry would, however, not accept the structure for subsea use because of its relatively weak construction. Therefore the number of supports were increased from one to three, which is the structure shown in Fig. 10.9. Again a simulation showed that the frequency response is good, with one clearly defined resonance, and a large distance to the next one. This eliminates the risk of peak confusion in a dynamic measurement situation. Because of the mechanical preferences and the good frequency response this structure was chosen for the Roxar WGM.

The resonance mode in the V-Cone with three supports (called V-Cone below) can be considered as a coaxial mode. The three supports act like a shorting end wall for the cone, which acts like the centre conductor. In the other end the resonator is open. The currents are axial and the lowest mode, which is the one used, is a circularly symmetrical quasi-TEM mode, with a radial electric field between the cone and the wall. The electric field maximum is at the point, where the cone is widest. Because of the varying diameter of the centre conductor the resonator is capacitively loaded in the gap between the broad end of the cone and the pipe wall, which lowers the resonant frequency considerably.

The resonant frequency is highest for the semisectorial structure and lowest for the conventional single-support structure. The structure with three supports has a higher resonant frequency than the single-support structure because the resonator is shorter. The electric field is zero at the meeting point of the three supports, which is therefore the location of the short, while the single-support structure is shorted at the attachment point of the support at the pipe wall. The three structures were simulated using HFSS. When the dimensions were identical except for the supporting structure, the simulation gave the following resonant frequencies for a typical 2" sensor: 1285 MHz (semisectorial), 774 MHz (three supports), and 570 MHz (single support). In addition to the method of attachment of the cone, the dimensions also affect the resonant frequency. Larger dimensions generally lead to a lower resonant frequency, and a narrower gap between the cone and the pipe wall also leads to a lower frequency because of increased capacitive loading. Because the permittivity of gas, condensate, and water is relatively independent of the frequency in the frequency range in question, the resonant frequency is not critical for the WGM. The dimensions can therefore be chosen to be optimal for the differential pressure flow measurement.

In a wet gas flow the liquid tends to form a film on the wall of the pipe. When the flow passes the V-Cone, it is accelerated due to the smaller cross sectional area, and pressed towards the pipe wall. It will therefore blow the film off the wall and produce a mist flow through the narrow gap. Because the microwave measurement is weighted by the distribution of the electric field, the most significant contribution also comes from where the gap is narrowest, and the mist flow is well developed.

Advantages with the V-Cone version of the WGM are:

- Compact design because the same structure is used as a mixer, microwave sensor, and differential pressure device. This is important especially in subsea applications.
- The electric field is even in the gap, where the mixture is good and the measurement is performed.
- The frequency response is good, and eliminates the risk of peak confusion in a dynamic measurement situation.

Figure 10.10 shows a photograph of the 5" Mikkel meter, and Fig. 10.11 the measured frequency response. It is worth mentioning that the simulated and the measured frequency responses showed an excellent agreement, and the resonant frequency was predicted correctly to within 1% by the simulation.

In many cases the ability to detect the salt content of the water is important for the inhibition of corrosion in the flow lines. As long as the well does not produce formation water the water in the flow is condensed water, which does not contain salt. Because salt gives the water ion conductivity it will also affect the microwave losses. By performing a two-parameter measurement [1, Chap. 2], i.e. by measuring the Q-factor in addition to the resonant frequency, it should be possible also to detect when the well starts to produce formation water [9]. The salt detection function is currently under development for the Roxar WGM.

Fig. 10.10. The 5" Roxar subsea WGM for Mikkel. On the left side are two differential pressure transmitters for redundancy. On the right is the electronics canister. In the front is the pressure and temperature transmitter. In the lower part is a gamma densitometer for cases of low GVF

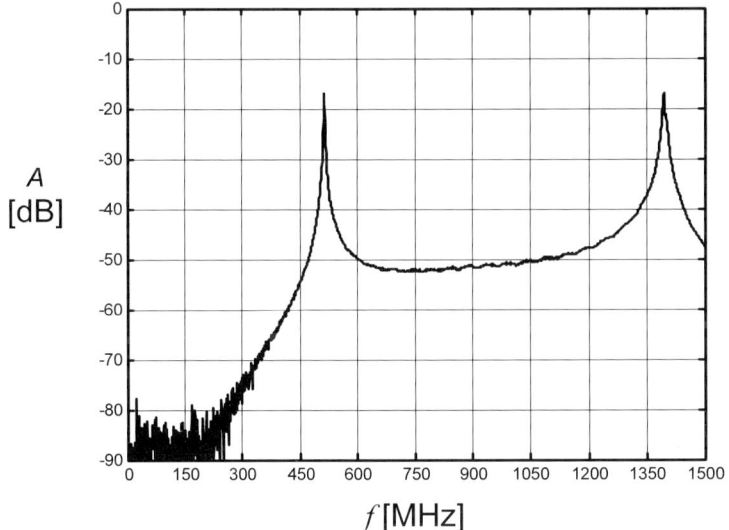

Fig. 10.11. The measured frequency response of the V-Cone WGM in Fig. 10.10, when filled with air

10.7 Watercut Meter for Measuring Downhole in an Oil Well

Oil well technology is moving towards longer or branched boreholes with production from several zones into the same main well. In order to optimise the production and recovery (percentage of the oil present in a reservoir that can be produced), the production from each zone must be measured and controlled separately. This is only possible by installing the measuring and controlling equipment in the borehole locally at the location of the producing zones. Such oil wells are called smart wells. Roxar has various downhole equipment, particularly also a Downhole WaterCut Meter (DWCM) for the measurement of the watercut of the fluid being produced [10].

The borehole in an oil well is lined by a pipe called the casing. Inside this is another one, called the tubing, which together with all chokes, safety valves, and instrumentation is called the completion. At a producing zone the oil enters the annulus (the annular space between the casing and the tubing) through perforations in the casing. It then joins the main flow in the tubing through a controlling choke. The DWCM is located in the annulus so that it measures the composition of the inflow from that zone. The area of the annulus with the perforations, DWCM, and the choke is sealed off by packers, as shown in Fig. 10.12. The environmental constraints on downhole equipment for the annulus are severe. Depending on the depth of the well, the pressure may be up to 1000 Bar, and the temperature up to 180°C. The space, and the electrical power available are limited, and the reliability must be high.

The DWCM is based on measuring the permittivity of the mixture of oil and water with a cavity resonator sensor. The watercut is then retrieved from the known relationship between the permittivity and the composition, and compensated for pressure and temperature, which are also measured by local downhole equipment.

The sensor of the Roxar DWCM is a semisectorial cavity resonator [11], which looks like a low rectangular resonator bent along the cylindrical surface of the tubing, as shown in Fig. 10.13. The resonance modes of such a sensor were described earlier in this chapter. Because of the relatively high ratio of radii r, they can also be estimated by regarding the resonator as a low rectangular cavity. The lowest mode is used, which has a radial electric field and a field distribution with a maximum at the centre of the resonator. The DWCM sensor is open in the ends allowing the flow to pass through the sensor. The short circuit for the microwaves is provided by 4 radial/axial plates that divide the end sections into 5 waveguides, with a cut-off frequency above the resonant frequency. The coupling devices are electrical probes. The steel-sheathed cable is electron beam (EB) welded to the probe housing, which is filled with a glass–ceramic mixture to stand the high pressure and temperature, and the chemically hostile environment. For best protection the probes are mounted on the concave side that faces the tubing. The resonant frequency of a resonator sensor can be measured using two different

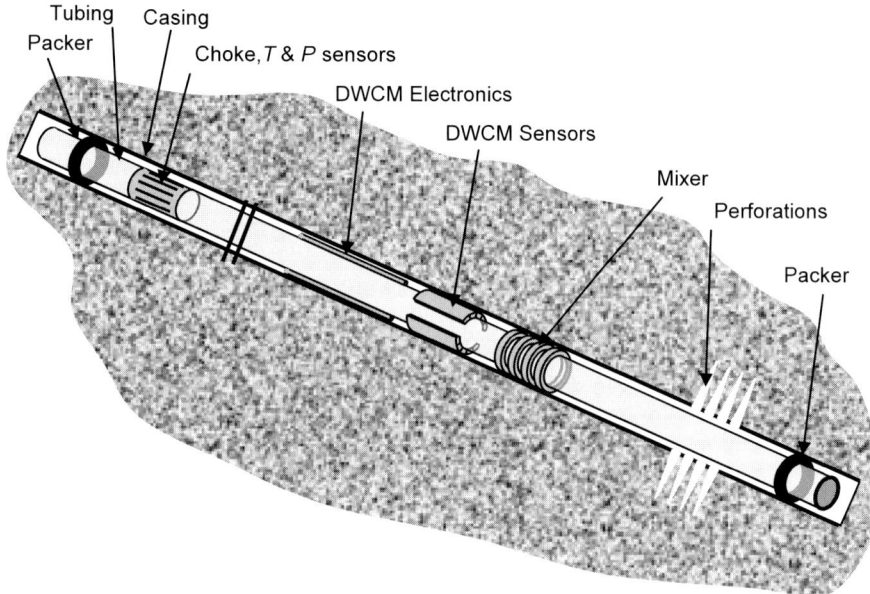

Fig. 10.12. The principal arrangement of the DWCM at the producing zone

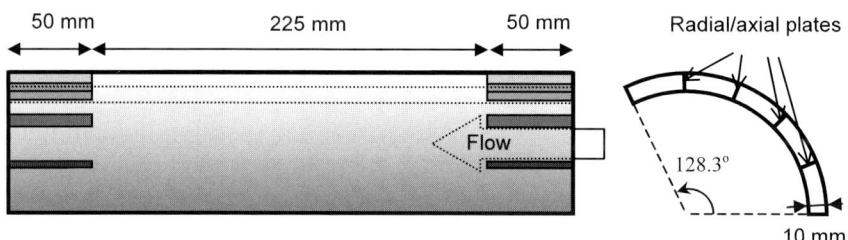

Fig. 10.13. The semisectorial cavity resonator sensor for the Roxar DWCM

methods: by performing a frequency sweep, or by locking an oscillator to the resonance peak. The former involves a lot of data. Either whole sweeps would have to be transferred to the software topside, which is not practical considering the limitations of the communication link, or the software would have to be located at the DWCM, which is also impractical. The latter method is based on an amplifier with positive feedback through the resonator, which then acts like a narrow filter. If the gain in the amplifier is higher than the insertion loss in the cables and the sensor at the resonance, the circuit starts to oscillate at a frequency near the resonant frequency, where also the phase criterion of $n \cdot 360°$ phase shift during one revolution is fulfilled. Only the frequency of oscillation then has to be counted and transferred to topside. Also, since the amount of electronics needed is very small, this method was chosen for the DWCM. It will here be called the feedback self-oscillating amplifier method (FSA) [2,11].

Because of the phase criterion the FSA circuit will jump from one frequency (n) to the next ($n \pm 1$), when the resonant frequency changes, and the frequency of oscillation is generally not exactly the same as the resonant frequency. This limits the resolution (and therefore the accuracy), but by using relatively long cables n becomes large and the resolution good. In the DWCM 5 m long cables with a ceramic insulator have been used, which gives a resolution of roughly 10 MHz, equivalent to roughly 1%$_{WC}$.

The FSA method requires that there is only one resonance peak in the range of the amplifier, or that the used peak is the highest. Otherwise the circuit may lock to the wrong peak. The electronics of the DWCM has a gain that rolls off with frequency favouring peaks at lower frequencies. The first resonance peak of the sensor is used, and the coupling to the next few modes has been minimized by the choice of locations for the coupling probes, as shown in Fig. 10.14. The crosses in

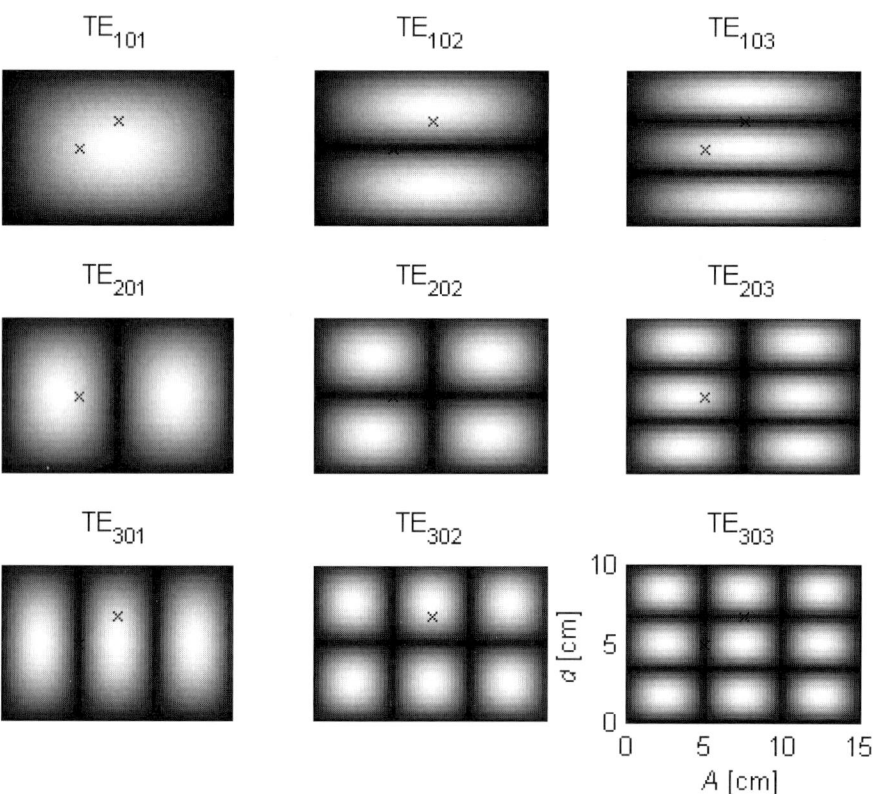

Fig. 10.14. The distribution of the electric field of the modes with n & $l \leq 3$ in the Roxar DWCM

the figure indicate the positions of the probes. Because only one or no cross is visible in the other figures, the arrangement eliminates the coupling to all but TE_{101} (for simplicity the indexes here and in the figure refer to the rectangular cavity modes that the semisectorial modes approach at high r). The characteristics of the gain and the frequency response of the sensor ensure that the circuit locks to the right peak. To fit into the annulus, the electronics is contained on a board that fits into a cylindrical enclosure, with an inner diameter of 19 mm.

Because of the high temperature the reliability of the electronics will be limited. To keep it acceptable it is therefore essential to keep the amount of components as small as possible, which is best achieved by using the FSA method. In addition, two sensors, with separate electronics units, are used for the improvement of the reliability by redundancy. In practice the sensors, a mixer (to ensure that the oil and water are well mixed), and the electronics units are mounted on a 3 m long threaded piece of tubing, called a carrier.

References

1. Nyfors E, Vainikainen P (1989) Industrial Microwave Sensors. Artech House, Norwood MA
2. Nyfors E (2000) Cylindrical Microwave Resonator Sensors for Measuring Materials under Flow. PhD Thesis, Helsinki Univ of Tech., Radio Lab, Report S243
3. Tiuri M, Nyfors E, Vainikainen P, Ståhl S (1986) Mittausmenetelmä ja –laite kiinteän, rakeisen aineen massavirtauksen ja kosteuden tai jonkin muun ominaisuuden määräämiseksi. (Method and device for measuring the material flow and the moisture, or some other property, of a solid, granular material; in Finnish), Suomi-Finland patent No. 69372
4. Lin F, Omar A (1989) Segment-sector waveguides. IEEE AP-S Digest
5. Collin RE (1966) Foundations for Microwave Engineering. McGraw-Hill, New York
6. Pozar D.M (1998) Microwave Engineering 2nd ed. Wiley
7. Arfken G (1970) Mathematical Methods for Physicists. Academic
8. Nyfors E, Wee A (2000) Measurement of mixtures of oil, water, and gas with microwave sensors. New developments and field experience of the MFI MultiPhase, and WaterCut Meters of Roxar. Proc Subsurface Sensing Technologies and Applications II, at SPIE's 45th Annual Meeting, San Diego, pp. 12–21
9. Nyfors E, Lund Bø Ø (2001) Compact flow meter. Norwegian patent application No. 2001.5132
10. Nyfors E (2001) Permanent downhole microwave sensor for the local measurement of the water content of the fluid being produced in an oil well. Proc. 4th Int Conf on Electromagnetic Wave Interaction with Water and Moist Substances, Weimar, Germany, pp. 293–300
11. Nyfors E (2000) Anordning for måling av egenskaper til en strømmende fluidblanding. (A Device for measuring the properties of a flowing fluid; in Norwegian), Norwegian patent No. 313.647

11 Microstrip Transmission- and Reflection-Type Sensors Used in Microwave Aquametry

Ferenc Völgyi

Budapest University of Technology and Economics, Hungary

11.1 Introduction

11.1.1 Microstrip Sensors Used in Microwave Aquametry

In this chapter we focus on microstrip sensors. Microstrip radiating sensors and microstrip antennas and arrays are attracting increasing interest due to their flat profile, low weight, ease of fabrication, and low cost. Because of these advantages, they are widely used in moisture sensors, in communications, in the nondestructive testing of materials, etc. With reference to the literature, we first describe some microstrip radiating sensors and antennas commonly used in microwave aquametry. Then the chapter summaries relevant R&D activities in microstrip antennas at BUTE/DBIS (Budapest University of Technology and Economics, Department of Broadband Infocommunication Systems) focusing on near-field experiments and monitoring of particleboards and concrete mixers. Furthermore, attention is given to the problems of new antenna technologies, with examples for active integrated antennas, photonic bandgap patch antennas, and silicon micromachined patch antennas, which show promise for microwave moisture measurement in the future.

The microstrip structure [1] is very popular as a sensor in microwave moisture measurement (MMM), because of its open fields (Fig.11.1a). As an application of an evanescent microwave probe (EMP) a microstrip resonator with a tapered tip (Fig.11.1b), capacitively coupled to a feed line and connected to a circulator, was reported in [2].

An EMP scan of a piece of balsa wood (2 cm ×1 cm) clearly shows regions containing moisture at a sub-micrometer lateral resolution. More widespread application of the microstrip structure is as a transmission line sensor placed near a moist substance, and the insertion loss, $|S_{21}|$, is measured, characterizing the moisture content, e.g., for oil palm fruits [3].

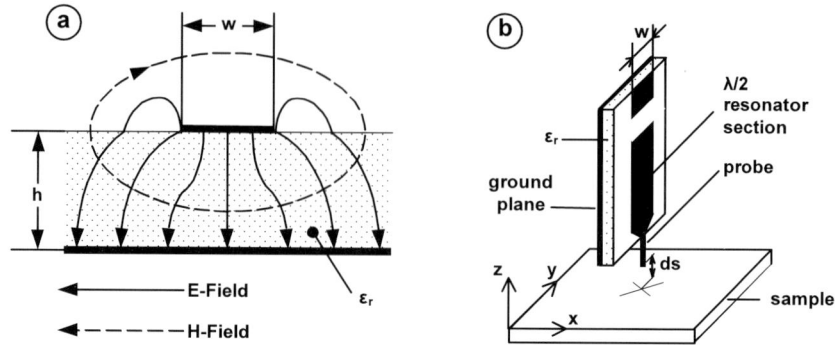

Fig. 11.1. a The E- and H-field lines of a microstrip transmission line at low frequencies with static approximation. **b** Microstripline resonator and the electric dipole probe assembly.. Evanescent waves extend out from the tapered tip of the resonator.(After [2])

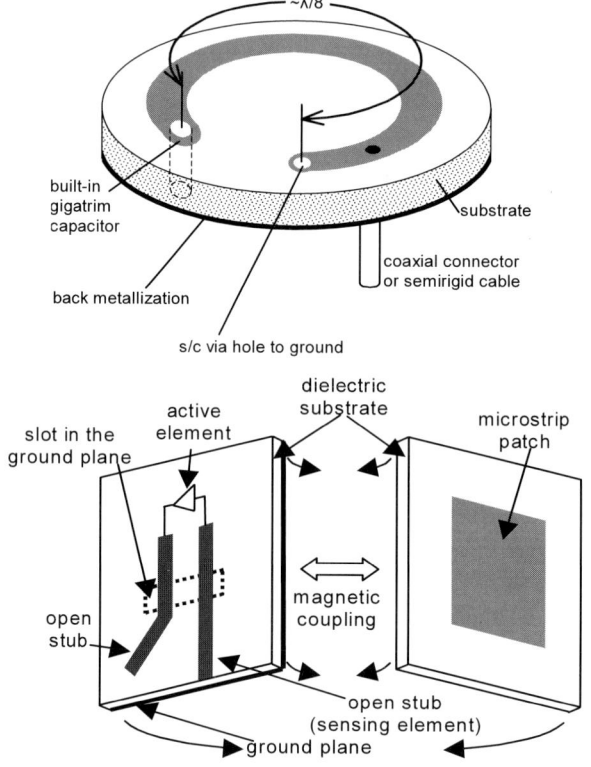

Fig. 11.2. a "Hula-hoop" microstrip resonator which was given originally with an airline [4]. **b** Microwave moisture sensor using an active integrated microstrip antenna. (Courtesy of G. Biffi Gentili [12])

The microstrip form of a "hula-hoop" resonator, which is a resonant open applicator "looking" into a dielectric half-space, is shown in Fig.11.2a. The influence of the moist material on the resonant behavior is weighted by the decay function, $1/r^3$, where r is the distance from the centre.

11.1.2 Microstrip Antennas as Sensors

Microstrip antennas are used as moisture sensors for near-field and radiating far-field measurements (see Fig.11.3 and [5-9]). The so-called super state loading (see Fig.11.4a and [10] p. 356) is suited well as a model of the MMM problem, and for measurement of the dielectric constant of homogeneous dielectric materials [11]. Recently, cylindrical (or conforming) microstrip antennas [10] also have been used in NDT and in microwave imaging (physical quantities, moisture content, etc.).

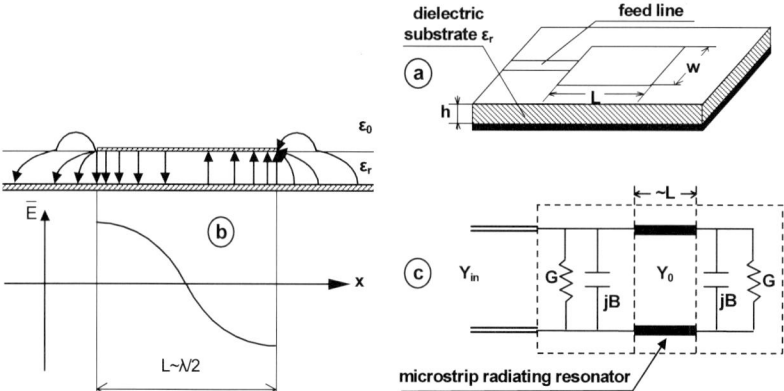

Fig. 11.3. a Rectangular microstrip antenna (RMSA) configuration. **b** E-field lines along the x-axis. **c** the equivalent circuit for the microstrip radiating element

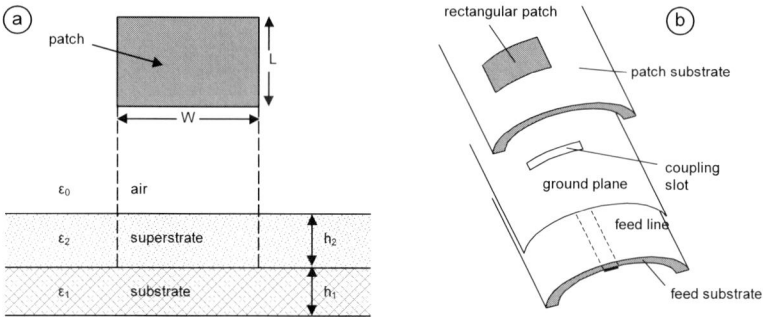

Fig. 11.4. a Geometry of a superstrate loaded RMSA structure. **b** Configuration of a slot-coupled cylindrical RMSA. (After [10])

The application of microstrip antennas (MSAs) in MMM and for NDT of materials is very popular. A unique solution of microwave moisture sensing was demonstrated in [12], and a schematic of the sensor is presented in Fig.11.2b. An integrated MSA, which was introduced in [13], is a multi-layer structure based on a patch antenna fed by a pair of microstrips through a non-resonant slot. The patch antenna represents the frequency-selective element of a phase shift transistor oscillator. The active integrated antenna frequency of operation is determined by both the patch geometry and the electrical loads connected at the two microstrip ends. A convenient termination is normally represented by two open-ended stubs whose electrical length is $\lambda/4$ at the patch resonant frequency. For operation as a moisture sensor, one of the open stubs is replaced by the open-ended coaxial transmission line, or a microstrip or coplanar waveguide element, which represents the sensing element. The automatic compensation of frequency drifts can be obtained by switching the amplifier input port between the sensing element and a reference load at a fixed rate, which results in a two-tone FSK-modulated waveform transmitted by the antenna.

A ready-to-use pocket-size moisture meter with interchangeable put on sensor heads was presented in [14], and is suitable for different recording depths in building materials. A linearly polarized single-element circular-patch antenna was used as volume probe. For a patch antenna placed above the sample surface with a definite air gap the resulting feed point impedance depends measurably on the dielectric constant of the samples. This is the basis for use as a moisture sensor. However, the feed point impedance of the antenna also depends on the received energy reflected from the rear side of the sample. Therefore the samples have to be thick enough to ensure a sufficient attenuation of back-reflected waves. This requires a sample thickness of 20 -50 cm, the lower value for larger densities and a moisture content of at least a few percent, and the higher value for dry low-density substances like heat insulation materials.

A combination of modulated scattering techniques and near-field microwave non-destructive evaluation techniques is used to determine the potential for evaluating the dielectric properties of a hardened mortar specimen [15]. This technique utilizes small, resonant PIN-diode loaded dipole scattering (see Fig.11.5a) embedded inside the mortar and an open-ended rectangular waveguide probe operating at 7 GHz to detect this dipole which is modulated at low frequencies. MSAs can surely be applied for this purpose, dipole scattering and substitution of the open-ended waveguide.

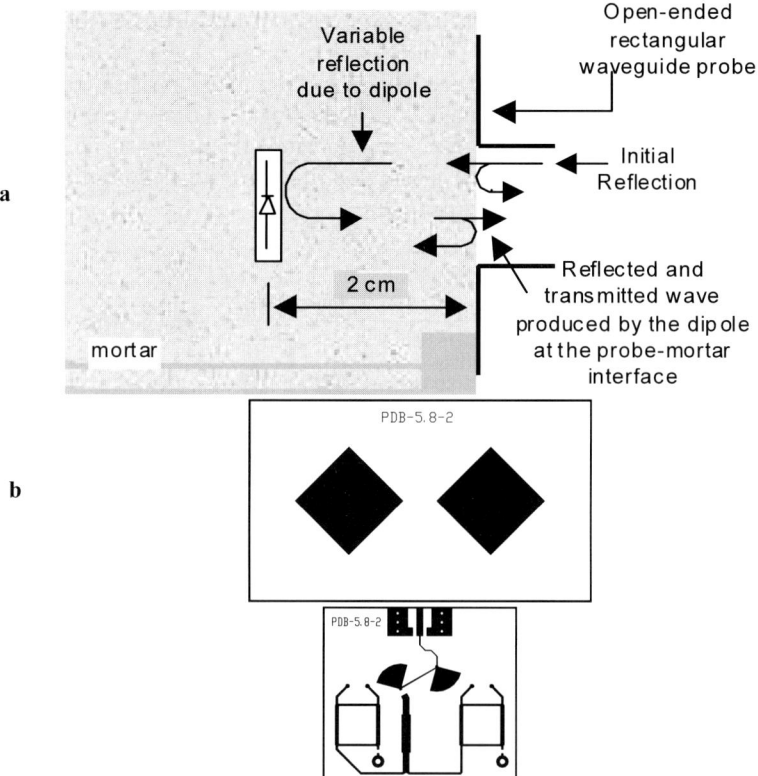

Fig. 11.5. a Bounce diagram representation of embedded modulating dipole scattering technique [15]. **b** Radiating-face and back-side of the PDB [18]. The 90° hybrids, λ/4 matching transformer, and LPF are shown (without a Schottky diode)

The classification of moisture sensors and recent research activity have been described in [16]. More details are given, respectively, for on-line set-ups in green tea production [17], in particleboard production [18], and in veneer production [19]. It is concluded that the most actively researched topics are the free space method using horn antennas or MSAs, and resonance methods using open-ended coaxial cable and MSAs.

Our dual-function microstrip circuit (passive detector/back-scatterer PDB [18]) is shown in Fig.11.5b. The appropriate functions are selected by the bias voltage applied to the Schottky diode. The measurement results for PDB are as follows: the gain of the dual patch is 9.8 dB; the 3 dB beamwidths are 82° and 44°; the axial ratio for circular polarization is 0.8 dB; the return loss of the OOK detector is 15 dB; the tangential signal sensitivity is -54 dBm; the reflection loss of the BPSK modulator is 1.8 dB; the amplitude error is 0.3 dB; the phase error is 4.3°; and the mono-static radar cross-section (RCS) is -17 dBm^2. We have designed, realized, and measured some other PDB models, too. More details are given in [6] and [9].

The disturbing effects of the reflected waves from the air- moist substance interface avoided by using circular polarization. We made four-element MSAs with circular polarization. By locating the input microstrip lines in opposite directions, the received relative power level as a function of range between transmitting and receiving antennas is as shown in Fig.11.6.

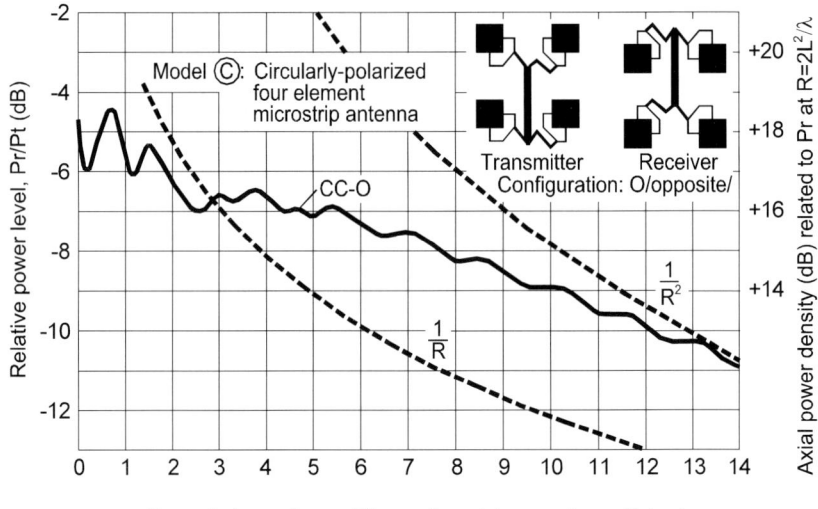

Fig. 11.6. Received relative power level as a function of range between transmitting and receiving apertures, using circularly polarized MSAs [21]

The best results are given by this configuration because of the high received level (−9 dB) at a distance of 10 cm, and low variation (± 0.15 dB) was observable. In our original solution, the microstrip power splitter network between radiating elements had a quasi-double symmetry [20].

We designed fixed beam MSA arrays and scanable phased arrays used as transmit/receive antennas for a reflection, aperiodic, and open (RAO, [22]) system configuration. Our circularly polarized phased array at a frequency of 5.8 GHz is described in [23], with circular patches as the radiating elements. Using 3-bit digital PIN-diode phase-shifters the maximum scanning angle was ± 49° with respect to the broadside direction. The three-layer structure consists of the radiator, phase-shifters (PHS), and the digital electronics. The second layer with the PHS consists of the input power splitter, three PHS using 3×2 PIN diodes, and a quadrature hybrid.

11.2 Transmission-Type Microstrip Sensors

11.2.1 Free-Space Set-up Using Modulated Backscatter Technology

For NDT of materials and for monitoring their parameters, MSAs can be advantageously applied. The basic measurement set-up for monitoring particleboards [18] is shown in Fig.11.7. This is a free-space, reflection/double transmission system in which the attenuation (ΔA) and phase ($\Delta \Phi$) are measured by the receiver.

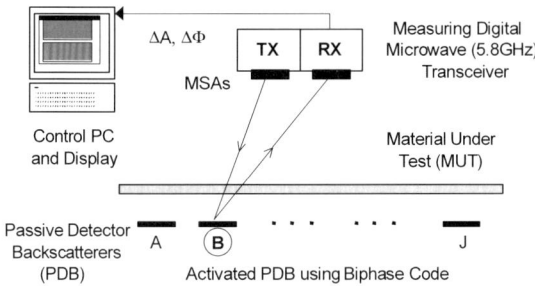

Fig. 11.7. The basic measurement set-up for monitoring particleboards [18]

The microwave transmitter (TX), which is working in one of the industrial-scientific-medical (ISM) frequency bands, radiates as a continuous wave (CW) during the measurement. After passing through the measured moist substance, this wave is reflected by one of the PDBs. At the microwave receiver (RX) the signal is converted to IF stage and then sampled by the DSP unit, which carries out the code correlation. The two parameters (ΔA and $\Delta \Phi$) can be measured, and from the correlated signal the relative complex permittivity ($\underline{\varepsilon} = \varepsilon' - j\varepsilon''$, moisture content (MC), wet and dry densities, etc., can be calculated from the measured data.

Circularly polarized MSAs are used in the system, eliminating the disturbing effects of the reflected wave from the air-slab interface. More details of this system are given in [8] and [24].

11.2.2 Bi-static Sensors for On-line Control

Our other measurement system was made for specifying the freshness of eggs, see Fig.11.8. We produced a microwave measurement set-up using three MSAs integrated with a microwave oscillator and Schottky diode detectors used as bi-static microwave sensors. We measured the microwave attenuation (water content) and the bi-static radar cross-section (dimensions) of eggs having different times of storage. The measured attenuation for fresh eggs is 8-10 dB at a frequency of 14.2 GHz; the variation with time is: dA [dB] = $-0.033 \times T$ (days), i.e., nearly -1 dB/30 days.

Our next program is to create microwave equipment for the automatic selection of old eggs. The basic arrangement and the results of our measurements are given in [25].

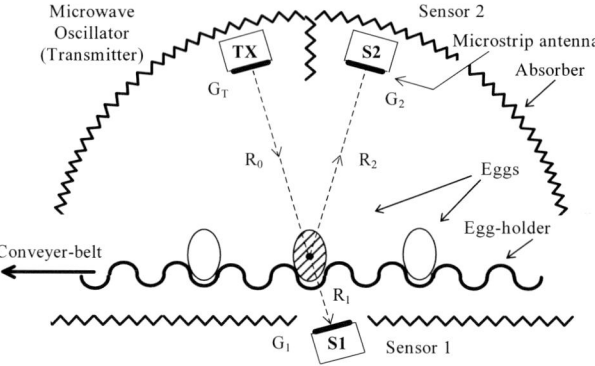

Fig. 11.8. Microwave measuring tunnel for specifying the freshness of eggs [25]

11.2.3 Emerging MSA and Sensor Technologies

Active integrated antennas (AIAs) provide a new paradigm for designing modern microwave and millimeter-wave architecture with desirable features such as compactness, light weight, low cost, low profile, minimum power consumption, and multiple functionality [26]. Optical controlling techniques for AIAs are also possible. A typical application of this emerging technology is in wireless sensors.

In recent years photonic crystals, which are artificial materials made of two-or three-dimensional periodic dielectrics, have been proved to have interesting characteristics not yet available with ordinary materials. They act on electromagnetic waves in a similar way to natural crystals acting on electron waves. These artificial materials, commonly called photonic band gap (PBG) materials, can significantly change electromagnetic wave propagation, providing forbidden frequency bands. High-performance passive structures based on the PBG concept were studied at University of California at Los Angeles UCLA [27]. This concept offers the possibility to control EM waves with additional degrees of freedom. Numerous applications of PBG structures have been attempted and successfully demonstrated, for example; a gain-enhanced patch antenna, at which reduced radiation along the substrate (with surface wave suppression) was observable. From input return loss measurements, the bandwidth was 5.4% (for a reference patch of 1.6%), and the measured gain at a frequency of 14.15 GHz was 6.77 dB (for a reference patch of 5.16 dB).

Recently, a silicon micro-machined patch (SMP) antenna for a frequency of 13 GHz has been reported [28]. This SMP had superior performance over conventional designs where the bandwidth and the efficiency have increased by as much as 64% and 28%, respectively.

This SMP consists of a rectangular patch centered over a cavity sized according to the effective index of the cavity region, and fed by a microstrip line. To produce the mixed substrate region, silicon micro -machining is used to laterally remove the material from underneath the specified cavity region resulting in two separate dielectric regions of air and silicone.

11.3 New Results Using Reflection-Type Sensors

11.3.1 New Concept for Near-Field Sensing

We have introduced reflection-type microwave-based moisture sensors for use in bins of raw materials such as sand, crushed stones etc., and in concrete mixers. The amount of water and cement present in a concrete mix greatly affects the physical characteristics of the concrete. The existing aggregate sensors are density-sensitive, which requires installation in bins of compacted material in order to attain sufficient accuracy. Our sensors are independent of density and measure to a good accuracy. The mode of operation is monostatic, in analogy with radar systems in which a single antenna is used for both transmitting and receiving purposes. First we studied the so-called density-independent functions, and a conceptual model was developed, using three microwave units with different frequencies. The basic equations and methods for calibration are given in an earlier paper [29]. Homodyne receivers [30] were used, because high dynamic range and measurement accuracy are needed.

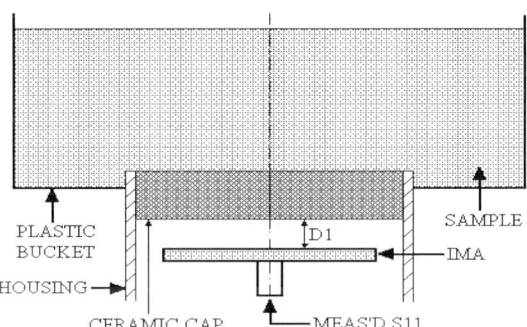

Fig. 11.9. Cross-sectional view of the microwave reflection sensor in direct contact with the sample (moist sand)

These receivers supply the in-phase (I) and quadrature-phase (Q) components of the reflection coefficients. Each receiver consists of a Wilkinson hybrid, a quadrature hybrid, two balanced mixers, bandpass filters and post-amplifiers. High-level mixers are used, thus decreasing the nonlinear distortion. The homodyne system offers a cheap solution due to the need for only one microwave source per frequency.

The radiating element is a three-frequency integrated microstrip antenna (IMA), which is applied in the reactive/radiation near-field region. An important part of the microwave reflection sensor is the ceramic cap, a faceplate material (Fig.11.9), which protects the microwave units and makes a significant trans-formation of S-parameters. In this context the distance D1 between the antenna and ceramic cap is an essential dimension, optimized to obtain high sensitivity for variation of the moisture content. Figure 11.10 shows the IMA.

The main challenge was the limited outer diameter $D = 78$ mm, which is only 64% of the free-space wavelength of the lowest frequency $F_1 = 2.45$ GHz. Be-cause of this limitation, only a stacked structure was conceivable. The central part of this innovative design is the microwave laminate (designation A10) itself, con-taining a barel-shaped patch (A11) for the frequency of F_1 on the back side, a rec-tangular patch (A12) for $F_2 = 5.857$ GHz and a four-element array (A13) for $F_3 = 10.502$ GHz on the upper side.

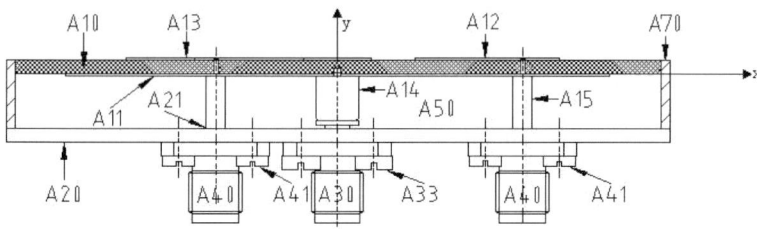

Fig. 11.10. Cross-sectional view of the three-frequency IMA. A30 is shifted by 8 mm in minus z-direction relative to A40

The resonant frequencies of antennas are tuned by the length of elements. "Electri-cally thick" antennas are realized, giving a broad bandwidth and good efficiency.

A cross-polarized arrangement is designed for minimizing the mutual cou-plings between antennas. The spacer material is a hard, closed cell foam A50 with a low dielectric constant and low loss. SMA jack-type connectors A30, A40 are mounted on a metal-plate A20 (a mechanical fixing element) and are used for in-put/output connections. The serial inductance of the central conductor of input A30 is resonated by a serially built-in mechanical capacitor A14, resulting in broad-band operation, which is important because of the de-tuning effect of the closely mounted ceramic disk. The I/O connections for the frequencies of F_2, F_3 are fed through an antenna for F_1 at its central line (zero field) by short, semi-rigid cables A15. For reasons of mounting, this was possible using only sliding contacts A21. The outer diameter and height of this antenna structure excluding the SMA connec-tors are, respectively, 76 mm and 9.6 mm. The IMA is mounted in a stainless-steel tube, which could create unwanted resonance. This is suppressed using a micro-wave absorber A70, made from silicon rubber, around the antenna. Humidity is kept out of the active region by a radome with material A50. Its thickness is equal to the optimal distance between the IMA and the ceramic cap.

All the elements are fixed by gluing. The new concept of the sensor is optimization at near field by the selection of the material for the protective hard ceramic cap and the distance from the IMA.

11.3.2 A Calibration Method Using Scattering Parameters

The calibration problem of the moisture sensing in the near field was analyzed using a scattering matrix representation and signal flow graphs. First the thickness of the sample was supposed high enough so that only the input surface reflection of the sample was taken into account. The main task was to convert the complex dielectric constant of the sample to a complex reflection coefficient measured at the input of the transmit -receive antenna (the direct problem).

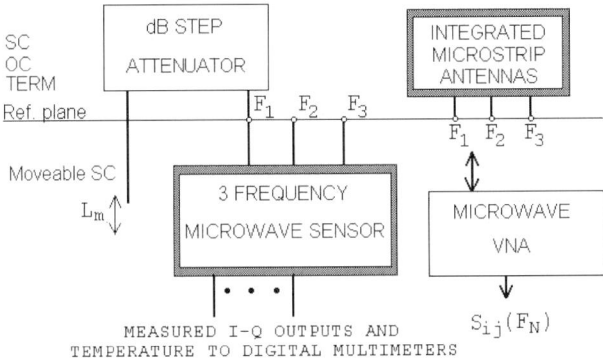

Fig. 11.11. Measurement set-up for calibration

We have also solved the inverse problem, namely, calculation of the moisture content and density from the measured input reflection at multi-frequency operation [29].

We have investigated an experimental calibration procedure (Fig.11.11), using a movable short circuit, a step attenuator, and a microwave vector network analyzer. The method is applicable at high signal levels for I -Q receivers, because the offset and peak amplitude versus the movable short-circuit position can be measured.

11.3.3 Experimental Results

A calibration result (measured I-Q outputs) is shown in Fig.11.12. Serial measurements were started using the conceptual model. Moist sand samples were measured first, with a moisture content between 0.5 and 16% and with a density range of 0.7-1.6 g/cm^3. The results are good enough to be used for network training of an artificial neural network (ANN) to a high accuracy in moisture measurement, which is independent of density [31].

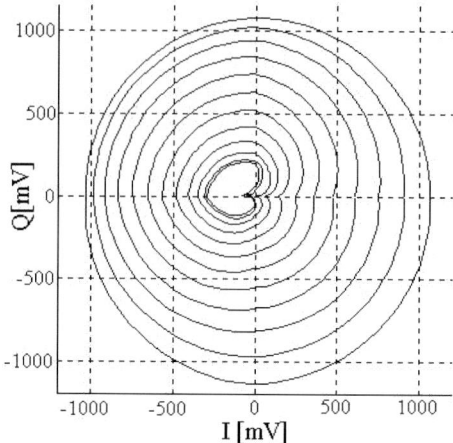

Fig. 11.12. Measured I - Q outputs versus attenuator in 1dB steps at a frequency of 2.45 GHz. Outer curve: attenuator at the 0 dB position

11.4 Conclusions

MSAs have several features that make them ideal not only for telecommunication applications, but also for special ones such as microwave aquametry. Based upon printed - circuit technology, they are very inexpensive to produce. Their inherent low profile allows them to be mounted directly on product housings. The examples given here show that many types of patch antennas can be used in MMMs and in moisture sensors, sometimes improving the system performance by the modulated scattering technique. Our experiments at BUTE/DBIS show the widespread application of MSAs, giving a high isolation between the transmitter and receiver, low sensitivity for near-field reflections, and a great number of realizations. New antenna technologies, such as active integrated antennas, photonic bandgap structures, and micro -machined patches, are promising, and provide compactness, multiple functionality, better efficiency, larger bandwidth, etc. New microwave moisture sensors presented here show some significant advantages for water determination in bins of raw materials and in concrete mixers. A sophisticated ANN technique was used in a reflection-type sensor.

Acknowledgements

I should like to thank Professor G. Biffi Gentili for Fig.2b, Professors Jean-Charles Bolomey and Reza Zoughi for Fig.5a and the related discussion. I should also like to extend my appreciation to my students Laszlo Nyul, Alajos Dlusztus, and Tibor Hegedus for their help in preparing the figures of this chapter. The invaluable assistance of Mr. Robin Shepherdson, President of Scale-Tron Inc., and Mr. John Burrows, President of JHB Electronics Inc., in developing the concept written in Ch. 11.3, is gratefully acknowledged.

References

1. Hoffmann RK (1987) *Handbook of microwave integrated circuits.* Artech House, Norwood, MA
2. Tabib-Azar M, Katz JL, Leclair SR (1999) Evanescent microwaves: novel super−resolution non-contact non−destructive imaging technique for biological applications. IEEE Trans Instrum Meas 48(6):1111–1116
3. Kaida b. Khalid, Zulkifly b. Abbas (1996) Development of microstrip sensor for oil palm fruits. In: Kraszewski A. (ed) *Microwave Aquametry.* IEEE Press, New York, pp 239–248
4. Assenheim JG (1993) UK Patent Appl. GB 2277803A, No. 9309221.1
5. Bahl IJ, Bhartia, P (1980) *Microstrip antennas.* Artech House, Dedham
6. Völgyi F (1996) Microstrip antennas used for WLAN systems In: Proc. of ISAP, Chiba, Japan, 24−27 Sept 1996, vol 3, pp 837–840
7. Völgyi F (1999) Microstrip antenna R&D in Hungary. In: 10th Microcoll, Budapest, Hungary, 21−24 Mar 1999, pp 249−252
8. Völgyi F (assoc. ed.) (1999) Special issue on WLAN and ISM applications of microstrip antennas, telecommunications. J C^5 L(12):76 pp
9. Völgyi F (2000) Microstrip antennas in subsurface sensing. Proc SPIE 4129: 47–58
10. Kin-Lu Wong (1999) *Design of nonplanar microstrip antennas and transmission lines.* Wiley, New York
11. Bogosanovich M (2000) Microstrip patch sensor for meas-urement of the permittivity of homogeneous dielectric materials. IEEE Trans Instrum Meas 49(5): 1144–114
12. Biffi Gentili G, Avitabile GF, Manes GF (1996) An integrated microwave moisture sensor. In: Kraszewski A. (ed) *Microwave Aquametry.* IEEE Press, New York, pp 215–221
13. Avitabile GF, Maci S, Biffi Gentili G, Roselli L, Manes GF (1992) A two-port active coupled microstrip antenna. Electron Lett 28(25), pp 2277–2279
14. Göller A, Handro A, Landgraf, J (1999) A new microwave method for moisture measurement in building materials. In: Third workshop on electromagnetic wave interaction with water and moist substances, Athens, GA, 11-13 Apr 1999, pp 84–88
15. Joisel A, Bois KJ, Benally AD, Bolomey JCh, Zoughi R (1999) Embedded modulating dipole scattering for near-field microwave inspection of concrete: preliminary investigation. Proc SPIE 3752: 208–214
16. Okamura S (2000) Microwave technology for moisture measurement. SSTA Int J 1(2): 205–227

17. Okamura S, Tomita F (1998) Microwave system for moisture content measurement of green tea leaves in drying process. In: Proc ISHM, London, 1998, vol 2, pp 204–210

18. Völgyi F (2000) Monitoring of the particleboard-production using microwave sensors. In: *Sensors update*, vol 7, Wiley-VCH, Weinheim, pp 249–274

19. Vainikainen PV, Nyfors EB, Fisher MT (1987) Radiowave sensor for measuring properties of dielectric sheets: application to veneer moisture content and mass per unit area measurement. IEEE Trans Instrum Meas 36(4):1036–1039

20. Völgyi F (1989) Versatile microwave moisture sensor. In: Conf. SBMO '89, Sao Paulo, Brazil, vol II, pp 456–462

21. Völgyi F (1996) Integrated microwave moisture sensors for automatic process control. In: Kraszewski A. (ed) *Microwave Aquametry*. IEEE Press, New York, pp 223–238

22. King RJ (2000) On-line industrial applications of microwave moisture sensors. In: *Sensors update* 7:109–170, Wiley-VCH, Weinheim

23. Völgyi F (2001) Microstrip sensors used in microwave aquametry. In: Kupfer K, Hubner C (eds) Proceedings of the 4th international conference on electromagnetic wave interaction with water and moist substances, Weimar, May 2001, pp 135–142

24. Völgyi F (2000) Multifrequency sensing of large-sized dielectric boards. Proc SPIE 4129:97–107

25. Völgyi F (2000) Specifying the freshness of eggs using microwave sensors. Subsurface Sensing Technol Appl 1(1): 119 –139

26. Quian Y, Itoh T (1998) Progress in active integrated antennas and their applications. IEEE Trans Antennas Propag 46(11): 1891–1900

27. Itoh T (1998) Active antennas - integration of microwave circuits with antennas. In: APMC'98 Short Course, Yokohama, 11 Dec 1998, p 66

28. Papapolymeru I, Drayton RF, Katehi LPB (1998) Micromachined patch antennas. IEEE Trans Antennas Propag 46(2):275–283

29. Völgyi F, Burrows J, Shepherdson R (2002) Calibration methods for non density-sensitive microwave-based moisture sensors. In: Proc ISHM, Taipei, Taiwan, 16–19 Sept 2002, CMS/ITRI, Hsinchu, pp 371–378

30. Bolomey JCh, Gardiol FE (2001) *Engineering applications of the modulated scatterer technique*. Artech House, Boston

31. Völgyi F, Shepherdson R, Burrows J (2003) New microwave moisture sensors for use in bins of raw materials and in concrete mixers. In: Proc of ISEMA, Rotorua, New Zealand, 23–26 Mar, 2003, pp 138–145

12 A Blind Deconvolution Approach for Free-Space Moisture Profile Retrieval at Microwave Frequencies

David Glay, Tuami Lasri

Institut d'Electronique de Microélectronique et de Nanotechnologie,
UMR CNRS 8520 IEMN-DHS, Avenue Poincaré - B.P. 69,
59652 Villeneuve d'Ascq Cedex, France

12.1 Introduction

An outstanding method of sensing moisture is the use of microwaves. Thus, the interaction of microwave radiation with moist substances is of general practical interest in many industrial areas. Consequently, electromagnetic aquametry is a domain that has been intensively studied in recent decades. Many characterization methods have been proposed, enabling large areas (agriculture, chemical, pharmaceutical, and coal industries, civil engineering, etc.) to be studied with this sensing technology [1–5]. Nevertheless, in most cases the test setups were designed to measure the moisture content in the frequency or time domains (reflection and transmission methods, coaxial probes and impedance methods, cavity methods, etc.), for static use in the laboratory. This is the reason why, today, there is a growing interest in proposing systems and methods that can be used outside the laboratories for on-line control. To that end we have developed microwave systems for non-contact moisture sensing applications [6]. These instruments measure the reflection coefficient and/or the transmission coefficient, in magnitude and phase, of a moist material to determine its complex permittivity. Different investigations have been carried out in various fields [7, 8].

This chapter addresses the problem of retrieving the moisture profile of a material presenting a non-uniform moisture distribution. The inhomogeneous dielectric medium, in terms of moisture content, is realized through the construction of a layered material that is made of layers of different moisture content. The configuration proposed permits us to simulate a large range of practical cases. To investigate the possibility of reconstructing, through a one-dimensional model, the moisture profile of the material under test, a free-space technique based on the determination of the reflection coefficient of an antenna is proposed. The collection of the data is performed at 2.45 GHz by using an antenna, fixed at a standoff distance, connected to an S-Parameters Measurement System (SPMS-2450). This apparatus is based on microstrip technology [6].

In the case of inhomogeneous materials characterization, the reflection coefficient is strongly influenced by the radiation pattern of the antenna. One solution to remedy to this problem is to make use of spot focusing antennas. Nevertheless, as far industrial applications are concerned, considerations such as the cost, the fragility, and the dimensions of this sensor have to be taken into account. The solution that has been chosen to overcome these problems is to make use, for example, of a rectangular horn antenna and to develop signal processing tools. Thus, in addition to the physical interacting mechanisms present, signal processing techniques are used to enhance the quality of the one-dimensional reconstructed profile. The proposed method considers that the measured data are the result of a convolution of the true data with a point spread function (PSF) that depends mainly on the operating frequency, the antenna, the characteristics of the structure under test, and the standoff distance between the antenna and the material. Consequently, a modeling effort is undertaken in order to solve the inverse problem by considering a blind deconvolution approach. The chapter describes the moisture profile reconstruction for different material configurations. In particular, subsurface moisture profiles are considered. Simulations and indoor experiments carried out on arrangements of cellular concrete samples are performed to discuss the effectiveness of the method.

12.2 Free-Space Characterization Method

12.2.1 Reflection Coefficient Modeling for a Two-Layer Material

Determination of moisture is very important in many branches of industry. Among the techniques available, microwave moisture measurement procedures are well suited, especially because of the great difference between the dielectric properties of water and most dry materials in the microwave frequency region [9]. Several methods, including transmission, reflection, and resonance techniques, are offered by this characterization means to measure moisture content through indirect methods. In this chapter, the determination is derived by making use of the reflection coefficient of the material under test.

Actually, in many measurement situations in industrial environments, only one side of the material to be characterized is reachable. So, to a great extent, the measurements are based on monostatic methods (one antenna). Another concern when we deal with industrial tests is the necessity to avoid, on the whole, disturbances to the process. To fulfill this requirement a non-contact method is used.

Finally, the measurement system has to be compatible with industrial use in terms of portability, sturdiness, cost, and ease of use. Taking into account all these demands, we propose in Fig. 12.1 a description of the method developed for materials characterization purposes. The material chosen to illustrate the technique is a two-layer material.

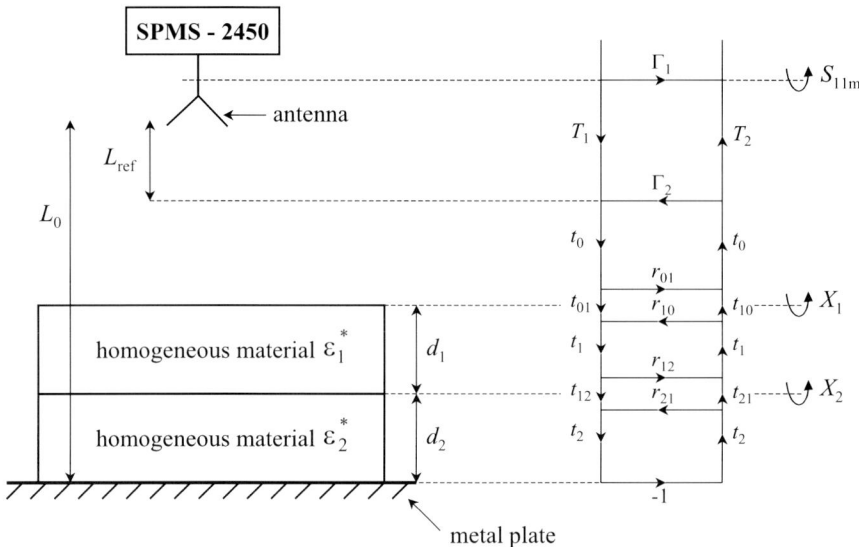

Fig. 12.1. Schematic diagram for a two-layer material characterization

The experimental set-up consists of a system operating at 2.45 GHz (SPMS-2450), realized in the laboratory [6], associated with an antenna (horn antenna) and a metal plate on which the material under test is laid.

The set-up and the measurement configuration chosen permit us to satisfy the expectations mentioned above. Besides the schematic view of the measurement environment we have also given the flow graph that corresponds to the modeling of the propagation phenomena. In this description of the flow graph the following parameters are mentioned:

- Γ_1, Γ_2, and T_1, T_2, which are respectively the reflection and transmission coefficients that take into account the antenna and the space contained between the antenna and the plane located at a distance L_{ref} chosen so that the far-field conditions are fulfilled,
- t_i which represents the transmission coefficient in the medium i (0 for air, 1 for the upper layer, and 2 for the lower one),
- r_{ij} and t_{ij} which define respectively the reflection and transmission coefficients at the interface between media i and j.

The explicit expression of the reflection coefficient S_{11m} calculated from the flow graph is given by

$$S_{11m} = \Gamma_1 + \frac{T_1 T_2 t_0^2 X_1}{1 - t_0^2 \Gamma_2 X_1} \qquad (12.1)$$

where

$$X_1 = r_{01} + \frac{t_{01}t_{10}t_1^2 X_2}{1 - t_1^2 r_{10} X_2}$$ (12.2)

and

$$X_2 = r_{12} - \frac{t_{12}t_{21}t_2^2}{1 + t_2^2 r_{21}} .$$ (12.3)

Simulations and measurements are conducted to determine the validity of the reflection coefficient model behavior. To make use of the model (Eq. (12.1)) the parameters Γ_1, Γ_2 and $T_1 T_2$ have to be determined. To that end an experimental procedure which includes the collection of the reflection coefficient for three distances, L_0 and $L_0 \pm \Delta L$, is performed. Thus, the unknown Γ_1, Γ_2 and $T_1 T_2$ are obtained from the resolution of a set of three equations. For this calibration step the material under test is a dry cellular concrete sample ($\varepsilon_{dry}^* = 2 - 0.1j$, dry bulk density $\rho_{dry} = 0.53$ g/cm^3, and thickness $d_1 + d_2 = 100$ mm) whereas the distances L_0, L_{ref}, and ΔL selected are respectively 250 mm, 120 mm, and 20 mm.

Knowing these parameters and the features of the material under test the reflection coefficient S_{11m} (Eq. (12.1)) can be determined. Comparisons between the model and measured data are made for three materials: air (no material on the plate), a block of dry cellular concrete whose thickness is 100 mm (homogeneous material), and a two-layer material. In this last case the lower layer is a moistened cellular concrete slab (moisture content $MC \sim 30\%$, $\varepsilon_{moist}^* = 7.0 - 0.8j$ and thickness $d_2 = 50$ mm) and the upper layer is a dry cellular concrete slab ($d_1 = 50$ mm). The moisture content MC is given by

$$MC = \frac{m_{water}}{m_{dry}} .$$ (12.4)

The results obtained are reported in Table 12.1.

Table 12.1. Reflection coefficients obtained from the model and from measurements

Reflection coefficient S_{11m}	Air (no material)	Homogeneous material	Two-layer material
Measurements	−7.0 dB/−0.4°	−11.3 dB/−1.9°	−11.1 dB/−10.1°
Modeling	−7.0 dB/6.0°	−11.3 dB/−1.9°	−10.8 dB/−9.7°

The findings exhibit a fair agreement between the two sets of data and suggest that the model correctly describes the measured reflection coefficient obtained.

In the next section, the composition of the layered material is changed to create a subsurface permittivity profile along the transversal axis (0x).

12.2.2 Construction of a Subsurface Moisture Profile

The previous arrangement (two homogeneous layers) allows us to respond to a particular class of problems in electromagnetic aquametry related to subsurface characterization. In order to expand the range of applications we now treat a slightly different case. The design of the structure proposed is given in Fig. 12.2.

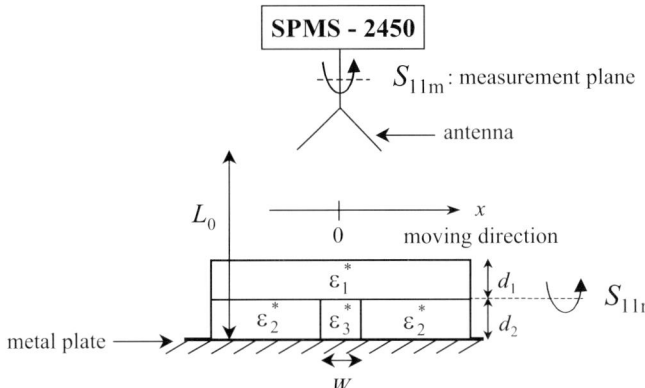

Fig. 12.2. Schematic diagram of the measurement of a subsurface moisture profile

The configuration of interest includes a three-layer material (a layer with a permittivity ε_3^* inserted between two pieces of the same material whose permittivity is denoted ε_2^*) coated with a homogenous material (permittivity ε_1^*). This arrangement allows us to simulate construction of a subsurface moisture profile, or a material presenting a non-uniform moisture content flowing in a pipe, or an unexpected collection of water inside a material, and many others situations. For this particular study we consider a material comprising an area whose moisture content is different from its surroundings ($\varepsilon_1^* \neq \varepsilon_2^* \neq \varepsilon_3^*$).

We propose to demonstrate how we can retrieve, more accurately, information such as permittivity and moisture content derived from the measurement of the one-dimensional reflection coefficient profile (the antenna is displaced above the material under test along the $0x$ axis at a fixed distance) by using signal processing techniques. In order to test a large range of arrangements we have first performed a simulation study based on the use of commercial software that has proven its wave modeling capabilities for such applications, namely Ansoft High Frequency Structure Simulator (HFSS) [10]. The principal purpose of this investigation was to evaluate the performance and the limitations of the proposed method.

The first test that is presented concerns a cellular concrete sample whose features are a dry permittivity $\varepsilon_{dry}^* = \varepsilon_1^* = \varepsilon_2^* = 2 - 0.1j$, a dry bulk density $\rho_{dry} = 0.53$ g/cm^3, and a thickness $d_1 + d_2 = 100$ mm, with a moistened area (moisture content $MC \sim 8.5\%$, moist permittivity $\varepsilon_{moist}^* = 3.15 - 0.14j$, width $W = 70$ mm, and thickness $d_2 = 50$ mm) located at a depth $d_1 = 50$ mm.

This sample is laid on a metal plate situated at a distance $L_0 = 250$ mm from the antenna so that the far-field conditions are fulfilled. The antenna used in the simulation study is an open-ended waveguide. This choice is made only for convenience reasons (simplicity of simulation design and economy of computational time). Prior to the construction of the reflection coefficient profile from the data collected by means of HFSS, we compared the findings obtained by HFSS to those given by the model in three cases. As has been shown in the previous section, knowledge of the parameters Γ_1, Γ_2 and $T_1 T_2$ is needed to use the model. The calculation of these coefficients follows exactly the same scheme. A calibration procedure including the test of a dry sample of cellular concrete, for three different distances ($L_0 = 250$ mm and $\Delta L = \pm 20$ mm), is carried out. The comparisons obtained between the HFSS simulations and the modeling (Eq (12.1)) are reported in Table 12.2.

Table 12.2. Comparison between reflection coefficients obtained from the model and from HFSS simulations

Reflection coefficient S_{11m}	Air (no material)	Homogeneous material	Two-layer material
HFSS simulations	−9.5 dB/−149.7°	−11.6 dB/−123.2°	−10.5 dB/−140.8°
Modeling	−9.1 dB/−144.6°	−11.6 dB/−123.2°	−10.7 dB/−140.0°

This simulation study yields comparable results. Thus, the findings presented in Tables 12.1 and 12.2 indicate that the model predicts the measured and the HFSS data to a good accuracy.

Nevertheless, the data reported in each table are not comparable in themselves because as mentioned above the antenna used in the HFSS simulations is an open-ended waveguide whereas the one selected for the measurement study is a rectangular horn. These two analyses demonstrate that the model behaves in a reasonable way when homogeneous or layered materials are investigated.

However, this description of the propagation phenomena is not sufficient when heterogeneous materials are investigated, especially when the choice of managing without the spot-focusing antenna is taken. The next section is devoted, in particular, to the emphasis of the antenna radiation pattern effects.

12.3 Blind Deconvolution Approach

This part of the chapter is essentially dedicated to HFSS simulations. Actually, as has already been noted, this simulation tool allows us to test a large number of configurations.

12.3.1 Simulation of Reflection Coefficient Profiles

For each configuration two sets of data are collected and organized to construct two different profiles. The first one, called the degraded profile, is established from the data obtained when the antenna is displaced along the material previously defined. The second one, called the true profile, is constructed from the width of the moist area and the reflection coefficient values simulated in the case of a dry material under test (homogeneous material) and in the case of the characterization of a two-layer material (W as large as the dry-part width).

The first example presented concerns the testing of an arrangement that includes a 70 mm wide inclusion. According to the description given above, the true profile is established by using the values in Table 12.2. For more clarity, if we consider, for example, the true magnitude profile, it has been constructed from the magnitudes obtained in the case of testing (Table 12.2) a dry sample (−11.6 dB) and a two-layer material (−10.5 dB).

The resulting profiles for S_{11m} (Fig. 12.2) are shown in Fig. 12.3.

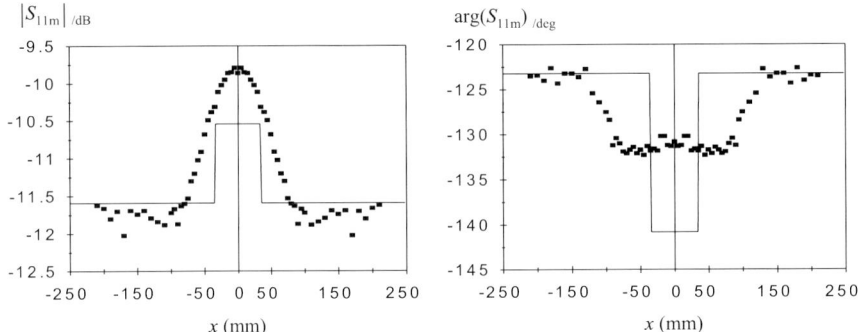

Fig. 12.3. Simulations of $S_{11m}(x)$ for $W = 70$ mm (■: degraded data; ——: true profile)

To take into account the influence of the antenna (matching, free-space propagation, etc.) and the material layer coating the inclusion, we calculate the reflection coefficient S_{11r} in the plane materialized by the dashed line in Fig. 12.2. The results are given in Fig. 12.4.

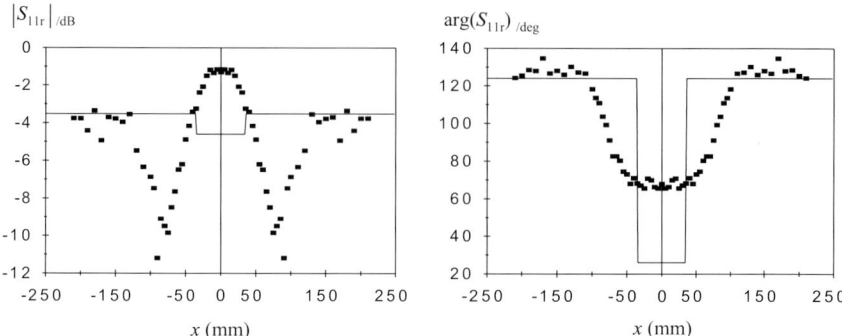

Fig. 12.4. Simulations of $S_{11r}(x)$ for $W = 70$ mm (■: degraded data; ——: true profile)

This figure shows that the magnitude of the reflection coefficient is strongly influenced by diffraction mechanisms. The compensation of such diffraction effects may require specific treatment and represents a limitation of the method. So, we have chosen to work with the phase of the reflection coefficient that clearly shows a convolution relationship between the true and degraded profiles. Thus, the phase degraded profile can be expressed as

$$\arg(S_{11r} \text{ degraded}) = \arg(S_{11r} \text{ true}) \otimes h(x) + n(x) \qquad (12.5)$$

where $h(x)$ is the PSF, $n(x)$ is additive Gaussian noise, and $\otimes$ is a discrete one-dimensional linear convolution operator.

So, knowing these two profiles (Fig. 12.4), the PSF is calculated from an inversion process of Eq. (12.5).

Before giving the result for this PSF we present simulation data obtained when smaller inclusions are tested. Investigations are performed for three different sizes, $W = 50$ mm, $W = 30$ mm, and $W = 10$ mm. All other parameters of the structure are kept unchanged.

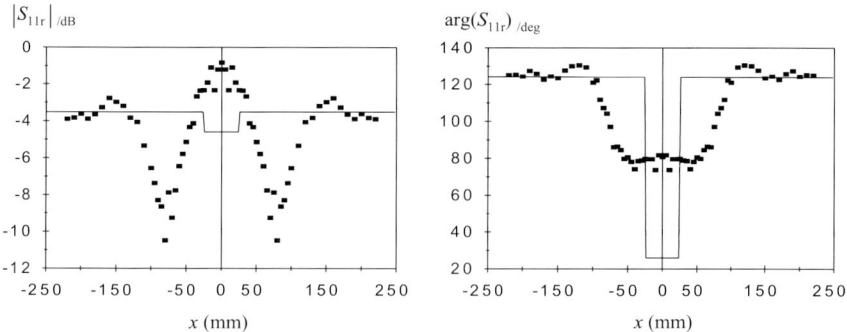

Fig. 12.5. Simulations of $S_{11r}(x)$ for $W = 50$ mm (■: degraded data; ——: true profile)

To simplify the presentation we give only the reflection coefficients $S_{11r}(x)$ deduced from the simulation of $S_{11m}(x)$. We present the simulation results for these three inclusions respectively in Fig. 12.5, 12.6, and 12.7.

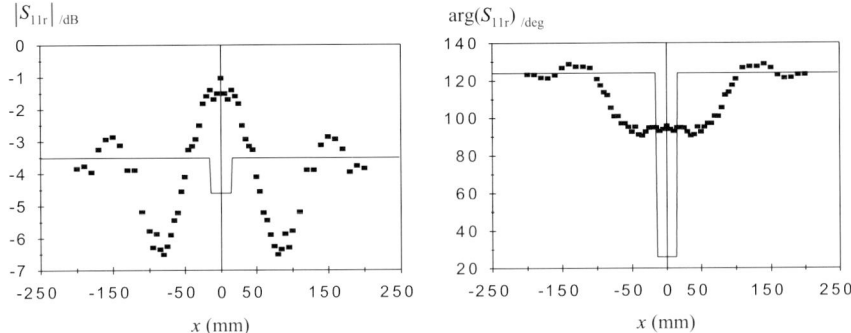

Fig. 12.6. Simulations of $S_{11r}(x)$ for $W = 30$ mm (■: degraded data; ──: true profile)

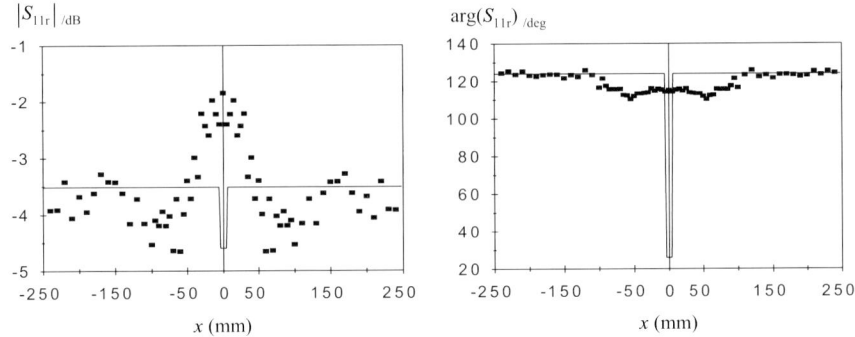

Fig. 12.7. Simulations of $S_{11r}(x)$ for $W = 10$ mm (■: degraded data; ──: true profile)

The simulations performed lead to magnitude and phase shapes relatively similar to those obtained for the case previously treated ($W = 70$ mm). Hence, the conclusions already drawn for the case $W = 70$ mm still stand. It can also be noted that the inclusions are still clearly detected down to $W = 10$ mm, especially if the magnitude is considered. Nevertheless, as we have chosen to take advantage of the reflection coefficient phase (Eq. (12.5)), the restoration of the moisture content is expected to be very difficult for the last configuration tested ($W = 10$ mm). Actually, the phase is not very sensitive to the presence of the smaller inclusion investigated (Fig. 12.7). This case is interesting to appreciate the limitations of the method. In order to illustrate the dependence of the PSF on the material under test, the information given in Figs. 12.4–12.7 is used to determine the PSF for all the configurations ($W = 70$ mm, $W = 50$ mm, $W = 30$ mm, and $W = 10$ mm) studied. The four shapes obtained are presented on the same graph (Fig. 12.8) to observe the degree of similarity.

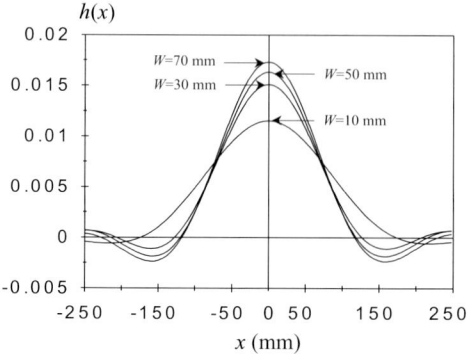

Fig. 12.8. PSF $h(x)$ for different values of W

This comparison shows the sensitivity of the PSF to the arrangement under test. It also demonstrates that the assumption that a PSF calculated in particular situations (frequency, standoff distance, material under test, etc.) remains unchanged is applicable under very restrictive conditions (very low variation of testing conditions). Thus, the first idea that consists in having a calibration procedure to calculate the PSF and assuming it is relatively unvarying in a certain range of experimental conditions is not very satisfactory [11]. So to overcome this difficulty we have made use of a blind deconvolution approach [12–14] for the restitution of the true profiles.

This technique has been successfully used in image processing [15–17] in many scientific and engineering disciplines.

12.3.2 Profile Estimate Results

The problem is that of reconstructing a reliable estimate of the true profile from the received degraded profile. To accomplish this task a blind deconvolution scheme is applied.

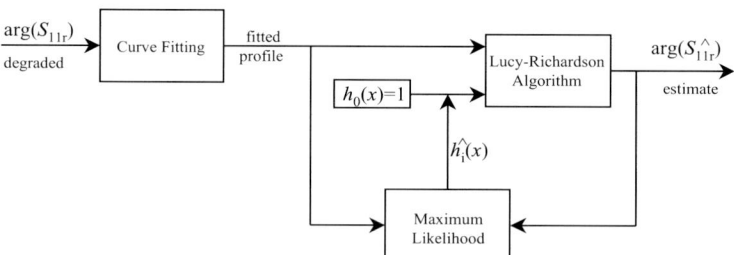

Fig. 12.9. Reconstruction algorithm based on a blind deconvolution approach

Indeed, the blind deconvolution problem is to estimate both the true profile and PSF simultaneously based on the degraded profile and a little a priori information such as non-negativity and support size of the true profile [17]. We give a description in Fig. 12.9 of the blind deconvolution algorithm used. As has been previously mentioned, the parameter of interest is the phase, $\arg(S_{11r})$. As the number of simulation points is not very large (~ 60 points over a [−200 mm; 200 mm] range) and as they are not regularly spaced (5 mm or 10 mm step), a curve fitting procedure is used to provide more data (512 points) presenting a narrow and regular spacing (2 mm). These functions are the input of an iterative algorithm that needs an initial value of the PSF ($h_0(x) = 1$) to retrieve an estimate of the restored profiles. To illustrate this deconvolution principle, in two examples ($W = 70$ mm and $W = 10$ mm) of a restoration profile we give Fig. 12.10.

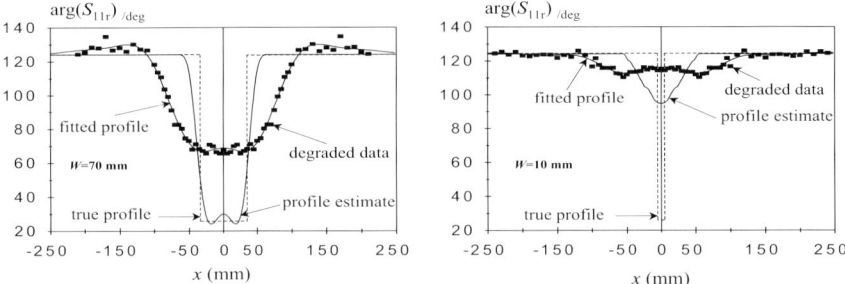

Fig. 12.10. Comparison of reflection coefficient phase profiles for different values of W

The fitted curves in the case of $W = 70$ mm and $W = 10$ mm are obtained respectively by using Eq. (12.6) and Eq. (12.7). The correlation coefficients are respectively equal to 0.996 and 0.968:

$$\arg(S_{11r} \text{ fitted}) = \left[11.2\cos\left(\frac{x}{19.6}\right) - 66.5\cos\left(\frac{x}{76.8}\right) \right] e^{\frac{-x^2}{14100}} + 124.1 \tag{12.6}$$

$$\arg(S_{11r} \text{ fitted}) = \left[3.5\cos\left(\frac{x}{19.8}\right) - 11.3\cos\left(\frac{x}{113.2}\right) \right] e^{\frac{-x^2}{22150}} + 124.1. \tag{12.7}$$

The results obtained before (fitted profile) and after (profile estimate) the deconvolution process exhibit a great improvement in the restitution of the profile in the case of the configuration including a 70 mm wide inclusion. The profile estimate is reasonably close to the true profile. On the contrary, the situation where an inclusion 10 mm wide is buried at a depth of 50 mm is more complicated. Although the enhancement of the retrieved profile is noticeable, the results are not satisfactory. The profile estimate remains appreciably far from the true profile. Consequently a good prediction of the moisture content from these results is not conceivable. The usefulness of the treatment of this configuration lies in its illustration of the method's limitation.

We make use of this technique to retrieve a permittivity profile. Among the parameters of practical interest that can be derived from this quantity we find the moisture content. Thus, the next section is devoted to a description of moisture profile restitution. Simulation and experimental studies are tackled to investigate the possibility of using a blind deconvolution approach to enhance the quality of the information retrieved taking into account the assumption made for the magnitude of the reflection coefficient.

12.4 Reconstruction Method of a Moisture Profile

12.4.1 Moisture Profile Simulation

It has already been shown [8] that the permittivity of a moist material ($\varepsilon^*_{\text{moist}}$) can be written as a function of its dry permittivity ($\varepsilon^*_{\text{dry}}$) and bulk density ($\rho_{\text{dry}}$), its moisture content (MC), and the water permittivity ($\varepsilon^*_{\text{water}} = 77 - 12\text{j}$):

$$\sqrt{\varepsilon^*_{\text{moist}}} = \sqrt{\varepsilon^*_{\text{dry}}} + MC\rho_{\text{dry}}(\sqrt{\varepsilon^*_{\text{water}}} - 1). \qquad (12.8)$$

The moist material permittivity is derived from the reflection coefficient: $S_{11r} = f(\varepsilon^*_{\text{moist}})$ [18]. For the reasons previously presented, the information at our disposal is the profile estimate of the phase. So, we have considered for the magnitude on the $0x$ axis a constant value equal to the one obtained in the case of the dry material testing (Fig. 12.4).

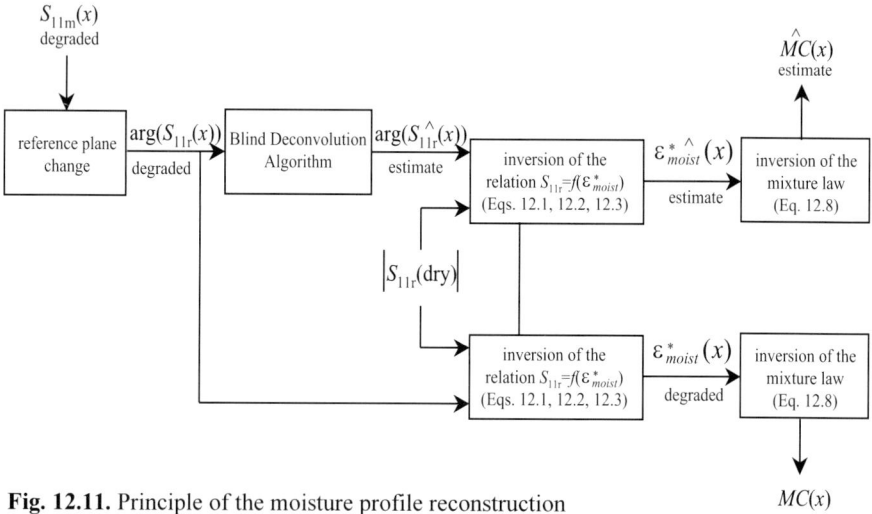

Fig. 12.11. Principle of the moisture profile reconstruction

The difficulty in retrieving the magnitude profile is one of the limitations of the method. Nevertheless, as the contrast permittivity investigated in this simulation is not very important, this approximation does not induce too hard a penalty.

The moisture profile reconstruction principle is detailed in Fig. 12.11 where all the stages of the method are given.

This technique is applied to the sample whose moist inclusion width is $W = 70$ mm. The different reflection coefficients (S_{11m}, S_{11r}, $S_{\hat{1}1r}$) appearing in the flowchart of this reconstruction algorithm have been presented, for $W = 70$ mm, in Figs. 12.3, 12.4, and 12.10. The permittivity profile is computed from these data according to the diagram given in Fig. 12.11.

Fig. 12.12 shows a comparison of the different complex permittivity profiles calculated. These results indicate that the profile of the real part of the complex permittivity is retrieved to an acceptable accuracy whereas the imaginary part is not restored in a good manner. This is not surprising considering the assumption made for the magnitude of the reflection coefficient ($\left|S_{11r}(x)\right| = \left|S_{11r}(\text{dry})\right|$).

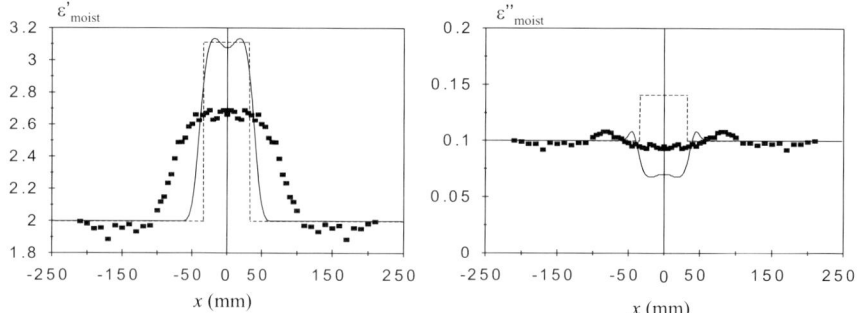

Fig. 12.12. Comparison of the complex permittivity profiles for $W = 70$ mm (■: degraded data; ----: true profile; ——: profile estimate)

Once again, these results ($\varepsilon''_{\text{moist}}$) are not given for their relevance but only to indicate the degree of validity of the method.

According to the description given in Fig. 12.11, we calculate the moisture content profiles from the data given in Fig. 12.12 and the Eq. (12.8).

A comparison of the different moisture profiles calculated, for the four configurations investigated, is presented in Fig. 12.13.

For $W = 70$ mm and $W = 50$ mm the findings show that the moisture content profile is comparable to the true profile. The data observed after the deconvolution process enhance the quality of the retrieved profile. When the antenna is just over the inclusion ($x = 0$) the degraded data (before deconvolution) indicate a moisture content of about 5.5% for $W = 70$ mm and 4.5% for $W = 50$ mm instead of 8.5%, whereas the value retrieved after the deconvolution treatment is correct. For $W = 30$ mm the shape is globally well restored; however, there are resolution problems, related to the diminution of the signal to noise ratio.

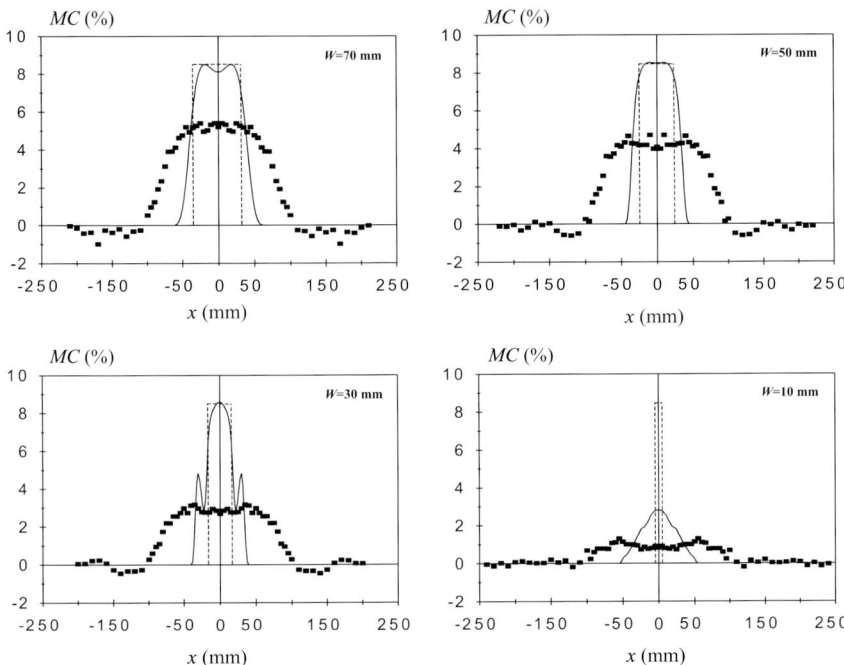

Fig. 12.13. Comparison between moisture profiles for different values of W
(■: degraded data; ----: true profile; —— : profile estimate)

But, once again, it can be noticed that for the position $x = 0$ the deconvolution process leads to a very good estimate moisture content. An examination of the last case, $W = 10$ mm, shows that the model is not valid for this size of inclusion and this level of moisture content. Thus, except for the last configuration, the different simulations performed clearly show that the method is very efficient when subsurface detection of a zone presenting a moisture content different from its surroundings is required. In addition, the moisture content of this zone is well estimated. Concerning the reconstruction of a moisture profile, it has been shown that the technique based on a blind deconvolution approach brings a noticeable improvement in the retrieved profile quality, at least for a moist inclusion with a width greater than 10 mm. After this theoretical analysis based on simulation results we propose to perform indoor experiments to analyze the performance of the method on real data.

12.4.2 Moisture Profile Measurement

The measurement study is performed by considering the same material investigated in the simulation analysis. Actually, the structure under test is made of four blocks of cellular concrete. Three of them are dry and the fourth is moist. The arrangement is realized according to Fig. 12.2. Taking into consideration the simulation results, we have chosen an inclusion whose thickness d_2 and width W are equal to 50 mm, buried at a depth d_1 equal to 50 mm. For the moisture content of this slab we have decided to increase the difficulty by considering a lower moisture level compared to the simulation tests. Therefore the moisture content presented by the inclusion is $MC = 4\%$.

The method used to reconstruct the moisture profile is copied from the development described in the simulation analysis. The first step consists in measuring the reflection coefficient S_{11m} of the structure under test. The results of this characterization are given in Fig. 12.14.

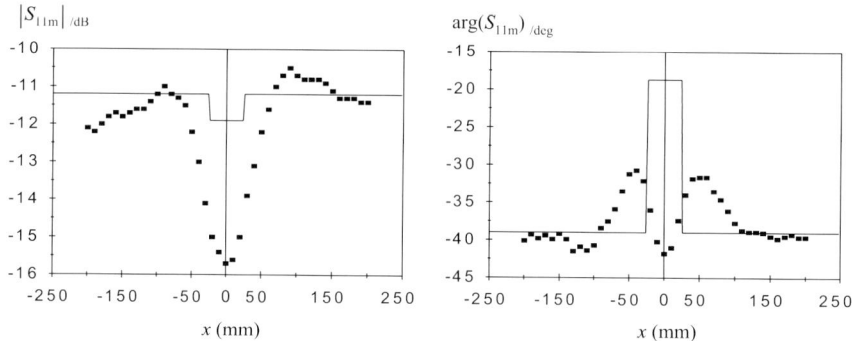

Fig. 12.14. Measurement of reflection coefficient profile $S_{11m}(x)$ for $W = 50$ mm (■: measured data; ——: true profile)

As for the previous simulation data we have also presented in this graph the true profile that should be obtained. This one is constructed from knowledge of the inclusion width and the true moisture content ($MC = 4\%$) that gives the true complex permittivity of the moist material ($\varepsilon^*_{moist} = 2.49 - 0.16j$) by using Eq. (12.7) and finally the true reflection coefficient S_{11m} (Eqs. (12.1–3)).

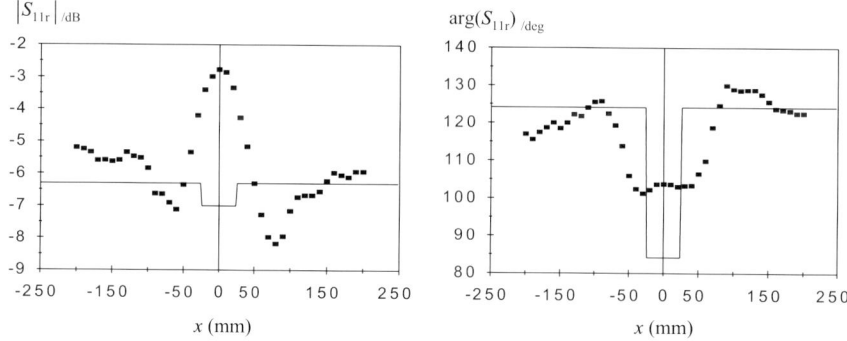

Fig. 12.15. Calculation of the reflection coefficient profile $S_{11r}(x)$ for $W = 50$ mm (■: measured data; ——: true profile)

The data collected for S_{11m} are the basis for the calculation of the reflection coefficient in the plane located at the interface between the upper block and the three others (Fig. 12.2). The findings for this parameter, S_{11r}, are presented in Fig. 12.15.

The results exhibit a shape that is not symmetrical. This is essentially due to a slight slope along the $0x$ axis and probably to a difference between the state of the blocks on each side of the moist slab. Actually, after verification it was shown that the block on the left ($x < 0$) was not entirely dry. A few tenths of a percent were present in certain zones of the sample. This residual moisture is the result of the drying of this sample that was initially moistened for other tests. Taking into account these experimental difficulties, we have obtained symmetry by considering only the data collected for positive values of x.

As already described in the simulation study, a curve fitting procedure is used (Eq. (12.9)). The correlation coefficient is equal to 0.994:

$$\arg(S_{11r}\ \text{fitted}) = \left[99.4\cos\left(\frac{x}{26.8}\right) - 118.9\cos\left(\frac{x}{31.5}\right)\right]e^{\frac{-x^2}{5520}} + 124.1. \tag{12.9}$$

The profile estimate is then recovered by applying the algorithm based on blind deconvolution given in Fig. 12.9. Fig. 12.16 shows all the profiles (fitted, estimated, and true) to make a comparison.

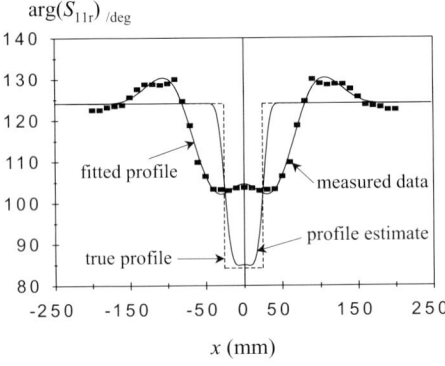

Fig. 12.16. Comparison of reflection coefficient phase profiles for $W = 50$ mm

The results show a sensitive enhancement of the retrieved profile.

According to the description given in Fig. 12.11 for the moisture profile reconstruction we calculate the permittivity profile. The comparison between the estimated and true profiles is shown in Fig. 12.17.

We can conclude that there is good agreement between the two kinds of data in the case of $\varepsilon'_{\text{moist}}(x)$. On the contrary the $\varepsilon''_{\text{moist}}(x)$ profile is not restored in a good manner. Bearing in mind the level of moisture investigated, this result was, however, expected (low-loss material).

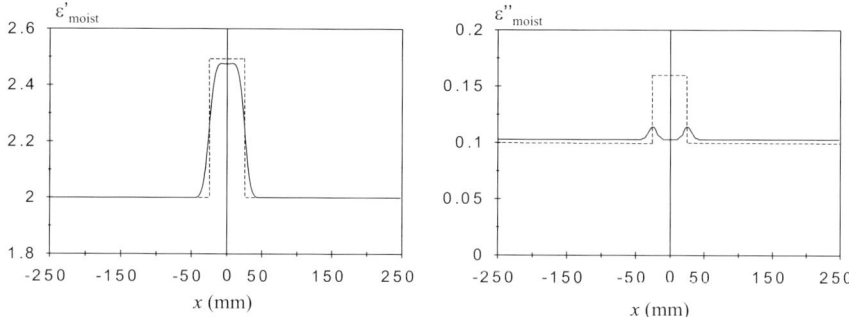

Fig. 12.17. Profiles comparison of the complex permittivity for $W = 50$ mm
(----: true profile; ———: profile estimate)

Finally, to complete the process we give in Fig. 12.18 a comparison of the different moisture content profiles.

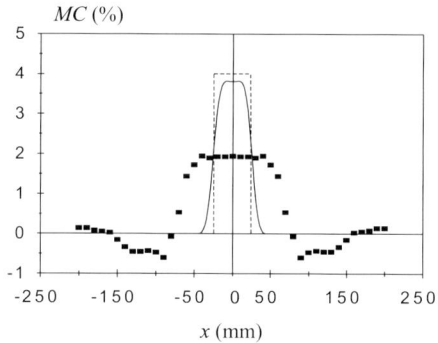

Fig. 12.18. Comparison of moisture content profiles for $W = 50$ mm
(■: measured data; ----: true profile; ———: profile estimate)

These results demonstrate the importance and the benefits of using signal processing tools to enhance the quality of a recovered profile. The method proposed seems to be promising, especially because the moisture content investigated is very low.

12.5 Conclusion

The use of microwave techniques has been investigated for many applications during the last few decades. The measurement of moisture content is one field where this technology has found a very large and successful development.

Nevertheless, for a long time many applications were based only on microwave capabilities. The performance of the characterization method was mainly related to the accuracy of determining the parameters measured by the microwave system (for example, magnitude and phase of the reflection coefficient or transmission coefficient) that lead to the quantity of interest such as moisture content. Today, we can show that with adequate signal processing treatment we are able to improve the quality of the information retrieved. The sensing problem has to be considered in terms of sensitivity with respect to the quantity of interest and also with aspects related to the development of signal processing tools, while maintaining acceptable costs. In this particular study the objective was to investigate the possibility of reconstructing a subsurface moisture profile for the purpose of detecting, locating, and qualifying a zone presenting a moisture content different to its surroundings. Simulation data have shown that the technique is relevant for a relatively small inclusion with regard to the wavelength. Confirmation of the good behavior of the proposed model has been obtained through indoor experiments. It can also be noted that these encouraging results have been obtained for inclusions presenting a low moisture content. In the simulation and experimental investigations the buried depth of interest was chosen equal to 50 mm, but the method is still applicable for greater burial distances.

References

1. Kraszewski AW, Trabelsi S, Nelson SO (2001) Broadband free-space microwave wheat permittivity measurements. In: Kupfer K (ed) Fourth international conference on electromagnetic wave interaction with water and moist substances, Weimar, pp 195–202
2. Kupfer K (2001) Material properties – measuring methods – applications. In: Kupfer K (ed) Fourth international conference on electromagnetic wave interaction with water and moist substances, Weimar, pp 517–527
3. Fischer M, Nyfors E, Vainikainen P (1996) On the permittivity of wood and the on-line measurement of veneer sheet. In: Kraszewski A (ed) Microwave aquametry - electromagnetic wave interaction with water-containing materials, IEEE Press, Piscataway, NJ, pp 347–354
4. Kent M, Knöckel R, Daschner F, Berger UK (1999) Measurement of composition and prior treatment of foodstuffs using complex dielectric spectra. In: Third workshop on electromagnetic wave interaction with water and moist substances, Athens, GA, pp 20–24
5. Zhang Y, Okamura S (2001) New function of dielectric properties for density-independent moisture measurement. In: Kupfer K (ed) Fourth international conference on electromagnetic wave interaction with water and moist substances, Weimar, pp 111–116
6. Lasri T, Glay D, Mamouni A, Leroy Y (2000) Development of microwave moisture measurement system around microstrip complex correlator. Sensors update, vol 7, Wiley-VCH, Weinheim, pp 233–248
7. Lasri T, Glay D, Mamouni A, Leroy Y (1996) A low cost microwave system for non destructive control of textile webs. J Microwave Power Electromag Energy 31(2):122–126

8. Lasri T, Glay D, Mamouni A, Leroy Y (1999) Free-space moisture measurements of cellular concrete. In: Third workshop on electromagnetic wave interaction with water and moist substances, Athens, GA, pp 184–188
9. Kraszewski A (1991) Microwave aquametry - needs and perspectives. IEEE Trans Microwave Theory Tech 39(5):828–835
10. Glay D, Lasri T, Mamouni A, Leroy Y (2000) A 35 GHz vector system for non destructive applications. In: Thompson DO and Chimenti DE (eds) Review of progress in quantitative non destructive evaluations, vol 19A. American Institute of Physics, New York, pp 579–586
11. Glay D, Lasri T, Mamouni A (2001) Non destructive permittivity profile retrieval of non-planar objects by free-space microwave techniques. Subsurf Sensing Technol Appl 2(4):391–409
12. Ayers GR, Dainty JC (1988) Iterative blind deconvolution method and its applications. Opt Lett 13(7):547–549
13. Miura N, Baba N (1992) Extended-object reconstruction with sequential use of the iterative blind deconvolution method. Opt Commun 89:375–379
14. Tsumuraya F, Miura N, Baba N (1994) Iterative blind deconvolution method using Lucy's algorithm. Astron Astrophys 282(2):699–708
15. Djafari AM, Qaddoumi N, Zoughi R (1999) A blind deconvolution approach for resolution enhancement of near-field microwave images. SPIE Proc 3816: 274–281
16. Jalobeanu A, Blanc-Féraud L, Zerubia J (2000) Study of the estimation of instrumental parameters in satellite imaging. INRIA Research Report RR-3957
17. Kundur D, Hatzinakos D (1998) Recursive blind deconvolution of still images based on nonnegativity and support constraints. IEEE Trans Signal Process 46(2):375–3
18. Glay D, Lasri T, Mamouni A, Leroy Y (2001) Free-space moisture profile measurement. In: Kupfer K (ed) Fourth international conference on electromagnetic wave interaction with water and moist substances, Weimar, pp 235–242

13 Sensors for Soil, Substrates, and Concrete Based on the MCM100 Microchip

Jos Balendonck[1], Max A. Hilhorst[1], William R. Whalley[2]

[1] Wageningen UR, Agrotechnology and Food Innovations, P.O. Box 17, 6700 AA Wageningen, the Netherlands.
[2] Soil Physics Group, Silsoe Research Institute, West Park, Silsoe, Bedford, MK45 4HS, UK.

13.1 Introduction

Monitoring dielectric properties through impedance measurements to characterize material composition is a commonly known sensing technique and is useful in a broad range of applications. Since the fundamentals of this principle were described [1], many applications based on this technique have been described in the literature. Among them are medical, industrial, agricultural as well as consumer-based applications. They span an enormous broad frequency range, from very low frequencies down to 1 mHz up to the microwave range above 10 GHz.

For research and development, mainly laboratory equipment like the Hewlett Packard impedance analyzers developed in the 1980s are used (Agilent Headquarters, Palo Alto, USA). To compensate for electrode polarization at low frequencies, instruments are available measuring with probes that have three or four electrodes (Solartron Analytical, Farnborough, UK). Recently broadband dielectric spectroscopy analyzers became available that go down to 3 μHz and up to 10 MHz, with good accuracy over an ultra-wide range of 16 decades for resistance and capacitance (Novocontrol GmbH, Hundsangen, Germany). This equipment is suitable for material analysis, since it covers a broad spectral range and even supports on-line temperature control.

In process monitoring as for instance in agriculture, the food industry, and construction engineering, there is an enormous need for low-cost sensors. Dielectric measuring equipment could fulfill this purpose, but the equipment described above is often too expensive and not suited for simple applications. In the 1950s, time domain reflectometry (TDR) became a favorite method to measure material properties with a dielectric method [2]. It is a special form of time domain spectrometry (TDS), while it is operated at a single frequency. In the beginning of the 1980s, cable analyzers were used for this purpose. Since data had to be interpreted visually, they were difficult to control, and they were too expensive to be used as a simple sensor. Only recently have cheaper TDR sensors become available that make use of advanced digital signal processing [3].

The impedance bridge [4] is one of the oldest applications of the frequency domain (FD) method. Over the years, this method has been applied with varying success. Due to innovations in electronics, stable circuitry has become available which made the use of the FD method at radio frequencies possible [5]. To make simple sensors for on-line process monitoring, the FD method suddenly has potential, especially for water content, but also for other parameters. Over the last two decades a large number of these sensors have come onto the market. Since dielectric properties are obtained at a single frequency, or at a limited number of discrete frequencies, the spectral range of these sensors is limited. We will refer to these techniques as small-band frequency domain spectroscopy (FDS). Sensors based on single frequency measurements are sometimes referred to as frequency domain reflectometry (FDR) sensors or FD sensors. In comparison to other material constituents, water has a high dielectric constant. Therefore, monitoring water content is the most widely spread application for on-line FD sensors. Many of them measure electrical conductivity (EC) as well, since this reflects the total amount of water-dissolved particles. FD sensors are tuned for a specific application. Each one has its own solution for the known problems of electrode polarization, temperature compensation, and the Maxwell–Wagner effect. In general users are not interested in the complex permittivity these sensors measure. They have to relate complex permittivity to the material properties they need, which involves the problem of calibration.

For application in agricultural and for automatic irrigation, numerous FD sensors have been developed to measure water content in soil and other growing media [6–9]. Over the last decade numerous manufacturers have introduced TDR or FD sensors. Examples are the Theta Probe (Delta-T Devices Ltd, Cambridge, UK), the Aquaflex (Streat Instruments, Bromley Christchurch, New Zealand), the TRIME-FM (IMKO Micromodultechnik GmbH, Ettlingen, Germany), and many others [10].

Dielectric measurements have great potential for monitoring moisture in construction materials like cement, sand, or asphalt. Numerous instruments measure moisture based on resistance or capacitance readings [11]. For concrete, this can be for new constructions to measure the hardening process [12], for precautionary purposes to monitor concrete aging [13], or to follow the drying process for curative measures after flooding accidents (TRIME-ES, Micromodultechnik GmbH, Ettlingen Germany).

In food processing, moisture could be measured with the FD method. However, since food texture is complex, the calibration is often a problem. Therefore, near-infrared or microwave technologies have taken over this application. Another promising application here is the monitoring of living cells. Living cells exhibit a very specific dielectric behavior – called the β-relaxation – in the LF and RF spectral range from 30 up to 300 MHz [14]. In this range, cell membranes are short-circuited, and as a result the cell's internal water and protein content can be seen. Many medical and biological sensor applications are based on this phenomenon. To control the fermentation process, in breweries for instance, an instrument is in use that measures the yeast cell concentration (Aber Instruments Ltd, Aberystwyth,

UK). Furthermore the Solartron 1260 (Solartron Analytical, Farnborough, UK) was used to study the behavior of yeast cells when toxic stimuli were applied [15].

All these applications vary in complexity and are more or less built with discrete electronics, which make them either expensive or inaccurate. Some years ago a mixed analog and digital integrated circuit for the measurement of complex impedances became available. It can be used in a frequency range from 10 to 30 MHz. With this microchip, reliable FD sensors can be built [16]. It can also operate at multiple discrete frequencies. So with it, a dielectric spectroscopy sensor with limited frequency range can be built. The small-band discrete spectrum can reveal much more information about the material than a single frequency measurement. In the following sections several applications based on this microchip will be described. They are based upon work from all authors. Much of this work has been published before, and is mentioned in the references. Part of the work was done together with others, who are mentioned in the acknowledgments.

13.2 A Microchip for Impedance Monitoring

In the 1980s, everyone started to use TDR to measure soil water content, working in the higher RF range between 100 and 200 MHz [6, 17]. Some ten years later it was revealed that FD sensors could be calibrated for soil water content in a lower RF range between 10 and 100 MHz as well. Although this calibration appeared to be more sensitive for soil texture, their accuracy was acceptable for practical applications. The FD method was used long before 1980, but the first FD sensors based upon discrete electronics became available around 1990. They measured a frequency shift as the electrical capacitance changed due to variations in water content [5, 7]. The tuning of the electronics was very tedious, and it showed that soon these sensors would only be successful if microchips were used. This would reduce the cost per sensor enormously. However, to ensure high-phase accuracy at the operating frequency (20 MHz), the microchip needed analog circuitry that could work up to 6 GHz. A microprocessor was needed to perform the signal analysis and the calibration task. Therefore, the chip needed an embedded processor or at least a digital interface.

Around 1992 this electronic microchip was designed as an application-specific integrated circuit (ASIC) by using a bipolar CMOS process (SGS-Thomson, Grenoble, France). Prototypes (see Fig. 13.1 left) of this microchip became available in 1994 [16], and shortly thereafter the first prototype FD sensors were built. Positive results were reported [18, 19]. Since only a few external components are needed, it is especially suitable to construct cheap, smart FD sensors to be used for on-line monitoring of water content in agricultural, environmental, and industrial processes. In this section the measuring principle of the microchip is described briefly; more detailed information can be found in other publications [20–22]. In the next sections the application of the chip as a sensor for water content in soils and substrate materials, pore water conductivity, and strength monitoring for young concrete is described.

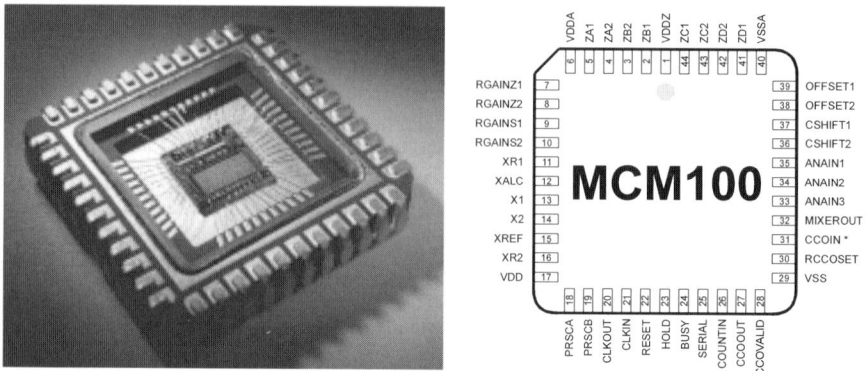

Fig. 13.1. Chip for dielectric measurements in open ceramic package (*left*). Chip pin configuration (*right*)

The microchip, named MCM100, is a vector voltmeter that measures the electric impedance at four differential inputs at a single frequency in the range from 10 to 30 MHz. From the digital data it generates, software can compute permittivity (ε) and conductivity (σ) of the material under test and subsequently volumetric water content. Electrodes can be connected to one of its inputs, via a set of decoupling capacitors that block DC currents through the measured medium, thus preventing electrolysis of the electrodes. Two other inputs can be used to measure a capacitor and resistor with known value as references. Many internal chip errors and even internal and external parasitic components can be compensated for with software. The fourth input can be left open or used for optional purposes (see Fig. 13.1 right).

The microchip has three additional single-ended analog inputs to which external analog or pulse-width-modulated sensors can be connected. Since many physical parameters are temperature dependent, a temperature sensor can be chosen, so measured values can be corrected for temperature. There is a TTL-level serial output that can connect directly to a microprocessor or with a simple RS232-level shifter to a PC or field bus system. Its baud rate can be derived from either the internal or an external clock source. An external prescaler may be used to tune the baud rate and timing of the chip. The output reveals data for the four differential and three analog inputs as well as for internal power and zero. For multiple sensor applications, several chips can be cascaded. This chip operates from a single 5 V power supply using approximately 35 mA. It has an oscillator and reset circuitry on-board, and is commercially available in a standard 44-pin PLCC package.

13.2.1 Operation of the Chip

Internally the chip contains a synchronous detector with a multiplier ($\times$) and low-pass filter (LPF), an analog to digital converter (ADC), a parallel to serial converter (PSC), and timing and controlling logic (see Fig. 13.2). Up to four impedances (Z_A, Z_B, Z_C, Z_D) can be measured at four differential inputs. A sine wave current (i_z) with a frequency determined by an externally connected crystal (f_0), comes from a stabilized oscillator (OSC). It develops a voltage (u_z) across the unknown impedances that are successively selected by S_1. This voltage is fed to one input of the analog multiplier. A second current (i_{shift}), equivalent to i_z, also comes from the oscillator. Its phase is shifted by respectively 0°, 90°, 180° and 270°, which is controlled by the switches S_2 and S_3. The voltage developed across the phase shifter (u_{shift}) is fed to the other input of the multiplier. The multiplier output (u) consists of a DC and an AC term with frequency $2f_0$. The DC term (U) is found at the output of the LPF. In the case of a 0° and 180° phase shift, U is a measure for the capacitance or inductance of Z_i. In the case of a 90° and 270° phase shift, U is a measure for the conductance of the impedances Z_A to Z_D. The output of the LPF is fed to the ADC and then converted into digital format by the PSC, whose data is outputted at the serial output. In order to compute Z_A to Z_D, this data must be processed further by a processor, which can be connected to the microchip externally.

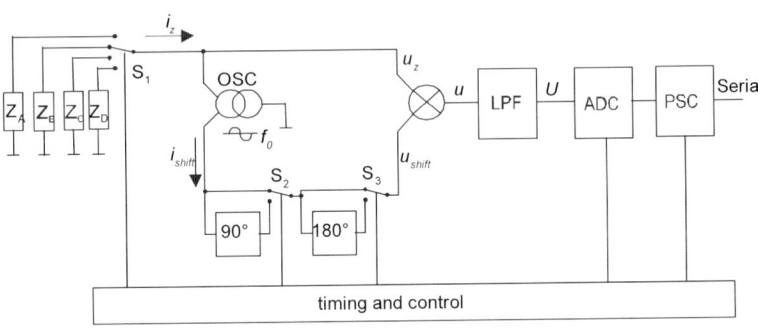

Fig. 13.2. Functional block diagram, showing the internal working of the chip

As long as the microchip is powered, it outputs serial data. Each single measurement (one cycle) involves 22 individual measuring steps. The duration of each step, as well as the baud rate, is dependent on f_0 and the set-up of the internal pre-scaler (PRSCA, PRSCB in Fig. 13.1, right). For $f_0 = 20$ MHz, typically a baud rate of 1200 Bd and a step time of 100 ms is obtained. There are 16 steps for the four impedances $Z_{A,B,C,D}$ at all four quadrants of the complex plane. Six other steps are used for the three analog inputs of which one is used for temperature and two for internal zero reference. To be able to compensate for a possible warming up of the chip during a measurement cycle, temperature is measured at the beginning and end of the measuring cycle. During every step the microchip outputs a 6-byte serial ASCII pattern that reflects the step identifier (A, B, ..., U, V) followed by the relative measured value, a five-digit BCD code in the range from 0 to 99999 (see Table 13.1). The serial data is sent as one package containing in total 134 characters starting with character "@" and ending with character "↔". In between these characters, there are the 22 packages belonging to the measuring steps.

Table 13.1. Overview of measuring steps for each measurement cycle

Step	Identifier, value	Comments
Start	@	No measurement
01	Annnnn	Analog input 3
02	Bnnnnn	Offset at beginning
03	Cnnnnn	Analog input 1
04–07	Dnnnnn-Gnnnnn	input Z_A:$U_{0°,90°,180°,270°}$
08–11	Hnnnnn-Knnnnn	input Z_B:$U_{0°,90°,180°,270°}$
12–15	Lnnnnn-Onnnnn	Input Z_C:$U_{0°,90°,180°,270°}$
16–19	Pnnnnn-Snnnnn	Input Z_D:$U_{0°,90°,180°,270°}$
20	Tnnnnn	Analog input 2
21	Unnnnn	Offset at end
22	Vnnnnn	Analog input 3
Stop	↔	No measurement

To build a sensor, only the microchip and a few extra components are needed (see Fig. 13.3). It needs a crystal (XTAL) for the internal clock, an operational amplifier (MAX480) for offset compensation, a reference resistor R_{ref} and a capacitor C_{ref}, two resistors for automatic gain control (R_1 and R_2), and (if needed) an external prescaler to set the baud rate. A set of measuring rods can be connected via two DC-blocking capacitors. As a temperature sensor, a common temperature dependent resistor (NTC), or an active element like the AD590 (Analog Devices) or the SMT160-30 (Smartec, Breda, the Netherlands), can be used. To test the working of the chip a prototype sensor was built. All electronics including the microchip, but excluding a microprocessor and memory, were placed on a printed circuit board (see Fig. 13.4).

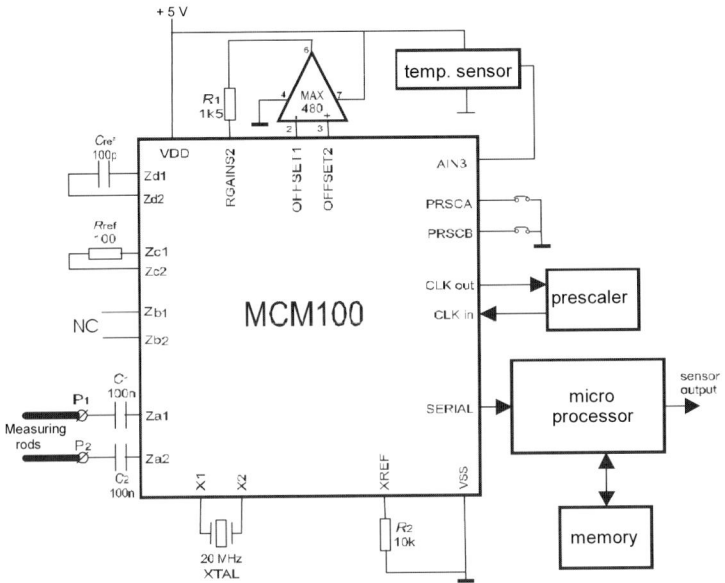

Fig. 13.3. Simplified schematic of sensor electronics (input $Zb_{1,2}$ is left open)

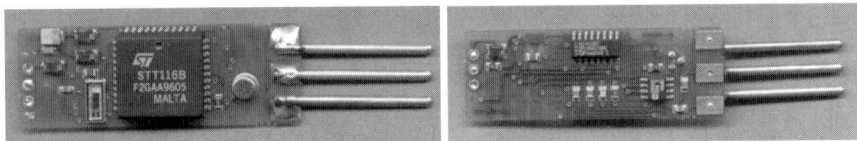

Fig. 13.4. Front side (*left*) of mounted PCB (20×55 mm) with chip, clock crystal, temperature sensor, and three measurement rods (25 mm). Back side (*right*) of PCB with prescaler and offset amplifier

13.2.2 Calibration Procedure and Computation of Dielectric Properties

Besides the unknown impedances $Z_{A,B,C,D}$, parasitic elements like electrodes, inputs, and internal chip circuitry contribute to the total measured impedance. The sensor must therefore be calibrated to measure the complex impedance properly. A simplified four-element model can be used for this. Although it is an approximation for a lumped model transmission line, this model has proven to be adequate for practical applications. It consists of a series inductor (L_s) and resistance (R_s) to model the electrical path length of the electrodes and a parallel capacitor (C_p) and resistor (R_p) that model the input circuitry of the chip (see Fig. 13.5). The values for L_s, R_s, C_p, and R_p are typical for each sensor and need to be obtained through calibration. The known external reference components (R_{ref} and C_{ref}) are used to compute C_p and R_p. Internal chip offsets are compensated for by subtracting two 180° phase-shifted signals for each input. Temperature readings (T) are linearized with a third-order polynomial and a single offset is used for calibration. Calibration data is typical for each sensor, and must be stored in memory to be retrieved at the time actual measurements are taken.

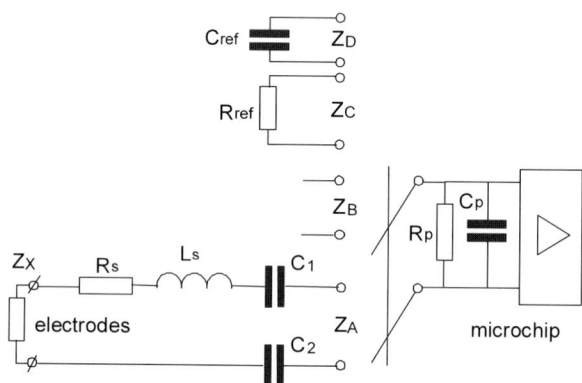

Fig. 13.5. Simplified electric model for the sensor measuring rods and chip input circuitry. Z_B is left open, and the DC-blocking capacitors C_1 and C_2 are considered to be much larger than the capacitive components in Z_x

Serial data coming out of the chip must be processed before the dielectric properties of the material under test become available. A multi-purpose microprocessor, a hand-held computer or a PC can do this job. From the serial data and previously stored calibration data for R_{ref}, C_{ref}, C_p, R_p, L_s, and R_s, and based upon complex arithmetic, the impedances Z_A to Z_D are computed first. Next the unknown impedance Z_x can be obtained. Finally, from this impedance, the dielectric properties ε and σ are computed [8, 22].

For a prototype sensor the ε- and σ-scales and the electrical path length compensation parameters were obtained by placing the electrodes in reference media. Air, tap water ($\sigma \approx 0.002$ S/m), and water of two other conductivities (0.1 S/m and 0.2 S/m) were used. The ε-scale was calibrated between $\varepsilon = 1$ for air and $\varepsilon = 80.3$ for tap water at 20°C. The σ-scale was determined for $\sigma = 0$ in air and $\sigma = 0.2$ S/m in water. L_s was found from the measurements in water at $\sigma \approx 0.002$ S/m and $\sigma = 0.2$ S/m. Water with $\sigma = 0.1$ S/m was used to adjust R_s such that ε-readings at all three conductivities were equal. The calibration software used a recursive approach to yield the dielectric calibration data. Tests with the sensor showed that ε could be measured with an accuracy of ±1 on a scale of 1–100 and a resolution of 0.1 in a temperature range from 15 to 25°C. Based on the temperature sensor SMT160-30 (Smartec, Breda, the Netherlands), readings have an accuracy of 0.5°C with a resolution of 0.1°C, in the operating range from −5 to 50°C.

13.2.3 Temperature Corrections

FD sensor readings are dependent on temperature. In the soil top layer, where temperature is very much dependent on sunlight, sometimes large temperature fluctuations are seen during the day. To allow for on-line correction, temperature is measured in the sensor. Little is found in the literature about the influence of soil texture and density on the temperature behavior of soil water content and EC. Recently it has been shown that the temperature behavior is dependent on soil texture [23]. Positive as well as negative effects have been seen for different soil types. This makes temperature corrections rather ambiguous. Nevertheless, for permittivity and conductivity separately we can perform some general corrections. The permittivity of pure water at a specific temperature can be obtained from [24]:

$$\log \varepsilon_{water}(T) = 1.94404 - 0.001991\ T. \tag{13.1}$$

This function was specified over a temperature range from 0 to 40°C with a maximum error of 0.3%. A simpler equation can be used, such as the following approximation of Eq. (13.1):

$$\varepsilon_{water}(T) = 80.327 - 0.368(T - 20). \tag{13.2}$$

EC depends on T and the dissolved ion types, which makes it impractical to handle this parameter just as it is measured. Growers use EC referred to a predefined reference temperature (T_{ref}), for instance 20°C or 25°C. For each water–salt mixture a specific temperature coefficient (α_i) should be used. This referenced conductivity (σ_{Tref}) can be computed from [25]:

$$\sigma_{Tref} = \sigma[1 - \alpha_i\ (T - T_{ref})]. \tag{13.3}$$

For average soil types a value of $\alpha_i = 0.0216$°C^{-1} [26], and for water with dissolved NaCl, $\alpha_i = 0.0225$°C^{-1} may be used.

13.3 FD Sensor for Water Content and Bulk EC in Soil

Based on the microchip described in the previous section, a water content FD sensor (see Fig. 13.6) was developed [18]. This sensor determines the complex permittivity of soil from the electric impedance at a single frequency (20 MHz). The complex permittivity can be related to bulk electrical conductivity (σ) and volumetric water content (θ) after calibration for a certain soil type [6, 18, 27]. By using the microchip, the sensor is solid state, robust, and needs no maintenance and no repeated calibrations. It has three in-line electrodes of which the outer two are electrically connected together and behave as a guard similar to the working of a coaxial probe. It has been shown that it can operate on a long-term basis and can be produced in large quantities at a low price [20, 28]. Currently it is available as the WET-sensor (Delta-T Devices Ltd, Cambridge, UK).

The validation of the FD sensor for permittivity was done by placing a number of FD sensors in reference liquids: pure water, a 1:2 water–ethanol mixture, and water-saturated glass beads (0.2 mm) at a constant temperature of 20°C. Conductivity was varied using increasing amounts of NaCl. It was shown that after calibration the sensors operated well and that the accuracy for permittivity was ± 1% over the full-scale range from 1 to 80. The sensors were also validated for conductivity. The accuracy found for conductivity was ± 1% for a full-scale range from 0 to 0.2 S/m. These observations were based on a limited number of sensors and were done for a full-scale calibration of 0.2 S/m [8].

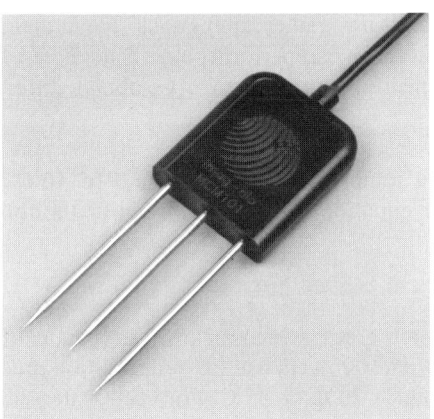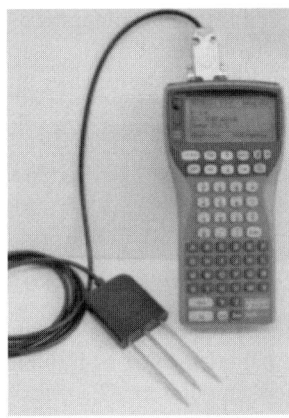

Fig. 13.6. FD sensor for soil water content, EC, and temperature (*left*). The dimensions of the housing are 46×55×12 mm³, and the electrodes have a length of 68 mm, each spaced 15 mm from each other. FD sensor connected to a PSION Workabout hand-held computer (*right*)

The sensors were used in several applications, and also calibrated for higher EC ranges up to 0.5 S/m. As will be shown in a later section, the sensors can be used to measure pore water conductivity as well. For that, σ is multiplied by a factor larger than one, which is proportional to the reciprocal value of the water content. Hence, errors in σ will be multiplied by this factor as well, so the accuracy of σ plays an important role. The factor becomes larger at lower θ levels, and the relative accuracy for σ gets lower, at lower σ-values, as is the case for dry agricultural soils ($\sigma < 0.1$ S/m, $\theta < 15\%$). For this reason extra attention was paid to the σ-linearity of the sensor, especially in the lower range ($\sigma < 0.1$ S/m).

The σ-linearity of the FD sensor was evaluated by measuring σ for eight different sensors in a number of water–salt mixtures. To see whether the calibration range influences the linearity of σ, this was done for three calibration ranges from 0 to σ_{max} (0.2, 0.4, and 0.8 S/m). First, the sensors were calibrated by using the procedure described in the previous section at three values for σ (0, $\frac{1}{2}\sigma_{max}$, and σ_{max}). Then, water–salt mixtures were made in the range from 0.01 to 0.5 S/m by mixing NaCl with water at a temperature of 20°C starting at the lower σ and adding salt to the mixture for each new experiment. Conductivity of the reference mixtures was measured using a four-point electrode LF conductivity meter from Profilab (WTW-LF597-S) working at 1 kHz with a standard conductivity cell, Tetracon 325 (Wissenschaftlich Technische Werkstätten GmbH, Germany).

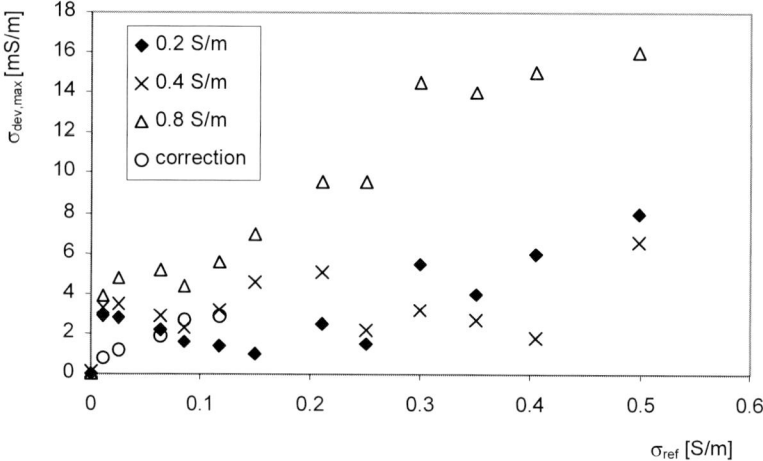

Fig. 13.7. Absolute maximum measured deviation in conductivity for eight FD sensors ($\sigma_{dev,\ max}$) versus reference conductivity (σ_{ref}) at three full-scale calibrations ($\sigma_{max} = 0.2$, 0.4, and 0.8 S/m)

Figure 13.7 shows the results of these experiments, where σ_{ref} indicates the conductivity measured with the LF meter in the reference mixtures, and $\sigma_{dev,max}$ is the maximum deviation for all 8 eight FD sensors found between σ_{ref} and the measured value (σ). We can see a sinusoidal behavior of the error that is dependent on the full-scale calibration range. This is due to the fact that a simplified model is used to compensate for the electrical path length of the electrodes. For the two calibration ranges 0.2 S/m and 0.4 S/m we see further that the errors become close to zero near the maximum conductivities at which the sensors were calibrated ($\sigma_{max} = 0.2$ and 0.4 S/m). Furthermore we see that the errors become larger outside the calibration range. Here obviously the model for electrical path length compensation does not work that well. Table 13.2 gives an overview of the maximum deviation found for all calibration ranges including the accompanying full-scale error. We see that the error is slightly more than the ± 1% found before [8]. Therefore, the sensors should be calibrated in the range where the expected measured values for σ will be.

Table 13.2. Maximum found deviation in conductivity for all 8 FD sensors within the calibration range, including the belonging full-scale error

Range (S/m)	$\sigma_{dev,max}$ (mS/m)	Error (% f.s)
0.2	3	1.50
0.4	5	1.25
0.8	16	2.00

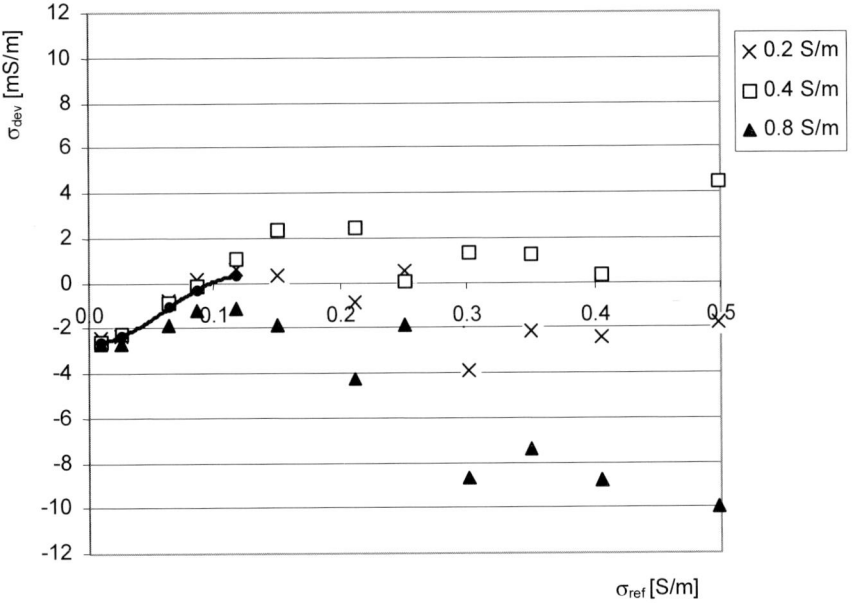

Fig. 13.8. Mean value of the measured deviation in conductivity for all FD sensors plotted against the reference conductivity.

Since the errors seem to correlate with conductivity due to the mismatch of the electrical path length compensation, we looked for ways to compensate for this systematic error. In Fig. 13.8 the mean value of the deviation for all 8 FD sensors for each used reference value is plotted against the reference values. We see clearly the sinusoidal behavior of the error, though it is different for each calibration range. For the two ranges 0.2 and 0.4 S/m the mean error stays within about 1.5% of the absolute value apart from conductivity values below 0.1 S/m. The errors are larger for the 0.8 S/m range. But, this range is rarely used in agricultural applications. At low conductivities ($\sigma < 0.1$ S/m) we see a significant and typical a-linear behavior, which seems to be independent of the full-scale calibration. Here, especially for $\sigma < 0.05$ S/m, the FD sensor systematically overestimates the conductivity by about 1–2 mS/m. Since this range is rather important for soils used in agricultural a correction is proposed for $\sigma < 0.05$ S/m, which leaves values above 0.05 S/m untouched. This correction can be used, irrespective the calibration scale. Figure 13.8 also shows a polynomial fit of the third order, common for the two calibrations for 0.2 and 0.4 S/m over the range from 0 to 0.1 S/m. Based on this fit the following equation was obtained for the corrected conductivity (σ^*):

$$\sigma^* = \sigma + a_0 + a_1\sigma + a_2\sigma^2 + a_3\sigma^3, \tag{13.4}$$

where we have assumed that $\sigma_{dev} \ll \sigma$, so we can replace σ_{ref} with σ as measured. For this polynomial fit the following coefficients were found: $a_0 = -2.6506\times10^{-3}$, $a_1 = -3.1322\times10^{-3}$, $a_2 = 0.64974$, and $a_3 = -3.4724$ ($R^2 = 0.999$). The results of this correction are showed in Fig. 13.7. We see that the errors for $\sigma < 0.05$ S/m are lower after correction. We can conclude that the full-scale accuracy for σ is 1.5%. After correction, at small values for $\sigma < 0.05$ S/m, for the two ranges of 0.2 and 0.4 S/m, an absolute accuracy better than $\pm 3\%$ can be obtained.

The relationships between ε, σ, and θ, strongly depend on soil density (ρ) and texture [6, 29]. Therefore, to obtain accurate readings, a soil-specific calibration is needed. This calibration is obtained by weighing soil samples during drying, while taking readings with the sensor. This gravimetric method gives reliable data under well-controlled conditions. For θ often the calibration Topp–curve is used [6]. This curve was obtained for sand and a number of sandy loam and clay–loam soils, by using the TDR principle at 150 MHz. It is used for an "average soil" and can be approximated with a third-order polynomial. Sometimes it is used in a simpler form:

$$\theta = 0.115\sqrt{\varepsilon} - 0.176. \tag{13.5}$$

For soils commonly used in horticulture, the accuracy found with this curve is $\pm 5\%$. To test the working of the soil-water-content FD sensor based on Eq. (13.5), an experiment was conducted with soil sampled from the top layer of a yellowish brown forest soil containing 18% clay and 3% organic matter [30]. The samples were air dried to reach weight equilibrium. Then θ was measured after drying in an oven at 105°C for 48 hours. A plastic container was filled with the dry soil and weighed to calculate its soil mass and volume. Then the amount of water to get a specific θ was calculated. Next, the dry soil was mixed with the water and poured into the plastic container. Then the soil volume in the container was recalculated and the final θ was determined. These samples were stored for 2 days to let the water redistribute and the temperature reach equilibrium. Thereafter the samples were measured with an FD sensor on three consecutive days, and average values were used to obtain the calibration curve, which were compared to Topp data (θ_{Topp}). The results for soil moisture from the experiment are given in Table 13.3. For this specific forest soil, the Topp curve slightly underestimates the actual moisture content (see Fig. 13.9). In the range from 0 to 35%, an accuracy of $\pm 1.1\%$ was achieved, which is nearly as low as the accuracy for ε.

Table 13.3. Results of ε measured with the FD sensor (ε_{FD}), volumetric water content as calculated with the oven dry method (θ), volumetric water content according to Topp (θ_{Topp}), and difference between θ and θ_{Topp} ($\Delta\theta$)

ε_{FD}	θ (%)	θ_{Topp} (%)	$\Delta\theta$ (%)
2.54	0.00	0.73	0.73
3.58	4.51	4.16	−0.35
5.50	10.35	9.37	−0.98
7.76	15.18	14.44	−0.74
12.85	24.74	23.62	−1.12
16.40	29.51	28.97	−0.54
18.79	32.64	32.25	−0.39
20.14	34.19	34.01	−0.18

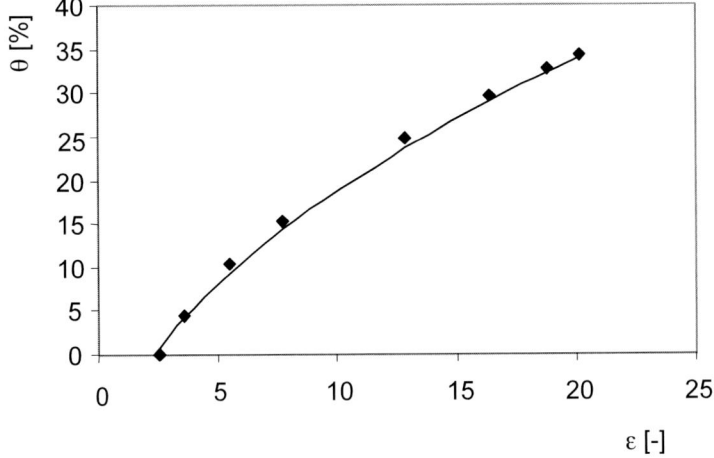

Fig. 13.9. Permittivity (ε) and soil water content (θ) measured for a forest soil (♦), related to the Topp-curve (—)

13.4 A Pore Water Conductivity Sensor for Growing Substrates

Traditionally TDR and FD methods measure permittivity (ε) and bulk conductivity (σ). Growers often refer to σ by using the term EC for electrical conductivity. EC here reflects the total electrical conductivity of the entire matrix containing substrate material, water, nutrients and air, which is strongly dependent on water content (θ) and subsequently ε. Since plants take up only water-dissolved fractions, growers are mostly interested in the EC of the water that can be extracted from the substrate matrix, for instance with a syringe. This so-called pore water conductivity (σ_p) relates to σ, but is dependent on ε. Straightforward measurement of σ with the FD sensor is therefore not useful for growers using substrates as for instance mineral wool mats or *Rockwool* (Grodan, Roermond, the Netherlands). Soil-based growers normally obtain σ_p by taking a soil sample. This sample is then mixed with a known volume of water to let the nutrients dissolve. Next they measure the EC in the aqueous solution with a standard EC meter. In case the amount of added water has the same volume as the soil sample, this method is referred to as "the 1:2 extract method." The pore water EC value is found by multiplying the measured value by a factor of two [31].

The FD sensor, based on the microchip, is capable of measuring σ, as well ε and temperature (T). Therefore, by measuring σ and correcting for ε and T, one can obtain σ_p with this sensor. This involves only a simple and straightforward model. This model was used to measure σ_p in situ in soil, for precision agricultural applications [32], and it describes the relationship between σ_p and the bulk soil values for ε and σ as measured with the FD sensor:

$$\sigma_p = \frac{\varepsilon_{water}\,\sigma}{\varepsilon - \varepsilon_{\sigma=0}} . \tag{13.6}$$

In this equation ε_{water} is the pure water permittivity corrected for temperature (see Eq. 13.2), and $\varepsilon_{\sigma=0}$ is an offset value. This offset can be obtained from ε and σ measured at two arbitrary free water content values and is not the value for ε when $\theta = 0$, but specifically for $\sigma = 0$. For a number of soils, empirically, values between 1.9 and 5.8 were found for $\varepsilon_{\sigma=0}$. These values are dependent on soil type, density, and the pin-type configuration [33]. For Eq. (13.6), it was assumed that the water is not bound to the soil matrix. Therefore, the model can not be used for bound water. Neither can it be used for conductivity due to ions moving through the lattice of ionic crystals in a dry or almost dry soil. For sand, the free water content corresponds to $\theta > 0.01$. For clay this is $\theta > 0.12$ [34]. As a rule of thumb the model applies for most normal soils if $\theta > 0.10$. Since it uses ε rather than θ, no calibration for θ is needed, and sensor working is not influenced by the soil–electrode contact and soil texture.

Since contact problems have only a minor effect on the measurements, a new sensor with a single and small sensor tip was developed (see Fig. 13.10, left). This sensor allowed for easy insertion into soils, but it was meant for measuring σ_p

only. It uses the same electronics as the FD sensor described in the previous section, but with a slightly higher measuring frequency of 30 MHz. The electronics are placed in a cylinder of a hard polyurethane molding at the top of the sensor rod. This rod is 10 cm long and has a diameter of 5 mm. It ends in a sharp point to facilitate insertion of the electrodes. The sensor tip is about 15 mm long and split into two metal electrodes separated from each other by a thin sheet of isolating material. The latter is a fringing field configuration where field lines concentrate just around the sensor tip. A temperature sensor is located close to the sensor tip to facilitate temperature measurements. A flexible polyurethane output cable contains the RS232 signal and power supply wires. This cable connects the sensor to a hand-held computer that runs signal-processing software.

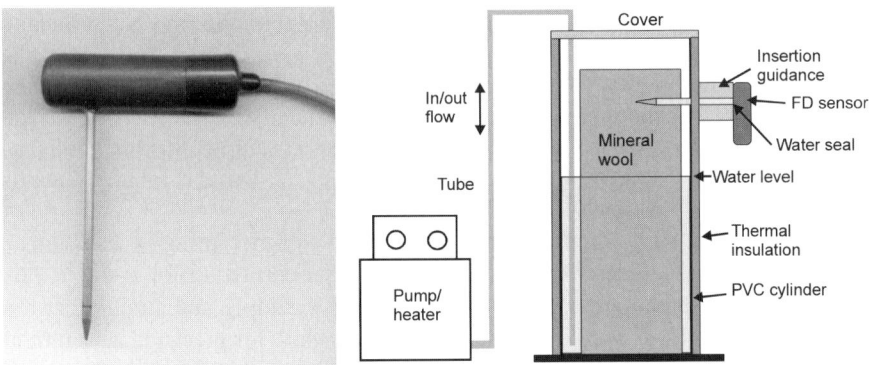

Fig. 13.10. Sensor for measurement of σ_p in soil (*left*). Experimental set-up (*right*)

It was shown that this sensor could be used for growing substrates as well, and for *Rockwool* a value for the offset $\varepsilon_{\sigma=0}$ of 4.1 was found [33]. For growers, this provides a new and more manageable way to measure in situ pore water EC, especially for hand-held application. In greenhouses, however, the encountered pore water EC levels in growing substrates are much higher than those found in soils. Furthermore the daily fluctuations of temperature are large. Therefore, the calibration of the model, the operating range, and the effect of temperature for application in mineral wool growing substrates were points of further study.

The performance of the sensor was studied in a *Rockwool* growing substrate at different values of EC, water content, and temperature, while it was connected to a PC [35]. The set-up (see Fig. 13.10, right) consisted of a vertical-placed PVC cylinder with an inner diameter of 15 cm and an height of 88 cm. The whole cylinder was thermally insulated. The top was covered to prevent evaporation and temperature gradients, and the bottom was sealed so as not to leak any water. Inside the cylinder a slab of growing substrate material was placed, filling about 80% of the inside of the cylinder. The sensor was inserted into this slab, horizontally through the wall of cylinder, about 10 cm under the top of the slab. An insertion guidance block was used to support the sensor to prevent it from leaning downwards. The sensor shaft was sealed with a silicone kit to make it watertight. Once inserted into the growing substrate the sensor was kept in place to keep the contact between the electrodes and the growing substrate intact.

The growing substrate around the sensor tip had to be brought at several levels of θ, σ_p, and T. Drying the sample through evaporation was not possible since the salt concentration in the pore water would then rise. Only a method that saturates and de-saturates the growing substrate sample in equal portions could be used. Therefore, water–salt mixtures were pumped in and out the container via a flexible tube fitted to a thermostatic bath containing a pump/heater combination (Ultra-Thermostat, COLORA, Germany) at a very slow rate.

Prior to the experiment the sensor was calibrated for ε and σ by measuring with reference values for air ($\varepsilon = 1$, $\sigma = 0$) and tap water ($\varepsilon = 80.3$, $\sigma \approx 2$ mS/m), and with water of two known conductivities (0.1 and 0.2 S/m) at 20°C. The temperature was calibrated using a single-point offset. Then, reference water–salt solutions were made by mixing NaCl with water at a reference temperature of 20°C. A Profilab WTW-LF597-S conductivity meter working at 1 kHz with a Tetracon 325 conductivity cell (Wissenschaftlich Technische Werkstätten GmbH, Germany) was used for this. Four EC values (σ_{ref} = 0.02, 0.05, 0.11, and 0.31 S/m) were taken by starting at the lower EC level and adding salts to the solution for each new experiment. Since salt and water have a different mobility, and to be certain of having the correct σ_p inside the growing substrate, the sample was slowly saturated and de-saturated several times to allow salt to fully penetrate the material. At the beginning of the measurement, the sample was completely immersed in water. Next, slowly and stepwise, it was de-saturated by pumping the water out of the cylinder. This was done until the sensor indicated a very low ε- or σ-value. The pump was stopped at water levels of +5 (saturated), –7, –10, –12, –15 and –18 cm, all referenced against the top of the substrate slab. During each stop, readings for ε, σ, and T were taken. Thereafter the cylinder was brought back to full saturation, and possible hysteresis effects were checked. A complete cycle took about 4 hours. Although exact values for θ could not be obtained, a broad range of θ-values was available. This procedure was repeated four times, by setting the thermostat subsequently at 10, 20, 30, and 40°C. Temperature and water content changes were applied four times at each EC value.

The experiment yielded 96 readings in total, for six water levels, four EC levels, and four temperatures. Since ε follows the wet–dry–wet cycles for θ, it behaves in a sawtooth manner. Readings for ε drop down to about 15 at low θ, and go up to about 70 at saturation (see Fig. 13.11, top). The temperature was set at a fixed level for each wet–dry–wet cycle (see Fig. 13.11, bottom). Based on Eq. (13.6) and the readings for σ, ε, and T, values for σ_p were computed. Figure 13.12 shows the conductivity for the original mixture (σ_{ref}), σ as measured directly with the sensor, and σ_p as computed, based upon $\varepsilon_{\sigma=0} = 4.1$, and referenced to 20°C with $\alpha_i = 0.0225$°C^{-1} for NaCl (see Eq. (13.3)). For a good working sensor, the curves for σ_p and σ_{ref} should be identical.

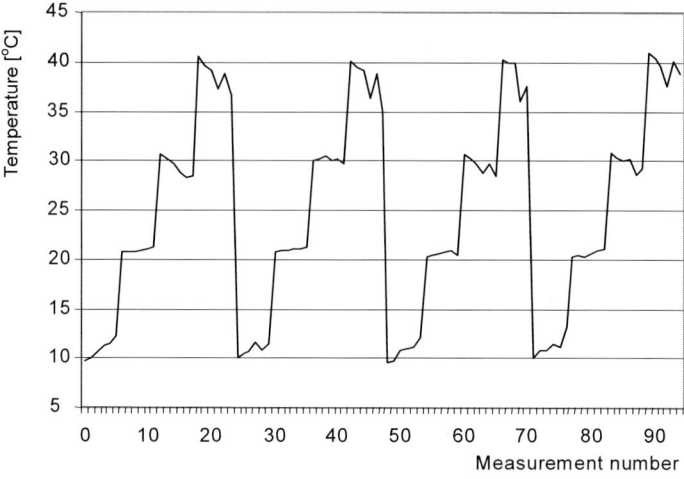

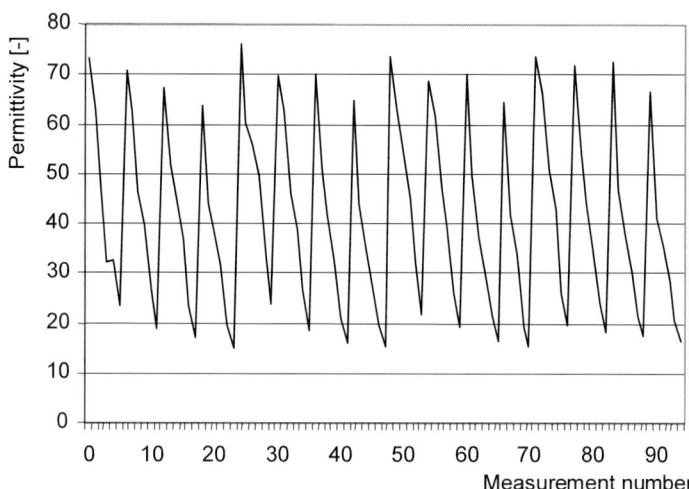

Fig. 13.11. Permittivity (*top*) and temperature readings (*bottom*) with measurement numbers on the *x*-axis

In spite of the temperature corrections for ε_{water} and σ, it can be seen that there is still a dependency on temperature (see Fig. 13.12). At higher temperatures, σ_p tends to decrease, while at lower temperatures there is a little overestimation of σ_p. A second-order fit could correct for this, but its effect was not fully analyzed. The performance of the temperature correction is considerably influenced by α_i. Although we used a fixed value for α_i of $0.0225°C^{-1}$ for NaCl, a better a correlation for σ_{ref} and σ could be obtained by using a coefficient of about $0.01°C^{-1}$. It seems that the impact of the medium, with respect to the salt-type mixture used, has a

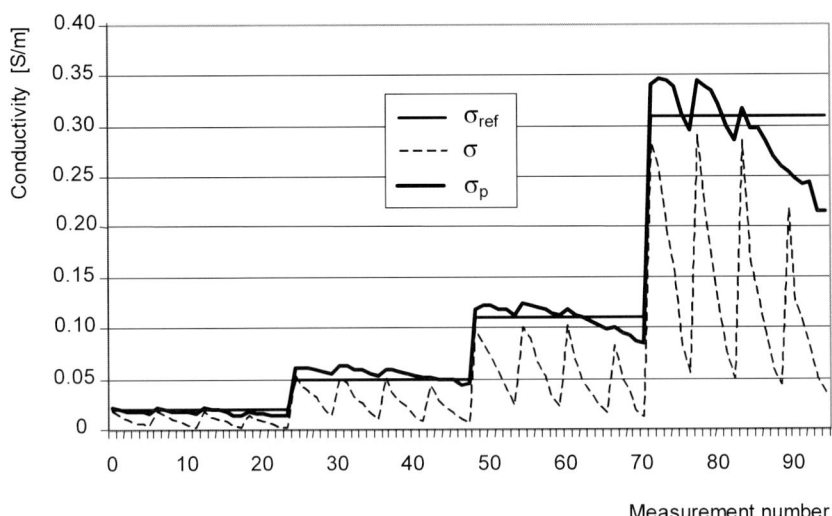

Fig. 13.12. Conductivity readings (σ), reference values (σ_{ref}), and computed pore water conductivity (σ_{p}) at each measurement

more complex nature than expected. Since we might expect that the medium used has no effect upon ε itself, probably pore binding effects influence $\varepsilon_{\sigma=0}$. There is a dependency on θ which leads to lower σ_{p} at lower θ. At high values of σ and at higher T, the absolute errors become larger. Further analyses should show whether the algorithm needs adaptation. This could be done by tuning $\varepsilon_{\sigma=0}$. At higher θ ($> 20\%$) the sensor performs well. For practical applications this is good, since growers keep their substrates normally at water content levels between 40 and 70%. The model works reasonably well over the range from 0.02 to 0.2 S/m; but nevertheless, the mean error is 2.1%, with a standard deviation of 14.8%. The experiment was performed with a *Rockwool* substrate, which might behave differently compared to other substrate media. Furthermore only a single salt (NaCl) was used. Different salts or salt mixtures might have other effects upon the behavior of the sensors. Further research should focus on different salt-type mixtures.

From the experiment we may conclude that the sensor has a reasonable performance over the range studied ($\sigma_{\mathrm{p}} = 0.02$–$0.31$ S/m), but not at low water contents ($\theta < 20\%$). There is a dependency on θ and T. At high temperatures ($T > 30°C$) and at high EC values ($\sigma_{\mathrm{p}} > 0.2$ S/m) absolute errors become large, even up to 0.09 S/m. The overall standard deviation error is 14.8%. Further research to enhance the model is likely to succeed and therefore certainly worthwhile to perform.

13.5 Monitoring the Strength of Young Concrete

Information about the development of the strength of young concrete during the first 28 days after pouring is the basis for deciding on formwork removal or the application of pre-stresses in construction engineering. Measuring the strength of young concrete may lead to faster formwork removal and fewer risks, and therefore has economical benefits [36]. Hydration of concrete causes a decrease of free water, an increase of compressive strength, and a temporal temperature rise in the mixture. To monitor hydration and strength development, currently the maturity method is applied. It uses in situ temperature monitoring in parallel with maturity and laboratory stress tests at the concrete manufacturer. In the Netherlands this method is standardized [37] and it works similar to, though in principle it is different from, the American standard procedure [38]. Both methods are commonly accepted, but time consuming and therefore costly when used under practical circumstances. The use of dielectric spectroscopy might be a good alternative for making in situ sensors.

During the hydration of concrete, the cement reacts with the water. This causes the concrete microstructure to grow and this in turn means that the concrete compressive strength increases. At the same time the amount of free water in the mixture decreases. In fact, more structure means less water and more strength. The decreasing amount of water and the increasing amount of structure influence the dielectric properties of the concrete. The conductivity (σ) and permittivity (ε) of the mixture therefore reflect the increasing strength of the concrete and can be obtained after a concrete mixture-specific calibration. Many authors have described the dielectric behavior of concrete [39–41]. Later it was shown that the FD sensor described previously offers a reliable non-destructive way to determine strength in situ, independent of the weather and other environmental conditions [42, 43]. This method was patented [44]. Based upon this principle, a sensor for monitoring the strength of young concrete was developed [12, 36] and made available commercially (ConSensor b.v., Rotterdam, the Netherlands).

The sensor consists of two stainless steel electrodes mounted in a watertight housing which includes the dielectric chip for measuring conductivity and temperature. A temperature sensor is placed in one of the rods. It connects via an RS232 connection to a hand-held computer (see Fig. 13.13). The sensor itself can be inserted into an electrode set that can be positioned in a predrilled hole in the concrete formwork by using an insertion tool. This insertion tool is in fact a sensor dummy that can temporarily be fixed to the formwork. Once the electrode set is cast in, it remains in the concrete permanently. The FD sensor can be used to monitor strength at multiple locations, just by plugging it into the electrode sets. A hand-held computer (PSION Workabout, Psion Teklogix Gmbh, Willich, Germany) computes the concrete strength parameters based upon the measured conductivity. PC software is used for further analyzing the collected data and for making calibration curves for the different concrete mixtures.

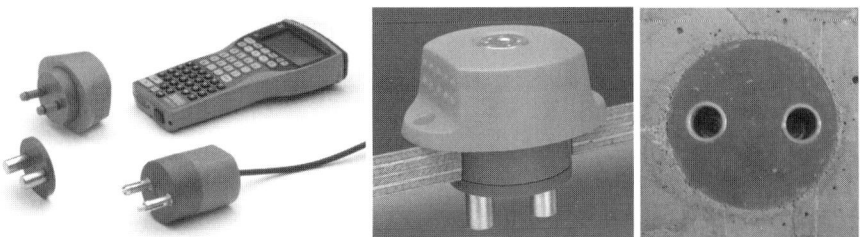

Fig. 13.13. The ConSensor system containing a PSION Workabout, sensor with cabling, electrode set, and an insertion tool with plastic cap (*left*); electrode set and insertion tool with plastic cap, inserted into a predrilled hole in formwork before molding (*middle*); electrode set enclosed in concrete after molding (*right*)

To evaluate the suitability of this FD sensor, its working was compared with the standard maturity method under practical circumstances. During the reconstruction of the Suurhof bridge in the Europoort region in the Netherlands, the concrete strength development in the new pillars of its foundations was monitored using both methods. In these foundations an exceptionally high amount of rebar steel was applied, which was a good opportunity to verify also the influence of rebar upon the dielectric measurements.

Before the in situ tests, a calibration for the specific concrete mixture was performed. Several small cubes of the mixture were made for this. One cube was used to measure dielectric data during the hydration process, which data were stored on a PC. Another cube was tested with the traditional maturity method by taking temperature readings. As reference for these measurements, tests were performed in the laboratory by pressing the cubes at regular time intervals and obtaining the compressive strength (*CS*) at the moment of collapse. The data was entered into a program, which calculated a calibration curve linking the dielectric data to the actual strength of that mixture of concrete. At the construction site this calibration curve was used to compute concrete strength with the hand-held terminal. At the site there were ten pillars. Two FD sensors were placed in every pillar. Two temperature sensors for the maturity method were placed nearby, one in the center of a pillar and one close to the outer site of the pillar. The measurement results, taken at two phases (Phase I and Phase II) of the project, four months apart in time, are given in Table 13.4.

Based upon this data a calibration curve (see Fig. 13.14) was obtained for the FD sensor (CS_{FD}) as well as the maturity method (CS_M):

$$CS_{FD} = 1.9184\sigma^{-0.8005} \quad (R^2 = 0.987) \tag{13.7}$$

$$CS_M = 13.415\ln(M) - 74.253 \quad (R^2 = 0.952). \tag{13.8}$$

Table 13.4. Calibration results with compressive strength (*CS*), conductivity (σ), and maturity (*M*) in arbitrary units

CS (N/mm^2)	σ (S/m)	M (—)
16.4	0.070	832
19.9	0.052	1,400
24.5	0.040	1,469
28.4	0.035	2,064
30.5	0.030	2,681
34.6	0.027	3,836
44.3	0.021	7,510
47.6	0.020	4,925
50.5	0.015	12,400
56.2	0.015	17,480

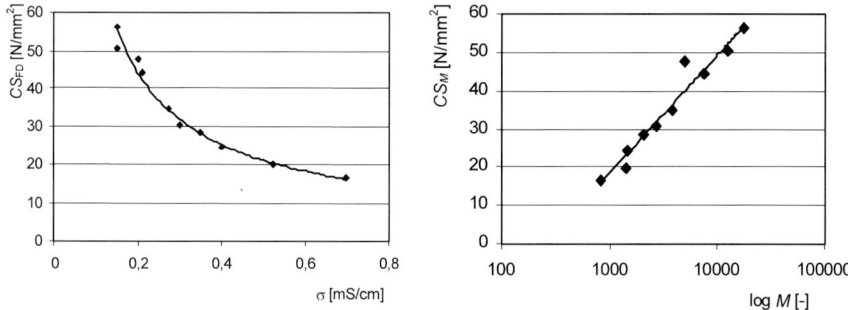

Fig. 13.14. Calibration curves for FD method (*left*), and maturity method (*right*)

Based on the calibration curves and by using the measured data (Table 13.4), for both methods strengths were recalculated and then plotted (see left part of Fig. 13.15). We can see that the two methods correlate rather well ($R^2 = 0.968$). Based upon the accuracy specifications for σ (see Fig. 13.8) for the FD method (resolution = 1 µS/m and 5% for $\sigma < 0.1$ mS/m), the confidence intervals were plotted as well. We can see that for higher strengths and at lower conductivities the accuracy is reduced. This is due to the power law curve that becomes rather steep in this range (see Fig. 13.14, left). Although the readings still fall within the confidence interval, the absolute deviation between CS_M and CS_{FD} is ±10%. The linearity stays within 0.05% over the measuring range. Next in situ measurements, first in Phase I and then in Phase II, were taken (see Fig. 13.15, right).

Looking at the in situ compressive strength data for phase I (see Fig 13.15, right), it can be noted CS_{FD} and CS_M differ substantially, up to 11 N/mm^2. It appears that the maturity method yields slightly higher strengths. Since the pillars contained a lot of rebar steel, it was suspected that the dielectric measurements were influenced by the rebar (see Fig. 13.16, left). After ruling out some possible sources of error such as initial water conductivity and variability due to sensor placement, new measurements were taken (Phase II), taking special care with the location of the FD sensor relative to the rebar (see Fig. 13.16, right). The length of the electrodes is 40 mm. The distances between the tip of the sensor electrodes and the rebar steel were classified as "large," "small," and "very small" respectively for distances of more than 50 mm, between 10 and 50 mm, and less than 10 mm (see Table 13.5). The distances sometimes became very small, even a few millimeters. In some cases the electrodes were even electrically shorted. The results of these measurements are plotted in Fig. 13.15 (right). For the three classes the mean values for the difference between CS_{FD} and CS_M were calculated (see Table 13.5).

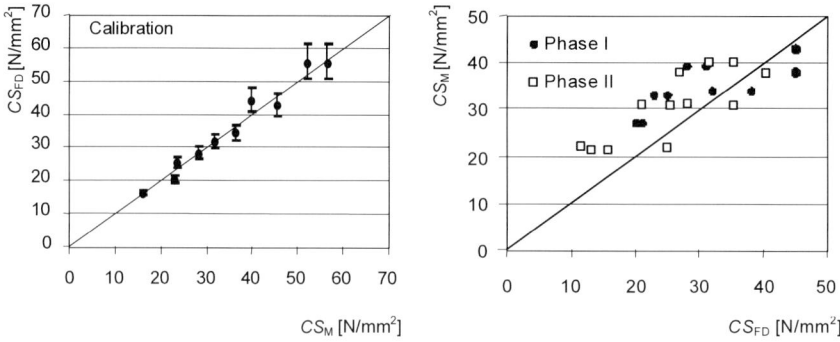

Fig. 13.15. Calibration for compressive strength (•), maturity method versus the FD method, including confidence intervals (—) for the FD sensor (*left*). Compressive strength measured in situ with both methods (*right*). The 1:1-line is shown in both graphs

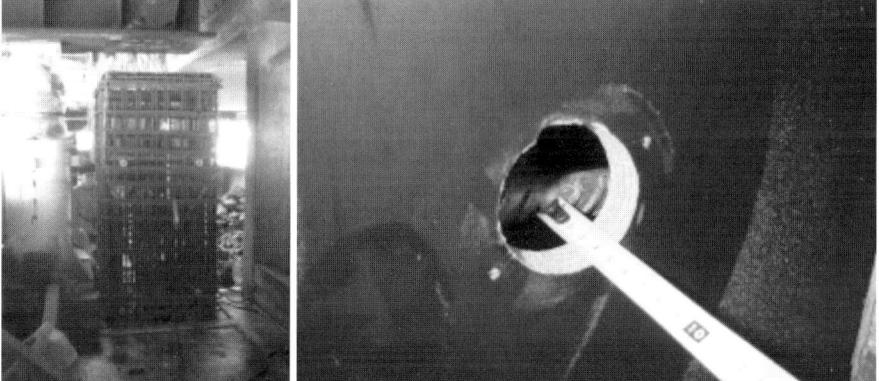

Fig. 13.16. Pillars contain a lot of reinforcement steel (*left*). At one measuring point even the sensor electrodes were in contact with the reinforcement steel (*right*)

Table 13.5. Strength differences in relation to sensor–steel rebar distances

Distance of sensor electrodes to steel rebar (mm)	Mean $(CS_M - CS_{FM})$ (N/mm^2)
> 50 (large)	−3
10–50 (small)	4
0–10 (very small)	9

Though the number of measurements is too small to draw any statistically significant conclusions, we can see that in those situations where the electrodes are close to the rebar ("very small"), there is a large difference in strength. When the electrodes have a distance larger than 1 cm, the errors are smaller and in the order of the measuring noise. It seems that the highly conductive steel leads to a higher conductivity and consequently a lower compressive strength when using the FD method.

We may conclude that for monitoring the strength development of young concrete, the FD sensor can be used as a practical alternative to the maturity method. The calibration was shown to be robust and comparable to the standard maturity method. No permanent measurement is needed; with portable equipment as many measuring points as required with one sensor can be taken. However it was seen that the near contact of the FD sensor electrodes with the rebar steel results in a higher value of the measured conductivity, and consequently an underestimation of strength. A practical solution for this is to keep a safe distance of at least 50 mm between the electrodes and the steel rebar.

13.6 A Dielectric Tensiometer to Measure Soil Matric Potential

Water content sensors, as described in the previous sections, measure the amount of water in the soil matrix and can indicate the amount of water that should be given to plants. On the other hand, tensiometers measure soil matric potential and indicate the moment at which plants should be given water. There is a relation between these two, which is described by the pf-curve [45]. For good water management both parameters are needed. If the pf-curve is known, only one parameter is needed. The pf-curve has an exponential behavior and depends on soil texture and density. Since both of these soil features vary greatly under practical circumstances, and soil behaves differently for wetting and drying, there is still a need for both types of water sensors. An FD sensor for soil water content was described in an earlier section; here we will discuss the use of the FD method to build a soil matric potential sensor.

For irrigation purposes normally water-filled hydraulic tensiometers are used [46, 47]. They consist of a ceramic cup placed in the soil, connected to a water tube with a mechanical or an electronic pressure gauge on top. These sensors are not so suited for automatic irrigation systems, since they need regular calibration and maintenance, such as refilling with water, and their working range is restricted for dry soils to about −85 kPa. In spite of the drawbacks, lacking a better alternative, and due to the need to save water, in semi-arid areas as well as in greenhouses hydraulic tensiometers are still being used. An alternative design, nowadays known as resistance blocks has been proposed [48]. Here, the AC resistance between two electrodes molded into a porous material is measured and calibrated against matric potential. Later, a conductivity sensor was used to measure water content inside a water permeable container filled with quartz sand [49]. These granular-type sensors came onto the market as the WATERMARK sensors (Irrometer, Riverside, CA, USA) and were optimized for use between −10 and −100 kPa. Though cheap, they were inaccurate and their calibration varied for different soils [50].

Hilhorst and de Jong [51] reported for the first time the use of a dielectric sensor to build a solid state tensiometer. This dielectric tensiometer measures matric potential indirectly by measuring the water content inside a ceramic material which is in equilibrium with the soil water. Once the sensor is inserted into soil, the water potential inside the ceramic will become equal to the soil water potential. Because the water retention characteristic (pf-curve) of the ceramic is known, matric potential can be calculated from the water content readings. These authors showed the results of a prototype based on the use of glass beads as the porous material. A decade later, the first version of this sensor was launched as the Equi-Tensiometer (UP Umweltanalytische Produkte GmbH, Cottbus, Germany), which was based upon the MR2 Theta Probe (Delta-T Devices Ltd, Cambridge, UK) working at 100 MHz. This sensor was also of the granular matrix type. This influences its accuracy, and its large size made its response slow. Recently, a newer version of the Equi-tensiometer was launched, which is based upon a new, more stable, and smaller substrate material [52]. It has a working range from 0 up to −1500 kPa. Or and Wraith [53] describe a TDR dielectric tensiometer with a coaxial transmission line embedded in a porous material. Their ceramic has a stable structure which does not change with time like other sensors [54, 55]. Since ceramics tend to have a narrow pore size distribution, they used a number of ceramics with different pore sizes and integrated them into one sensor to extend the measurement range. The dielectric tensiometer principle [51] was used to design a new experimental prototype [64]. It used the MCM100-chip described in Sect. 13.2 to measure permittivity inside the ceramic. The aim of this work was to explore the design criteria for dielectric tensiometers based on porous media over the range from saturation to −60 kPa, which is suitable for automatic irrigation control applications. One of the major concerns was the effect of hysteresis.

The new prototype (see Fig. 13.17) used a ceramic (Coralith, grade C0) consisting of glass-bonded aluminum particles (Fairy Industrial Ceramics, Staffordshire, UK). It had a mean pore size of 11 μm, a porosity of 35%, and an air entry potential of −27 kPa. Three holes were drilled into this ceramic. The holes were slightly filled with electrically conducting silver-loaded epoxy glue (RS Components, Northants, UK) to prevent an air gap between the electrodes and the ceramic. Three stainless steel electrodes with a diameter of about 1 mm were carefully pushed into these holes. Finally the electrodes were soldered to an electronic circuit board (see Fig. 13.4). Just above the ceramic a small air gap was kept to allow air to flow in and out the ceramic freely. This air gap is kept at atmospheric pressure by a tube with an outlet to the open air.

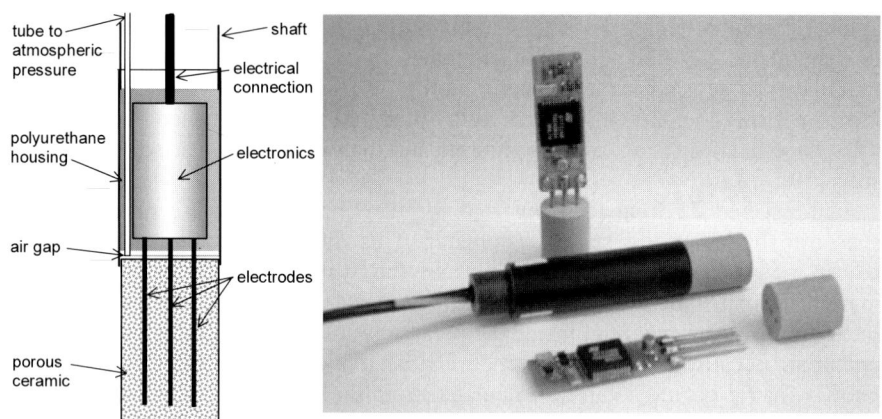

Fig. 13.17. A schematic drawing of a dielectric tensiometer (*left*), and a prototype with electronics, housing, and ceramic (*right*)

By using the FD method, small electrodes can be applied, contrary to similar TDR probes. The FD sensor, operating at 30 MHz, also measures conductivity, which might be a useful means to measure pore water salinity inside the ceramic. The electronics were embedded into a hard polyurethane molded cylinder of about 20 mm in diameter and a length of 5 cm. A flexible polyurethane output cable was connected to the electronics. The ceramic was 25 mm long and 19 mm in diameter. The complete sensor can be mounted on one end of a long tube, containing the cable and the air tube. In this way the dielectric tensiometer looks much like the hydraulic tensiometer, and it can be installed and removed from the soil very easily.

The FD sensor essentially measures the degree of saturation (*S*) of the ceramic, expressed as

$$S = \frac{\sqrt{\varepsilon} - \sqrt{\varepsilon_d}}{\sqrt{\varepsilon_s} - \sqrt{\varepsilon_d}},$$

(13.9)

where ε is the dielectric permittivity in the ceramic in equilibrium with the surrounded soil and ε_d and ε_s are dielectric constants for the air-dry (d) and the water-saturated (s) ceramic. To find the relation between *S* and the soil matric potential (*h*) we use the inverted version of the van Genuchten equation [56], which is also used to characterize soil pf-curves:

$$h = \frac{1}{\alpha} \left[\left(\frac{S - S_r}{S_m - S_r} \right)^{-\frac{1}{m}} - 1 \right]^{\frac{1}{n}} ,$$

(13.10)

where S_m and S_r are the maximum and residual values of S, and α, m, and n are the shape parameters of the curve. To quantify these parameters for the ceramic used, its pf-curve was obtained from a drying curve cycle by using a conventional tension table. A sample of the ceramic was placed on the tension table immersed in silica paste. This sample was exposed to several matric potentials. Each time S was calculated from an oven-drying experiment. The curve found is in accordance with the model (see Fig. 13.18). This curve shows that the specific ceramic has an operating range that is suitable for irrigation management (–20 to –60 kPA). With parameter fitting the following van Genuchten parameters were found: $S_r = 0.111$; $S_m = 0.996$; $m = 0.405$; $n = 9.95$; and $\alpha = 0.041$.

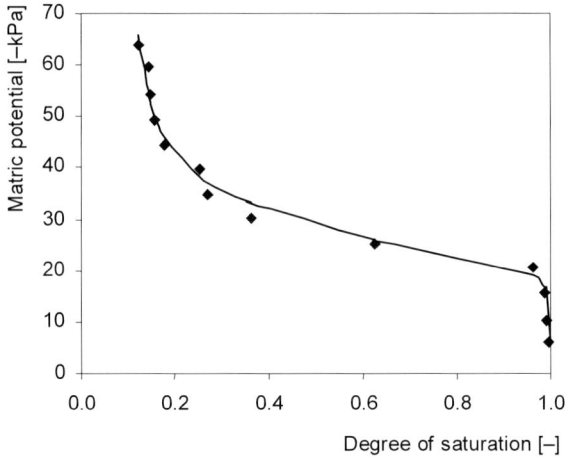

Fig. 13.18. pf-curve for the used ceramic with measured values during drying (♦) and fitted van Genuchten curve (—)

Thirteen prototype sensors were built. To observe the effect of drying and wetting, all 13 sensors were placed on a tension table in a saturated kaolin mixture. They were allowed to equilibrate at 0 kPa (saturation) for at least 2 days. The water potential of the tension table was controlled using a vacuum pump and the actual value was monitored using a hydraulic tensiometer. All measurements were carried out at a constant temperature of 20°C. Water potentials were decreased in small steps, typically –5 kPa. Values for each sensor were recorded following an equilibration period of 24 hours. This procedure was repeated until a water potential of –60 kPa was obtained. The water potential was then increased in small steps until saturation (0 kPa) was reached. The results obtained from this experiment are shown in Fig. 13.19.

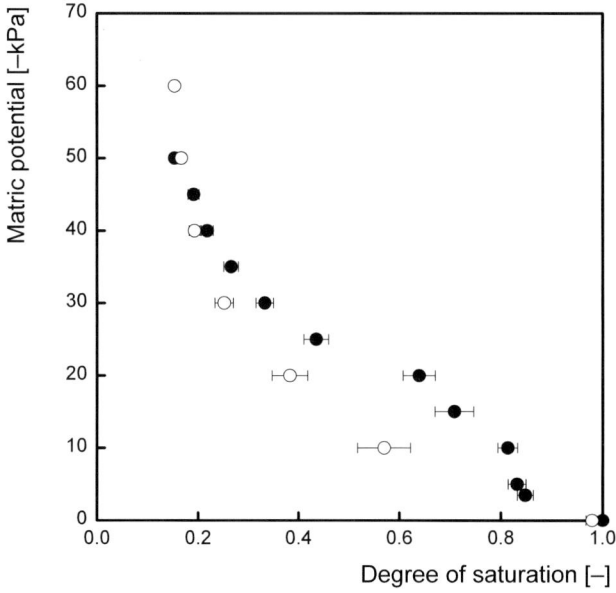

Fig. 13.19. Data obtained with 13 prototype sensors for the drying (●) and wetting (○) cycle. Horizontal lines indicate the standard error of the mean degree of saturation (Reproduced from Whalley et al. [64] with the permission of Blackwell Publishing)

The main wetting and drying curves clearly show hysteresis. We also see that for S between 0.2 and 0.8 these curves are flatter than the ceramic pf-curves found in the previous experiment. Furthermore, for S between 0.85 and 0.95 the behavior is also different. Because of this, the main drying curve alone is not suitable to obtain the matric potential. By not accounting for hysteresis effects, even by using a mean pf-curve, large errors up to ± 5 kPa may occur. When the sensor is not operated along the main drying or wetting curve an accurate value for h can never be obtained. Therefore it was decided to explore a hysteresis model that could correct for this effect in the area between the main wetting and drying curves.

Several hysteresis models are known [57, 58]. We have chosen the Kool and Parker model [59], which combines the empirical model of van Genuchten [56] for the moisture characteristic curve and the hysteresis model of Scott et al [60]. This model is capable of calculating matric potentials from the water content or saturation of the porous ceramic substrate, provided the wetting history is known. It requires that the main drying and wetting curves be known and expressed in terms of the van Genuchten equation (Eq. (13.10)). For all prototype sensors these parameters were obtained by using a curve-fitting program. The results are shown in Table 13.6.

Table 13.6. Van Genuchten and dielectric parameters for 13 prototype dielectric tensiometers.

No.	S_r	S_m	α (d)* α (w)**	n (d) n (w)	m (d) m (w)	ε_r	ε_s
1	0.119	0.818	0.00371	3.356	1.784	7.5	31.9
			0.02018	205.0	0.007		
2	0.128	0.913	0.00038	1.961	117.48	9.4	35.3
			0.02147	74.6	0.027		
3	0.121	0.845	0.00395	4.035	1.065	10.3	36.9
			0.01715	32.96	0.371		
4	0.084	0.695	0.00673	5.690	0.436	7.8	30.9
			0.02003	741.3	0.003		
5	0.108	0.795	0.00493	4.481	0.674	10.0	36.2
			0.01423	8.395	0.153		
6	0.135	0.862	0.00411	4.935	0.879	8.1	30.5
			0.00609	2.400	1.114		
7	0.114	0.828	0.00429	4.536	0.740	4.3	27.0
			0.00669	2.495	0.798		
8	0.156	0.915	0.00380	4.718	1.065	4.8	26.4
			0.00245	2.157	3.692		
9	0.105	0.790	0.00506	18.88	0.134	4.4	27.5
			0.00585	2.876	0.745		
10	0.118	0.835	0.00510	6.299	0.391	4.7	27.4
			0.00685	2.764	0.709		
11	0.153	0.900	0.00500	6.720	0.431	4.7	26.9
			0.00507	2.542	1.101		
12	0.113	0.889	0.00541	12.57	0.149	4.8	25.1
			0.00645	2.901	0.574		
13	0.135	0.742	0.00411	3.837	0.938	4.7	31.2
			0.00454	2.380	1.237		

*(d) represents parameters determined from the drying cycle (0 to –60 kPa).
**(w) represents parameters determined from the wetting cycle (–60 to 0 kPa)

To test the hysteresis model, one prototype tensiometer (number 5) was exposed to a series of random changes in matric potential in the range 0 to –60 kPa. This was achieved by using a tension table. The sensor was immersed in wet silica paste together with a hydraulic tensiometer, which was used to monitor the matric potential of the silica paste. Both the drying pf-curve for the ceramic (see Fig. 13.18), as well as the Kool and Parker model using the van Genuchten parameters (Table 13.6, number 5), were used to obtain the matric potentials. Although the Kool and Parker model works unsatisfactorily at matric potentials lower than –30 kPa (see Fig. 13.20, bottom), in general it behaves clearly better than the single drying pf-curve (see Fig. 13.20, top).

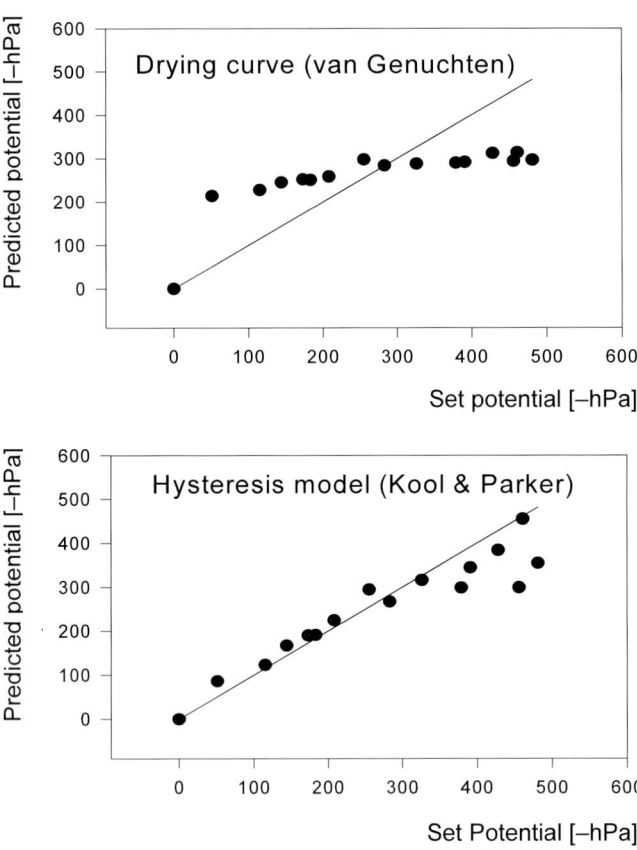

Fig. 13.20. Comparison between the output of a conventional tensiometer and a dielectric tensiometer by using the pf-curve for drying (*top*) and the Kool and Parker hysteresis model (*bottom*) (Reproduced from Whalley et al. [64] with the permission of Blackwell Publishing)

In the time series data plot (see Fig. 13.21) we see that the dielectric and hydraulic tensiometers show a similar shape. The time constant for both tensiometers seems to be in the same range, so they have a similar response time. Here also, in the dry range (−30 to −60 kPa), we see a discrepancy between the two types of tensiometers. In the wet range (0 to −30 kPa), both tensiometers have similar outputs. In general the Kool and Parker model gives more accurate results, but for longer measuring periods and smaller matric potential differences we observed some drift in readings. This is due to the fact that the hysteresis model loses track of the historic data due to measuring errors. Fortunately the model synchronizes again after the tensiometer is brought back to (near) saturation or to the dry end of the curve. However, for practical irrigated crop production systems, where water is given more often, this is not the case. Here the differences in matric potentials will probably be so low that the hysteresis effect may be neglected and a mean pf-curve, somewhere between the wet and dry cycling curves, could be used instead.

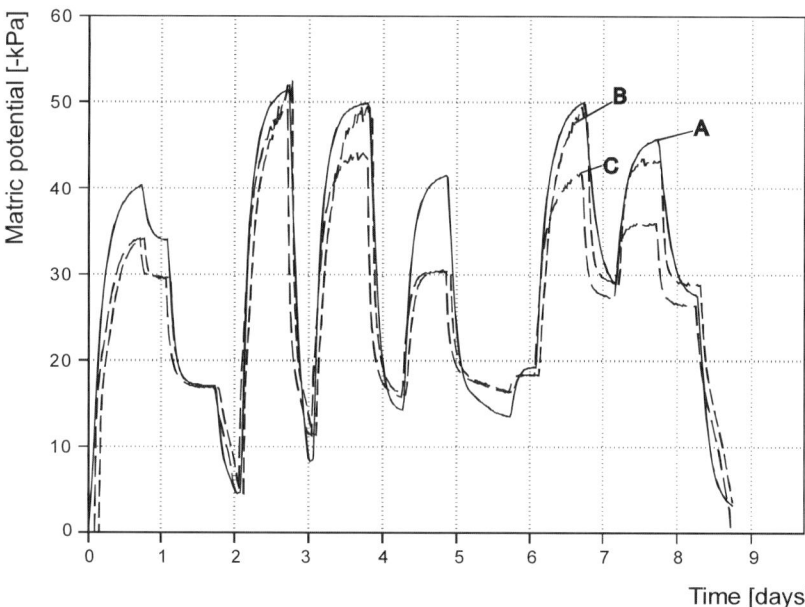

Fig. 13.21. Time series plot for the number 5 tensiometer. Readings from the hydraulic tensiometer (**A**) are compared with the calculated matric potential based on the Kool and Parker hysteresis model (**B**) and the drying pf-curve (**C**) (Reproduced from Whalley et al. [64] with the permission of Blackwell Publishing)

The development of our experimental tensiometer has raised a number of important issues that have general relevance to the class of matric potential sensors based on the use of porous materials. In particular the importance of air access into the porous material is relevant. The success of our experimental sensor, which was embedded in silica paste with much lower air entry potential, following the output of a tensiometer, demonstrates that the provision of access to air to invade the draining porous material is needed. An advantage of providing for air access into the ceramic is that it is then possible to install the sensors in the field in exactly the same way as a conventional tensiometer. Surrounding the sensor with a paste with a very low air entry potential also has the advantage of providing a good connection between the soil and the sensor. We have shown that a model of hysteresis can be used to track changes in matric potential in a porous medium in equilibration with soil.

13.7 An FD Sensor Auto-Calibration Method for Volumetric Water Content

Monitoring soil water content (θ) with FD sensors that work in the lower RF range beneath 50 MHz involves a soil-specific calibration defined as $\theta = f(\varepsilon)$, similar to Eq. (13.5). In this frequency range a raised permittivity is found due to the Maxwell–Wagner effect. This effect is soil-texture dependent since it is influenced by the grain size of the soil particles. For clay this effect is larger than for sandy soils. For frequencies above 50 MHz this effect becomes smaller and can be neglected at even higher frequencies. Therefore dielectric soil-water content sensors that operate at frequencies above 100 MHz can generally do without a soil-specific calibration. Examples for this are the TDR sensors used by Topp [6, 17, 27] and the Theta probe (Delta-T Devices Ltd, Cambridge, UK) which is an FD sensor working at 100 MHz.

Generally, a gravimetric approach is used to calibrate these sensors. This is done by taking soil samples manually and measuring θ under wet and dry conditions while taking readings with the FD sensor (see Table 13.2 and Fig. 13.8). This method yields the most reliable calibration. However, under practical circumstances, soil texture and density have a large spatial variability. Inherently, FD sensors, when calibrated under laboratory conditions, exhibit in practice an error due to the local variation in soil texture and density. For this reason it would be nice to have an in situ calibration procedure for θ under practical conditions that could compensate for the Maxwell–Wagner rise automatically.

Little is found on this topic in the literature, but one paper reveals the fact that it might be possible to estimate the Maxwell–Wagner rise ($\Delta\varepsilon_{MW}$) by measuring ε at multiple, at least three, discrete frequencies in the range from 10 to 30 MHz [8, p 26]. Based on $\Delta\varepsilon_{MW}$ the more reliable permittivity at higher frequencies ($\varepsilon(f\to\infty)$) can be estimated. This idea is based upon the fact that soil water is either free or bound. Free water reacts with the electrical field at all frequencies up to its relaxation frequency, whereas bound water reacts better at lower frequencies. The Maxwell–Wagner rise in the dielectric spectrum at a specific measuring frequency is defined as:

$$\Delta\varepsilon_{MW}(f) = \varepsilon(f) - \varepsilon(f \to \infty),\qquad(13.11)$$

where $\varepsilon(f)$ is the permittivity at the measuring frequency and $\varepsilon(f\to\infty)$ is the constant permittivity at a higher frequency where the Maxwell–Wagener effect can be neglected (e.g., $100 - 150$ MHz). When $\varepsilon(f)$ is known at several measuring frequencies, $\varepsilon(f\to\infty)$ and $\Delta\varepsilon_{MW}(f)$ can be obtained through extrapolation toward a higher frequency. Since FD and TDR sensors usually yield a more accurate result in this frequency range, the common Topp curve can than be used for calibration. In order to reduce this "bound water fault," here a simpler model is proposed to obtain an estimate for $\Delta\varepsilon_{MW}$ (f), based on taking an extra reading for ε at a slightly lower frequency (f^*) as can be seen in Fig. 13.22.

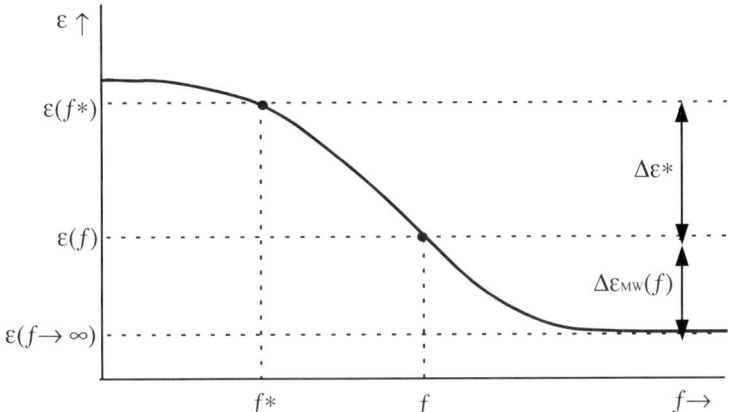

Fig. 13.22. Principle of automatic self-calibration based on estimation of the Maxwell–Wagner rise in the spectrum by taking readings for permittivity at two nearby frequencies

For this, it is assumed that $\Delta\varepsilon_{MW}$ (f) can be computed through a function F based on the difference in permittivity $(\Delta\varepsilon^*)$ measured at two nearby frequencies f and f^* with $f^* < f$ defined as

$$\Delta\varepsilon^* = \varepsilon(f^*) - \varepsilon(f) , \qquad (13.12)$$

and that this function F can be described as a Taylor polynomial:

$$\Delta\varepsilon_{MW} (f) = F (\Delta\varepsilon^*) \approx a_0 + a_1\Delta\varepsilon^* + a_2(\Delta\varepsilon^*)^2 + \ldots . \qquad (13.13)$$

Next, $\varepsilon(f\rightarrow\infty)$ can then be obtained by using Eq. (13.11), and subsequently θ can be calculated from the standard Topp curve.

To explore this automatic soil-type calibration, some experiments were performed [61–63]. FD sensors were modified to measure permittivity at two frequencies $(f = 10$ and 20 MHz). This was achieved by successively switching two crystals to the microchip and taking readings with a PC. The sensors were used to measure ε, σ, and T inside five containers with a volume of 5 liters. Each container was filled with a different type of soil: sand, sandy loam, or loess all sampled from the field. For loess, three densities of which two were known ($\rho_{pot4} = 1.1$ g/cm³ and $\rho_{pot5} = 1.4$ g/cm³) were used. The samples were exposed to a constant temperature of 15°C, by using a water bath. They were wetted between air dry and field capacity ($\theta \approx 5$–40%) in six or more steps for θ. The wetting water had low conductivity. For equilibration, after each wetting, three days were taken for the water to redistribute. The values for water content were calculated from the amounts of wetting water. Due to inhomogeneous water distribution in the relatively huge pots, the readings obtained were not very accurate. For each sample, the permittivity was measured at the two frequencies. To show the Maxwell–Wagner rise, these values were plotted for a frequency of 20 MHz (see Fig. 13.23).

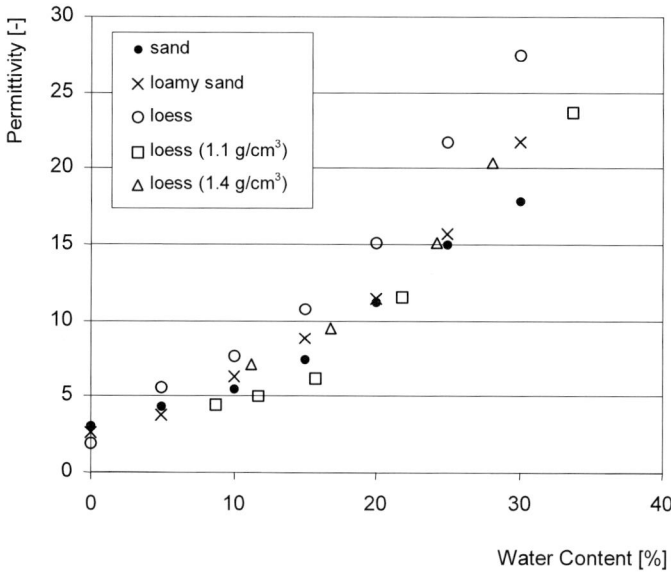

Fig. 13.23. Permittivity versus water content for three soil types (sand, loess, loamy sand) and for loess at two known densities ($\rho_1 = 1.1$ g/cm³ and $\rho_2 = 1.4$ g/cm³) taken with the FD sensors at a single frequency (20 MHz)

It can be seen that higher values for ε are found for loess (small soil particles) and that lower values are found for sand (large soil particles). Sandy loam shows a curve somewhere in between. This finding reflects nothing more than the fact that smaller particles provide a larger part of bound water and therefore a larger $\Delta\varepsilon_{MW}(f)$. Looking at the results for loess at two different densities ($\rho_1 = 1.1$ and $\rho_2 = 1.4$ g/cm³), we see that the Mawell–Wagner effect is larger at higher densities.

Next the 10 MHz readings from the FD sensors were also taken into account. The difference in permittivity ($\Delta\varepsilon^*$) was calculated for each reading based on Eq. (13.12). For each reading a value for $\varepsilon(f \rightarrow \infty)$ was obtained from the reference values for water content and the Topp curve. Then for each reading $\Delta\varepsilon_{MW}(f)$ was calculated by using Eq. (13.11). Now having a series of values for $\Delta\varepsilon_{MW}(f)$ and

$\Delta\varepsilon^*$ belonging together the function F and the coefficients a_i were obtained through a polynomial fitting program. By using this function for all readings, corrected values for θ were calculated. These corrected values, as well as θ results based on the Topp curve, are compared with the reference values for water content used in the experiment (see Fig. 13.24).

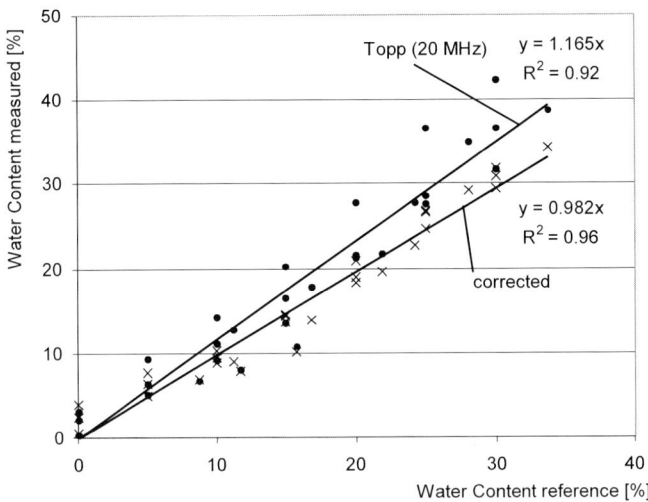

Fig. 13.24. Water content (θ) based on the auto-calibration algorithm (×) and water content based op the Topp curve applied at the 20 MHz permittivity data (●), compared to the reference water contents

It was shown that the correction scheme works fine for the used samples. The 1:1-line in Fig.13.24 for the auto-calibration values has a good correlation ($R^2=0.96$) and only deviates 2%. As expected, the Topp curve yields higher values since it is used on a low frequency of 20 MHz. It deviates 16.5% and has a lower correlation because of the soil-type dependency. Indeed, through a soil-specific calibration a better match than this 16.5% could have been obtained, but nevertheless this would as well lead to a higher spreading due to soil-type variation. It may be concluded that, by measuring at two frequencies, an interesting and important step toward self-calibration for soil type can be achieved. In practical circumstances, however, not only do soil type and density vary, but also temperature and conductivity have an influence on the dielectric spectrum. Hilhorst [8] showed that at higher conductivities the Maxwell–Wagner relaxation frequency shifts to the right. This suggests that the above relation is only valid at a specific conductivity or temperature. Further research is needed to explore these effects and to see whether the two-frequency approach will still work in practical situations, even after correction for temperature and conductivity.

Acknowledgment

The work presented in this chapter was mainly funded from the EC project WATERMAN (FAIR4-PL0681). We thank Wolfgang Paul from the Federal Agricultural Research Center, Institute of Technology and Biosystems Engineering, Braunschweig (Germany), for sharing his work on self-calibrating sensors from the WATERMAN project [65] with us. We are grateful to Kalman Rajkai from the Soil Science Department of the Research Institute for Soil Science and Agricultural Chemistry of Budapest (Hungary) for performing the water content calibration for our FD sensor and the willingness to share experimental data from the EC project QLK5-CT-1999-01349. Furthermore we have found Wim Stenfert Kroese from CONSENSOR b.v. (Rotterdam, the Netherlands) and Ton van Beek from Delft University (the Netherlands) willing to share their results from experiments performed with the concrete sensor, carried out in close collaboration with the Dutch Rijkswaterstaat. The work performed within our own laboratory at Wageningen-UR was co-funded by the Dutch DWK agricultural research program on Water and Nutrient Management. We also thank Max Wattimena, Henk van Roest, and Peter Nijenhuis for their work on the electronic design, software programming, and testing of the FD sensor. *Rockwool* is a trademark of Rockwool Grodan (Roermond, the Netherlands).

References

1. Debye P (1929) Polar molecules. Rheinhold, New York
2. Davidson DW, Cole RH (1951) Dielectric relaxation in glycerol, propylene glycol, and n-propanol. J Chem Phys 19:1481–1493
3. Stacheder M, Blume P, Fundinger R, Koehler K, Ruf R (2001) Reliability of Trime-TDR Sensors for moisture determination in pure and contaminated concrete. In: Proceedings of the fourth international conference on electromagnetic wave interactions with water and moist substances, Weimar, 13–16 May 2001, pp 266–273
4. Ferguson JG (1953) Classification of bridge methods of measuring impedances. Bell Syst Tech J 12:452–459
5. Hilhorst MA (1984) A sensor for the determination of the complex permittivity of materials as a measure for the moisture content. In Bergveld P (ed) Sensors & actuators. Kluwer Technical Books, Deventer, pp 79–84
6. Topp GC, Davis JL, Annan AP (1980) Electromagnetic determination of soil water content: measurements in coaxial transmission lines, Water Resour Res 16(3):574–582
7. Hilhorst MA, Groenwold J, De Groot JF (1992) Water content measurements in soil and rockwool substrates: dielectric sensors for automatic in situ measurements. In: Sensors in horticulture, Acta Hortic 304:209–218
8. Hilhorst MA (1998) Dielectric characterisation of soil. Doctoral-thesis, Wageningen University and Research Center, Wageningen, the Netherlands, ISBN 90-5485-810-9
9. Bratton WL, Pluimgraaff DJMH, Hilhorst MA (1995) CPT sensors for bio-characterization of contaminated sites. In: International symposium on cone penetration testing, Sweden, Oct

10. SOWACS website: www.sowacs.com, february 2004
11. Hadjar A (1997) Zerstörungsfreie Feuchtemessverfahren für Beton. In: Kupfer K (ed) 9 Feuchtetag, 7/18 Sept, MFPA an der Bauhaus-Universität Weimar, pp 301–316
12. van Beek A, Hilhorst MA (1999) Dielectric characterization of young concrete. Heron 44(1), pp 3–17
13. Sokoll T, Jannsen B, Jacob AF (2002) A novel sensor for measuring ion concentration in concrete structures. In: Kupfer K (ed) 11 Feuchtetag, 18/19 Sept 2002, pp 36–46
14. Foster KR, Schwan HP (1986) Dielectric permittivity and electrical conductivity of biological materials. In: Polk C, Postow E (eds) Handbook of biological effects of electromagnetic fields. CRC Press, Boca Raton, FL, pp 27–98
15. Nacke T, Frense D, Göller A, Beckmann D (2001) Impedance spectroscopy – a tool for in situ biomass analyses and for the study of the toxic sensitivity of cells in suspension cultures. In: Proceedings of the fourth international conference on electromagnetic wave interactions with water and moist substances, Weimar, 13–16 May 2001, pp 93–100
16. Hilhorst MA, Balendonck J, Kampers FWH (1993) A broad-bandwidth mixed analog/digital integrated circuit for the measurement of complex impedances. IEEE J Solid-state Circuits 28(7):764–769
17. Topp GC, Davis JL, Annan AP (1982) Electromagnetic determination of soil water content using TDR: II. Evaluation of installation and configuration of parallel transmission lines. Soil Sci Soc Am J 46:678–684
18. Hilhorst MA, Dirksen C (1994) Dielectric water content sensors: time domain versus frequency domain. In: Proc of the symposium on TDR in environmental, infrastructure and mining applications, Evanston, Illinois, Sept 1994, pp 23–33
19. Dirksen C, Hilhorst MA (1994) Calibration of a new frequency domain sensor for soil water content and bulk electrical conductivity. In: Proceedings of the symposium on TDR in environmental, infrastructure and mining applications, Evanston, Illinois, Sept 1994, pp 43–153
20. Balendonck J (1997) Smart sensor chip for dielectric measurements In: Proceedings 8th international congress, transducers & systems, Sensor 97, Nürnberg, May 1997, vol 1, pp 253–258
21. Balendonck J, Hilhorst MA (1998) MCM100 Smart sensor interface for complex impedance measurement. Datasheet and application note. Report IMAG-DLO, Wageningen, Note P98–50, 45 pp
22. Balendonck J, Hilhorst MA (2001) Application of an intelligent dielectric sensor for soil water content, electrical conductivity and temperature. In: Proceedings of the 18th IEEE instrumentation and measurement technology conference, IMTC-2001, Budapest, 23–25 May 2001, pp 1817–1822
23. Seyfried MS, Murdock MD (2002) Effects of soil type and temperature on soil water measurement using a soil dielectric sensor. In: I.C. Paltineau (ed.), First International Symposium on Soil Water Measurement using Capacitance and Impedance, Beltsville, MD. 6–8 November 2002, pp 1–13
24. Kaatze U, Uhlendorf V (1981) The dielectric properties of water at microwave frequencies. Z Phys Chem, Neue Folge, 126:151–165
25. Balendonck J, Hilhorst MA (2001) WET sensor application note. IMAG Report 2001-07, Wageningen
26. Heimovaara TJ (1993) Time domain reflectometry in soil science: theoretical backgrounds, measurements and models. PhD thesis, University of Amsterdam

27. Topp GC, Ferré PA (2001) Electromagnetic wave measurements of soil water content: a state-of-the-art. In: Fourth international conference on electromagnetic wave interaction with water and moist substances, Weimar, 13–16 May 2001
28. Kuyper MC, Balendonck J (1997) Application of dielectric soil moisture sensors for real-time automated irrigation control. In: Sensors in horticulture, Tiberias, Israel, August 1997
29. Perdok UD, Kroesbergen B, Hilhorst MA (1996) Influence of gravimetric water content and bulk density on the dielectric properties of soil. Eur J Soil Sci 47:367–371
30. Kalman Rajkai (2002) Personal communication, Soil Science Department of the Research Institute for Soil Science and Agricultural Chemistry of Budapest, Hungary
31. Sonneveld C, van den Ende J (1971) Soil analysis by means of a 1:2 volume extract. Plant Soil 35:505–516
32. Hilhorst MA, Balendonck J (1999) A pore water conductivity sensor to facilitate non-invasive soil water content measurements. In: Staffort JV (ed) Proceedings of the 2nd European conference in precision agriculture, Society of Chemical Industry, Odense, pp 211–220
33. Hilhorst MA (2000) A pore water conductivity sensor. Soil Sci Soc Am J 64(6), pp 1922–1925
34. Dirksen C, Dasberg S (1993) Improved calibration of time domain reflectometry for soil water content measurements. Soil Sci Soc Am J 57:660–667
35. Balendonck J, Hilhorst MA, van Roest H (2002) Water content and temperature dependency of pore water conductivity for the FD sensor in growing substrates. In: 11 Feuchtetag, 18/19 Sept 2002, MFPA an der Bauhaus-Universität Weimar, pp 67–76
36. van Beek A (2000) Dielectric properties of young concrete, non-destructive dielectric sensor for monitoring the strength development of young concrete. Dissertation, Delft University
37. NEN 5970 (1999) Bepaling van de druksterkteontwikkeling van jong beton op basis van de gewogen rijpheid, oktober
38. ASTM C-1074-93 (1998) Revised standard for testing young concrete, defined by the Am Soc for Testing and Materials
39. Tobio JM (1957) A study of the setting process: dielectric behaviour of several Spanish cements. In: Silicates Industrials, Comunication présentée aux Journées Internationales d'études, Liant hydrauliques 1957, de l'Assoiation belge pour favoriser L'étude des Verres et Composés siliceux, pp 30–35, 81–87
40. De Loor GP (1953) Method of obtaining information on the internal dielectric constant of mixtures. Appl Sci Res pp 479–482
41. Al-Qadi IL, Hazim OA, Su W, Riad SM (1995) Dielectric properties of Portland cement concrete at low frequencies. J Mat Civ Eng 7:192–198
42. van Breugel K, Hilhorst MA, van Beek K, Stenfert-Kroese W (1996) In situ measurement of dielectric properties of hardening concrete as a basis for strength development. In: Proceedings of the 3rd conference on non-destructive evaluation of civil structures and materials, Sept 1996, pp 7–21
43. Hilhorst MA, van Breugel K, Pluimgraaf DJMH, Stenfert Kroese W (1996) Dielectric sensors used in environmental and construction engineering. Mat Res Soc Symp Proc 411:404–406
44. Stenfert Kroese WH, Hilhorst MA (2000) Method for determining the degree of hardening of a material. Patent WO9642014 and US 6023170, 2 Aug 2000

45. Dirksen C (1999) Soil Physics Measurements. Geo-Ecology, Catena Verlag, Reiskirchen, Germany, 1999
46. Richards LA (1949) Methods for measuring soil moisture tension. Soil Sci 68:95–112
47. Mullins CE, Mandiringana OT, Nisbet TR, Aitken MN (1986) The design, limitations, and use of a portable tensiometer. J Soil Sci 37:691–700
48. Bouyoucos GJ, Mick AH (1940) Electrical resistance method for the continuous measurement of soil moisture under field conditions. Michigan Agricultural Experimental Station, Tech Bull no 172
49. Thomson SJ, Armstrong CF (1987) Calibration of the Watermark Model 200 Soil Moisture Sensor. Appl Eng Agric 3(2):186–189
50. Spaans EJA, Baker JM (1992) Calibration of the Watermark soil-water sensors for soil matric potential and temperature. Plant Soil 143:213–217
51. Hilhorst MA, de Jong JJ (1988) A dielectric tensiometer. Agricultural Water Management 13:411-415, technical note
52. Liu Jin-Chen (2002) Ein neues Verfahren zur Messung des Matrixpotenzials im Bodem. In: 11 Feuchtetag, 18/19 Sept 2002, MFPA an der Bauhaus-Universität Weimar, pp 77–84
53. Or D, Wraith JM (1999) A new soil matric-potential sensor based on time-domain-reflectometry. Water Resour Res 35:3399–3407
54. Bouyoucos GJ (1953) More durable plaster of Paris blocks. Soil Sci 76:447–451
55. Liu Jin-Chen (1999) Device and method for determining properties of a soil. US Patent 5.898.310, 27 Apr 1999
56. van Genuchten MT (1980) A closed-form equation for predicting the hydraulic conductivity of unsaturated soils. Soil Sci Soc Am J 44:892–898
57. Jaynes DB (1984) Comparisons of soil-water hysteresis models. J Hydrol 75:289–299
58. Otten W, Raats PAC, Kabat P (1999) Hydraulic properties of root-zone substrates used in greenhouse horticulture. In: MTh van Genuchten et al (eds) Characterization and measurement of the hydraulic properties of unsaturated media, proceedings of international workshop, 22–27 Oct 1997, Riverside, California, pp 477–489
59. Kool JB, Parker JC (1987) Development and evaluation of closed form expressions for hysteretic soil hydraulic properties. Water Resour Res 23:105–114
60. Scott PS, Farquhar GJ, Kouwen N (1983) Hysteretic effects on net infiltration. In: Advances in infiltration, Publication 11–83, American Society of Agricultural Engineering, St. Joseph, MI, pp 163–170
61. Paul W, Hilhorst MA, Münstermann C, Schmitz M (1998) Neue Meßtechniken zur gleichzeitigen Bestimmung von Wassergehalt, Wasserspannung und verfügbaren Düngersalzen im Boden, Vortrag Internationale Tagung Landtechnik, Garching 1998, VDI-MEG Verlag, Düsseldorf, 1998, S.223–228
62. Balendonck J et al (2001) Waterman. Final report, EC-project FAIR1–CT95–0681, CD-ROM, 13 Jan 2001
63. Paul W (2002) Prospects for controlled application of water and fertiliser, based on sensing permittivity of soil. Comput Electron Agric 36:51–163
64. Whalley WR, Watts CW, Hilhorst MA, Bird NRA, Balendonck J, Longstaff DJ (2001) The design of porous material sensors to measure the matric potential of water in soil. Eur J Soil Sci 52:511–519
65. Paul W (1998) Sensors for soil attributes, plant transpiration and water stress. In: Int Conf on Agr Eng; part 2, Oslo 1998, 98-C-010, pp 850–853

Measurement Methods
and Sensors in Time Domain

14 Advanced Measurement Methods in Time Domain Reflectometry for Soil Moisture Determination

Christof Huebner[1], Stefan Schlaeger[2], Rolf Becker[3], Alexander Scheuermann[4], Alexander Brandelik[5], Wolfram Schaedel[3], Rainer Schuhmann[5]

[1] Fachhochschule Mannheim – University of Applied Sciences
[2] Research Center Karlsruhe, Institute of Technical Chemistry, Water and Geotechnology Section
[3] University of Karlsruhe, Institute of Water Resources Management, Hydraulic and Rural Engineering
[4] University of Karlsruhe, Institute of Rock Mechanics and Soil Mechanics
[5] Research Center Karlsruhe, Institute of Meteorology and Climate Research

14.1 Introduction

Knowledge of soil moisture is essential to many applications in hydrology, agriculture, and civil engineering. Among the various electromagnetic or dielectric moisture measurement methods time domain reflectometry (TDR) has become one of the most used methods. This is due to the early establishment of simple approximate relations between soil dielectric properties and water content and the availability of field portable instruments [1]. Sensors used for TDR usually consist of two- or three-wire transmission lines which are buried into the soil and connected to a TDR set-up (Fig. 14.1). A fast rise time (about 200 ps) voltage step is launched into a coaxial cable and propagates through the sensor system. At the transition between the coaxial cable and the two-wire transmission line a part of the pulse is reflected due to impedance mismatch. The remaining pulse travels along the two-wire line until it is totally reflected at the open end. An oscilloscope records the sum of the incident signal and the reflected signal (step response), also called the TDR trace, from which the travel time and the mean wave velocity in the two-wire line section can be determined. The permittivity of the surrounding soil can be calculated from the wave velocity and is related to the mean soil water content. For many applications this basic TDR measurement method is not sufficient for user requirements. The most serious limitations are

1. the restricted length of the transmission line (usually not more than 30 cm),
2. the lack of spatial water content resolution along the transmission line, and
3. the lack of a low-cost high-precision TDR instrument.

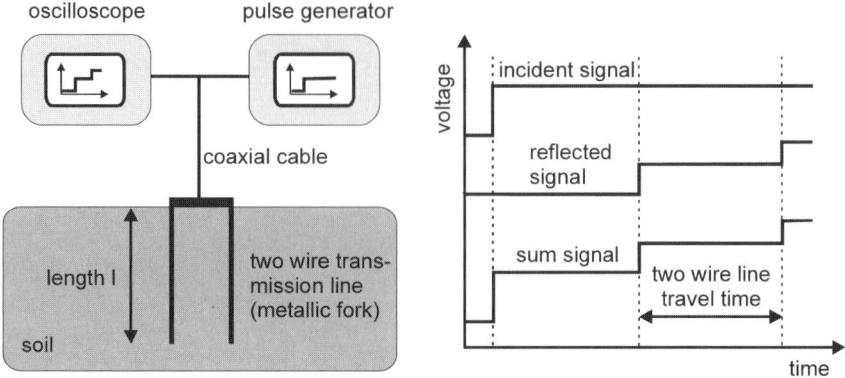

Fig. 14.1. Basic TDR set-up and signals. Oscilloscope and pulse generator are usually integrated in a single TDR instrument

Therefore a new transmission line technology, a reconstruction algorithm for determining the water content profile along the transmission line and a new TDR instrument have been developed and will be presented in this work. Experimental results, e.g., obtained in a project dealing with water transport processes in a full-scale levee model, demonstrate the applicability of the proposed methods for practical problems.

14.2 Dielectric Properties of Soils

TDR as an electromagnetic moisture measurement method is an indirect way to determine water content. The relation between water content and electrically measured permittivity of the soil has to be established by calibration. Though this is not the focus of this work, some of the most important definitions and dielectric properties of soils are reviewed.

Definitions of water content are different in various scientific or industrial fields, e.g., a weight-related definition (dry or wet) is preferred in soil mechanics and civil engineering whereas hydrologists ask for volume-related water content to evaluate the transport equations. As the dielectric measurement is a volumetric phenomenon, we agree on a volumetric definition of the water content W. Conversion into gravimetric water content or saturation requires additional information on dry density or pore fraction.

The straightforward but most laborious way to find the relation between the water content and permittivity of a soil is to perform a site-specific calibration with gravimetric sampling. This is often impracticable for operational use and does not allow sensor performance prediction prior to installation. Therefore, several empirical, semi-empirical, and theoretical mixing rules with different degrees of experimental justification have been developed and applied. One of the most often used empirical equations was given by Topp et al [2] for the soil permittivity ε_m:

$$\varepsilon_m = 3.03 + 9.3 \; W + 146 \; W^2 - 76.7 \; W^3 . \tag{14.1}$$

This approach with a real-valued permittivity is valid as long as losses in the measurement frequency range can be neglected, which is not true for soils with high clay and/or salt content. To apply a reconstruction algorithm for water content profile determination it is crucial to know the frequency-dependent complex permittivity of the soil. One approach is a simple model based on the refractive mixing rule developed by Birchak et al [3]:

$$\sqrt{\varepsilon_m} = W \sqrt{\varepsilon_w} + \; S \; \sqrt{\varepsilon_s} + \; A \; \sqrt{\varepsilon_a} . \tag{14.2}$$

S, W, and A are the volumetric fractions of solid particles, water, and air with their corresponding permittivities ε_s, ε_w, and ε_a. The permittivity of water is considered to be complex valued due to its ionic conductivity σ_w and therefore depends on the frequency f. ε_{ws} is the static permittivity of water. Thus

$$\varepsilon_w = \varepsilon_{w,real} - j\varepsilon_{w,imag} = \varepsilon_{ws} - j \frac{\sigma_w}{2\pi \; f \varepsilon_0} . \tag{14.3}$$

This simple model is compared to measurements on sand and clay materials in Fig. 14.2 and Fig. 14.3. The sand exhibits very low losses and a nearly frequency-independent real part of the permittivity except at very low frequencies. It can be expected that this trend will remain up to about 1 GHz until water relaxation smoothly starts to decrease the real part of the permittivity and increase dielectric losses.

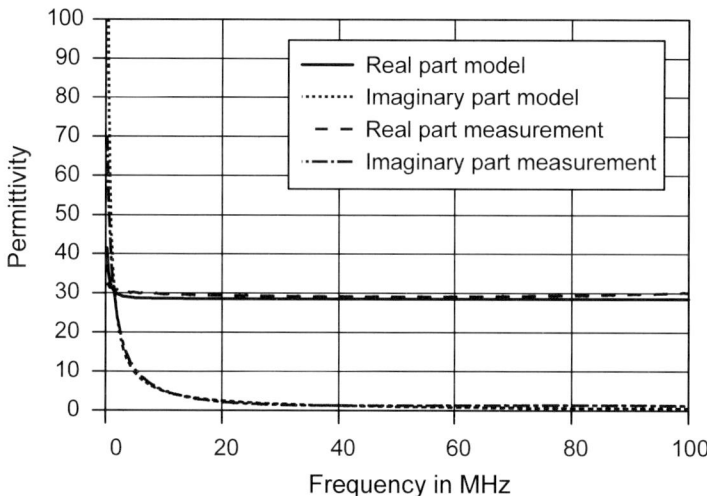

Fig. 14.2. Frequency-dependent complex permittivity of a sand material (S = 0.5, W = 0.455, A = 1-W-S, ε_w = 80, ε_s = 6, ε_a = 1, σ_w = 0.01 S/m)

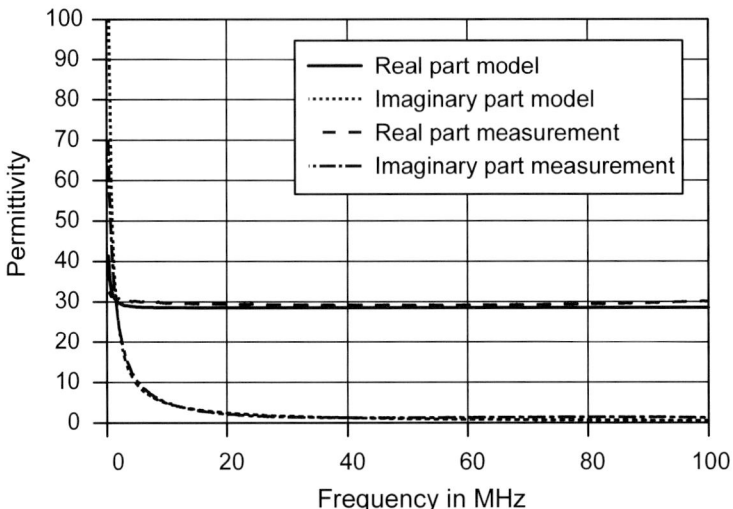

Fig. 14.3. Frequency-dependent complex permittivity of a clay material ($S = 0.6$, $W = 0.35$, $A = 1$-W-S, $\varepsilon_w = 80$, $\varepsilon_s = 3$, $\varepsilon_a = 1$, $\sigma_w = 0.8$ S/m)

The frequency dependence of the clay is much more pronounced due to the high conductivity. Furthermore the quality factor $\varepsilon_{real}/\varepsilon_{imag}$ is less than 1 even at frequencies as high as 100 MHz.

Though this simple model is in close agreement with the measurement results, it does not take into account the relaxation of water, temperature-dependent permittivities, nor bound water which is especially found in clayey soil. Nevertheless the general trend is clearly visible. More sophisticated models of the dielectric properties of soils have been investigated by Dobson et al [4] and Heimovaara et al [5] and compared to experimental results.

Soil moisture measurements should be preferably carried out at frequencies where losses are small. TDR, however, is an inherently broadband method with a frequency range from a few kilohertz up to several gigahertz. Fortunately the standard travel time analysis of TDR traces can approximately be considered as a measurement in the frequency range between about 100 MHz and 1 GHz. The relatively small dependency of the TDR method on soil type is attributed to this high-frequency measurement [6].

The use of a reconstruction algorithm to determine the water content profile along the transmission line implies the analysis of the amplitude variations of the TDR trace. Therefore losses cannot be neglected and the frequency-dependent permittivity has to be accounted for. Losses also limit the maximum length of the transmission line, because the electromagnetic pulses are attenuated and disappear on longer lines. The upper limit is about 1 m depending on the conductivity [7]. In order to overcome this limitation a new transmission line technology has been developed and will be discussed in the next section.

14.3 Transmission Lines

For longer transmission lines we propose insulated flexible flat band cables. They show much less pulse attenuation than non-insulated metallic forks in the same media. Several cables with different geometries have been developed and manufactured, from simple concentric insulation to sophisticated multiwire structures with unilateral sensitivity [8, 9]. The flat band cable used for most experiments here is shown in Fig. 14.4 together with its electrical field distribution in cross-section. The cable consists of three copper wires covered with polyethylene insulation. The electrical field is concentrated around the conductors and defines the sensitive area of 3 to 5 cm around the cable. The spatial weighting of the measurement in the cross-section of the cable is directly related to the energy density distribution. In order to predict the time domain response of the cable the electromagnetic properties have to be measured or calculated by numerical methods. The well-known equivalent circuit for an infinitesimal section of a transmission line is shown in Fig. 14.5. The equivalent circuit parameters are the series resistance R, inductance L, shunt conductance G and capacitance C.

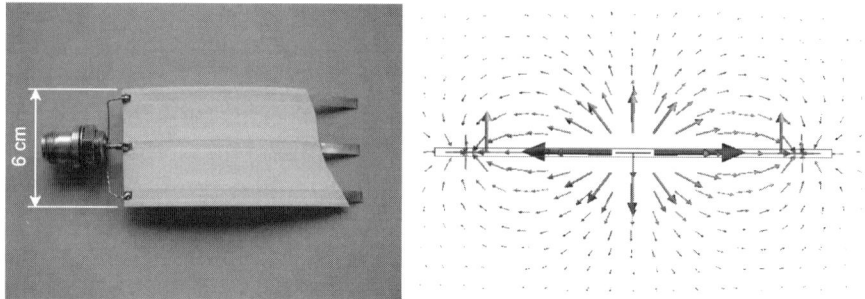

Fig. 14.4. Flat band cable and electrical field distribution in cross-section for symmetric excitation (inner conductor positive, outer conductors negative)

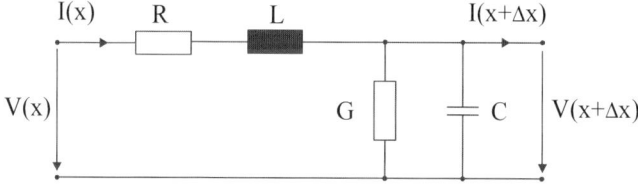

Fig. 14.5. Equivalent circuit of an infinitesimal section of a transverse electromagnetic (TEM) transmission line. $V(x)$ and $I(x)$ are the voltage and the current at the beginning and at the end of the line

A limitation of this model is that radiation and higher order modes are not included. Open-wire lines exhibit electromagnetic radiation when the electrical spacing between the wires is large. Then they behave as a hybrid transmission line/antenna mixture. Radiation strongly increases with frequency and may be a major source of loss and rise time degradation. Future investigations will focus on this important aspect but require full-wave numerical field simulations. Nevertheless the equivalent circuit in Fig. 14.5 is a reasonable approach for calculating the wave propagation along the transmission line.

14.3.1 Resistance R and Inductance L

The resistive losses are due to the skin effect. At low frequencies there is a uniform distribution of the current through the cross-section of the copper conductors. As the frequency increases, current concentrates at the edges of the conductors, and at the highest frequency most of the current is flowing in a very thin region. This behavior is illustrated in Fig. 14.6 and Fig. 14.7 which show the cross-section of the cable with its magnetic field distribution at 1 MHz and the current density in the outer left conductor at four different frequencies.

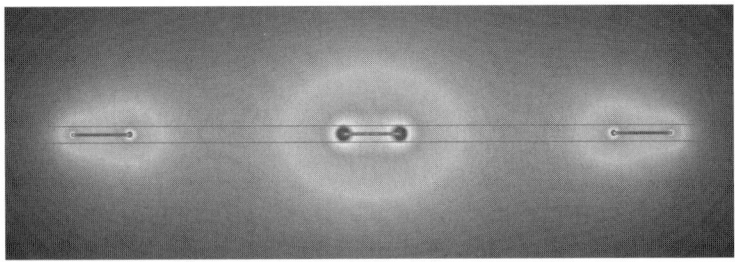

Fig. 14.6. Magnetic field distribution in the cross-section of the cable at 1 MHz

Fig. 14.7. Current density in the outer left conductor for four different frequencies. As frequency increases the current concentrates at the edges of the conductor. The dark color in the middle of the conductor represents low current density

As the frequency increases, the cross-sectional area available for current flow decreases and causes an increase in series resistance (Fig. 14.8). In most practical cases resistance losses are small compared with dielectric losses (shunt conductance) except for long cables in nearly lossless materials like snow. For the same reason that resistance is frequency dependent, the inductance will be too. As the current distribution changes, the internal self-inductance decreases with frequency (Fig. 14.8). At the highest frequency only the external inductance will remain. The transition frequency is around 100 kHz. This means that within the usual time window of TDR measurements (< 100 ns) no significant influence of the inductance increase at low frequencies will be expected.

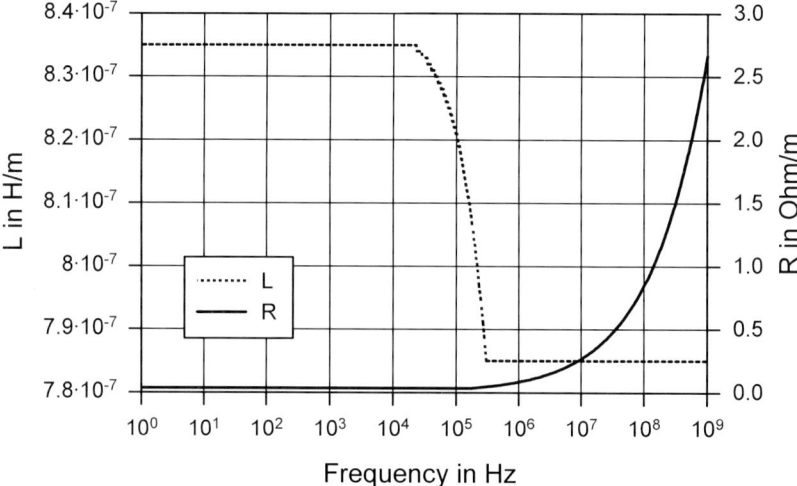

Fig. 14.8. Series resistance and inductance of the flat band cable for symmetric excitation (stepwise interpolated data based on numerical field simulation)

14.3.2 Capacitance C and Conductance G

Unlike resistance and inductance, capacitance and conductance are dependent on the dielectric properties of the soil surrounding the cable. Both C and G can be determined with numerical field calculation. For the lossless case ($G = 0$) the relation between real permittivity ε_m and capacitance C shown in Fig. 14.9 has been obtained. A simple model with three capacitances is able to explain this behavior. The capacitance C may be replaced by C_1, C_2 and $\varepsilon_m C_3$ as shown in Fig. 14.10. The total capacitance is given by

$$C(\varepsilon_m) = C_1 + \frac{C_2 \varepsilon_m C_3}{C_2 + \varepsilon_m C_3}. \tag{14.4}$$

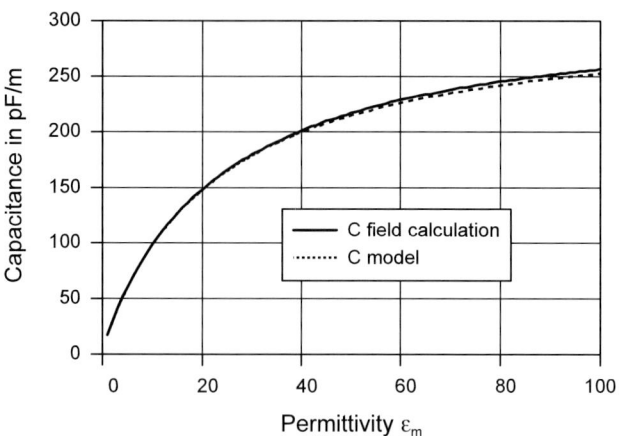

Fig. 14.9. Relation between permittivity and capacitance for the lossless case and symmetric excitation of the cable (cf. Fig. 14.4)

The three unknown capacitances C_1, C_2, and C_3 were derived form numerical field calculation and calibration measurements with three different media around the cable (Table 14.1).

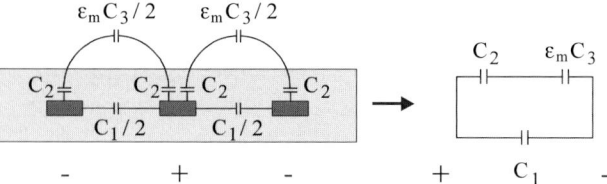

Fig. 14.10. Capacitance model of the flat band cable

Table 14.1. Parameters of the three-capacitance model

Circuit element	C_1	C_2	C_3
Measured	3.4 pF/m	323 pF/m	14.8 pF/m
Calculated	4.0 pF/m	303 pF/m	13.7 pF/m

The differences between measurement and calculation are attributed to the uncertainty in measuring the thickness of the insulation, which is a very sensitive parameter, especially at high permittivities. With some limitations at high conductivities, Eq. (14.4) is a good approximation for complex permittivities as well. The DC conductivity of the soil, σ_m, can be combined with real permittivity and frequency with a complex permittivity:

$$\varepsilon_m = \varepsilon_{m,real} - j\varepsilon_{m,imag} = \varepsilon_{m,real} - j\,\frac{\sigma_m}{2\pi\,f\varepsilon_0}. \qquad (14.5)$$

Then C and G are calculated according to the following equation:

$$C(\varepsilon_m) + j\frac{G(\varepsilon_m)}{2\pi f} = C_1 + \frac{C_2 \varepsilon_m C_3}{C_2 + \varepsilon_m C_3} \qquad (14.6)$$

C_1 and C_2 contribute to G as well due to dielectric losses in the polyethylene (tan $\delta \approx 2 \times 10^{-4}$), but these are usually very small compared to losses caused by soil conductivity and can therefore be neglected. Even when the soil has a frequency-independent real permittivity and conductivity the equivalent circuit parameters C and G of an insulated transmission line are dispersive, which is a kind of Maxwell-Wagner effect [10]. Figure 14.11 shows the frequency dependence of C and G for real permittivity $\varepsilon = 1$ and conductivity $\sigma = 0.1$, 0.01, and 0.001 S/m.

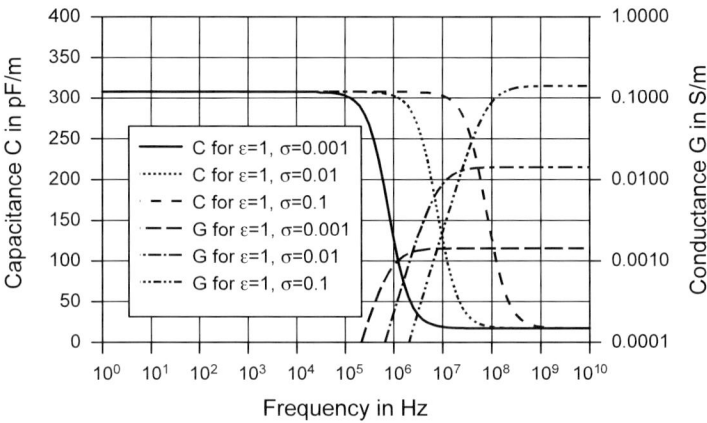

Fig. 14.11. Capacitance C and conductance G of the flat band cable embedded in a material with real permittivity $\varepsilon = 1$ and conductivity $\sigma = 0.1$, 0.01, and 0.001 S/m

At low frequencies the capacitance is strongly enhanced and conductivity is near zero. Lower conductivity corresponds to lower transition frequencies and consequently has less influence on the TDR trace in the usually short time range. In order to study these effects the transient response of a transmission line under various conditions will be calculated in the next section.

14.3.3 Transient Analysis

The transient response of a transmission line can be calculated in the time domain or by Laplace/Fourier transformation from the frequency to the time domain. The advantage of a frequency to time domain transformation is the easy incorporation of dispersive equivalent circuit parameters. Time domain methods on the other hand are usually limited to non-dispersive L, C, R, and G. Their advantage is the availability of fast and robust inversion algorithms for the reconstruction of inhomogeneous C and G profiles along transmission lines, which are required for many soil moisture measurement applications. Fortunately inversion algorithms usually analyze only once or twice the time range to the reflection at the open end of the transmission line. Within this time range the assumption of non-dispersive equivalent circuit parameters may be a good approximation. Figure 14.12 shows the transient response of a 1 m long flat band cable embedded in two materials with different conductivities. Within the time range up to the reflection at the open end at about 18 ns the agreement for the low-loss case is good. The deviation for the high-loss case is greater. For longer times the TDR traces diverge more and more. At infinity the TDR trace in the dispersive case reaches a level of 2 V due to zero conductivity of the insulated flat band cable at DC.

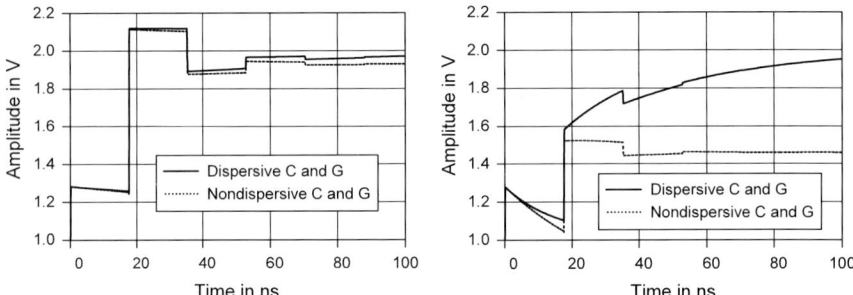

Fig. 14.12. Simulation of the TDR trace of a 1 m long flat band cable; 1 V step as incident signal. *Left:* dispersive C and G ($\varepsilon_{m,real} = 10$, $\sigma_m = 0.001$ S/m, Eqs. (14.5-6)), non-dispersive C and G ($C = 98.6$ pF/m, $G = 0.00074$ S/m) *Right:* dispersive C and G ($\varepsilon_{m,real} = 10$, $\sigma_m = 0.01$ S/m, Eqs. (14.5-6)), non-dispersive C and G ($C = 98.6$ pF/m, $G = 0.0074$ S/m)

14.4 Reconstruction Algorithm

The standard TDR measurement procedure based on travel time analysis delivers only the mean water content along the transmission line. On the other hand many applications require soil moisture profiles to be determined. One approach to satisfy these demands has been developed by Hook et al [11]. The transmission line is divided into several sections by remote-controlled switches. This solution only allows a very coarse spatial resolution of the soil moisture distribution. A high spatial resolution can be achieved by exploiting the full information in the reflected electromagnetic signal. Each change in water content along the transmission line means an impedance change and a partial reflection. A three-step algorithm has been developed to reconstruct the soil moisture profile from these reflections [12].

In the first step the transmission line parameters $C(x)$ and $G(x)$ are reconstructed with two independent time domain measurements from both ends of the flat cable by solving the inverse problem for the telegraph equations. If there are only one-sided reflections available it is possible to reconstruct $C(x)$ with a given distribution of $G(x)$. The second step transforms $C(x)$ into the dielectric coefficient $\varepsilon_m(x)$ by the inversion of Eq. (14.4). In the third step $\varepsilon_m(x)$ is converted into the water content profile $W(x)$ by standard transformations based on Eq. (14.1) or material-specific calibration functions. The first step as the key component for the reconstruction will be discussed in detail.

14.4.1 The Telegraph Equations

The telegraph equations describe the variation of the voltage $V(x,t)$ and the current $I(x,t)$ in time along the transmission line. By applying Kirchhoff's voltage and current laws to the equivalent circuit in Fig. 14.5 the following is obtained:

$$\frac{\partial}{\partial x}V(x,t) = -R(x)I(x,t) - L(x)\frac{\partial}{\partial t}I(x,t) \tag{14.7}$$

$$\frac{\partial}{\partial x}I(x,t) = -G(x)V(x,t) - C(x)\frac{\partial}{\partial t}V(x,t). \tag{14.8}$$

In order to reconstruct the two parameters $C(x)$ and $G(x)$, two independent measurements are needed. Consequently, the task is divided into two parts, one part dealing with an incident wave from one side and the other part dealing with an incident wave from the other side of the cable.

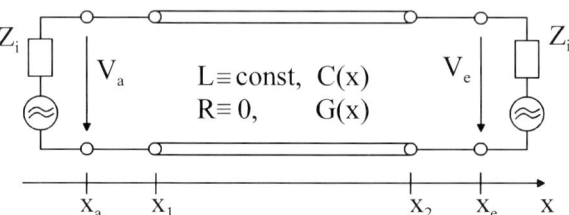

Fig. 14.13. The non-uniform flat band cable, situated between x_1 and x_2, is connected to two lossless uniform coaxial cables with matched impedances Z_i at their endpoints and capacitances C_a, C_e and inductances L_a, L_e.

Figure 14.13 describes the set-up for receiving the reflection data from both sides of the unknown material. Therefore two separate measurements must be carried out with the external currents

$$F_{ex}^1 = \delta(x - x_a)f(t) \text{ and } F_{ex}^2 = \delta(x - x_e)f(t). \tag{14.9}$$

The input data $f(t)$, which describes the incident pulse, can be easily extracted from the measurements $V_a(t)$ of the coaxial cable between x_a and x_1 with an open end at x_1.

The inverse method presented in the next section is based on an iterative search for the electrical parameters of the non-uniform flat band cable with the full-wave solution of both direct problems. During the search the solutions of the line need to be calculated many times. It is therefore important to use a technique that is computationally efficient, and which provides a simple mapping from the parameters to the direct solution, to guarantee a fast convergence.

Equations (14.7-8) can easily be transformed into a partial differential equation (PDE) of second order for $V_i(x,t)$ representing the voltages of each separate experiment with regard to F_{ex}^i, $i=1,2$

$$\left[LC \frac{\partial^2}{\partial t^2} + LG \frac{\partial}{\partial t} + \frac{\partial L / \partial x}{L} \frac{\partial}{\partial x} - \frac{\partial^2}{\partial x^2} \right] V_i(x,t) = 0, \quad i = 1,2 \qquad (14.10)$$

with the initial conditions

$$V_i(x,t)\big|_{t\leq 0} = 0, \quad \frac{\partial}{\partial t} V_i(x,t)\bigg|_{t\leq 0} = 0, \quad x_a \leq x \leq x_e, \ i = 1,2. \qquad (14.11)$$

The boundary conditions depend on the location of the external current $F_{ex}^i(x,t)$, $i=1,2$. For the source at $x = x_a$, there is an absorbing boundary condition at $x = x_e$

$$\left[\frac{\partial}{\partial x} + \sqrt{L_e C_e} \frac{\partial}{\partial t} \right] V_1(x,t)\bigg|_{x=x_e} = 0, \ t \geq 0 \qquad (14.12)$$

and an absorbing boundary condition for the reflected wave at $x = x_a$ in connection with the external current:

$$\left[\frac{\partial}{\partial x} - \sqrt{L_a C_a} \frac{\partial}{\partial t} \right] V_1(x,t)\bigg|_{x=x_a} = L_a \frac{\partial}{\partial t} F_{ex}^1(x,t)\bigg|_{x=x_a}, \ t \geq 0. \qquad (14.13)$$

For the other PDE for $V_2(x,t)$ with external current $F_{ex}^2(x,t)$ the boundary conditions are exchanged.

14.4.2 The Optimization Approach

The aim of the investigation is the determination of the unknown distributions of $C(x)$ and $G(x)$ with measurements of input data $f(t)$ and output data in the time domain. The output data $\lambda_i(t)$, $i = 1,2$ ($\lambda_1(t) = V_a(t)$, $\lambda_2(t) = V_e(t)$) can be received from the two separate measurements with the external current F_{ex}^1 and F_{ex}^2, respectively.

A cost function $J(C,G)$ is defined which measures the difference in the L_2-norm between the solutions $V_i(x_i,t)$, $i = 1,2$ of the direct problems (Eqs. (14.10-13)) corresponding to the parameters $C(x)$ and $G(x)$ and the given measurements

$$J(C,G) = \sum_{i=1}^{2} \int_0^T [V_i(x_i,t) - \lambda_i(t)]^2 dt \qquad (14.14)$$

with $T = 2\tau(x_1,x_2)$, where $\tau(x_1,x_2)$ is the travel time between x_1 and x_2. The cost function refers to the error in the solutions $V_1(x_1,t)$ and $V_2(x_2,t)$, respectively. The concept of the method is to find the parameter distributions $C(x)$ and $G(x)$ that minimize the cost function J. If the inverse problem has a solution the theoretical minimum of J is zero. One more important reason for choosing the L_2-norm is that this makes it possible to derive exact expressions for the gradient of J.

14.4.3 Exact Expression for the Gradient of the Cost Function

It is advisable to derive the gradient of J with respect to C, J_C, by $\lim_{\|\delta\ C(x)\| \to 0}(J(C+\delta C)-J(C)) =: (J_C,\delta C(x))$ where $\delta C(x)$ denotes the differential of $C(x)$ and (f,g) is the inner product defined by

$$(f(x), g(x)) = \int_{x_1}^{x_2} f(\xi)g(\xi)d\xi . \qquad (14.15)$$

Thus, the gradient $J_C(x)$ is the direction in which J increases most rapidly with respect to the norm of the change of the function $C(x)$. $J_C(x)$ and $J_G(x)$ are given by

$$J_C(C,G) = -\sum_{i=1}^{2} \int_0^T \left(\frac{\partial}{\partial t}\Psi_i\right)\left(L\frac{\partial}{\partial t}V_i\right)dt \qquad (14.16)$$

$$J_G(C,G) = \sum_{i=1}^{2} \int_0^T \Psi_i\left(L\frac{\partial}{\partial t}V_i\right)dt \qquad (14.17)$$

where the dual functions Ψ_1 and Ψ_2 satisfy the following PDEs:

$$\left[LC\frac{\partial^2}{\partial t^2} - LG\frac{\partial}{\partial t} - \frac{\partial L/\partial x}{L}\frac{\partial}{\partial x} - \frac{\partial^2}{\partial x^2}\right]\Psi_i(x,t)$$

$$= 2\delta(x-x_i)[V_i(x,t) - \lambda_i(t)], \quad i = 1,2 \qquad (14.18)$$

with the initial conditions for $t > T$ and boundary conditions similar to Eqs. (14.11-13).

14.4.4 Reconstruction of the Parameters

Only reflection data is used to determine uniquely the parameters $C(x)$ and $G(x)$ [13]. The optimization is carried out with a conjugate gradient method using Fletcher-Reeves update formulas [14]. To determine the optimal step size α for the search direction a parabolic approximation technique is used in each conjugate gradient step.

The initial capacitance $C_0 = \tau^2(x_1,x_2)/(L(x_2-x_1)^2)$ can easily be determined by simple travel time measurements along the probe [15]. To ensure the invariability of this sensor travel time during the conjugate gradient algorithm one has to find a constant shift C_α for every given α during the optimization to fulfill

$$\sqrt{C_0}\,(x_2 - x_1) = \int_{x_1}^{x_2}\sqrt{C(x)+\alpha J_C(C,G)+C_\alpha}\,dx \ . \qquad (14.19)$$

The result of the minimization of Eq. (14.14) is the spatial distribution of the total capacitance C and total conductance G. $C(x)$ can easily be transformed into the dielectric properties $\varepsilon_m(x)$ of the surrounding material. Furthermore, the volumetric water content $W(x)$ can be determined using standard transformations or a special calibration function for the material used. The total conductance $G(x)$ describes the conductivity of the material between the copper wires, i.e., the system of polyethylene insulation and the surrounding material. The determination of the water content distribution of the surrounding material does not require knowledge of the conductivity distribution of the material, but it cannot be neglected during the reconstruction of $C(x)$.

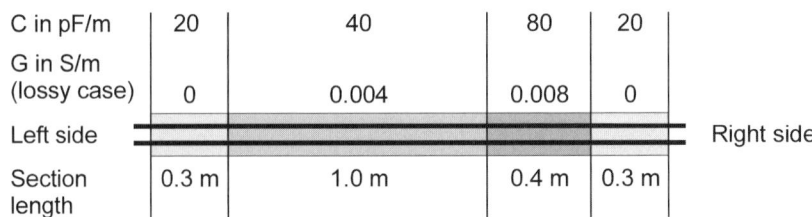

C in pF/m	20	40	80	20
G in S/m (lossy case)	0	0.004	0.008	0
Left side				Right side
Section length	0.3 m	1.0 m	0.4 m	0.3 m

Fig. 14.14. Inhomogeneous transmission line with four sections. Lossless case: C as indicated, $G \equiv 0$. Lossy case: C and G as indicated

14.4.5 Reconstruction Examples

The performance of the reconstruction algorithm will be demonstrated with an example. Consider the inhomogeneous transmission line shown in Fig. 14.14. The transmission line consists of four sections. Two different arrangements will be investigated. In the lossless case G is zero everywhere whereas in the lossy case G is non-zero in the two sections in the middle. The incident signal is shown in Fig. 14.15.

14.4.5.1 Lossless Case

The TDR reflections from the left side and the right side are shown in Fig. 14.16. For each simulation the other side of the cable is left open, which results in a total reflection occurring at about 22 ns. In this simple example, the variation of the capacitance can roughly be estimated from the amplitude variations. High amplitude means low capacitance and vice versa.

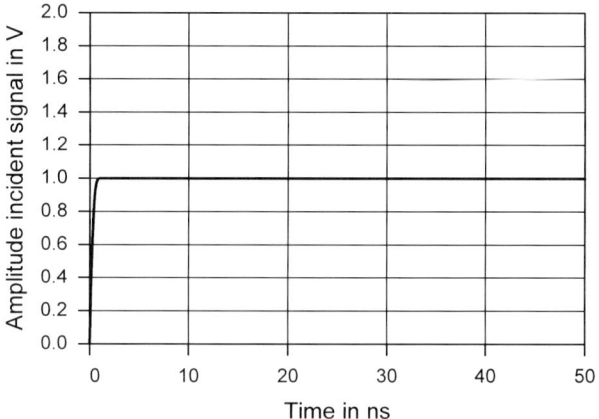

Fig. 14.15. Incident signal (step with a 10% to 90% rise time of about 460 ps)

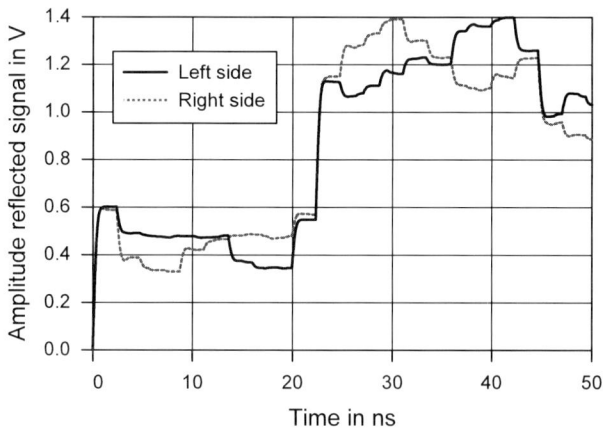

Fig. 14.16. TDR reflections from left and right side for the lossless case

Of course the reconstruction algorithm is more accurate, because it takes into account multiple reflections. In the lossless case only one parameter, namely, $C(x)$ has to be reconstructed. Therefore it is sufficient to use only one-sided reflection data. The reconstruction shown in Fig. 14.17 is in close agreement with the true capacitance profile. Only at the sharp edges of the profile is some overshoot visible.

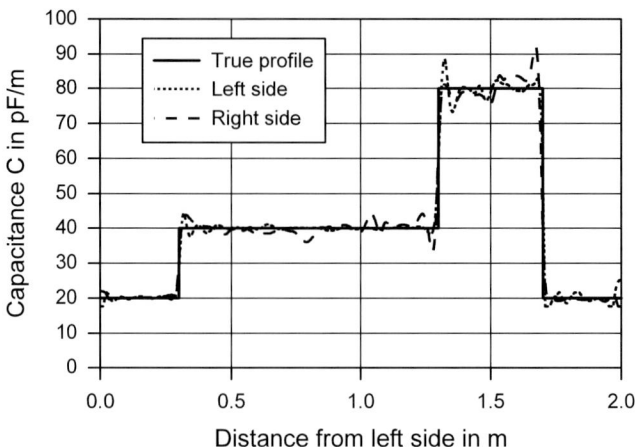

Fig. 14.17. Reconstruction of $C(x)$ from one-sided data (left and right) in the lossless case

14.4.5.2 Lossy Case

Figure 14.18 shows the TDR reflections for the lossy case. Losses reduce signal energy and decrease amplitude. The pronounced downward slope prevents prediction of the capacitance profile by simple visual inspection of the amplitude variation.

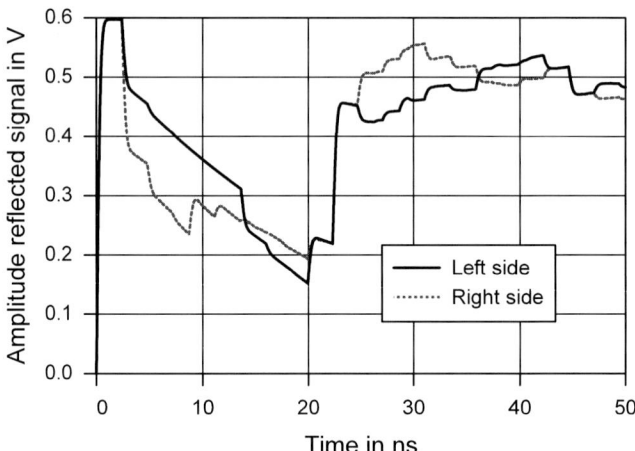

Fig. 14.18. TDR reflections from left and right side for the lossy case

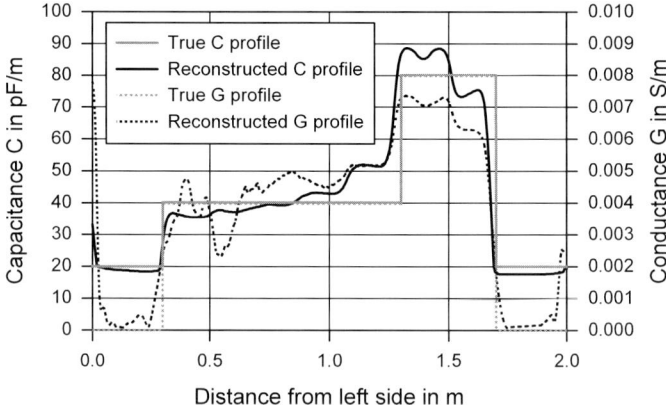

Fig. 14.19. Reconstruction of $C(x)$ and $G(x)$ from two-sided data in the lossy case

In the lossy case, our reconstruction algorithm requires reflection data from both sides of the cable to achieve reasonable accuracy. The results shown in Fig. 14.19 approximate the true profile $C(x)$ and $G(x)$ well.

14.5 A Novel TDR Measurement Principle

The overall costs of a soil moisture measurement system strongly depend on the cost of the TDR instrument. Commercially available TDR instruments are too expensive for widespread use in operational monitoring of soil moisture. Also they often lack the necessary robust design for long-term application under field conditions. To improve this situation a new TDR instrument "Observer" has been developed in collaboration with the ADD Automation company (Fig. 14.20).

14.5.1 The "Binary Sampler", a Delta Modulator in Equivalent-Time Sampling

The Observer is a unique TDR, because it differs significantly from conventional instruments. Its sampler is based on the recently developed "Binary Sampler" [16]. The Binary Sampler is a differential pulse code or delta modulator [17]. A delta modulator is a codec (encoder/decoder), which encodes the error between the input signal and a prediction with 1-bit resolution, e.g., the signal is greater or less than the last prediction. It tries to minimize this error continuously by correcting the prediction. An integrator is used as first-order predictor. If the input at one sample instant is greater than the last prediction, the integrator will predict the input signal to be even larger at the next sample instant, and vice versa. The result of a delta modulator is a differential pulse code. It is transmitted to a receiver, e.g., a microcontroller, where it has to be digitally integrated to reconstruct the sampled signal. The delta modulator itself has a local analog integrator as decoder, whose output is coupled back and used as a prediction for the input signal at the next sample instant. Therefore it is a control system which is different to the usual sampling bridge or gate circuits found in conventional samplers [18].

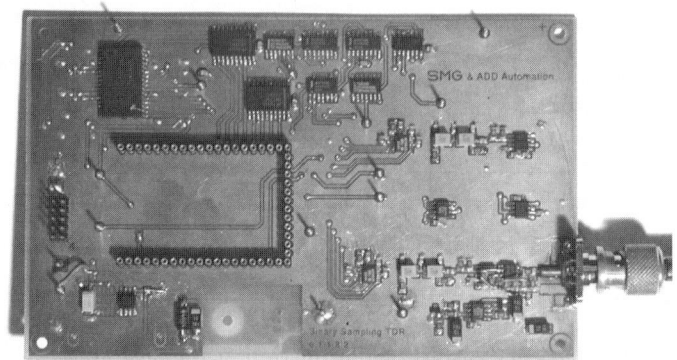

Fig. 14.20. Prototype of the new TDR instrument "Observer"

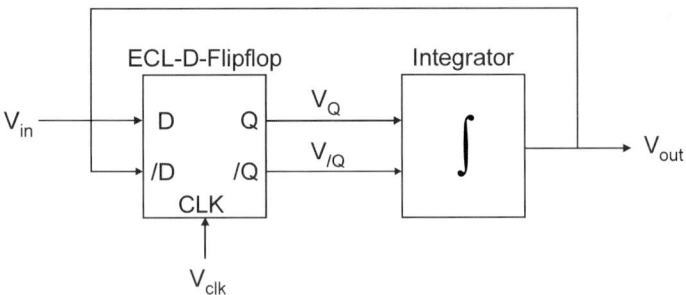

Fig. 14.21. Simplified block diagram of the Binary Sampler

The core of the Binary Sampler consists of two main functional blocks: an edge triggered D-flip-flop (D-FF) of a fast emitter coupled logic family (ECL), and an integrator, the output of which is coupled back to the input of the D-FF (Fig. 14.21). A D-FF is a simple storage element. At each rising edge of the D-FF clock signal the logic input state is shifted to the output and latched. A change at the input does not influence the output, unless a clock pulse is fired. The ECL family uses differential input and output stages. The differential input of the D-FF consists of the two signals D and /D (not D). A logic "1" or "0" is represented by a positive or negative voltage difference between D and /D, respectively. The high-frequency signal under investigation, V_{in}, is connected to D, whereas /D is connected to the output V_{out} of the integrator. The commonly available, cost-effective ECL D-FF acts in principle like a specialized strobed comparator [19]. It compares V_{in} and V_{out} at each rising edge of a clock pulse, which is the sampling strobe. If

$$\Delta V_{in} = V_{in} - V_{out} \qquad (14.20)$$

is positive or negative, the output voltage difference

$$\Delta V_Q = V_Q - V_{/Q} \qquad (14.21)$$

will have a positive or negative constant value, respectively. It is important that even a very small input voltage difference leads to a full-scale output voltage swing. In the next stage of the control system the integrator integrates ΔV_Q, which is a fixed positive or negative value between two consecutive clock pulse edges occurring at time t_{n-1} and t_n:

$$V_{out}(t_n) = \frac{1}{t_c} \int_{t_{n-1}}^{t_n} \Delta V_Q(t)dt + V_{out}(t_{n-1})$$

$$= \frac{1}{t_c} \Delta V_Q(t_{n-1})(t_n - t_{n-1}) + V_{out}(t_{n-1}) = \frac{\pm |\Delta V_Q|}{t_c} \frac{1}{f_{CLK}} + V_{out}(t_{n-1}). \qquad (14.22)$$

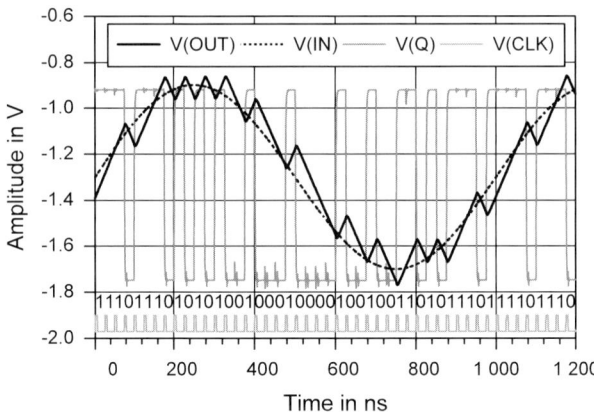

Fig. 14.22. Real-time sampling of a sinusoid by a delta modulator. V_{in}: input sine wave, $f_{in} = 1$ MHz, V_{clk}: sample clock, $f_{clk} = 40$ MHz, V_{out}: feedback, piecewise linear approximation of V_{in}, V_Q: Q-output of the D-FF. The sequence of the D-FF's logic states is the pulse-coded signal (PSPICE simulation results)

Integration up to the next clock pulse time t_n adds a linearly rising or falling voltage ramp to the former voltage level $V_{out}(t_{n-1})$. The slope of the ramp is determined by the integrator's time constant t_c and the constant voltage ΔV_Q. The resulting voltage V_{out} is fed back to the /D input of the D-FF.

The working principle of this feedback can be best explained with an example (Fig. 14.22). Assume V_{in} to be greater than V_{out} at sampling time t_0: $\Delta V_{in} = V_{in} - V_{out} > 0$. Thus the output ΔV_Q of the D-FF is set positive. This leads to a rising voltage ramp at the output of the integrator, which drives V_{out} toward V_{in}. The output voltage does not change direction until at some sample strobe the feedback voltage V_{out} exceeds V_{in}, i.e., $\Delta V_{in} = V_{in} - V_{out} < 0$. Then the D-FF changes its output state and ΔV_Q becomes negative, which leads to a falling feedback voltage correcting the error between input and feedback.

Thus the circuit always tries to minimize the input voltage difference as far as possible. In this respect it resembles the Sampling Voltage Tracker (SVT) of the American National Institute of Standards and Technology [20], but differs significantly from the SVT in the way the signal under test is coded. The SVT samples the same instant of the repetitive input signal over many periods and digitizes the analog feedback voltage with a high-resolution analog-to-digital converter, e.g., 18 bits per sample, giving the absolute voltage level. In contrast the Binary Sampler is a delta modulator and uses the logic state of the D-FF to code the signal, i.e., 1 bit per sample, giving just differential information, i.e., step up (logic 1) or step down (logic 0). The encoding of the signal leads to a bit stream, which has to be decoded by integration and filtered to reconstruct the signal. The differential pulse code leads to very effective data compression, since only one bit per sample is needed.

14.5.2 Slope Overload, Amplitude Range, and Quantization Noise

The example of Fig. 14.22 shows a fairly rough approximation of the input sinusoid. To reduce the quantization error, the swing or step size of the integrator's output has to be reduced by an increase of its time constant. It is obvious that, without an increase of the sampling rate, the output signal cannot follow the steep edges of the sinusoid. When the slope of the input signal exceeds the maximum slope of the feedback signal a "slope overload" condition has been reached. Especially in TDR applications where a precise recording of steep rising edges of incident and reflected waves plays the essential role, slope overload is strictly to be avoided. To investigate the avoidance of slope overload, the input is assumed to be sinusoidal with amplitude A and frequency f_s:

$$V_{in}(t) = A\sin(2\pi f_s) . \tag{14.23}$$

Its maximum slew rate is given by the first derivative around its zero crossings. The condition of slope overload is generally avoided if the slew rate of the feedback (cf. Eq.(14.24)) is larger than the maximum slew rate of the input signal:

$$2\pi f_s A < \frac{1}{t_c} |\Delta V_Q| . \tag{14.24}$$

If the frequency of the input signal increases, the integrator's time constant has to be decreased to fulfill the condition for non-overload. Equation (14.24) can also be interpreted as a condition for the maximum allowable amplitude A_m of an input signal with bandwidth f_s. This determines the theoretical upper value of the sampler's amplitude range. The lower value is the smallest amplitude of a sinusoid which still disturbs the idling pattern of the sampler. The idling pattern is a regular pattern of alternating rising and falling slopes, which occurs when the input signal is a constant. The amplitude of the idling pattern is

$$A_I = \frac{1}{2} \frac{|\Delta V_Q|}{t_c} \frac{1}{f_{clk}} .$$

(14.25)

The larger the input signal amplitude becomes with respect to Eq. (14.25), the better the tracking. According to Eqs. (14.24 and 14.25) the amplitude range covered by the delta modulator is:

$$R = \frac{A_m}{A_I} = \frac{f_{clk}}{\pi f_s}$$

(14.26)

This ratio is not really the dynamic range, since the minimum amplitude, which gives an acceptable decoded signal-to-noise ratio, is a subjective choice.

Each sampling process is accompanied by quantization noise which is due to quantized time and amplitude. The calculation of the quantization noise produced by delta modulators is not a trivial task. Only the simple formula for quantization noise derived by DeJaeger [21], is given here:

$$N_q^2 = K_q \frac{f_c}{f_{clk}} \gamma$$

(14.27)

γ is the step size of integrator after one sampling period:

$$\gamma = \frac{|\Delta V_Q|}{t_c} \frac{1}{f_{clk}} .$$

(14.28)

f_c is the bandwidth of the message band of the codec, K_q is an arbitrary constant. A detailed analytical derivation of an improved formula describing the quantization noise is given in [17] and a summary of this analysis in [22].

14.5.3 Equivalent-Time Sampling

Among other applications delta modulators have been used to encode speech in realtime [21, 23]. The Binary Sampler is a delta modulator running not in realtime but equivalent-time mode, like most conventional samplers used in oscilloscopes or TDRs. In equivalent-time sampling the complete sample of a repetitive, stationary signal is composed of interleaved samples taken over many signal periods. The Binary Sampler for example takes one sample per period and sweeps the sample instant slowly across the signal. Equivalent-time sampling enables the sampler to acquire signals up to the analog bandwidth of the sampling system regardless of the sample rate. Real-time sampling of signals with a bandwidth of a few gigahertz is a much higher technical and economical expense.

14.5.4 "Observer", a Novel TDR Instrument Based on the Binary Sampler

The TDR prototype "Observer" is an application of this sampling technology. Figure 14.23 shows the block diagram. It mainly consists of a 1-bit sampler (delta modulator, Binary Sampler), two similar pulse generators, and two channels with different coding alternatives. In channel A the differential pulse code of the delta modulator is read. Every 8 bits, 1 byte is formed by means of a shift register. These bytes are buffered in a RAM, which is not shown in Fig.14. 23. The second channel B has just been added for test purposes. Here an analog-to-digital converter is used to digitize the feedback of the sampler, which yields a representation of absolute voltage levels rather than a differential pulse code.

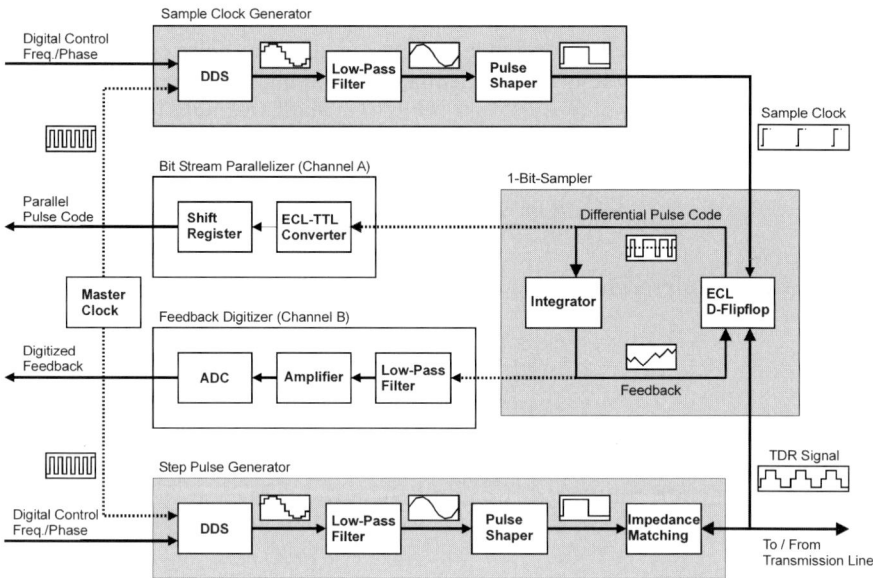

Fig. 14.23. Block diagram of the Observer

The two signal generators are driven by direct digital synthesis (DDS). The output frequency of the DDS can be programmed and controlled very precisely. The DDS produces a staircase approximation of a sinusoid, which has to be filtered to get the desired sine wave as pure as possible. The filtered sine wave enters a Schmitt trigger with hysteresis ("pulse shaper") which puts out a square wave. Thus both signal generators produce square waves with programmable frequency. The sample clock generator generates the sampling strobes for the sampler. The step pulse generator is connected to the transmission line. The superposition of emitted and reflected pulses traveling along the transmission line reaches the input of the sampler.

To achieve equivalent-time sampling, the step generator and clock generator have to be detuned slightly. The clock frequency can be expressed in terms of the step generator frequency f_{step} and a small offset:

$$f_{clk} = f_{step} - f_{offset} .$$
(14.29)

The offset frequency f_{offset} is chosen to be six to eight orders of magnitude less than f_{step} or f_{clk}. During each TDR signal period one sample is taken. Since the clock is a little bit slower than the step signal, the phase of the sample clock sweeps slowly across the TDR signal from sampling pulse to sampling pulse. After a time

$$t_s = f_{offset}^{-1}$$
(14.30)

step and clock pulse are in phase again, indicating that one complete period of the TDR signal has been sampled. The result is a sampled approximation of the TDR signal on a much longer time scale. The input signal period $t_{step} = f_{step}^{-1}$ is dilated to the time scale t_s. The dilation factor is

$$s = \frac{t_s}{t_{step}} = \frac{f_{step}}{f_{offset}}$$
(14.31)

with $10^6 \le s \le 10^8$, typical for the application of soil moisture measurement. The sampling of one TDR period takes t_s seconds. During this time, $n_{clk} = t_s f_{clk}$ sampling pulses occur. Thus one TDR signal of period t_{step} is sampled with n_{clk} points, yielding a theoretical temporal resolution of

$$\Delta t_r = \frac{t_{step}}{n_{clk}} = \frac{f_{offset}}{f_{step} f_{clk}} \approx \frac{f_{offset}}{f_{step}^2} .$$
(14.32)

This theoretical resolution can easily be set down to $\Delta t_r = 10^{-12}$ s. In reality the corresponding bandwidth cannot be reached, since it is limited by timing jitter, noise, and low-pass effects of the electronics.

14.5.5 Application and Comparison to Other Instruments

A wide range of different requirements for such a device exists even in the limited field of soil moisture measurement. In one application a moisture profile along a rod probe of a few centimeters may be needed, in another case the moisture distribution along a very extended system consisting of a flat band cable connected to a coaxial cable may have to be determined. To demonstrate the capabilities of the Observer, comparative measurements with two commercial TDRs have been performed. These instruments are the "1502B metallic cable tester" by Tektronix and the "TDR100" by Campbell Scientific. Both TDRs use conventional technology with sampling gates. The time base in these instruments is given by delay generators, which compare a linearly rising voltage ramp with variable slew rate to a given voltage threshold.

A 60 cm three-rod-probe has been vertically installed in a laboratory lysimeter. All three TDRs are connected to the probe via a multiplexer. The aim of the hydrological study was to demonstrate the possibility to track the temporal evolution of infiltration fronts propagating in the soil with TDRs. Figure 14.24 shows the TDR signal after 30 minutes of precipitation. The rising edge at 3 nanoseconds is the partial reflection at the probe's head. Approximately 10 nanoseconds later a strong reflection due to the probe's end is detected. In between the signal propagates along the rods buried in soil. The depression at the beginning of this part of the signal indicates the wet zone in the first centimeters of soil.

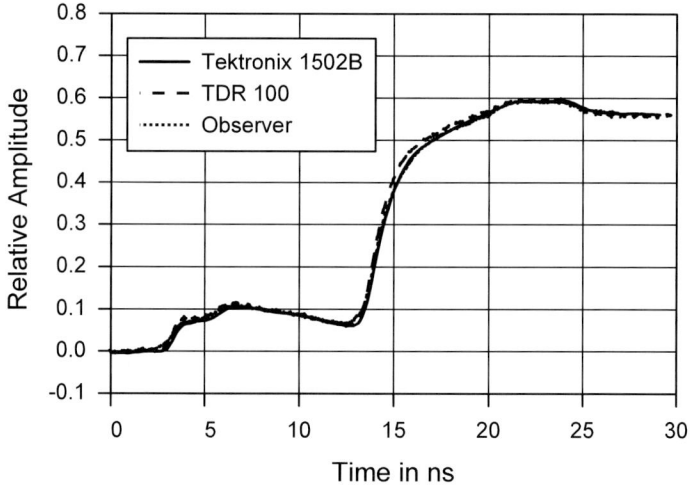

Fig. 14.24. First measurement results of the Observer in comparison to Tektronix cable tester 1502B and Campbell TDR100.

The good agreement between the simple prototype and the other TDR instruments is very promising. The main difficulty with the current design of the Observer is signal integrity issues, e.g., due to the noise generated in the crystal oscillator and in the power supply. Crosstalk disturbs the sampler feedback path and probably also increases the phase noise of the signal generators because of the high sensitivity of the zero cross detectors and the resulting uncertainty of the switching instant. It may be advisable to change the current design from DDS to PLL to avoid the sensitive pulse shaping. Another problem is a slow voltage drift of the TDR step generator, which looks like the charging and discharging of a capacitor, an effect not yet understood. Another concern is the phase stability between clock and step signal at low frequencies. Nevertheless we assume that with the Observer's concept an adequate performance for the desired applications can be achieved.

14.6 Experimental Results

14.6.1 Measuring of Soil Water on a Full-Scale Levee Model

The percolation of levees due to a hydraulic load from a flood or the infiltration of water during a precipitation event is currently being investigated [24]. For this investigation a full-scale levee model is available (cf. Fig. 14.25). The levee is built up homogeneously with uniform sand (grain size 0.2 to 2 mm) and it is based on a waterproof sealing of plastic, so that infiltrating water in the dike will flow to a drain at the landside slope. A basin has been included to provide the opportunity to simulate flood events.

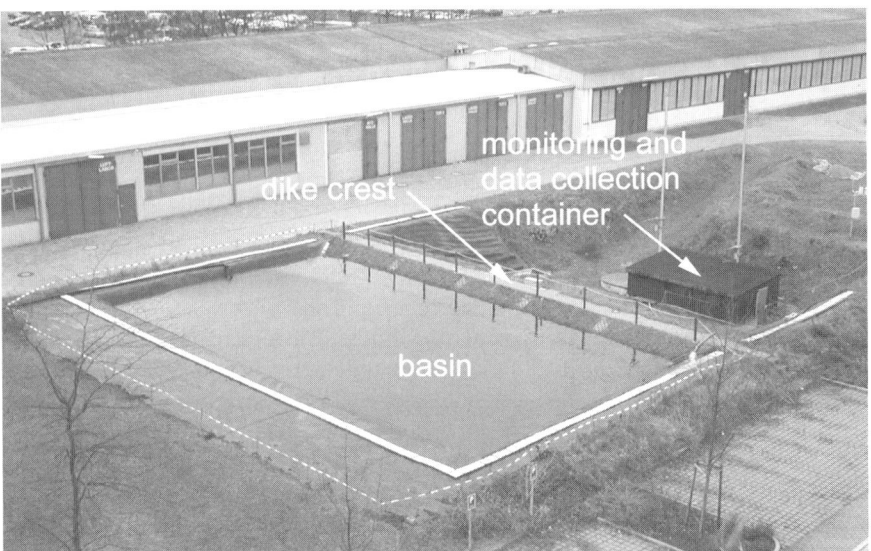

Fig. 14.25. Full-scale levee model at the Federal Waterways and Research Institute in Karlsruhe during a flood simulation test in December 2000 (steady state of seepage condition)

The aim of the project is the quantitative description of the influence of initial soil water distribution within the levee body on the progression of the percolation of water during a flood event. For this, the levee was equipped with 12 vertically installed flat band cables from 1 to 3 m in length, connected from both sides with coaxial cables to a multiplexer, and TDR device in a box on the crest of the levee. The data collection and controlling equipment (PC) of the multiplexer and the TDR device are placed in a measuring container at the toe of the landside slope (Fig. 14.26). With this system the data acquisition time for the whole cross-section is reduced to only 5 minutes. Subsequent processing of the time-dependent data into a spatial distribution, e.g., for the water content, requires several hours on a desktop PC. For the purpose of comparing the results there is the possibility of

measuring soil water profiles with conventional probes for tubes based on TDR. In
addition to these TDR systems, pore water gauges are installed on the sealing at
the base of the model in order to measure the hydraulic head inside the levee body.

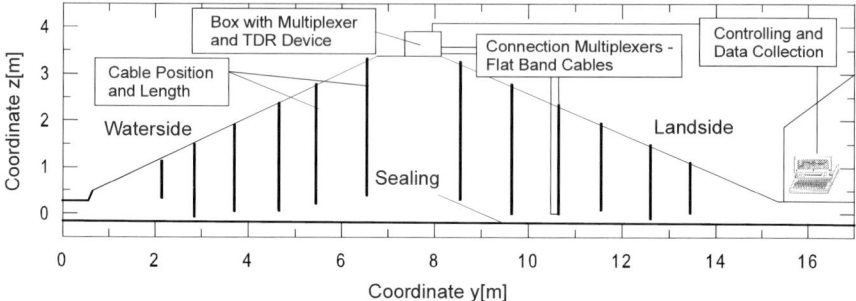

Fig. 14.26. Location of the flat band cables in the cross-section of the levee and schematic
description of the measuring system

Figure 14.27 shows the results of measurements during a flood simulation test
carried out with the levee model in December 2000. The top picture represents a
result measured with a TDR probe for tubes and the picture below is measured
with the flat band cables as transmission lines. The dotted lines represent the
positions of the individual flat band cables together with the volumetric water
content at these locations (see also the gray-scale bar in Fig. 14.27). For better
visualization, the single measurements were interpolated over the observed area in
the cross-section. The dark gray shows wet zones whereas the light gray
represents the more dry zones. In addition, the levels of the hydraulic head are
given in the figures representing the approximate position of the phreatic line
inside the levee body. The independently measured water level in the basin is also
marked.

At first glance one big advantage of the measuring system with the flat band
cables can be recognized. While the flat band cables are able to measure the water
content in the levee body under the water level of the basin, the tubes cannot be
reached in order to make any manual measurements, which require about 1 hour.
The measurement with the flat band cables needs only 5 minutes.

The discrete measurements with the probe for tubes determine a mean value for
the soil water content over a length of about 20 centimeters. The results along the
flat band cables are available in a resolution from 1 to 3 centimeters. In spite of
these differences, the results of both measurements correspond well with the
independently determined phreatic surface. The differences in the water content
between both observation methods vary by about 3 vol %. The porosity of the
levee material – this means the ratio of the pore volume to the whole volume – is
about 37%, but the measured volumetric water content underneath the phreatic
surface is lower. This means that there are air bubbles in the area percolated.

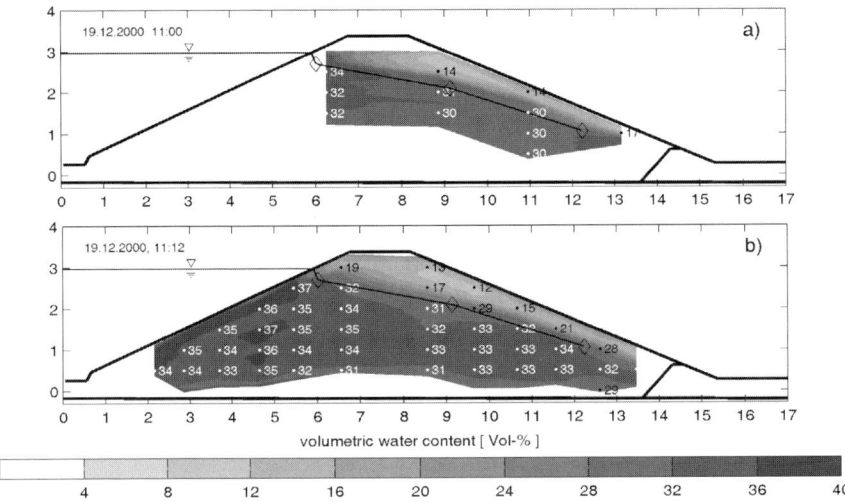

Fig. 14.27. Results of the volumetric water content distribution measured with a probe for tubes (top) and flat band cables as transmission lines (middle) with an independently measured phreatic surface inside the levee body and water level in the basin (steady state of seepage condition during a flood simulation test in December 2000)

Two results of a sprinkler irrigation test are shown in Fig. 14.28. In this experiment a precipitation event which occurs in Karlsruhe around every 100 years was simulated. The resulting rainfall of 140 mm was subdivided over three days each with 9 hours sprinkler irrigation from midnight to 9 a.m. In Fig. 14.28 (top) the initial soil moisture distribution can be seen before the first sprinkler irrigation and Fig. 14.28 (middle) shows the measurement result after the last one. It can be recognized that almost over the whole cross-section the volumetric water content has increased besides the area in the middle of the levee body. This non-uniform distribution of the water content with a declining value from the top down to the base of the levee was verified by independent measurements with tensiometers in the levee body. For the first time, the new TDR method using flat band cables as transmission lines has offered the possibility of a quantitative observation of transient hydraulic processes to a high resolution in both space and time.

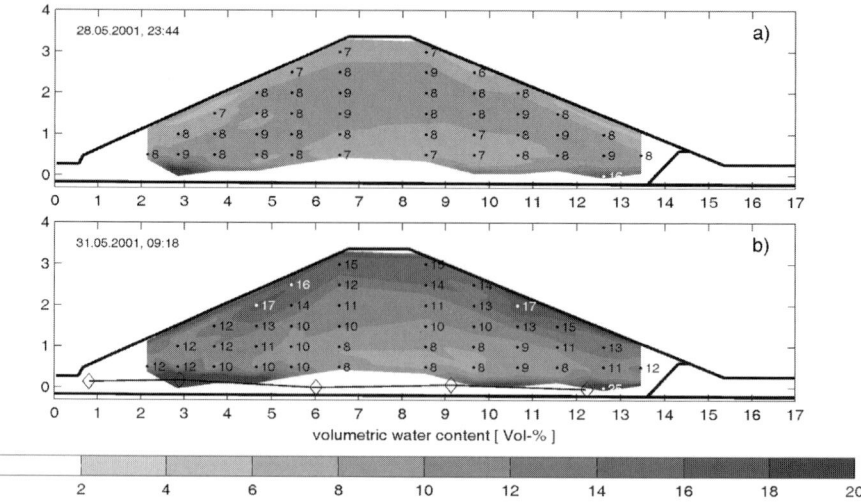

Fig. 14.28. Results of the volumetric water content distribution measured with flat band cables as transmission lines before a simulation of a precipitation event (top) and after the last sprinkler irrigation in May 2001 (middle)

14.6.2 Monitoring of Levees and Dams

At the Institute of Soil Mechanics and Rock Mechanics, University of Karlsruhe, large-scale infiltration tests were carried out on a cable sensor 30 m long, in order to test the suitability of the TDR measuring technique for assessing the water side sealing elements of levees or dams. For this investigation a sensor was inserted into a sand-filled pipe, which had been slit open at the top and perforated at the bottom. Water was sprinkled over the sensor at different positions and at varying widths. At the same time reflection measurements were carried out from both ends of the sensor, in order to improve the resolution accuracy. The aim is not to reconstruct the water content along the cable, as in the levee model at the BAW (cf. previous section), but to directly evaluate the measuring signals to localize any water content anomalies.

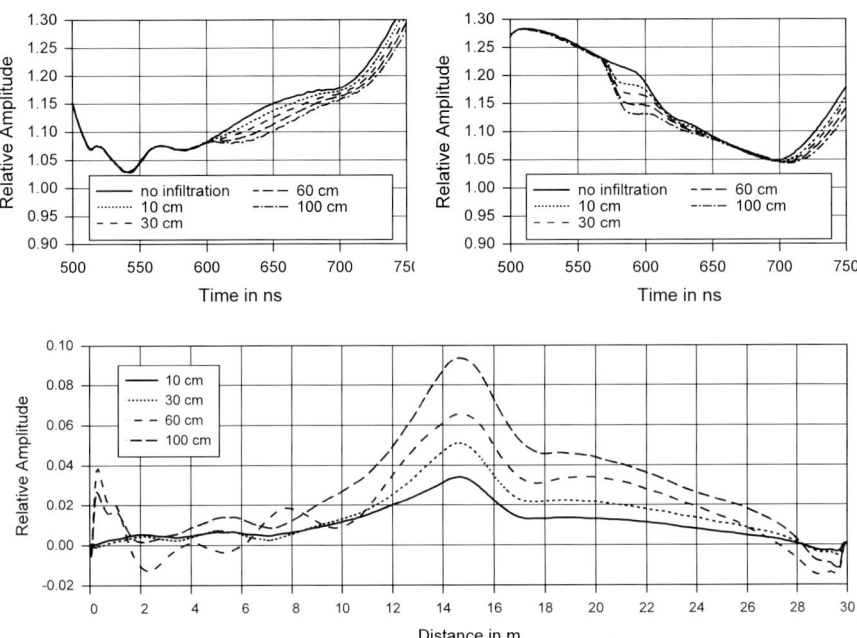

Fig. 14.29. Top: reflection measurements on both sides of a flat band cable 30 m long during an infiltration at $x = 15$ m (i.e., in the middle of the cable) over different widths (10, 30, 60, and 100 cm). Bottom: transformation result of reflection measurements

At the top of Fig. 14.29 the reflection measurements are shown, which are carried out at both ends for infiltration over different widths (10, 30, 60, and 100 cm) at $x = 15$ m. As can be seen from the curve in the left hand diagram (measurement from $x = 0$ m), the area between $x = 0$ m and $x = 15$ m (corresponding to the time 500-600 ns) is hydraulically disturbed by prior infiltrations at $x = 2$ m (510 ns), $x = 5$ m (540 ns), and $x = 10$ m (570 ns). The soil between $x = 16$ m and $x = 30$ m retained its original dry condition. As opposed to the measurement without infiltration, the changes in this section in the reflection signals originate only from the infiltrations at $x = 15$ m. In the left hand diagram several of these changes start at 600 ns; however, in the right hand diagram they already start at 570 ns. This difference results from the slower velocity of the signal in a medium with a higher dielectric constant ,i.e., with a higher soil water content. An exact localization of the soil water anomaly from a one-sided reflection measurement is thus hardly possible. However, if use is made of the information from both reflection measurements by

1. subtracting the initial signal from the time signals (without infiltration),
2. reflecting the corresponding right hand curve and superimposing it with the left one (by addition), and
3. transforming these resulting curves to the cable length,

a very good estimate of the position of the water content anomaly can be obtained (cf. Fig. 14.29, bottom). In comparison to a reconstruction of the water content, this estimate can be calculated quickly and without any great effort. Thus it is very well suited as an evaluation algorithm for a monitoring system. This monitoring system will first be used on the River Rhine in a zoned levee with mineral sealing on the water side and a drain at the toe of the land side slope [25].

14.7 Conclusion

Advanced measurement methods in TDR are presented that focus on three areas. First, a new transmission line technology has been developed which is based on insulated flat band cables and extends the maximum length of TDR transmission lines up to 10 m or more. The electrical parameters of the transmission lines have been investigated with numerical field calculations and incorporated into an electrical equivalent circuit. Second, a new algorithm for determining the water content distribution along transmission lines has been developed. It is based on TDR measurements from both sides of the transmission line and an optimization approach for determining the parameters of the telegraph equations. Third, a novel TDR instrument with binary sampling has been presented. It is a low-cost, high-performance alternative to conventional TDR devices.

The new reconstruction algorithm has been integrated into a soil moisture measurement system for investigating the water transport processes in a full-scale levee model. For this purpose flat band cables were installed in a levee and connected to a cable tester and data acquisition equipment. A simulation of a flood event was carried out. The spatial resolution had an accuracy of about 3 cm and the volumetric water content profiles along the cables were determined with an average deviation of ±2% compared to independent measurements. Due to the data acquisition time for the TDR data of about 5 minutes for the complete cross-section of the levee, fast-running water transport processes could be monitored for the first time at this spatial resolution. Future improvements of the measurement system will include optimized processing of the TDR data and better soil specific calibration.

References

1.   Topp GC (2001) Electromagnetic wave measurements of soil water content: a state-of-the-art. In: Kupfer K, Hübner C (eds) Proceedings of the fourth international conference on electromagnetic wave interaction with water and moist substances, Weimar, Germany, May 13-16, pp 327-335
2.   Topp GC, Davis JL, Annan AP (1980) Electromagnetic determination of soil water content: measurement in coaxial transmission lines. Water Resour Res 16:574-582
3.   Birchak JR, Gardner CG, Hipp JE, Vicor JM (1974) High dielectric constant microwave probe for sensing soil moisture. Proc IEEE 62:93-98

4. Dobson MC, Ulaby FT, Hallikainen MT, El-Rayes MA (1985) Microwave dielectric behaviour of wet soil, II Dielectric mixing models. IEEE Trans. Geosci, Remote Sensing 23:35-46
5. Heimovaara TJ, Bouten W, Verstraten JM (1994) Frequency domain analysis of time domain reflectometry waveforms: 2. A four-component complex dielectric mixing model for soils. Water Resour Res 30:201-209
6. Heimovaara TJ (1994) Frequency domain analysis of time domain reflectometry waveforms: 1. Measurement of the complex permittivity of soils. Water Resour Res 30:189-199
7. Dalton FN, van Genuchten MTh (1986) The time domain reflectometry method for measuring soil water content and salinity. Geoderma 38:237-250
8. Huebner C, Brandelik A (2000) Distinguished problems in soil and snow aquametry. In: Baltes H, Goepel W, Hesse J (eds) Sensors update, vol 7, Wiley-VCH, Weinheim, pp 317-340
9. Huebner C, Brandelik A (2000) Near subsurface moisture sensing. In: Nguyen C (ed) Subsurface sensing technologies and applications, Proc SPIE 4129:88-96
10. Nyfors E,Vainikainen P (1989) Industrial microwave sensors. Artech House, Norwood, MA
11. Hook WR, Livingston NJ, Sun ZJ, Hook PB (1992) Remote diode shorting improves measurement of soil water by time domain reflectometry. Soil Sci Soc Am J 56:1384-1391
12. Schlaeger S (2002) Inversion von TDR-Messungen zur Rekonstruktion räumlich verteilter bodenphysikalischer Parameter. PhD thesis, University of Karlsruhe, Institute of Rock Mechanics and Soil Mechanics
13 He S, Romanov VG., Stroem S (1994) Analysis of the Green function approach to one-dimensional inverse problems. Part II: Simultaneous reconstruction of two parameters. J Math Phys 35:2315-2335
14. Fletcher R, Reeves CM (1964) Function minimization by conjugate gradients. Comput J 7:163-168
15. Heimovaara TJ, Bouten W (1990) A computer-controlled 36-channel time domain reflectometry system for monitoring soil water content. Water Resour Res 26:2311-2316
16. Monett MR (2004) http://www3.sympatico.ca/add.automation/sampler/intro.htm Feb 12
17. Steele R (1975) Delta modulation systems. Pentech Press, London
18. Mulvey J (1970) Sampling oscilloscope circuits. Tektronix, Beaverton, OR
19. Laug OB, Souders TM, Flach DR (1992) A custom integrated circuit comparator for high-performance sampling applications. IEEE Trans Instrum Meas 41:850-855
20. Souders TM, Hetrick PS (1988) Accurate RF voltage measurements using a sampling voltage tracker. IEEE Trans Instrum Meas 38:451-456
21. DeJaeger F (1952) Delta modulation – a new method of p.c.m. transmission using the 1 unit code. Philips Res Rep 8:442-446
22. Steele R (1980) SNR formula for linear delta modulation with band-limited flat and RC-shaped Gaussian signals. IEEE Trans Commun 28:1977-1984
23. Pohlmann KC (1995) Principles of digital audio. McGraw-Hill, New York
24. Scheuermann A, Schlaeger S, Hübner C, Brandelik A, Brauns J (2001) Monitoring of spatial soil water distribution on a full-scale dike model. In: Kupfer K, Hübner C (eds) Proceedings of the fourth international conference on electromagnetic wave interaction with water and moist substances, Weimar, Germany, May 13-16, pp 343-350
25. Scheuermann A, Brauns J (2002) Die Durchfeuchtung von Deichen – Modellversuche und Analyse. In: DGGT 12, Donau-Europäische-Konferenz, Proceedings, Passau, 197-200

15 Simulations and Experiments for Detection of Moisture Profiles with TDR in a Saline Environment

Klaus Kupfer, Eberhard Trinks

Materialforschungs- und -prüfanstalt an der Bauhaus Universität Weimar
Amalienstr. 13, D-99423 Weimar, Germany

15.1 Introduction

In Germany hazardous and non-recyclable wastes are permanently disposed of in salt mines. The long-term isolation from the outer environment (biocycle) should be realized by geotechnically sealed barriers like sealed shafts, road ways, and boreholes. Therefore clay and bentonite are used as sealing materials. By the state of the art the long-term workability and reliability of bentonite sealings are insufficiently proofed in the case of fluid ingress or water inrush. This means that bentonite sealings under the influence of moist saline conditions not yet been satisfactorily investigated. The measurement of moisture distribution in structures designed for free monitoring after site closure is stringently necessary for the indefinable trend of fluid ingress and for data recording of long-term security analyses. Because of the absence of long-term experiences, measurement control under the influence of disturbances is imperative over a long period of observation in test barriers. Such measurement monitoring not only will be required for quality assurance during the final building inspection, but also it will determine the discharge over the monitoring period and shorten it essentially. Measurements showed for this application that the accuracy and reliability of most moisture measurement methods are limited or even unacceptable. A time domain reflectometry (TDR) moisture measurement system using flexible electrodes will be presented here for bentonite as the sealing material in a saline environment. Influences of chemically bound water, ionic conductivity, pressure, bulk density, and temperature should also be investigated in detail. Electromagnetic field simulation of the sensor and of the surrounding material will help to optimize the cable sensor, to increase the accuracy and reliability of the measurement method. By the application of TDR with a cable sensor it was possible to record the moisture profiles across a cable sensor in a trough with different moist chambers, and to determine the moisture front in a small bentonite layer in a pressure test stand at pressures > 40 bar [1, 2].

15.2 Fundamentals of Time Domain Reflectometry

TDR served originally for the detection of cable defects (cable radar). It was the first method which was applied for moisture determination in soils. Parallel-wire lines as sensors were carried out as two- or three-line forks. The method is based on the detection of the travel time of a pulse at the end of an open line which is surrounded by a moist material [3–6].

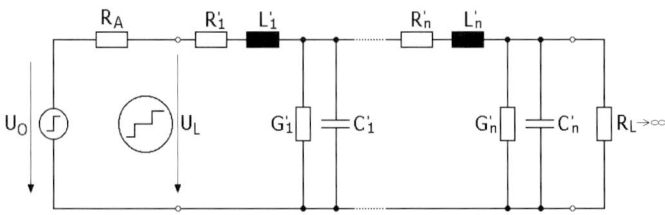

Fig. 15.1. Equivalent circuit of an open cable for determination of travel time

Figure 15.1 shows an equivalent circuit of an open cable which consists of n partial cable pieces. A cable test set-up at the input of the circuit generates a pulse with a rise time of 200 ps. The pulse response and the determination of pulse travel time can be evaluated at the resistor R_A using appropriate software. The open sensing lead was realized as a three-conductor-band-cable where both the outer conductors were grounded (Fig. 15.7). The connection to the cable test set-up was realized with a non-matched coaxial cable. The experimental set-up consisting of a cable tester, a cable sensor, and a computer is shown in Fig. 15.2.

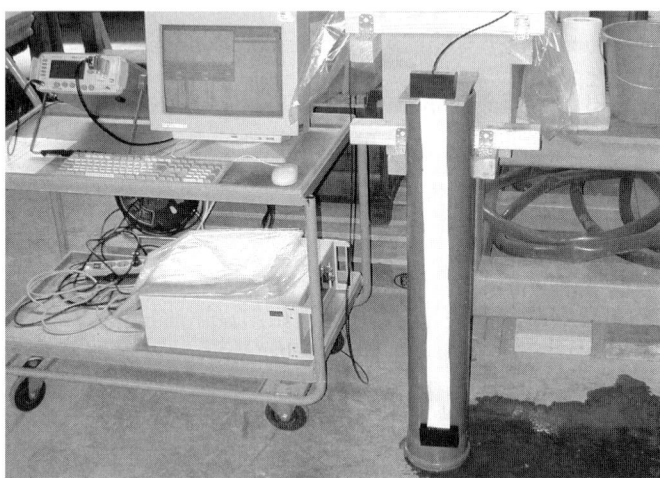

Fig. 15.2. Measuring set-up – cable tester, band cable, and computer

The propagation of an electromagnetic wave is characterized by the propagation constant γ which is determined by the attenuation constant α and the phase constant β. Equation (15.1) shows the propagation constant dependent on the cable parameters:

$$\gamma = \alpha + j\beta = \sqrt{(R' + j\omega L')(G' + j\omega C')} \approx \sqrt{j\omega L'(G' + j\omega C')} \qquad (15.1)$$

At a zero-loss line the loss components of the surrounding material influence the values of capacity and conductivity, each per unit length. Using the loss factors for dielectrics $\tan \delta_e$ and for magnetics $\tan \vartheta_m$,

$$\tan \delta_e = \frac{G'}{\omega C'} \qquad (15.2)$$

$$\tan \vartheta_m = \frac{R'}{\omega L'} \qquad (15.3)$$

the attenuation and phase constants in Eq. (15.1) can be written as

$$\alpha = \omega \sqrt{L'C'} \sqrt{\frac{1}{2}\left[\sqrt{\left(1 + \tan^2 \delta_e\right)\left(1 + \tan^2 \vartheta_m\right)} + \tan \delta_e \tan \vartheta_m - 1\right]} \qquad (15.4)$$

$$\beta = \omega \sqrt{L'C'} \sqrt{\frac{1}{2}\left[\sqrt{\left(1 + \tan^2 \delta_e\right)\left(1 + \tan^2 \vartheta_m\right)} + 1 - \tan \delta_e \tan \vartheta_m\right]} \qquad (15.5)$$

The following approximations are valid for the penetration of an electromagnetic wave through a dielectric with high losses in a big volume: $\tan \delta_e \gg \tan \vartheta_m$, $\mu_r = 1$, and $\tan \vartheta_m \ll 1$; $(L'C')^{1/2} = (\varepsilon_0 \varepsilon_r' \mu_0 \mu_r')^{1/2}$. The Eqs. (15.4 and 15.5) can be derived in the following form:

$$\alpha = \omega \sqrt{\varepsilon_0 \mu_0} \sqrt{\frac{\varepsilon_r'}{2}\left[\sqrt{1 + \tan^2 \delta_e} - 1\right]} \qquad (15.6)$$

$$\beta = \frac{\omega}{v} = \omega \sqrt{\varepsilon_0 \mu_0} \sqrt{\frac{\varepsilon_r'}{2}\left[\sqrt{1 + \tan^2 \delta_e} + 1\right]} \qquad (15.7)$$

A step pulse, which is given at the input of the cable, travels a distance l_L to the end of the waveguide and back. This allows the determination of pulse travel time of the step response using Eq. (15.7) and $c = 1/(\varepsilon_0 \mu_0)^{1/2}$.

$$t_p = \frac{2l_L}{v} = \frac{2l_L}{c} \sqrt{\frac{\varepsilon_r'}{2}\left[1 + \left(1 + \tan^2 \delta\right)^{\frac{1}{2}}\right]} \qquad (15.8)$$

The step responses dependent on the surrounding air and on the sand water ratio is depicted in Fig. 15.3. With an increasing sand-water ratio the permittivity of the mixture and the pulse travel time measured at the embedded cable sensor will be increased. The pulse travel time (see Eq. (15. 8)) can be determined by double differentiation at the reversal points (Fig. 15.4). The second derivative intersects the neutral axis at the reversal points.

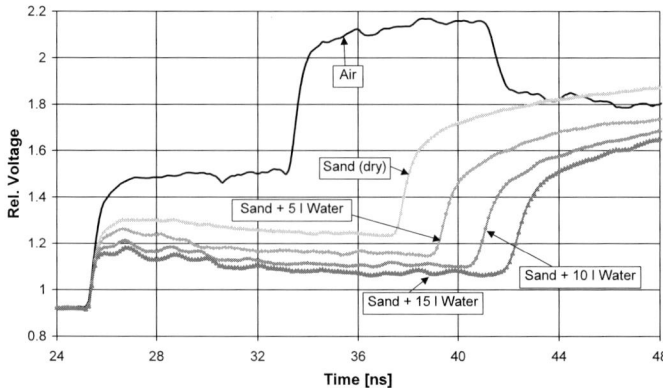

Fig. 15.3. Pulse travel time dependent on the sand-water ratio of a moist sand mixture measure with a cable sensor of 2 m length

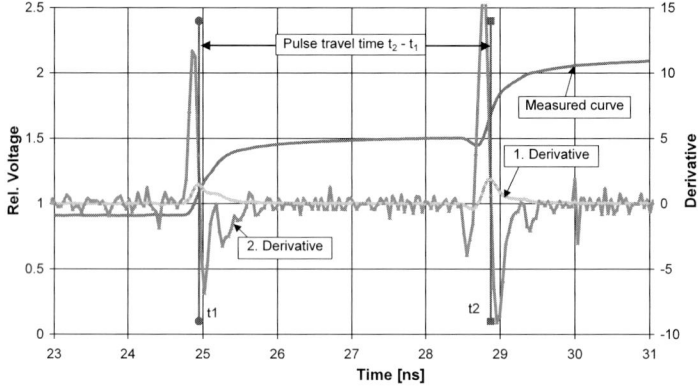

Fig. 15.4. Determination of the pulse travel time using the second derivative

The loss factor is characterised by the tan δ (see Fig. 15.10b):

$$\tan\delta = \frac{\varepsilon_r^{''}(\omega) + \dfrac{\sigma(\omega)}{\omega_0\varepsilon_0}}{\varepsilon_r^{'}(\omega)}$$

with the following approximations:

$$\tan\delta \approx \frac{\sigma(\omega)}{\omega_0\varepsilon_0\varepsilon_r^{'}} \quad \text{for frequencies} < 3\ \text{GHz} \qquad (15.9)$$

$$\tan\delta \approx \frac{\varepsilon_r^{''}(\omega)}{\varepsilon_r^{'}(\omega)} \quad \text{for frequencies} > 3\ \text{GHz}$$

where σ is the conductivity.

It is shown in Fig. 15.10b and Eq. (15.9) that the losses of conductivity dominate at lower frequencies < 3 GHz; the dielectric losses dominate at frequencies > 3 GHz. This approximation is valid at a low concentration of salt content. TDR pulses with short rise times and materials with high losses (salt solutions) require that losses of conductivity and dielectric losses have to be considered.

The attenuation constant α can be derived in the form Eq. (15.10), where $Z_0 = (\mu_0 / \varepsilon_0)^{1/2} = 120\pi \ \Omega \approx 377 \ \Omega$ is the intrinsic wave impedance:

$$\alpha = \frac{\frac{1}{2}\sqrt{\frac{\mu_0}{\varepsilon_0}}\left(\omega\varepsilon_0\varepsilon_r'' + \sigma\right)}{\sqrt{\frac{\varepsilon_r'}{2}\left[1+\left(1+\tan^2\delta\right)^{\frac{1}{2}}\right]}} \qquad (15.10)$$

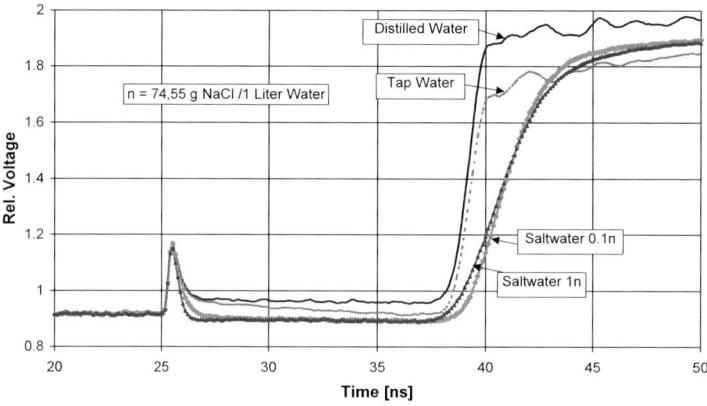

Fig. 15.5. Pulse rise time of step response dependent on the salt content of the water

A pulse will be reduced along a cable by the factor $\exp(-\alpha \times length)$. The decrease of amplitude can be used to determine the attenuation constant for a test cable of defined length. The reflected voltage U_2 can be measured at the source resistor R_A (Fig. 15.1):

$$\frac{U_2}{U_1} = e^{-\alpha 2 l_L} \qquad (15.11)$$

The conductivity can be obtained by combining of Eq. (15.10 and 15.11):

$$\sigma = \frac{1}{l_L}\sqrt{\frac{\varepsilon_0}{\mu_0}}\ln\frac{U_1}{U_2}\sqrt{\frac{\varepsilon_r'}{2}\left[1+\left(1+\tan^2\delta\right)^{\frac{1}{2}}\right]} - \omega\varepsilon_0\varepsilon_r' \qquad (15.12)$$

The pulse rise time of step response dependent on the salt content is shown in Fig. 15.5. With increasing salt content of solution the pulse rise time will be extended.

15.3 Simulation of the Measurement Set-up

Calibrations of bentonite are relatively time consuming, because of the simultaneous changing of several material parameters such as moisture, density, and ionic conductivity. Small changes of the sensor for adaptation to the material involve costly design and testing; often the principle of trial and error dominates. With an EM-field-simulator it is possible to investigate specific material properties and to optimize sensors in relatively short times. Here, simulations of the cable and surrounding materials were performed using Ansoft-HFSS (High Frequency Structure Simulator), based on a finite element method and operating in the frequency domain.

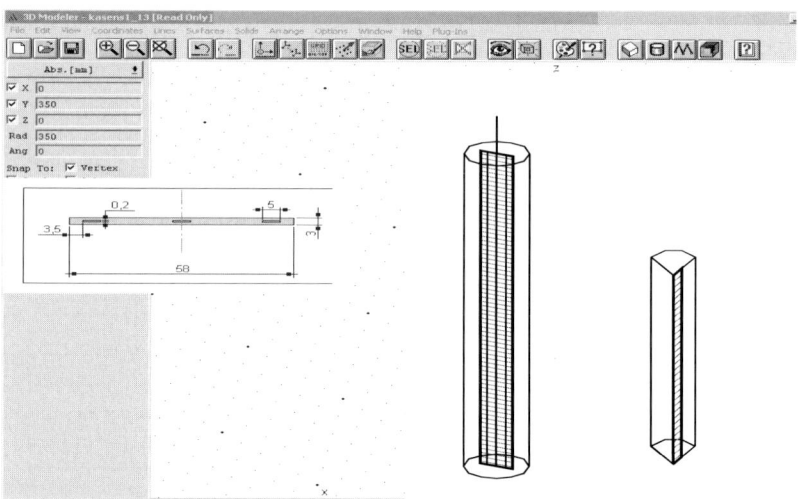

Fig. 15.6. Design of the band cable on 3D-Modeler

The cable sensor – as a polyethylene cable – has been designed on the 3D-Modeler using the original dimensions (Fig. 15.6).

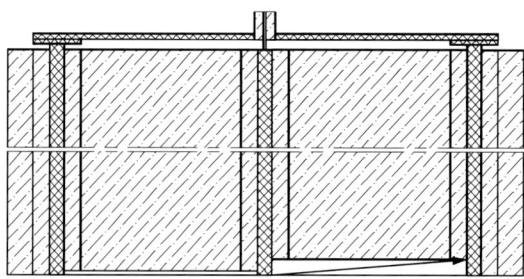

Fig. 15.7. Coaxial connector and load resistances of the cable as vector current flow lines with orthogonal position

It will be surrounded by an outer cylinder which can be filled with a lossy dielectric. To save calculation time the cable and the cylinder were cut into length of two or four parts and the symmetry mode was used. A coax connector as the input (port) was adapted to the band cable without any final mismatching. The ends of the band cable were connected with load resistances of 1 MΩ (Fig. 15.7). The materials of the cable sensor and the data of the lossy dielectric were entered in the data-input of the HFSS Material Manager and assigned to the related objects.

The conductors, ports, symmetry planes, load resistances, and radiation boundaries are entered in the HFSS 3D-Boundary/Source-Manager. The starting parameters of the calculation are defined in the HFSS-Solution Setup (Tab. 15.1).

Table 15.1: Start parameters for calculation

Start frequency	1 MHz	Required passes	8
Stop frequency	12,5 GHz	Max. delta S[1]	0.02
Single frequency	6 GHz	Error toleranz	0.2
Number of steps	2999	Max. solutions	100
Interpolating sweep		Tetrahedra-refinement	20
Mesh refinement	adaptive		

[1]) Amount of relative change of all S-Parameters

The resulting S-parameters can be plotted versus frequency or on a Smith chart in the HFSS-Matrix Plot. The display of different fields, vectors, or phase animations can be realized using the HFSS-Field Display. Figure 15.8 shows the magnitude versus frequency of the TDR system surrounded by different materials with constant parameters ε_r' and $\tan\delta$.

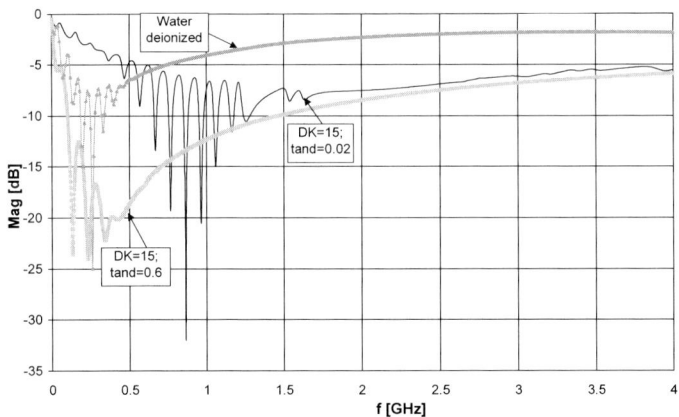

Fig. 15.8. Magnitude via frequency of the TDR system surrounded by different lossy dielectrics

The calculated frequency-dependent functions (Fig. 15.8) will be sent to the HFSS-Spice Simulator. There the step responses will be calculated using FFT and circuit calculation with a transient analysis. The model of an open band cable is shown as a two-terminal network. A pulse with a rise time of 200 ps is given at the input. Figure 15.9 shows both the step responses of the TDR system surrounded by materials with relative low and high permittivity ($\varepsilon_r' = 4$ and 15), and also relative low and high losses ($\tan\delta = 0.1$ and 1). These separations can be realized using one simulation only. The change of the ratio $\varepsilon_r'/\tan\delta$ from 4/0.1 to 15/0.1 shows a parallel shift in the step responses to a higher pulse travel time; the rise time remains constant. At an increasing of the loss factor from 0.1 to 1 the gradient of the curve is strong reduced. The modification of the ratio $\varepsilon_r'/\tan\delta$ from 4/1 to 15/1 also shows a parallel shift of the curves. Compared to the first case, the rise time is essentially increased.

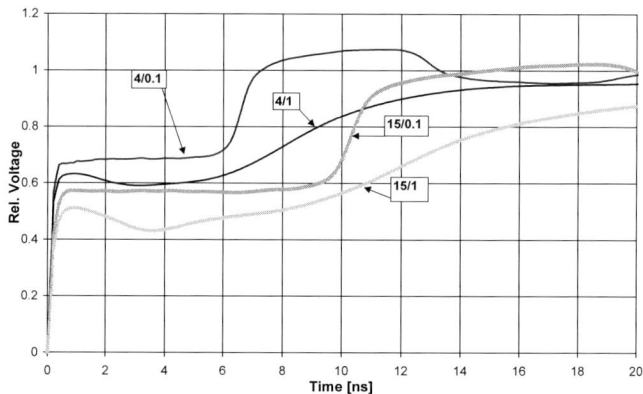

Fig. 15.9. Step responses of the TDR system surrounded by different lossy dielectrics (Parameters: ε_r' /$\tan\delta$)

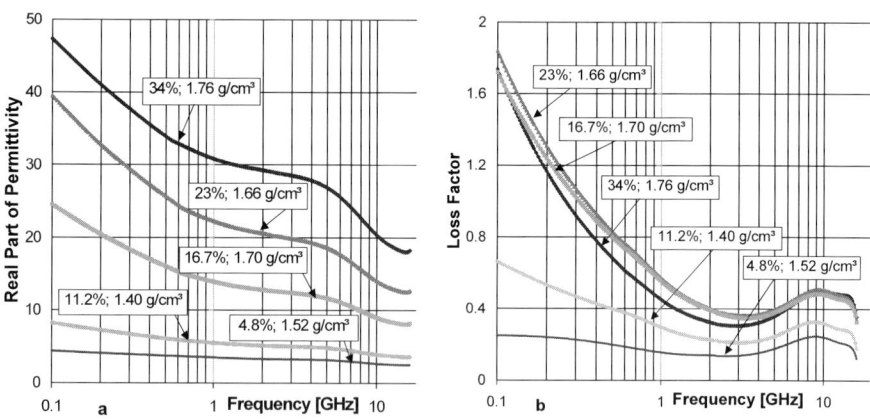

Fig. 15.10. Real part of permittivity (**a**) and loss factor of bentonite (**b**) versus frequency at different moisture content

The dielectric material properties of bentonite (ε_r' and $\tan\delta$) dependent on frequency will be determined using the network analyzer in a frequency range of 1 MHz to 15 GHz (Fig. 15.10). The frequency-dependent material characteristics of ε_r' and $\tan\delta$ are entered into the HFSS-Material Manager.

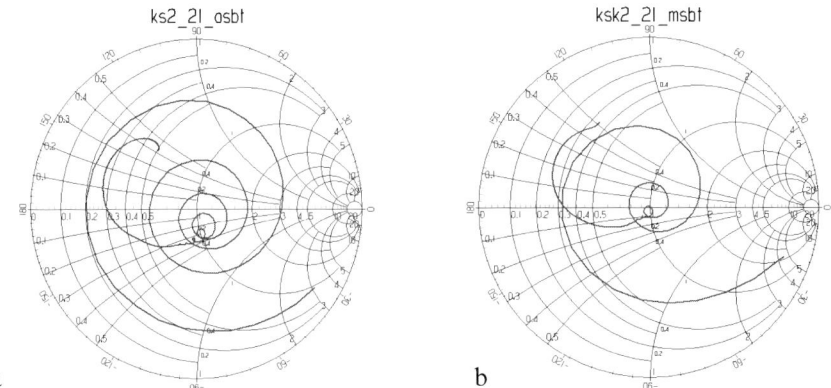

Fig. 15.11. Calculated Smith chart of a bentonite and tap-water mixture (**a**) and of a bentonite and salt-water mixture (250g/l) (**b**)

A comparison of the calculated Smith charts of a bentonite and tap-water mixture and a mixture of bentonite and salt solution is shown in Figs. 15.11a and 15.11b respectively. The salt solution has a strong attenuation in the lower frequency range.

The calculated curves of the step responses of the pulse travel time of different moisture for a bentonite and tap-water mixture and for a bentonite and salt-solution mixture are shown in Figs. 15.12a and b. Based on the different values of pulse travel time the calibration curves dependent on the moisture content or on the content of salt solution can be depicted. By comparison the step responses of a tap-water mixture show shorter rise times than that of a salt-solution mixture.

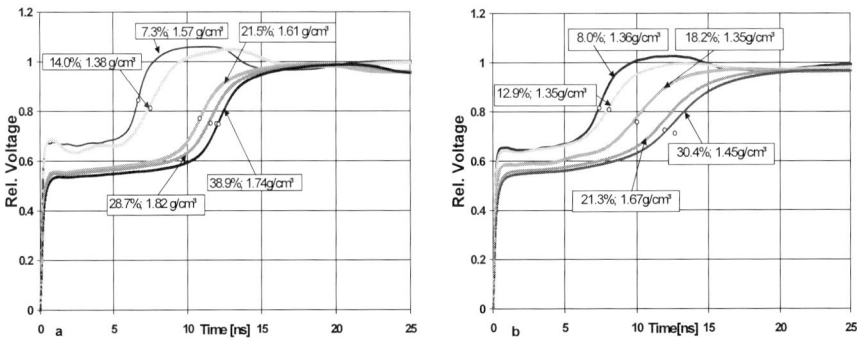

Fig. 15.12. Calculated step responses for (**a**) bentonite-sand and tap-water mixtures of different moisture content and for (**b**) bentonite-sand and salt-solution mixture

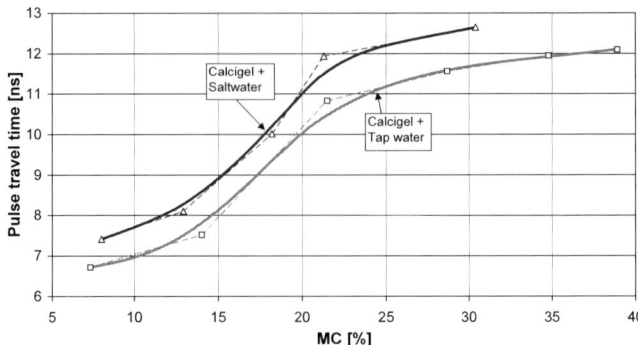

Fig. 15.13. Calibration curves of pulse travel time dependent on the moisture content (MC) / content of salt solution

The calibration curves (Fig. 15.13) show in the range of 7 – 12 % MC / content of salt solution a relativ slow rise, which is determined by the bound water of intermediate layers of bentonite. Free water appears in the range of 12 – 21 % MC, above 21 % the saturation range is reached. The calibration curve of the bentonite-sand and salt-solution mixture has an approximately constant distance to the calibration curve which was recorded for a bentonite-sand and tap-water mixture. The reason is that the salt content which was in the mixture after reference drying was not calculated separately from the mixture.

A very important safety-related problem can occur during the installation of cable sensors in bentonite barriers. A thin air layer and infiltrating water can produce a water leading layer over the band cabel and may destroy the sealing. Compared to a sensor which is fully covered by bentonite (MX 80), Fig. 15.14 shows the changes in the measuring signals, by introducing different layers filled with air, water, and salt solution. Air layers 0.5 and 1 mm thick indicate a shorter pulse travel time compared to layers which are filled with water or salt solution. The installation of cable sensors has to be carried out very carefully, to reduce these problems.

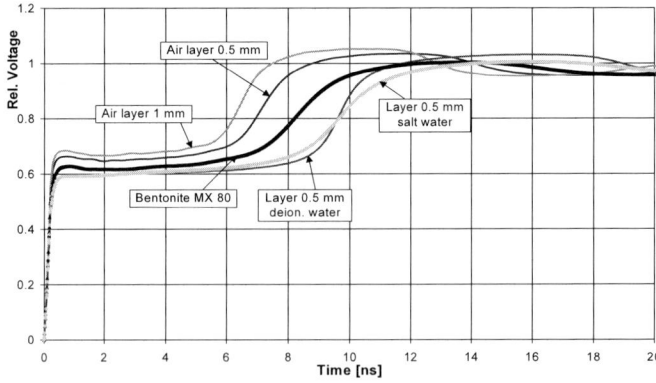

Fig. 15.14. Influence of thin layers filled with air, water or salt solution on the pulse travel time

15.4 Calibrations

A bentonite-sand mixture with different values of MC or content of salt solution was produced to record the pulse travel time of the cable sensor dependent on the MC or on the content of salt solution (see Figs. 15.15 and 16).

Fig. 15.15. Embedded cable sensor in a Calcigel-sand mixture

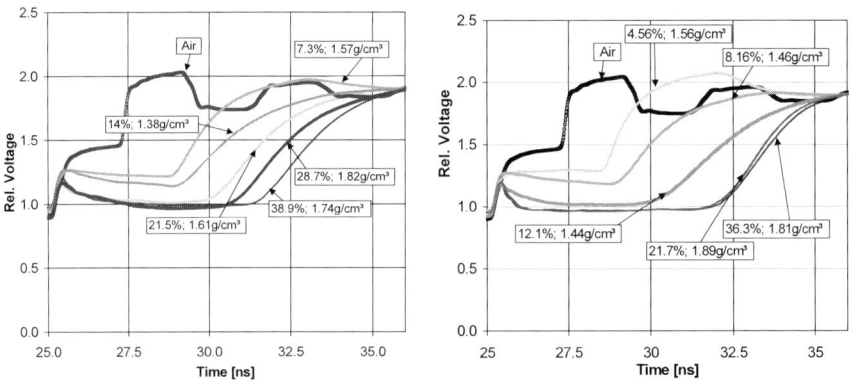

Fig. 15.16. Measured pulse travel time of a Calcigel-sand mixture dependent on moisture content (**a**) and content of salt solution (**b**)

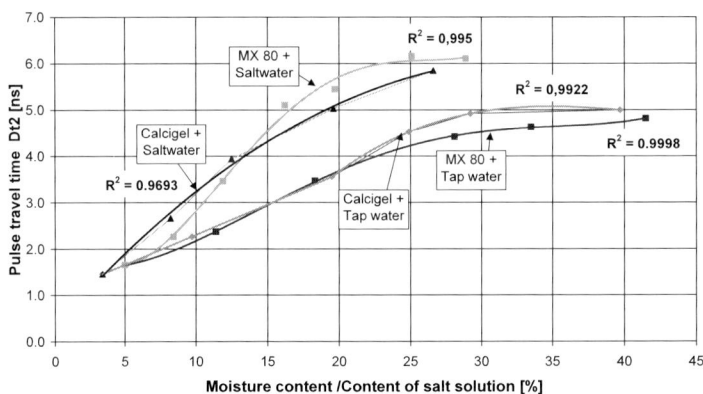

Fig. 15.17. Calibration curves of pulse travel time, measured with a cable sensor, dependent on MC or content of salt solution at a Calcigel-sand-mixture and a MX 80-sand mixture

For the comparison of curves and for the determination of pulse travel time of different MC the step responses has to be normalized to a standardized value of relative voltage (i. e. 1.9) at a steady state (Time > 50 ns) (see Fig. 15.16). The amplitude values of step responses has to be taken into account during the determination of ionic conductivity according to Eq. (15.12). The step responses of a Calcigel-sand-water-mixture show a wider dynamic range (Fig. 15.16a) compared to the same mixture dependent of the content of a salt solution. At high contents of salt solution (> 20%) the step responses only show small right shift or indicate an inverse behavior. The pulse travel time calibration curves of calcium- and natrium bentonite dependent on MC and content of salt solution (Fig. 15.17) show the same characteristics as depicted in Fig. 15.13. In the diagram Fig. 15.17 the pulse travel time Dt_2 shown at the y-axis is the difference between the pulse travel time measured with a cable sensor in the material and in air $Dt_2= t_p$ (mat) - t_p (air).

15.5 Moisture Profiles

A trough, subdivided into eight sections, was used to investigate moisture profiles (Fig. 15.18). The cable was installed vertically to avoid water accumulating on its surface

Fig. 15.18. Trough with cable sensor and divisions for dielectrics (seramis) of different MC

During the experiment the section 1 was filled with seramis (MC = 0.55%) - as result a curve with a pulse form 1 was measured. After that the section 2 was filled etc. The curve labeled 6 displays the actual state; curves 1–5 are the results from earlier measurements. The change in the pulse form (Fig. 15.19) shows the moist packet moving across the trough.

.

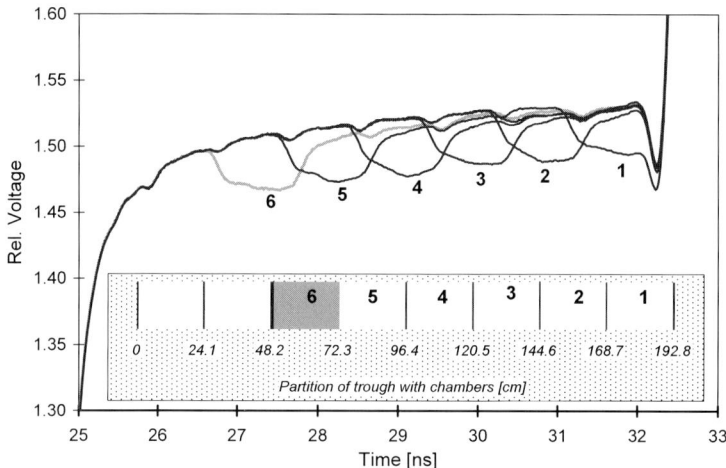

Fig. 15.19. The change in pulse form shows the transportation of a packet of seramis (0.55 % MC) across the trough

Figure 15.20 shows the change of pulse form dependent on filling the chambers with material of different MC. The pulse travel times and their differences Δt depend proportionally on MC.

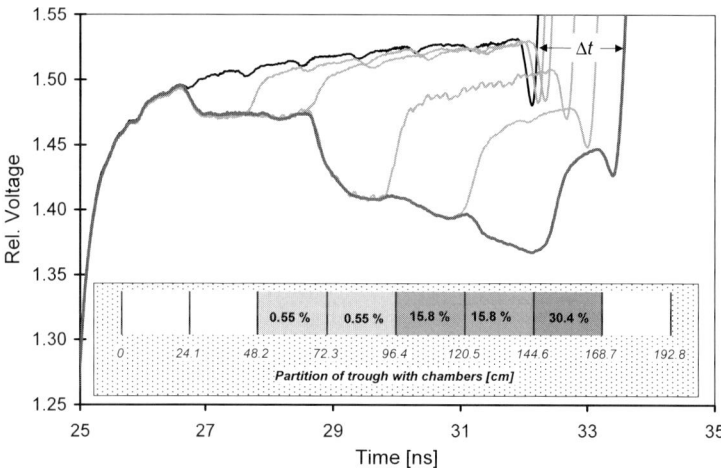

Fig. 15.20. Pulse form dependent on filling the chambers with material of different MC

15.6 Bentonite in a Pressure Test Stand

Investigations influenced by pressure are necessary to simulate the conditions in the depths of a mine. A pressure test stand (Fig. 15.21) was used to detect moisture profiles and to investigate the sealing of bentonite.

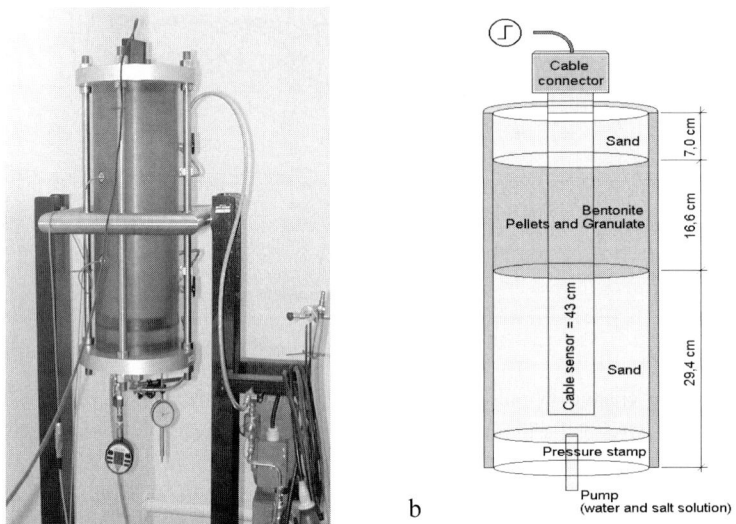

a   b

Fig. 15.21. Detection of moisture profiles in bentonite using a pressure test stand (**a**); and the schematic representation of filling (**b**)

The pressure vessel, which is open at the upper end, is filled from the bottom with deionized water or salt solution. The schematic representation shows the material layers and the embedding of the cable sensor. The experiments were carried out at pressures up to 40 bar. This pressure corresponds to a depth in salt mines of 300-400 m.

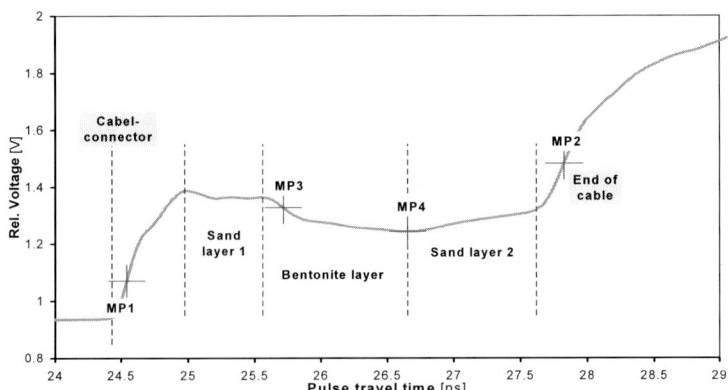

Fig. 15.22. Truncation of locations in bentonite layers and measuring points in the pulse form

The gradual progress of the moisture front in the bentonite was detected using TDR with a cable sensor. The truncation of locations in bentonite layers can be determined using reversal points in the pulse form (Fig. 15.22). The pressure is active at the end of the cable in the first instance. The step response starts at the measuring point MP1 at the beginning of the cable connector. The voltage is increased up to the beginning of the upper sand layer 1 and then remains relatively constant. The increase of ε_r' is connected with the increase of pulse travel time, and the drop in voltage is shown over the bentonite layer between MP3 and MP4. Thereafter that the voltage is increased up to the end of the cable MP2 (cf. the schematic representation in Fig. 15.21b).

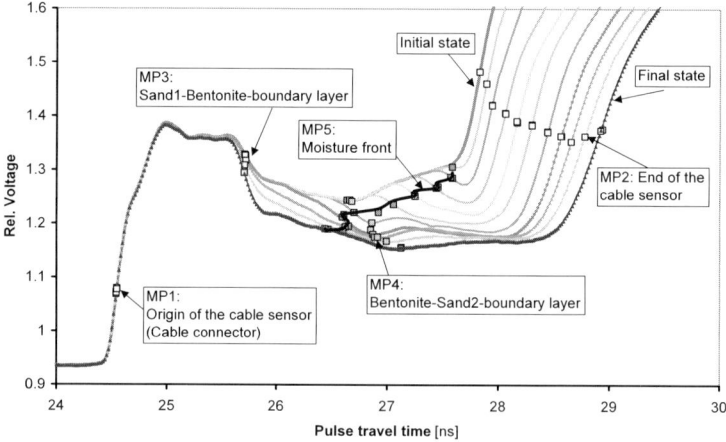

Fig. 15.23. Progress of moisture front in bentonite in the pressure test stand

Figure 15.23 indicates the progress of the moisture front in bentonite as a result of many TDR measurements over 200 days. The upper curve shows the initial state and the bottom curve the final state of measurement. The water comes from the pump (Fig. 15.21a) at the end of the cable sensor (MP2), and penetrates the sand layer very quickly (1 day) up to the sand2-bentonite boundary layer (MP4). The water can penetrate into the bentonite only very slowly. The bentonite swells with the injected water and produces a sealing effect (measuring time 2 months). The pressure is increased in steps up to 40 bar. Higher pressures produce better barrier properties in bentonite. The dot series at MP5 show the moisture front first in the sand layer and then, after penetrating the boundary MP4, that in the bentonite layer. The widening of the group of curves behind the moisture front between MP4 and MP3 results from density changes and not from water leakage along the cable sensor. Salt solutions were used to get practical knowledge for applications. The swelling pressure of bentonite is decreased and the front of the solution penetrates the bentonite layer faster than that from using water.

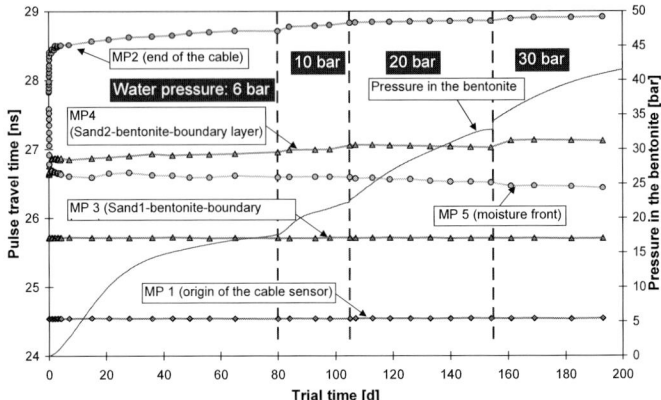

Fig. 15.24. Pulse travel time at increasing pressure during the trial time: at the origin of the cable sensor MP1; at the sand1-bentonite boundary layer MP3; at the sand2-bentonite-boundary layer MP4; at the end of the cable MP2; and the change of moisture front at MP5

The pulse travel time at increasing pressure during the trial time is shown in Fig. 15.24. At the origin of the cable sensor MP1, at the sand1-bentonite boundary layer MP3, at the sand2-bentonite boundary layer MP4, and at the end of the cable MP2 the values of the pulse travel time remain relatively constant. The change of moisture front is depicted at MP5. The pressure in the bentonite layer is increased continuously up to 41bar.

15.7 Summary

The chapter describes the recording of moisture profiles in sand and bentonite using the TDR method. Several influences vary during the experiments, such as moisture, density, pressure, salt content, and temperature. EM-field simulations have been used to virtually change some selected characteristics of the sensor and the material. Essential topics are the calibration of the TDR measurement set-up recording the moisture profiles in bentonite sealing systems, the evidence of stability, and the repeatability of the measurements.

Acknowledgement

The authors grateful acknowledge the project agency for Water Technology and Waste Management of the BMBF and BMWA for support of the project 02C0800.

References

1. Kupfer K, Trinks E, Schäfer Th, Keiner Th (2002) Theoretische und experimentelle Untersuchungen zur Detektion von Feuchteprofilen mittels TDR-Messleitung. 11. Feuchtetag Weimar, S.23–35
2. Kupfer K, Trinks E (2003) Measurement method for long-term monitoring of geotechnically sealed barriers in salt mines. In: 5th conference on electromagnetic wave interaction with water and moist substances, Rotorua, Mar 2003, pp 181189
3. Ferre PA, Topp GC (2000) Time-domain reflectometry sensor techniques for soil water content measurements and electrical conductivity measurements. In: Sensor update, vol 7, Wiley-VCH,   Weinheim, pp 277–300
4. Stacheder M, Koehler K, Fundinger R (2000) New time-domain reflectometry sensors for water content determination in porous media. In: Sensor update, vol 7, Wiley-VCH, Weinheim, pp 301–316
5. Hübner Ch, Brandelik A (2000) Distinguished problems and sensors in soil and snow aquametry. In: Sensor update, vol 7, Wiley-VCH, Weinheim, pp 317–340
6. Scheuermann A, Schlaeger St, Huebner Ch, Brandelik A, Brauns J (2001) Monitoring of the spatial soil water distribution on a full-scale dike model. In: Proceedings of the 4th conference on electromagnetic wave interactions with water and moist substances, Weimar, May 2001, pp 343–350

16 Combined TDR and Low-Frequency Permittivity Measurements for Continuous Snow Wetness and Snow Density Determination

Markus Stacheder[1], Christof Huebner[2], Stefan Schlaeger[3], Alexander Brandelik[1]

[1]Research Center Karlsruhe, Institute of Meteorology and Climate Research
[2]Fachhochschule Mannheim – University of Applied Sciences
[3]Research Center Karlsruhe, Institute of Technical Chemistry

16.1 Introduction

Measuring snow wetness and density is essential for many applications in snow hydrology like avalanche warning, flood prediction, optimization of hydro power generation, and investigations of glacier melting due to global warming and climate change. Seasonal snow cover is highly variable in both space and time, especially when melting occurs. Deposition and depletion represents sporadic rather than continuous processes varying with meteorological conditions. After initial deposition, snow layers change over time and display a wide range of physical characteristics, from metamorphism of snow crystals and grains to melting processes and liquid water transport. The vertical arrangement of different snow layers and their properties is most important for avalanche prediction. The total snow water equivalent represents the available supply for filling the reservoirs of hydro power stations. Monitoring the variations of temporal snow wetness is essential for determining water percolation through the snow pack and the assessment of flood danger. Therefore several measurement stations at representative sites within a hydrological basin or area of interest are required. The sensors themselves have to provide vertically resolved mean snow properties of a sufficiently large area at these sites.

So far snow sensors are not suitable for long-term continuous measurements when disturbing melting processes occur around the sensor, nor are they capable of measuring snow wetness and density at the same time with sufficient accuracy. The performance of these sensors, moreover, suffers from their small measurement volume which is not adequate to achieve representative values for natural snow cover with its large spatial variability. These sensors usually consist of coaxial lines, two-wire lines, microstrip lines, or capacitor plates (Fig. 16.1). They measure the permittivity of the snow in their surroundings from which the water content and/or the density of the snow can be derived.

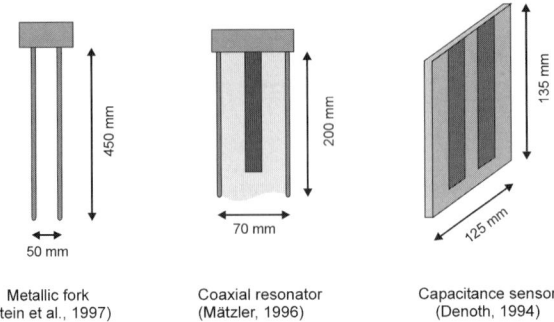

Fig. 16.1. Conventional snow moisture sensors [1], [2], [3]

One of these sensors has been used during a measurement campaign to determine the spatial variability of wetness in natural snow cover. A trench 16 m long and about 0.5 m deep was excavated in the snow. The water content along the trench was measured with the Denothsensor [4] at two consecutive days at a depth of about 0.5 m below the surface and with 1 m spacing (Fig. 17.2).

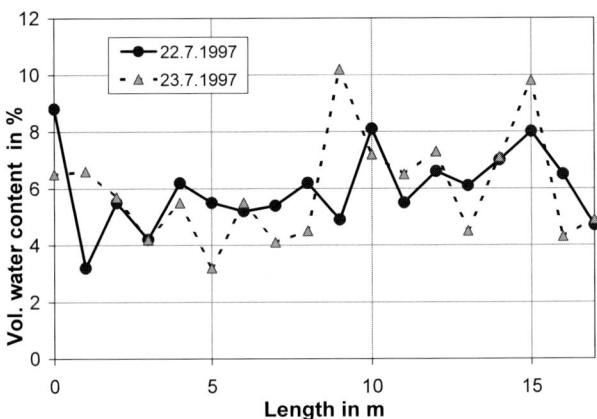

Fig. 16.2. Water content of a high Alpine snow pack at two consecutive days measured with a Denothsensor

The measurement results show strong fluctuations of water content which are typical of the time of the year with daily melting and nightly freezing. When water flow starts, rapid changes in the snow pack occur and preferential percolation paths emerge. A high temporal and spatial variability of the snow wetness is the result. Hence, on one hand, destructive methods are not practical for studying temporal evolutions. On the other hand representative mean values of snow wetness require a large number of measurements and consequently an enormous expenditure of work.

Therefore a new sensor has been developed. It consists of a flat band cable as a transmission line up to about 100 m long which is enclosed by snowfall. TDR measurements in combination with low-frequency measurements are suitable for the determination of both snow wetness and density at the same time. Besides integral measurements along the flat band cable, the reconstruction of snow parameter profiles with a sloping configuration is possible as well.

16.2 Dielectric Properties of Snow

Dry snow is a mixture of ice crystals (volumetric fraction I) and air (volumetric fraction A). Wet snow contains an additional fraction of liquid water (W). The relative dielectric permittivity of ice, ε_{ice}, and water, ε_w, are frequency and temperature dependent, whereas the permittivity of air, ε_a, is equal to 1. Figure 16.3 shows the relaxation spectra for water and ice at a temperature of 0°C.

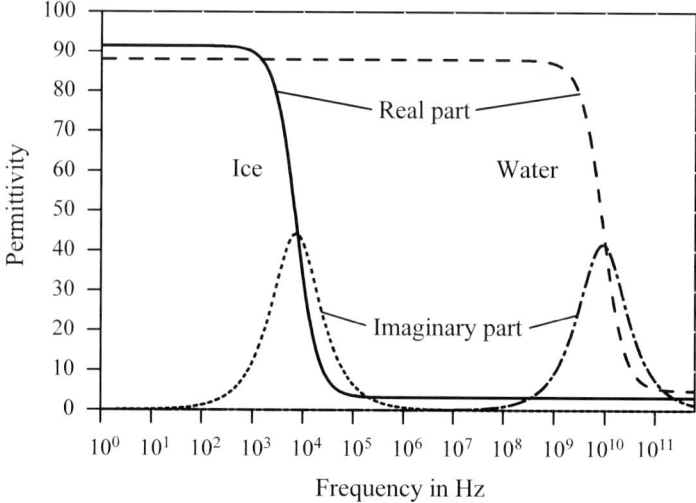

Fig. 16.3. Relaxation spectra of water and ice at a temperature of 0°C

The relaxation frequency of ice is much lower than that of water due to the strong binding forces within the ice crystal. Wet snow has a constant temperature of 0°C, whereas dry snow has varying temperatures below or equal to 0°C. Therefore only the temperature-dependent permittivity of ice has to be accounted for.

Experimental evidence indicates that the permittivity of ice may be considered independent of both temperature (below 0°C) and frequency in the microwave region and may be assigned the constant value 3.15 [5]. But at lower frequencies, e.g., in the kilohertzrange, the permittivity of ice depends considerably on temperature (Fig. 16.4).

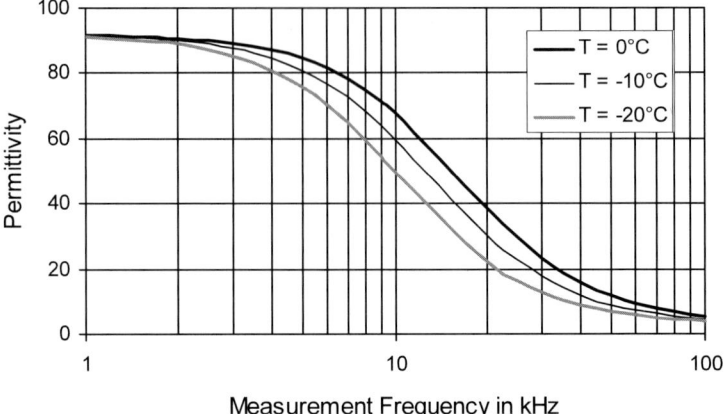

Fig. 16.4. Temperature-dependent permittivity of ice

The complex permittivity of ice, ε_{ice}, is given by

$$\varepsilon_{ice}(f,T) = \varepsilon_\infty + \frac{\varepsilon_{ice,static} - \varepsilon_{ice,\infty}}{1 + jf / f_{rel}(T)} \qquad (16.1)$$

with temperature-dependent relaxation frequency

$$f_{rel}(T) = \frac{1}{2\pi}10^{-\left(\frac{664.873}{T} - 7.447\right)}. \qquad (16.2)$$

The static permittivity is $\varepsilon_{ice,static} = 92.0205$, whereas the high-frequency limit is $\varepsilon_{ice,\infty} = 3.16$ [5]. The permittivity of water is considered to be real and independent of frequency within the measurement range below 1 GHz:

$$\varepsilon_w(f < 10^9 Hz, T = 0°C) = 88. \qquad (16.3)$$

The permittivity of snow, ε_{snow}, is related to the volumetric fractions of the constituents ice, water, and air and their dielectric properties. This relation can be described by an adequate mixing rule. Looyenga's formulae [6] for spherical intrusions ($\alpha = 0.3$) is in close agreement with experimental results:

$$\varepsilon_{snow}(f,T) = \left(W\varepsilon_w^\alpha + I\varepsilon_{ice}^\alpha(f,T) + A\varepsilon_a^\alpha\right)^{1/\alpha} \qquad (16.4)$$

with $W + I + A = 1$ for the unity volume.

16.3 Measurement Principle

In order to determine the two unknowns, water content and density, two independent equations are required. These are

$$\varepsilon_{snow}(f_1) = \left(W\varepsilon_w^{\alpha} + I\varepsilon_{ice}^{\alpha}(f_1,T) + A\varepsilon_a^{\alpha} \right)^{1/\alpha} \tag{16.5}$$

$$\varepsilon_{snow}(f_2) = \left(W\varepsilon_w^{\alpha} + I\varepsilon_{ice}^{\alpha}(f_2,T) + A\varepsilon_a^{\alpha} \right)^{1/\alpha} \tag{16.6}$$

with $\varepsilon_{snow}(f_1)$ and $\varepsilon_{snow}(f_2)$ as permittivity measurements at two frequencies with different water to ice permittivity ratios.

The equation set can be solved for water content, W, and ice fraction, I, using $W + I + A = 1$ as follows

$$I = \frac{\varepsilon_{snow}^{\alpha}(f_1) - \varepsilon_{snow}^{\alpha}(f_2)}{\varepsilon_{ice}^{\alpha}(f_1,T) - \varepsilon_{ice}^{\alpha}(f_2,T)} \tag{16.7}$$

$$W = \frac{\varepsilon_{snow}^{\alpha}(f_1) - 1 - \dfrac{\varepsilon_{snow}^{\alpha}(f_1) - \varepsilon_{snow}^{\alpha}(f_2)}{\varepsilon_{ice}^{\alpha}(f_1,T) - \varepsilon_{ice}^{\alpha}(f_2,T)} \left(\varepsilon_{ice}^{\alpha}(f_1,T) - 1 \right)}{\varepsilon_w^{\alpha} - 1}. \tag{16.8}$$

The snow density, D, can be derived from water content, W, and ice fraction, I, as follows

$$D = W \cdot 0.9999 + I \cdot 0.9150 \tag{16.9}$$

taking into account the different densities of liquid water (0.9999) and ice (0.9150). In order to achieve maximum accuracy the differences between the water to ice permittivity ratio at the two frequencies should be as large as possible. This can be achieved by using a low frequency (smaller than the relaxation frequency of ice) and a high frequency (higher than the relaxation frequency of ice and lower than the relaxation frequency of water). Typical frequencies are below 10 kHz for f_1 and frequencies between 100 kHz and 1 GHz for f_2.

16.4 Flat Band Cable Sensor

Instead of the small-scale sensors (Fig. 16.1) with their rigid and thus inadequate constructions, a flexible flat band cable up to about 100 m in length is proposed, which can follow the settlement of the snow cover. A picture of the cable together with a sketch of the cross-section is shown in Fig. 16.5.

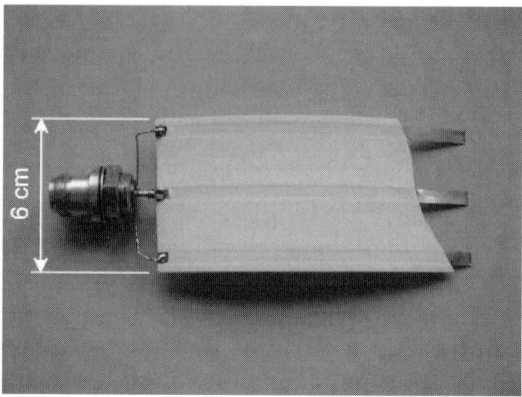

Fig. 16.5. The PE-insulated flat band cable (short section with uncovered conductor to show connection and geometry)

The electrical field concentrates around the conductors and defines a sensitive area of 3 to 5 cm around the cable. The spatial weighting of the measurements in the cross-section of the cable is directly related to the energy density distribution. The electric properties of the flat band cable used in this study can be calculated and measured. One can assume that the well-known equivalent circuit for the infinitesimal line section as shown in Fig. 16.6 fully describes the electrical properties of the line.

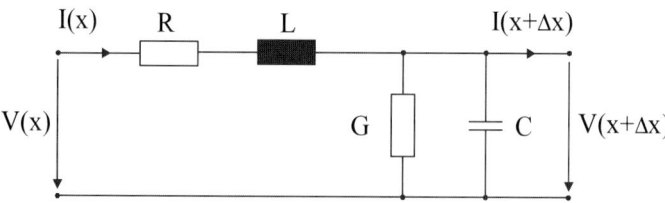

Fig. 16.6. Electric equivalent circuit of an infinitesimal section of a TEM transmission line

The white polyethylene (PE) insulation reduces heating due to solar radiation and the thin copper conductors have an advantageous low thermal capacity. Nevertheless, air gaps may develop around the flat band cable, e.g., due to multiple freezing and thawing cycles.

These air gaps cause under-prediction of the permittivity of the surrounding snow. To detect air gaps and correct the measurement results the three-wire cable is measured twice, both with small and with large spacing, leading to different measurement volumes as shown in Fig. 16.7. Thus an air gap has different effects on the volume and it is possible to correct it. A correction equation has been derived for calculating air gap size and true permittivity of snow [7].

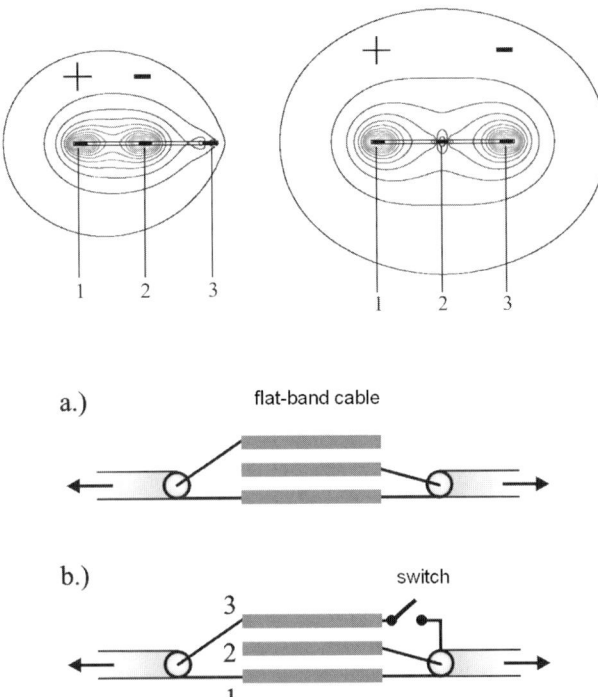

a.)   flat-band cable

b.)   switch

Fig. 16.7. Three-wire flat band cable: extension of the electromagnetic field for different connection possibilities. The cable is connected once from the left (a) and once from the right side (b)

16.5 Field Experiment Set-up

The new flat band cable sensor was tested at the high-elevation field site "Weissfluhjoch" at Davos (Switzerland) at 2550 m a.s.l. Fig. 16.8 shows an overview of this well-known avalanche test site which is equipped with numerous meteorological and nivological instruments and has a heated measurement shelter.

Fig. 16.8. Overview of the test-site at Davos with Parsenn cable car in the background

Two different set-ups were examined:

1. One cable was mounted sloping (Fig. 16.9) at an angle of $30°$ from the bottom to a mast with the aim of measuring the vertical properties of the snow pack (snow depth, density profile, snow water equivalent).

2. Three cables were placed horizontally on the snow surface at different stages of the winter in order to measure the spatial variability of the liquid water content and snow density, and especially to detect water conducting zones during the main snowmelt.

In the following only the evaluation of the horizontal cables is shown. These cables followed the natural settlement of the snow pack. The electronic measurement devices of the system were placed in the shelter approximately 10 m away from the cable sensors (Fig. 16.9). For the high-frequency measurements (100 MHz to 1 GHz) a TDR cable tester "Tektronix 1502B" was used the low-frequency measurements in the range of 1 kHz to 300 kHz were carried out with an impedance analyzer "HP-4192A". A PC and a home-made multiplexer completed the system. The measurements started on December 21, 2001 and ran automatically until the end of the winter season (June 21, 2002).

Fig. 16.9. The cable sensor system at "Weissfluhjoch", Davos (2550 m): the electronic devices in the shelter (*left*) and the test field with sloping cable sensor (*right*).

With the exception of some weeks in December, when an erroneous setting prevented the impedance analyzer from running, and in April, when the breakdown of an internal electronic component forced us to replace the impedance analyzer, the electronic system ran more or less failure-free throughout the entire winter season. The performance of the measurements and the position of the horizontal cables in the snow pack are shown in Fig. 16.10.

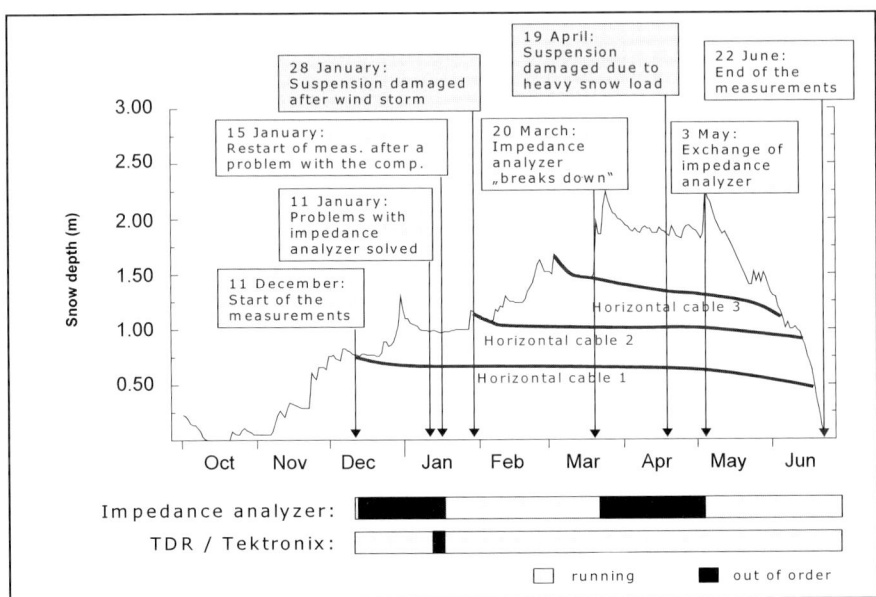

Fig. 16.10. Summary of the test measurements at Davos during winter 2001/2002 indicating problems and measures taken in the run over the season. Also, snow depth and position of the horizontal flat band cables are indicated

16.6 Experimental Results

The sloping flat band cable was able to withstand the harsh weather conditions quite well. However, with regard to the suspension and the supporting poles, which bent considerably when the snow load was heaviest, more sophisticated solutions need to be found. There were no clear indications of an air gap during the accumulation phase, but during melting an air gap was noticed at the sloping cable. The evaluation of the sloping cable is still under development, so only the results of the horizontal cables are shown.

The capacitance and dielectric permittivity of these cables for both the high- and low-frequency range were calculated from the raw signals. Figure 16.11 shows the capacitance in picofarads per meter of the lowest horizontal cable for both the small and large spacing measurement mode.

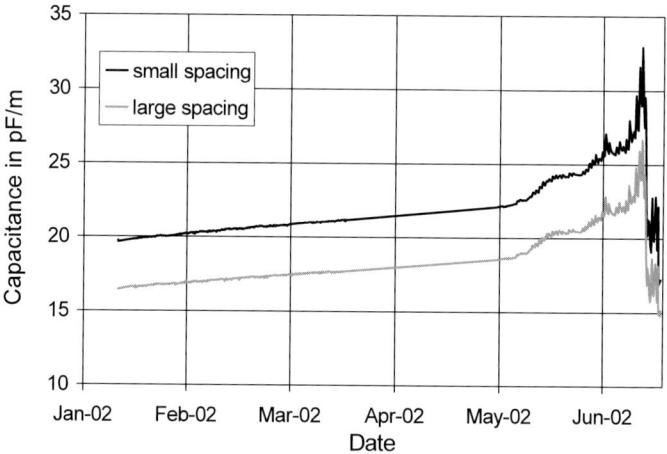

Fig. 16.11. Capacitance at high frequency of lowest horizontal cable

The clear difference between the two spacing modes is mainly caused by the different influence of the PEinsulation and is corrected by a calibration of the cable in well-defined materials.

Figure 16.12 shows the capacitance of the horizontal cable for the low-frequency measurements at 10 kHz. As expected the capacitance is much higher, thus leading to higher permittivity so that the differences in the water to ice permittivity ratio at the two frequencies is large enough to use the proposed evaluation method.

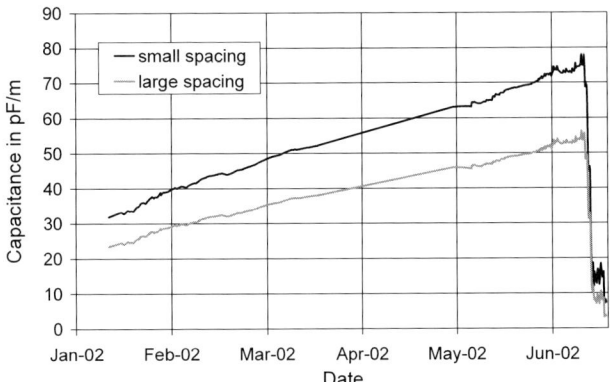

Fig. 16.12. Capacitance at low frequency (10 kHz) at lowest horizontal cable

The calculated high-frequency permittivity is shown in Fig. 16.13. It can be seen that the calibration was successful and the small and large spacing modes nearly yield identical results. We did not find any evidence for significant voids around the horizontal cables when we excavated dummy flat band cables, so the presence of air gaps around the horizontal cables can be excluded.

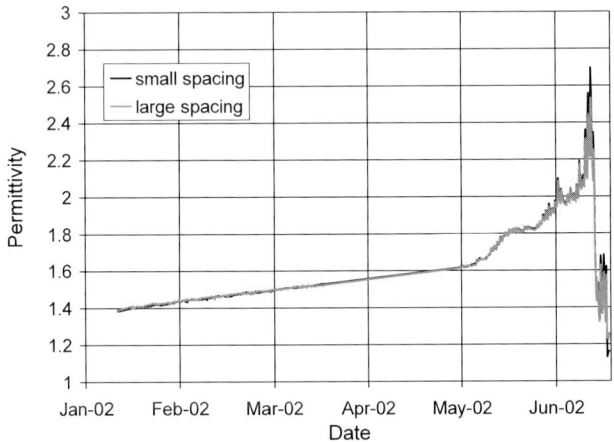

Fig. 16.13. High-frequency permittivity at lowest horizontal cable

From the curve it can be seen that the measurements started at the end of January 2002 and the cable was successively covered with snow. The slight increase in permittivity is caused by the increase in density of the dry snow pack. At the beginning of May, the sharp rise of the curve indicates the start of the melting period, where liquid water penetrates the snow pack. Although interrupted by a short cold period at the end of May when the capacitance stayed constant for a few days, the melting intensified in June. In this state also, variations in permittivity between day and night times due to melting and refreezing processes are clearly visible. At the end of June, the sharp drop in permittivity is caused by the melt-out of the cable and its exposure to air. The determined dimensions of the high-frequency permittivity are in the same order of magnitude as that found by Tiuri et al [8] for similar snow packs.

The low-frequency permittivity (at 10 kHz) is given in Fig. 16.14. It is approximately five times higher than the high-frequency permittivity and thus gives the possibility to use the equations above to determine the density and the liquid water content of the snow pack. The steeper increase of the curve in the dry snow phase compared to the high-frequency results is due to the much higher permittivity of ice at the low frequency (ε_{ice}(10 kHz) = 38.11) used, which influences the permittivity of the ice–air mixture. For the same reason, the variations during the melting phase are less pronounced. The bigger difference between the small and large spacing modes for the low-frequency permittivity is still a question of adequate calibration in the low-frequency range and will be improved in future.

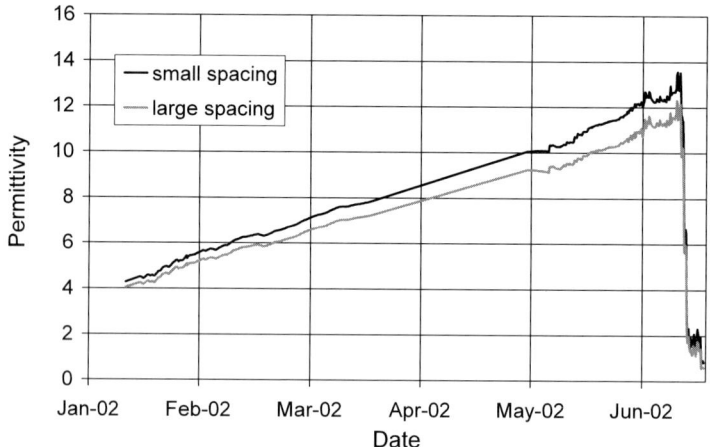

Fig. 16.14. Low-frequency permittivity at lowest horizontal cable

The combination of high- and low-frequency permittivity measurements finally led to the liquid water content along the horizontal cable shown in Fig. 16.15. Practically no liquid water was detected with the horizontal cable until the end of April, when the snow pack had reached its maximum height. This is supported by the snow temperature measurements taken at the same height as the cable (Fig. 16.16), which indicate dry snow conditions due to temperatures well below zero before this period. If the small amounts of 0.1 to 0.2% water content that were measured during the dry snow phase are realistic and possible is still a topic of the current discussion and remains unclear at the moment.

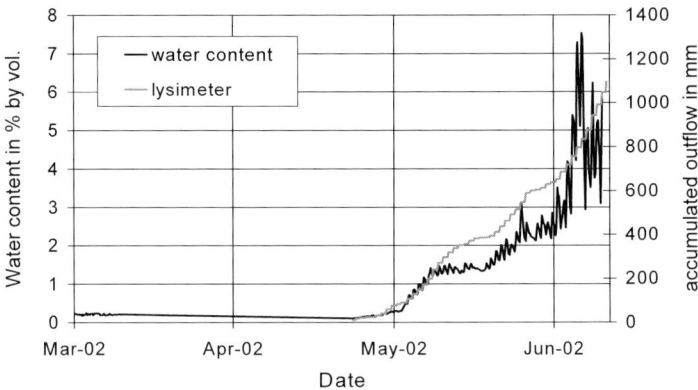

Fig. 16.15. Calculated liquid water content at lowest horizontal cable compared to accumulated outflow of lysimeter

Once the snowmelt set in at the end of April, a steadily increasing liquid water content was measured with the horizontal cable, indicating the downward penetration of the wetting front. The cable even reacted remarkably to the diurnal variation by detecting slightly higher water contents (0,1%-0,5%) during the day when melting occurs and lower water contents during the night, when the liquid water refreezes again due to lower temperatures.

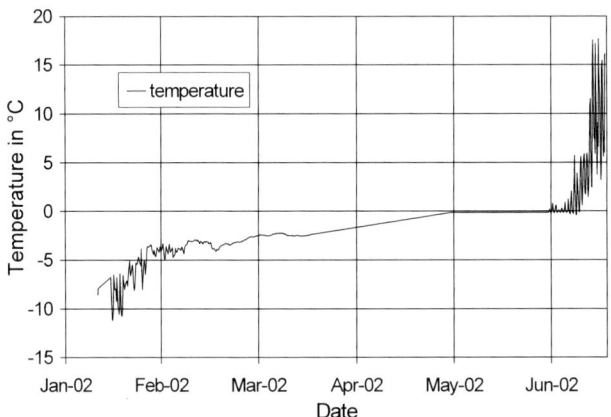

Fig. 16.16. Temperature of snow pack at same height than horizontal cable during winter

Figure 16.16 shows the temperature of the snow pack during winter at the same height as the cable. The part of the dry snow pack with temperatures below 0°C and the wet snow pack with temperatures of 0°C and above can be clearly distinguished and corresponds to the results found with the capacitance and permittivity measurements. It can also be seen that the snow temperature sensor is no longer covered with snow in the middle of June and is thus influenced by air temperature. This corresponds also with results of the horizontal cable, which melted out in the middle of June. The slight difference is caused by the settlement of the cable, being covered by snow a little longer than the temperature sensor, which is mounted on the ground like a candle and thus cannot settle with the snow cover.

The natural settling of the snow cover was reflected nicely in the horizontal cable measurements. The snow density (Fig. 16.17) increased from initially 200 kg/m^3 to approximately 450 kg/m^3 at the end of the winter season, which was in accordance with manual snow density measurements of the corresponding snow layers taken in the field every fortnight. We have no explanation for the high manual values at the beginning of the measurement period. The other horizontal cables located at different depths of the snow pack also correctly reproduced a lower density in the upper part of the profile and a faster compaction during the snowmelt.

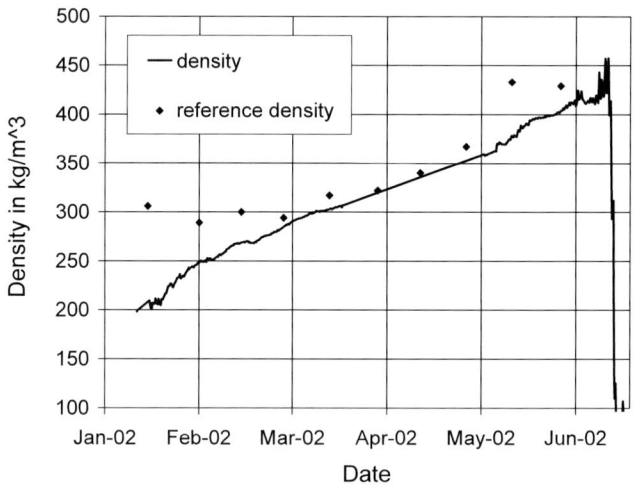

Fig. 16.17. Density of snow pack during winter 2002 at horizontal cable 2

With the TDR signal reconstruction algorithm developed by Schlaeger [9], the distribution of liquid water content along one of the horizontal cables was also reconstructed for two different stages during the melting period in May.

With regard to the spatial variation of liquid water, the horizontal flat band cable nicely demonstrated the formation of preferential water flow paths (Fig. 16.18), which is the natural process of water transport in a melting snow pack [10]. The horizontal cable at about 1 m below the snow surface indicated an emerging water conducting zone in mid May at 14 m from the beginning of the cable, as well as a newly developing flow finger at 8 m. These results will open up new possibilities for snow hydrology research and especially for the forecast of wet snow avalanches.

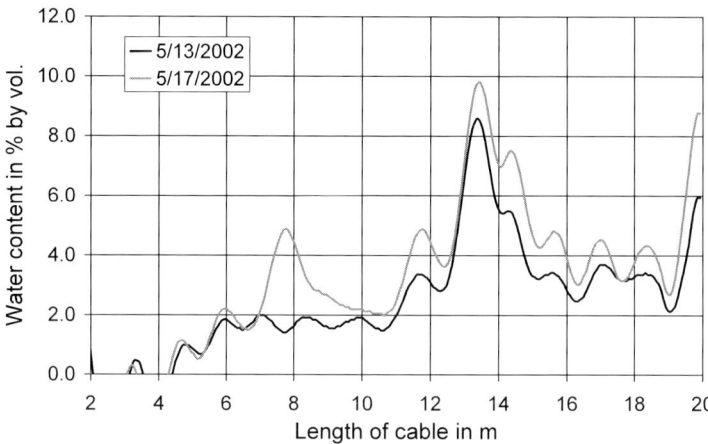

Fig. 16.18. Spatial variation of liquid water content along the upper horizontal cable at two days during the melting period

Following these promising results, a second field test of the new flat band cable has been started at Davos and will go on throughout winter 2002/2003.

16.7 Conclusion

It has been demonstrated that dielectric methods are very well suited for snow moisture determination. A combination of high- and low-frequency measurements gives a good opportunity of determining both the liquid water content and the density of the snow pack.

The measurement results yielded good correspondence of the snow pack density with manual reference measurements taken twice a month at the measurement site. Also, the determination of liquid water content from measurements of high- and low-frequency dielectric permittivity gave plausible results both compared to lysimeter data taken on the test field and with regard to the spatial variation of flow fingers that are normally experienced in a natural snow pack.

In light of these first results, the sensors seem to have the promise to become operational tools for continuous and large-scale monitoring of snow cover properties. The devices' suitability for the measurements has been demonstrated, but there is a dire need for an improvement of the instrumentation set-up to withstand the harsh conditions of Alpine winter seasons. This will be a task for the future.

Acknowledgments

Parts of this work have been carried out within the EUproject SNOWPOWER (5th framework, NNE5/2000/251).

References

1. Stein J, Gaetan L, Levesque D (1997) Monitoring the dry density and the liquid water content of snow using time domain reflectometry. Cold Regions Science and Technology 25:123-136
2. Maetzler C (1996) Microwave permittivity of dry snow. IEEE Trans Geosc Remote Sens 34:573-581
3. Denoth A (1994) The monopole antenna: A practical snow and soil wetness sensor. IEEE Trans Geosci Remote Sens 35:1371-1375
4. Denoth A (1989) Snow dielectric measurements. Adv Space Res 9 (1):233-243
5. Ulaby FT, Moore RK, Fung AK (1986) Microwave dielectric properties of natural earth materials. In: Ulaby FT, Moore RK, Fung AK (eds) Microwave remote sensing, vol III, From theory to applications. Artech House, Norwood, MA, pp 2017-2027
6. Looyenga H (1965) Dielectric constant of heterogeneous mixtures. Physica 31:401-406
7. Huebner C (1999) Entwicklung hochfrequenter Messverfahren zur Boden- und Schneefeuchtemessung. Wiss Ber FZKA 6329, Forschungszentrum Karlsruhe
8. Tiuri M E, Sihvola AH, Nyfors EG, Hallikainen MT (1984) The complex dielectric constant of snow at microwave frequencies. IEEE J Ocean Eng 9(5):377-382
9. Schlaeger S (2002) Inversion von TDR-Messungen zur Rekonstruktion räumlich verteilter bodenphysikalischer Parameter. Veröffentlichungen des Instituts für Boden- und Felsmechanik, vol 156, Karlsruhe
10. Marsh P, Woo M-K (1984) Wetting front advance and freezing of meltwater within a snow cover: 1. Observations in the Canadian Arctic. Water Resour Res 20(12):1853-1864

17 Principles of Ultra-Wideband Sensor Electronics

Jürgen Sachs

Technische Universität Ilmenau, Ilmenau, Germany

17.1 Introduction

The ultra-wideband (UWB) technique will open a new perspective in microwave moisture sensing. This chapter intends to introduce the readers to this technique and to familiarize them with the appropriate electronics and related aspects.

Microwave moisture sensors take advantage of the strong influence of water on the propagation of an electromagnetic field. Thus the moisture content can be determined by measuring the propagation and development of an alternating electromagnetic field. However, not only water but also other substances, the material constitution (homogeneity, granular size, particle inclusions, material texture, etc.) and the geometry of the arrangement under test (size and shape of the test object, surface properties of the test object, object sensor distance, etc.) influence the electromagnetic sounding fields. One tries to exclude these troubling effects in classical microwave moisture sensors by calibration referring to the material under test and by a fixed geometry of the exposed volume. But increasing claims referring to application flexibility and robustness against troubling factors require improved sensor principles.

The measurement of moisture by microwaves is an indirect method. The moisture content is deduced based on the behavior of the electromagnetic field which is generated by a known source and influenced by an unknown object within its propagation path. That means it can be concluded from knowledge (measurement) of the resulting field as to the reason (i.e., the moisture) which made the field behave as it did. By generalization, this kind of question is called an inverse problem. It is known that the solution of such problems gives ambiguous results under most experimental conditions. From a mathematical point of view, inverse problems are ill-conditioned and ill-posed, since the quantity and quality of the available data are not sufficient to solve the problem in an unambiguous way. But practical experience gives rise to some optimism. The probability of finding an adequate solution of an inverse problem increases with the availability of better information (measurement data) about the test objects. Furthermore, a situation-specific model (adapted to the present test scenario) is needed in order to be able to extract the required information from the measurement data.

Translated to the task of microwave moisture sensing, the message from the above details can be summarized by the following three points:

1. *Need for a sufficient quality and quantity of measurement data*: Starting from a general viewpoint, the information content gained from a pure microwave measurement is determined by the frequency diversity, by the diversity of sensor locations, and by the polarization diversity. That means, for example, that by measuring over a large bandwidth there is a better chance to separate different substances due to their specific dispersion or to separate volume effects (moisture, grain size, etc.) and boundary effects (size and shape of test arrangement, surface structure). Or it could be useful to gather data from several sensor positions (i.e., sensor arrays) under different polarizations in the case of inhomogeneous objects, objects with limited dimensions, or a rough surface. Consequently a sophisticated sensor arrangement might be a wideband system and it should cover lots of measurement channels to operate with sensor arrays which could be fully polarimetric.

2. *Need for an adequate model of the test scenario*: In order to be able to extract the required information from the measurement data, a model of the test scenario is necessary. The model can be a purely formal or physical-based system of mathematical relations. The key issue of modeling is a set of (desirably few) parameters which should describe the behavior of the real object as it was measured. The values of these parameters are the information searched for, i.e., moisture content, material density, etc. Thus, the task of the model is to link the available (measurement) data with the model parameters of interest.

3. *Need for methods to extract the model parameters from the measurement data*: UWB models are of course more complex than the usual narrowband models. With the further improvement of algorithms to solve inverse problems and the permanent growth of computing power, both the methodical and technical prerequisites are given as a matter of principle to extract the required information from a huge amount of data. The selection of an appropriate solver depends on the data throughput, the model structure, and the restrictions regarding computing power (costs, size, and power consumption of the system) and the available computing time (e.g., the control of a fast industrial process).

In summary, the improvement in sensor performance is based on the technical improvements of measurement and computing electronics as well as the improved capabilities of modeling and parameter extraction. Currently, the major handicap in the volume application of the UWB method for moisture sensing is seen by the lack of economical solutions for the appropriate electronics. In what follows, this chapter will concentrate on this topic in order to give a perspective for future sensor developments.

The UWB term is defined by the fractional bandwidth b, which has to be between 25% and 200% for a UWB system

$$b = \frac{f_\mathrm{u} - f_\mathrm{l}}{\frac{1}{2}(f_\mathrm{u} + f_\mathrm{l})} 100\% . \tag{17.1}$$

Here f_u and f_l refer to the upper and lower 10 dB bound of the occupied spectrum for the stimulus signal. Following FCC rules [17.1], the UWB term (cf. Eq. (17.1)) is limited to an instantaneous spectrum. This fact may be important for regulation issues but not for the function of moisture sensors. Hence it will disregarded in what follows.

The classical device which covers a large bandwidth is the network analyzer. For moisture sensing, it is a useful device for research and development purposes but it is useless in industrial volume applications due to its cost, size, weight, and power consumption. The goal of this chapter is to show that there are some interesting concepts which can replace this device with economically interesting solutions and to give the readers the opportunity to select the best principle for their purpose.

Before some of these device concepts are introduced, a short review of UWB system theory will be given. It will be followed by a summary of practical constraints which UWB arrangements underlie. The basic concepts of UWB systems architecture will be explained and three of the most suitable (concerning volume applications) device concepts will be presented. Finally some examples will give an impression of the utility of the UWB technique for moisture sensing.

17.2 Basics of UWB System Theory

The basic parts of an arrangement for microwave moisture sensing are built from the material under test (MUT), the applicators – the real sensor elements providing and capturing the sounding fields – and the RF-electronics as well as the signal processing unit. There are lots of different measurement arrangements and applicator principles in use. But in all cases, the MUT applicator arrangements may be considered as a one-, two-, or n-port system which we call the device under test (DUT). Figure 17.1 schematically symbolizes such an arrangement.

At any port, two different signals are needed to describe completely the energetic interaction with the DUT. These signals are either a voltage and a current or an ingoing and outgoing wave each referred to a defined location (measurement plane). Half of these signals can be chosen independently. They are called input signals, stimulation signals, or perturbation signals. The ingoing waves a_i are mostly considered as these stimulation signals in the microwave technique. The remaining signals, the so-called system response (usually the outgoing waves b_j), depend naturally on the stimulation signals but also on the behavior of the DUT. Thus there is some possibility to determine the properties of the DUT by measuring these signals.

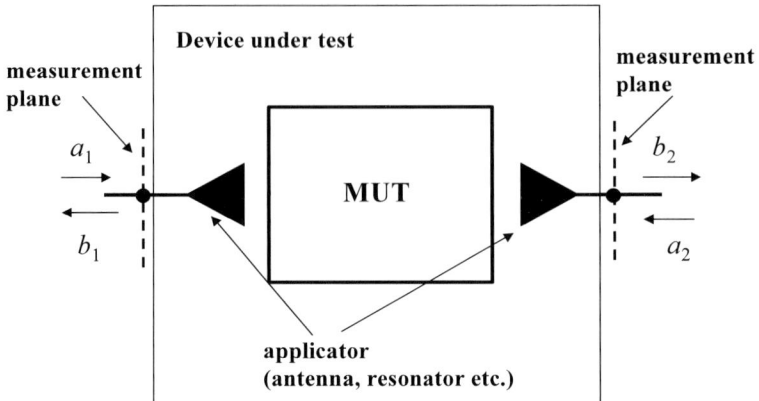

Fig. 17.1. Schematic arrangement of a microwave moisture sensor forming an electrical two port. The port signals applied here are the normalized waves *a* and *b* as usual in microwave techniques.

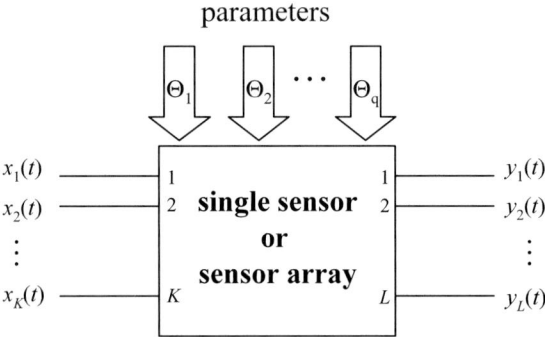

Fig. 17.2. Sensor arrangement symbolized as MiMo system having *K* inputs (stimulation points) and *L* outputs (measurement points). In the case of microwave moisture sensors, the input variables x_k usually correspond to the ingoing waves a_k and y_l correspond to b_l. The parameters Θ_q refer to the quantities which influence the properties of the DUT. Examples of such parameters are the moisture, the material density, the temperature, the salt content, etc.

   Mathematically, the relation between the stimulation signals and the response signals via the DUT behavior is described by a multiple-input – multiple-output (MiMo) system as symbolized in Fig.17.2. In order to abstract from a specific physical meaning of the stimulation signals, they are described as $x_i(t)$ and the response signals as $y_j(t)$. An arrangement of *n* applicators provides a maximum of *n* stimulation and *n* response signals. But all these signals must not or can not be used so, in general, one has to deal with an *L·K* system having *L* output and *K* input variables.

   In general, every output signal $y_l(t)$ of a MiMo system depends on every input signal $x_k(t)$ and the behavior of the DUT which is influenced by the parameters Θ_q. This can be expressed by the following generic relation:

$$y_l(t) = f_l\big(t, x_1(t), \cdots, x_k(t), \Theta_1 \cdots \Theta_q\big).$$ (17.2)

Equation (17.2) simplifies greatly if time invariance and linearity of the DUT can be supposed. Time invariance means that the dynamic behavior of the system merely depends on an "internal propagation time" but not on the ("absolute") observation time. Of course, this is not a completely exact assumption but for practical purposes it can be accepted as long as the DUT does not (only slightly) change its behavior during the measurement time. Therefore, highly variable phenomena require a short measurement time. Linearity holds if the superposition of signals is valid. This is generally the case as long as limiting or hysteresis effects do not occur and if no semiconducting pn-transitions are involved within the DUT. For the sensors we consider, the use of low peak power stimulus signals usually prevents non-linearities.

Under the constraints mentioned above, an arbitrary signal $x_k(t)$ may be decomposed into a series of standard functions such as Dirac pulses, sine waves, or step functions. Thus the stimulation effect of every sub-component of every variable $x_k(t)$ on the DUT may be considered independently of each other. In the case of a Dirac-pulse decomposition – often described as "time domain consideration" – Eq. (17.2) transforms into a set of convolutions which may be expressed in matrix form:

$$\begin{bmatrix} y_1(t) \\ y_2(t) \\ \vdots \\ y_L(t) \end{bmatrix} = \begin{bmatrix} h_{11}(t, \Theta_1 \cdots \Theta_q) & h_{12}(t, \Theta_1 \cdots \Theta_q) & \cdots & h_{1K}(t, \Theta_1 \cdots \Theta_q) \\ h_{21}(t, \Theta_1 \cdots \Theta_q) & h_{22}(t, \Theta_1 \cdots \Theta_q) & \cdots & h_{2K}(t, \Theta_1 \cdots \Theta_q) \\ \vdots & \vdots & \ddots & \vdots \\ h_{L1}(t, \Theta_1 \cdots \Theta_q) & h_{L2}(t, \Theta_1 \cdots \Theta_q) & \cdots & h_{LK}(t, \Theta_1 \cdots \Theta_q) \end{bmatrix} * \begin{bmatrix} x_1(t) \\ x_2(t) \\ \vdots \\ x_K(t) \end{bmatrix}$$ (17.3)

$$\mathbf{y}(t) = \mathbf{h}(t, \Theta_1 \cdots \Theta_q) * \mathbf{x}(t).$$

Here the asterisk $*$ symbolizes the convolution product and $h_{lk}(t, \Theta_1 \cdots \Theta_q)$ is called the impulse response function (IRF). The IRF $h_{lk}(t, \Theta_1 \cdots \Theta_q)$ corresponds to the system response $y_l(t)$ at output l by stimulating the input k with $x_k(t)$. Consequently, a MiMo system is characterized by a set of $L \cdot K$ time functions that have to be measured.

The determination of the whole number of IRFs requires a minimum of K steps. Successively, within every step j, only one of the input channels will be stimulated by a short pulse $x_{kj}(t)$. The corresponding output signals are recorded. If they are additionally indexed by the number j of measurement, a new matrix equation (17.4) can be built, which provides all the required IRFs of the MiMo system. Particularly if the stimulation signals are close to Dirac pulses $\delta(t)$, the time shape of the IRFs directly correspond to the measured output signals $y_{lj}(t)$ (see Eq. (17.4), where $\mathbf{I}$ is the identity matrix):

$$\begin{bmatrix} y_{11}(t) & y_{12}(t) & \cdots & y_{1K}(t) \\ y_{21}(t) & y_{22}(t) & \cdots & y_{2K}(t) \\ \vdots & \vdots & \ddots & \vdots \\ y_{L1}(t) & y_{L2}(t) & \cdots & y_{LK}(t) \end{bmatrix} = \begin{bmatrix} h_{11}(t) & h_{12}(t) & \cdots & h_{1K}(t) \\ h_{21}(t) & h_{22}(t) & \cdots & h_{2K}(t) \\ \vdots & \vdots & \ddots & \vdots \\ h_{L1}(t) & h_{L2}(t) & \cdots & h_{LK}(t) \end{bmatrix} * \begin{bmatrix} x_{11}(t) & 0 & \cdots & 0 \\ 0 & x_{22}(t) & \cdots & 0 \\ \vdots & \vdots & \ddots & \vdots \\ 0 & 0 & \cdots & x_{KK}(t) \end{bmatrix}$$

$$\mathbf{y}_m(t) = \mathbf{h}(t) * \mathbf{x}_m(t) \tag{17.4}$$

$$\mathbf{y}_m(t) \propto \mathbf{h}(t) \qquad \text{with} \quad \mathbf{x}_m(t) = \delta(t)\,\mathbf{I}\,.$$

The set of $L{\cdot}K$ IRFs covers all accessible information about the behavior of the DUT, since a Dirac pulse is able to stimulate all system states which may be externally stimulated. These functions form the database to determine the required parameters, i.e., the moisture etc. via an appropriate model and inversion algorithm. However, since this chapter is mainly directed toward measurement problems rather than inversion problems, the parameter dependence of the IRFs will be omitted in what follows (and even in Eq. (17.4)) in order to simplify the notation of the equations. But it should be stressed that the actual goal is directed toward the extraction these parameters.

A second possibility to characterize the MiMo behavior is to decompose the time signals at the different ports into a series of sine waves whose amplitude and phase are given by the complex values

$$\underline{X}_k(f) = X_k(f)e^{j\varphi_k(f)} \text{ and } \underline{Y}_l(f) = Y_l(f)e^{j\gamma_l(f)}$$

for each frequency component. In equivalence to Eqs. (17.3) and (17.4), this results in a simple matrix product omitting any convolution:

$$\underline{\mathbf{Y}}(f) = \underline{\mathbf{H}}(f) \cdot \underline{\mathbf{X}}(f) \tag{17.5}$$

and for the representation of the K measurements

$$\underline{\mathbf{Y}}_m(f) = \underline{\mathbf{H}}(f) \cdot \underline{\mathbf{X}}_m(f) \tag{17.6}$$

which simplifies to

$$\underline{\mathbf{Y}}_m(f) \sim \underline{\mathbf{H}}(f) \quad \text{if} \quad \underline{\mathbf{X}}_m(f) = X_0\,\mathbf{I}\,. \tag{17.7}$$

Equation (17.7) indicates the usual measurement procedure, in which every input channel is individually stimulated by a sine wave of constant power. Its phase is considered as a reference phase and its frequency is stepped over the band of interest. The function $\underline{H}_{lk}(f)$ is called the frequency response function (FRF). It includes all electrically accessible information about the DUT as does the IRF. In microwave measurements the FRFs often relate to the scattering matrix.

Equations (17.3) and (17.5) as well as (17.4) and (17.6) can be transformed mutually by the Fourier transform. Thus there is no difference between both measurements concerning the information content about the DUT. Finally a third method is partially in use, which applies a decomposition into step functions. Correspondingly the characteristic system function is called the step response function (SRF). It can be measured in an analogous way to the IRF. But it can also be calculated from the IRF by integration; thus the SRF also does not provide any new information on the DUT.

In summary, the starting point to determine the parameters of the MUT like the moister etc. is knowledge of the characteristic functions of the DUT. These are the IRF, the FRF, and the SRF. Theoretically there is no preference of any kind with these functions since they all include the same information. But sometimes it is usual to have them all, because some effects can be better interpreted by the IRF while others are easier to see with the FRF or SRF. In practice this means measuring one of them and providing the others by transformation. This is a straightforward method. But some attention has to be paid in order to avoid errors. Some problems which can arise will be discussed shortly in the following section.

Finally it should be mentioned that a sensor arrangement having n applicators can provide n^2 characteristic functions in total, from which $(n-1)n/2$ functions are redundant because of the reciprocity of the DUT.

17.3 Practical Constraints of UWB Measurements

The previous considerations are based on very theoretical assumptions which usually cannot be fulfilled exactly in reality. In order to minimize errors, some basic rules should therefore be followed. In what follows, they will be considered using a single-input – single-output DUT for simplicity.

Assume a bandwidth B_{DUT} of the DUT and a length of its IRF of T_{DUT}. Note that the B_{DUT} value in our sense does not refer to the 3 dB value, rather it means the highest frequency component which is not yet buried by noise. This also holds for T_{DUT}, which implies that the IRF cannot be considered to be terminated before all signal components have died out, i.e., they are covered by random noise. Under the following measurement conditions, the characteristic functions will be well represented:

$$B_x \approx \frac{1}{\tau_x} \geq B_{\text{DUT}} \tag{17.8}$$

$$f_s \geq 2\, B_{\text{DUT}} \tag{17.9}$$

$$T_y \geq T_{\text{DUT}} . \tag{17.10}$$

Here B_x is the 3 dB bandwidth of the stimulus and τ_x refers to its impulse width, f_s is the sampling rate of the receiver, and T_y the record length of the captured signal. From this it can be seen that the data from any the characteristic functions covers

$$N = T_y\, f_s \geq 2 T_{\text{DUT}}\, B_{\text{DUT}} \tag{17.11}$$

data points. Thus the IRF or the SRF is a "chain" of N (real-valued) data points sampled at a time interval $t_s = 1/f_s$, whereas the FRF represents a series of $N/2$ complex-valued (non-redundant) data points determined at a regular frequency grid spanned from DC to $f_s/2$ by increments of $\Delta f = 1/T_y$.

For time domain measurements within the microwave range, the sampling theorem Eq. (17.9) burdens the electronics with extremely high measurement rates. But, fortunately, the system response can also be measured piecewise by repetitive stimulation with the same signal as long as the DUT does not change its properties. Such a procedure is described as under-sampling where only a subset of data points is gathered by one stimulation. Referring to n data points captured at every stimulation, the actual sampling rate f_{sa} is given by

$$f_{\text{sa}} = \frac{f_s}{u_{\text{sf}}} \quad \text{with} \quad u_{\text{sf}} = \frac{N}{n} . \tag{17.12}$$

The value u_{sf} represents the under-sampling ratio and f_s is now called the equivalent sampling rate. In order to avoid aliasing, the equivalent sampling rate must respect the sampling theorem Eq. (17.9). However, the real sampling rate f_{sa} may be quite low and easy to handle with technical systems. This will expand the time T_{obs} which is necessary to observe the DUT, i.e., to complete the data set. Supposing a repetition rate f_0 of the stimulus, the observation time will be

$$T_{\text{obs}} = T_0\, u_{\text{sf}} \quad \text{with} \quad T_0 = 1/f_0 \geq T_y \geq T_{\text{Dut}} . \tag{17.13}$$

Note that the period length T_0 of the stimulus must be larger than the settling time T_{DUT} of the DUT, otherwise time aliasing arises. The maximum value of the under-sampling ratio u_{sf} is limited by the non-stationarity of the DUT. The most critical non-stationarity in moisture measurements is certainly given by fast-moving MUTs. The wideband ambiguity function is one way to evaluate the sensor performance under such conditions. Corresponding considerations are showing that the measurement arrangement behaves nearly stationary, i.e., time invariant if

$$4\,T_{\text{obs}}\left|v_{\text{max}}\right| < \frac{c}{B_{\text{DUT}}}. \qquad (17.14)$$

Thus the under-sampling factor u_{sf} must be less than

$$u_{\text{sf}} < \frac{c}{8\left|v_{\text{max}}\right|N}. \qquad (17.15)$$

Here v_{max} and c stand for the maximum speed of the MUT and the speed of light respectively. Equation (17.14) holds independently of the measurement principle. That means the sine wave sweep technique underlies the same requirements. Here, T_{obs} corresponds to the time needed for one sweep over the frequency band of interest.

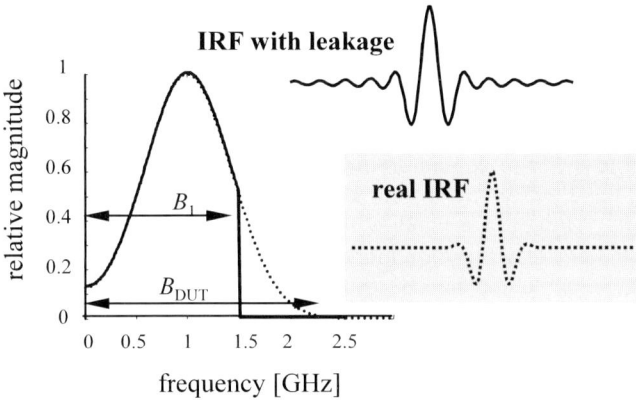

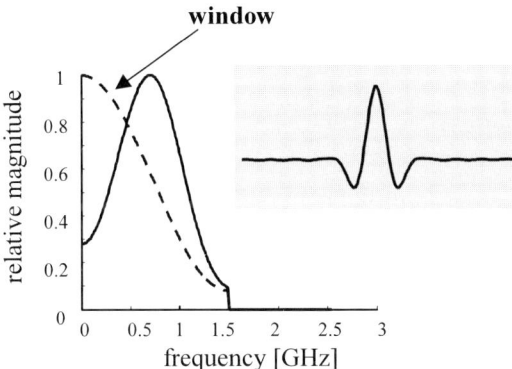

Fig. 17.3. Leakage effect. *Top*: ringing on the IRF by leakage (dotted line: real FRF and IRF of the DUT; solid line: measured part of FRF and corresponding IRF gained from the Fourier transform). *Bottom*: reduced ringing by windowing (dotted line: applied window function; solid line: windowed measurements and resulting IRF)

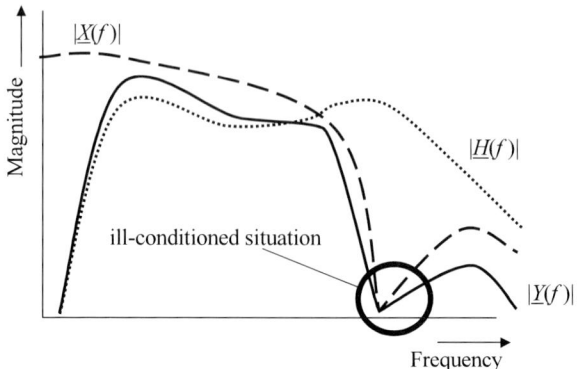

Fig. 17.4. Ill-conditioned deconvolution caused by zeros in the stimulation spectrum (dotted line: real FRF; dashed line: spectrum of stimulus; solid line: spectrum of measured signal)

By disregarding the condition in Eq. (17.8) two problems arise. Consider first a sine wave, i.e., a frequency domain measurement which covers the bandwidth $B_1 < B_{DUT}$ (see Fig. 17.3). Thus some data of the FRF missed. By transforming this reduced data set to the time domain in order to get the IRF, the abrupt cut will be "interpreted by the Fourier transform" as a steep filter flank causing a sin x/x ringing in the IRF. However, a window function can be applied to the frequency data to smooth the cut resulting in a more friendly IRF. But note that windowing cause data loss, reducing the time resolution of the IRF by widening the main lobe. So a reasonable compromise must be found.

The second problem arises if the deviations of the stimulus signal referring to a Dirac pulse cannot be tolerated. Then the stimulus $x_m(t)$ must be deconvolved from the measurement signal $y_m(t)$ in order to gain the IRF $h(t)$. The easiest way to do this is to transform the stimulus and measurement signal into the frequency domain, to divide their (complex) spectra, and to re-transform the quotient back to the time domain. Thus,

$$h(t) = \text{IFT}\left\{\frac{\text{FT}\{y_m\}}{\text{FT}\{x_m\}}\right\}. \tag{17.16}$$

FT{} represents the Fourier transform and IFT{} the inverse Fourier transform.

But unfortunately deconvolution often represents an ill-conditioned problem. Figure 17.4 demonstrates the difficulties. Suppose that the spectral energy of an (arbitrary) stimulus is not regularly spread over the frequency band of interest (i.e., B_{DUT}). Obviously it is difficult to divide the spectra at those spectral segments where their values are close to zero. Mathematically the term 0/0 is indefinite, which practically means that noise will determine the result. Although no energy appears in the process at these frequencies, there may be huge values in the FRF caused by the uncertainties of the division. By transforming such an FRF back to the time domain, usually those quantities dominate the shape of the whole IRF. They must be suppressed by appropriate methods.

17.4 Measurement Errors

Electronic measurement devices are always associated with errors. One distinguishes several types of errors which differ by their temporal behavior. Additive noise and jitter are short-term perturbations, drift refers to medium-term variation of the system properties, and systematic errors are stable deviations from the desired conduct. Particularly due to the large bandwidth, the real challenge of the UWB system design is the error handling. In what follows, the different error types will be discussed and some general measures will be given to suppress them.

17.4.1 Additive Random Noise

Random noise is considered as a stochastic short-term variation of the amplitude values of a signal. Caused by the large bandwidth, the noise rejection of UWB systems is very poor. That is why proper design and signal processing must suppress the noise as much as possible. The first measure is to restrict the bandwidth to the actual needs by short time integration of the measurement signal (see Fig. 17.5). The integration window either continuously slides over the waveform (low-pass filter) or is moved in steps (S&H circuit with an appropriate sample time). A short time integrator reduces noise at the expense of bandwidth. In an ideal case it should limit the bandwidth of the receiver to the bandwidth of the DUT. The remaining variance of the captured waveform is approximately

$$\sigma_T^2 \approx \frac{N_x}{T} \approx N_x B. \tag{17.17}$$

Here σ_T is the standard deviation of the captured signal, and N_x the power spectral density of the noise (which was supposed to be white). The bandwidth B of the receiver is given by the inverse of the integration time.

A second measure particularly in the case of a periodic stimulation is synchronous averaging as demonstrated in Fig. 17.6. It does not affect the resulting bandwidth of the recorded waveform as long as only additive noise is perturbing the signal, though in the presence of jitter or time drift, the resulting bandwidth will decrease (see Chap. 17.4.2). The improvement of the noise performance is proportional to the square root of the averaging number m:

$$\sigma_{av}^2 \approx \frac{\sigma^2}{m}. \tag{17.18}$$

Note that the observation time T_{obs} is stretched by the same factor which slows down the measurement speed.

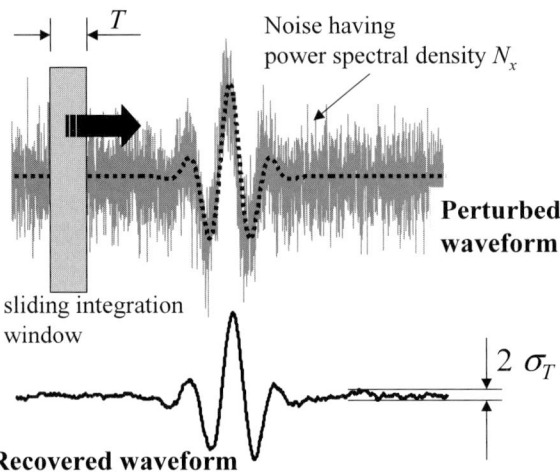

Fig. 17.5. Noise reduction by short time integration. An integration window is moved over the waveform smoothing out short noise peaks

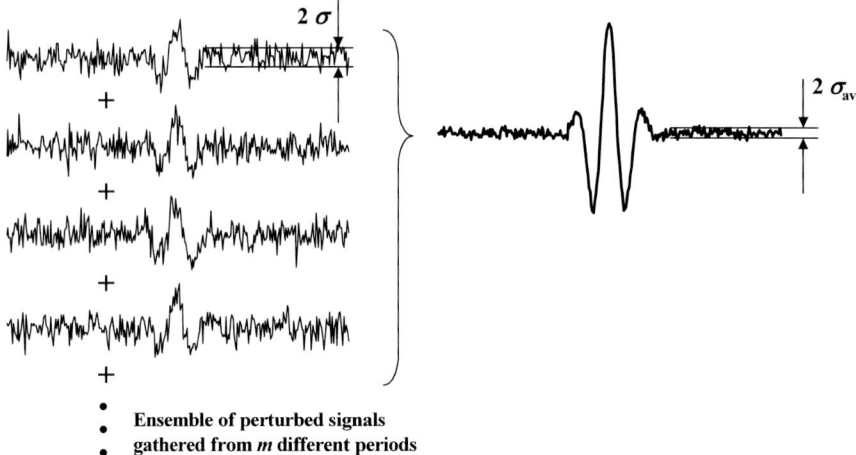

Fig. 17.6. Noise reduction by synchronous averaging

The problem of random noise should, however, not be reduced to the question of how to suppress it. Rather the question is more how to maximize the signal-to-noise ratio *SNR*. This quantity represents a central point. It influences all the other aspects of the electronic design and behavior. The *SNR* value is given by

$$SNR = \frac{E}{N_x}. \tag{17.19}$$

This states that the receiver must accumulate a certain quantity E of coherent energy connected with the measurement process under consideration. The accumulated energy depends on the average signal power P_{av}, the accumulation time τ_{ac}, as well as on an efficiency factor η of the receiver:

$$E = \eta \, P_{av} \, \tau_{ac}. \tag{17.20}$$

The higher the average signal power P_{av} of the stimulus signal, the shorter the accumulation time τ_{ac} can be for a given SNR. Accumulation time in that sense is another expression for measurement and observation time T_{obs}. Among other things, the receiver efficiency η depends on the under-sampling ratio $\eta \sim 1/u_{sf}$. Only a small part of the available signal energy is really included in the measurement data if under-sampling is applied. Thus a high under-sampling ratio may have a negative impact on the SNR value.

Whereas the average power accounts for the SNR value, the peak power P_{peak} is usually the limiting factor due to the electronics. The crest factor CF of a signal is a measure to characterize the uniformity of the instantaneous power distribution:

$$CF = \frac{\hat{X}}{X_{rms}} = \sqrt{\frac{P_{peak}}{P_{av}}}. \tag{17.21}$$

Here $\hat{X}$ represents the peak value of the signal and X_{rms} is its RMS value. The lower the crest factor, the more evenly the instantaneous signal power is distributed. This means that fewer the systems are stressed by high-power "shocks." Thus, thanks to the low crest factor signal more energy can be handled by the systems without going beyond their limits.

Measurement devices based on sine the wave technique respect this fact. Pulse measurement systems do not. This is why they usually have a poor noise performance. One way out is to spread the energy of a wideband signal over a large time and to compress it again in order to get the IRF of a DUT. The convolution of Eq. (17.3) with the time-inverted and transposed stimulation vector permits such an approach:

$$\begin{aligned}
\mathbf{y}(t) * \mathbf{x}^{T}(-t) &= \mathbf{h}(t) * \mathbf{x}(t) * \mathbf{x}^{T}(-t) \\
\mathbf{C}_{yx}(t) &= \mathbf{h}(t) * \mathbf{C}_{xx}(t) \\
\mathbf{C}_{yx}(t) &\propto \mathbf{h}(t) \quad \text{if } \mathbf{C}_{xx}(t) = \delta(t)\,\mathbf{I}.
\end{aligned} \tag{17.22}$$

Taking $\lim_{T \to \infty} 1/T$ over the first line in Eq. (17.22) results in two matrices of correlation functions $\mathbf{C}_{xx}$ and $\mathbf{C}_{yx}$. The matrix $\mathbf{C}_{xx}$ forms a diagonal matrix of functions like a Dirac pulse if the stimulus signals at the different input ports are mutually un-correlated and if they occupy a large spectrum. Thus the set of IRFs is approximately given by the matrix of the cross-correlation functions $\mathbf{C}_{yx}$ which must be determined by the measurement system. Obviously, the actual signal shape of the stimulus is of no further interest. This opens up a great deal of different device architectures.

The determination of the cross-correlation functions $\mathbf{C}_{yx}$,i.e., the IRF, is performed by a correlator and a matched filter. Figure 17.7 schematically demonstrates the procedure. A wideband waveform of any shape stimulates the DUT. Its response is processed by a matched filter or correlator having the stimulation waveform as reference. The compressed waveform represents the IRF of the DUT as long as the bandwidth of the stimulus is larger than the bandwidth of the DUT under consideration.

For estimation of the *SNR* improvement, let us consider a DUT of infinite bandwidth. Referring to Fig. 17.7 and Eq. (17.19), and taking into account the energy conservation Eq. (17.23) by the matched filter (respectively correlator)

$$E = x_{rms}^2 T_0 \approx \hat{Y}^2 t_m \approx \frac{\hat{Y}^2}{B}. \tag{17.23}$$

the improvement of the *SNR* value of the compressed waveform can be approximated by

$$SNR_{out} = \frac{\hat{Y}^2}{n_{rms}^2} = \frac{E}{N} = \frac{x_{rms}^2}{n_{rms}^2} B T_0 = SNR_{in} B T_0. \tag{17.24}$$

Here B is the bandwidth of the matched filter (respectively the waveform), and $n_{rms}^2 = NB$ is the total noise power. The improvement factor BT_0 is also called the correlation gain.

The aim of impulse compression is to load critical system components only by low peak power signals and to regain high peak power signals by impulse compression at less critical points within the systems. The intention is to put as much energy as possible into the measurement arrangement. Impulse compression is performed if the receive filter has an IRF which is a time inverse replica of the transmit waveform. A second possibility is to perform a cross-correlation which applies a reference signal corresponding to the desired input waveform. The output waveform is as short as the bandwidth $B \approx 1/t_m$ of the input signal is wide and the energy E stored in the filter is as big as the time T_0 of the input signal is long. Both factors determine the compression and the correlation gain.

It should be noted that impulse compression by analog circuit principles is always limited in the dynamic range because of saturation effects of the components. A compression in the software domain is less critical since an appropriate number representation can always be used. Furthermore, the quantity of energy which can be stored in an analog matched filter for the purpose of compression is also quite limited, particularly if it is spread over a wide spectrum. Finally, analog correlator principles must always fight against the inadequacy of programmable wideband delay lines. As may be seen in the subsections below, a digital principle has benefits also with respect to these two aspects.

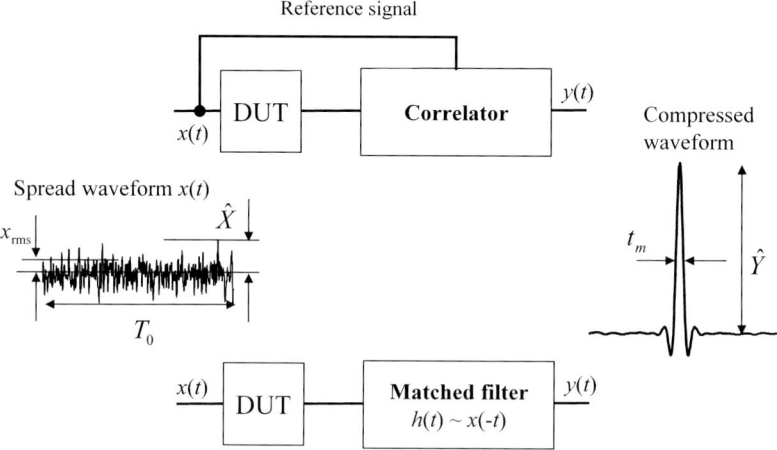

Fig. 17.7. Improvement of the *SNR* by impulse compression. If the signal $x(t)$ is misshapen by a linear system (DUT) before it enters the impulse compression system, its output signal will represent the IRF of the DUT as long as the bandwidth of $x(t)$ is wide enough

17.4.2 Jitter

Jitter is a stochastic short-term instability of the temporal position of a signal and of a sampling moment. Jitter affects signal flanks by increasing the noise but not flat signal parts. Averaging of jittered waveforms flattens impulse flanks, which means it reduces the available bandwidth (see Fig. 17.8 and Eq. (17.25), where t_r is the rise time).

$$t_{r,av}^2 \approx t_r^2 + t_j^2 \qquad (17.25)$$

Jitter is caused by additive random noise within timing control circuits or oscillators. Figure 17.9 demonstrates one possibility of its origin. The quantity of trigger jitter can be estimated by Eq. (17.26). Obviously it can be largely avoided by steep flanks.

$$t_j \approx \frac{n_{rms}}{V_0} t_r. \qquad (17.26)$$

A second countermeasure is to omit VCOs for clock or stimulus generation, since monochromatic frequency sources provide less phase noise, i.e., jitter, than frequency tunable circuits due to their high-quality resonators.

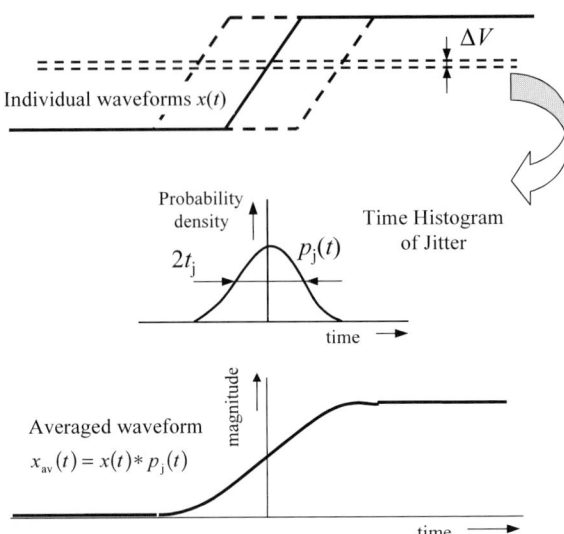

Fig. 17.8. Effect of averaging on jittered waveforms. The jitter behavior may be determined by recording a histogram over a small voltage gap

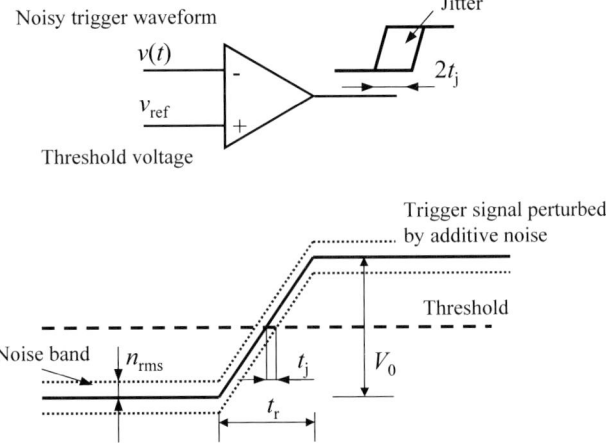

Fig. 17.9. Generation of jitter by triggering at a threshold

17.4.3 Drift

Drift is a medium-term instability which acts on both amplitude and time values. Drift limits the maximum usable number of averages and prevents correction of systematic errors in the sensor arrangement. Thus the drift behavior is a critical parameter of UWB systems. Usually it is caused by variations of threshold or offset values and gain due to changes in temperature or supply voltage or by aging. Three types of drift can be distinguished as follows.

Time drift: Comparable to the jitter performance, time drift can be largely suppressed by the steep flanks of the control signals (compare Fig. 17.9) and a temperature-stabilized high-Q oscillator.

Offset drift: This may be critical in connection with additive sampling (see Fig. 17.14). The stability of the threshold value as well as the amplitude stability of the sampling generator translate immediately into amplitude errors. Since the threshold and sampling pulse must be larger than the measurement signal, even small, relative variations of their values cause considerable additive errors. Homodyne receivers are also sensitive to drift caused by self-mixing effects (see Fig. 17.11 and the following text).

Gain drift: Gain errors are proportional to the actual measurement value. They can be eliminated by referring the measurement signal to a reference signal (usually the stimulus). It is important that the reference signal is captured by an identically constructed receiver, because then it can be supposed that they behave in the same way.

17.4.4 Systematic Errors

Systematic errors can be classified into linear and non-linear effects. Since such errors do not change with time, they can be removed theoretically from the measurement result. The success of such an error correction largely depends on the stability of the measurement system (drift, noise, jitter), the error model, and the quality of the calibration standards.

The causes of systematic errors are manifold (mismatched ports, frequency-dependent receivers, crosstalk, limited directivity, saturation effects of amplifiers, etc.) The simplest and most promising way to deal with such errors is to consider them by a formalized strategy, i.e., to model the real measurement system and sensor arrangement by an ideal device cascaded by an error box including all errors independently of their cause. Following Fig. 17.10, a sensor arrangement having n applicators represents an n-port system resulting in an error box having $2n$ ports. Such a system is characterized by $4n^2$ characteristic functions as long as the errors behave linearly. However, one of them is redundant, thus $4n^2 - 1$ error terms may appear in total. If the error terms are known, one can remove them from the measured values (a_{ei}, b_{ei}) by matrix inversion.

The determination of the error terms is based on a calibration routine which uses DUTs with a known behavior – the so-called calibration standards. The simplest way is to use commercial calibration standards. But they usually refer only to standardized RF connectors. Unfortunately, that kind of calibration excludes the influence of the applicators on the systematic errors. In order also to deal with them, either a numeric applicator model is needed or the applicators must be included into the calibration routine by using appropriate calibration materials.

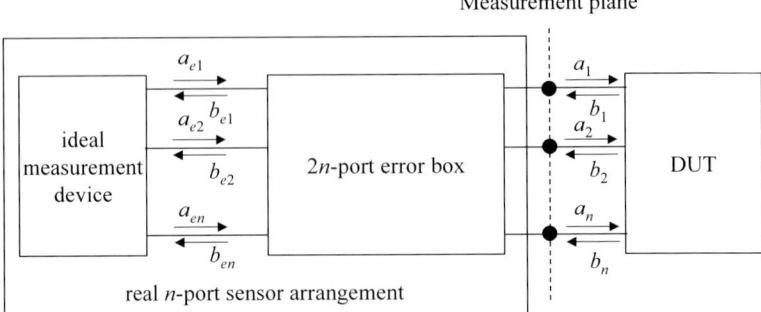

Fig. 17.10. Generalized modeling of systematic errors

Finally it should be mentioned that except for the simplest error correction methods, the full *n*-port matrix must be measured. Nowadays, this is not usual in microwave moisture sensors. For future UWB moisture sensors, however, it is recommended to equip their electronics with a sufficient number of measurement channels in order to be able to determine the complete *n*-port matrix of the DUT.

Even if it seems theoretically possible to remove all systematic errors by the above-mentioned procedure, only the most important one can be actually eliminated in practice, because the more error terms are respected the more (different) calibration standards are needed and the more reference measurements have to be done. Since every additional measurement increases the random error level of the final result, there exists an optimum number of systematic error terms that should be removed.

17.5 Architecture of UWB Systems

Concerning the electronics, a UWB sensor system is built from three parts – an RF unit which provides the stimulus signals, an RF unit which captures the measurement data, and a usually digital unit which is charged by the data processing.

The stimulation of the test objects can be done either sequentially by a swept or stepped narrowband (sine wave) source, or instantaneously by a periodic or random wideband signal. The nature of the stimulus signal has a determinant influence on the architecture and the behavior of the system.

The task of the measurement unit is to determine the response of the DUT due to its stimulation. In the case of moisture sensing, the time constants which fix shape and duration of the response function are to be found in the nano- and sub-nanosecond range. Thus a typical record length T_y of the measurement data is in the order of tens of nanoseconds. The gathering of such data streams in real time requires an extreme technical effort. Fortunately the characteristic parameters change only slowly, thus signal recording may be distributed over a longer time. This fact can be used to reduce the data gathering rate which enables less critical and hence more cost-effective technical system implementations. There are three basic approaches to reduce the real-time requirements, which will be summarized below shortly.

Nowadays, sensor measurements include a more or less sophisticated level of digital signal processing. The burdens of the digital unit and hence its complexity and costs depend on the extent of the algorithms which extract the desired information from the measurement data as well as from the data rate. Though a low data rate system often permits cost-effective technical implementations, it should be noted that the stochastic confidence in the measurement results will degrade if data rate reduction is simply done by losing data (as in case of under sampling). Consequently, an optimum must be found for every application between the performance of the UWB system and its costs. Such an optimum solution mainly depends on the overall architecture of the UWB system and the available technology for the electronic components.

An assessment of the usability of a specific UWB principle for a certain task should cover not only technical parameters (bandwidth, dynamic range, stability, measurement speed, etc.) but also features concerning power consumption, aspects of technical implementation (size, costs, weight, level of integration) as well as the capability to form multi-channel sensor systems or to permit systematic error corrections. Certainly, in deciding in favor of a specific principle, the various features weight differently. But this consideration must be left to the user, corresponding to the actual task.

In what follows, some examples for a UWB system layout will be given including some general assessments.

17.5.1 Down-Conversion

Down-conversion is a method to shift the RF measurement signals into a lower frequency band where they can be easily handled. However, a simple down-conversion is not able to reduce the data rate. Down-conversion is often applied in connection with sine wave excitation, which finally results in a considerable reduction of measurement speed. The idea is to determine the real and imaginary parts of the FRF $\underline{H}(f_x)$ of the DUT referred to the applied excitation frequency f_x. In order to record the full FRF the sine wave sources must be (slowly) stepped and swept over the entire band of interest.

There are two basic methods – the homodyne and the heterodyne approach (see Fig. 17.11). In the case of the technically simpler homodyne receiver, the response of the DUT is mixed by the two channels of a quadrature modulator with the replica and a 90°-shifted version of the stimulus sine. If the stimulus source keeps its frequency constant, both output channels of the quadrature demodulator will provide a DC voltage whose values correspond to the real part (inphase component) and the imaginary part (quadrature component) of the FRF of the DUT. The noise suppression and hence the receiver sensitivity is performed by the low-pass filters included in the quadrature demodulator. At first glance, the homodyne concept seems to be ideal for volume applications due to its simplicity and the absence of bulky off-chip components which are amenable to monolithic integration. However, the homodyne approach exacerbates a number of issues [2]:

- The required 90° phase shift of the reference signal within the IQ demodulator can only by handled for narrow band sweeps.

- The measurement values are superimposed by a varying DC offset originating from self-mixing of the reference signal due to leakage of the mixers.

- Even-order distortions also initiate DC perturbations.

- Any tolerance between both branches of the quadrature modulator leads to an I/Q-mismatch causing systematic errors.

- Flicker noise (1/*f*) may have a profound effect since the down-converted spectrum is located close to DC.

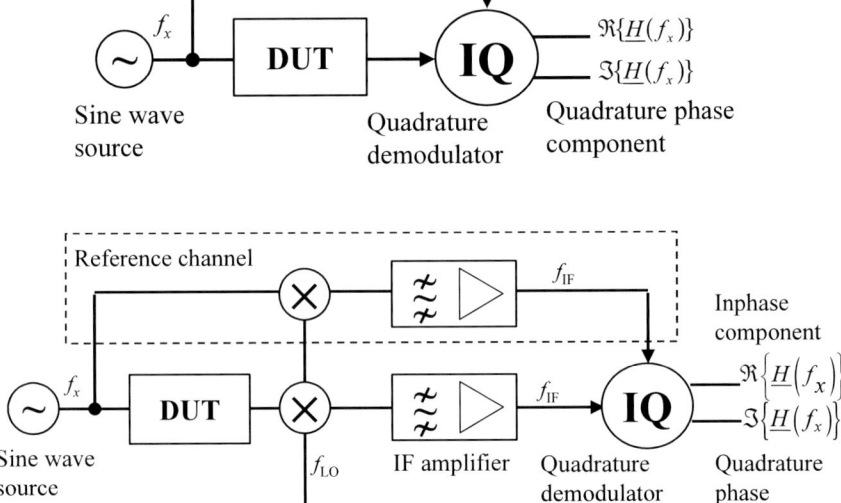

Fig. 17.11. Down-conversion by a homodyne (top) and heterodyne (bottom) approach. Note that the heterodyne concept usually includes an image rejection filter in front of the mixers. It can be dropped only if single tone signals exists as shown

The heterodyne concept avoids most of the inadequacies of the homodyne receiver by introducing an intermediate stage (see Fig. 17.11). The RF signal is mixed by the local oscillator to a fixed intermediate frequency $f_{\mathrm{IF}} = |f_x - f_{\mathrm{LO}}|$.

Thus narrow band filtering can be performed which effectively rejects noise and distortion. The quadrature demodulator must be designed only for a fixed frequency.

High-precision, multi-purpose UWB measurement devices as network analyzers are based on such heterodyne principles. However, their excellent features must be paid for by system complexity. Narrowband systems as IF amplifiers require high-Q elements which prevent monolithic integration. But the most critical parts in this sense are the sine wave sources (stimulus generation, local oscillator), which must be sophisticated synthesizers in order to gain both a wide sweep range and an excellent frequency stability (accuracy and phase noise).

Even if the down-conversion in connection with a swept/stepped sine wave excitation is used to leave the microwave range and to have a modest data rate, the measurement rate is often unacceptably low. Indeed, most of the time cannot be used for data gathering in waiting for synthesizer and the narrowband filters to settle down.

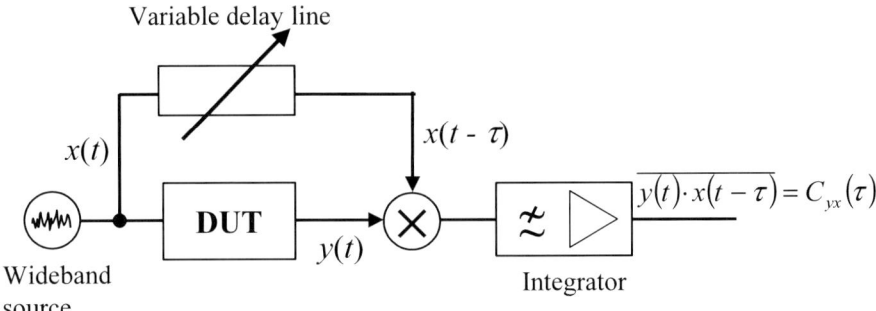

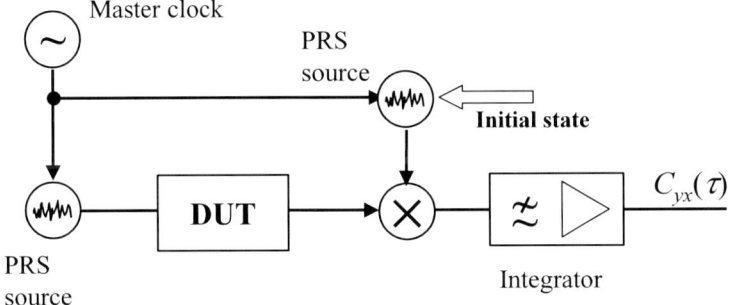

Fig. 17.12. Sliding correlator with variable delay line (top) and a reference source (bottom) which is shifted by its initial state

17.5.2 Sliding Correlation

Sliding correlation is a further method to record successively the shape of a response function of the DUT from DC values. Following Fig. 17.12 (top) and supposing a fixed delay τ_0, the correlator forms the mean value of the product between the DUT response $y(t)$ and a delayed version of the stimulus $x(t - \tau_0)$. As demonstrated by Eq. (17.22), this value corresponds to the cross-correlation $C_{yx}(\tau_0)$ at τ_0 which is equal to the IRF $h(\tau_0)$ if the DUT is stimulated by any wideband source. In order to capture the whole IRF, the delay time must be successively swept over the whole record length T_0. This is the bottleneck of the principle, since no satisfactory method is known that does this in an automatic and cost-effective way. An alternative is to use two identical pseudo random signal sources. One of them successively shifts its initial state, which has the same effect as a delay line (Fig. 17.12, bottom).

The bandwidth of the integrator affects the sensitivity, i.e., the noise rejection, and limits the measurement speed. Note that at every delay step the integrator must have settled down before the next measurement value can be captured. Furthermore the sliding correlator is sensitive to DC offset (by self-mixing and even-order distortion) and flicker noise as in the homodyne principle. Some examples of sliding correlators are to be found in [3] – [5].

17.5.3 Bandwidth Compression by Under-Sampling

Under-sampling is strictly limited to periodic (repetitive) signals. Without violating of the Nyquist theorem, the gathering of a complete set of data samples may be distributed over several periods. Hence the data gathering rate can always be broken down to a speed which does not overload the more or less slow and cost-effective digital signal capture. The sampling method reduces the number of analog components and opens up the full flexibility of the digital world as close as possible to the RF measurement ports.

Theoretically, under-sampling is nothing but a multiplication of the input signal having a repetition rate f_0 with a pin-pulse train having the rate f_{sa}. If properly done (see Fig. 17.13) this results in a signal with reduced bandwidth which can be sampled at a lower speed without disregarding the Nyquist theorem. Suppose the bandwidth of the measurement task requires an actual sampling rate of f_s. In the case of real-time sampling this would give $N = f_s/f_0$ samples per period of the measurement signal. However, it can be shown that, corresponding to Eq. (17.27) any other sampling ratio f_{sa} may also be used to collect the same samples as long as N and the under-sampling factor u_{sf} do not have a common divider.

$$u_{sf} = \frac{f_s}{f_{sa}} = N\frac{f_0}{f_{sa}} \quad (u_{sf}, N \text{ integer numbers}) \tag{17.27}$$

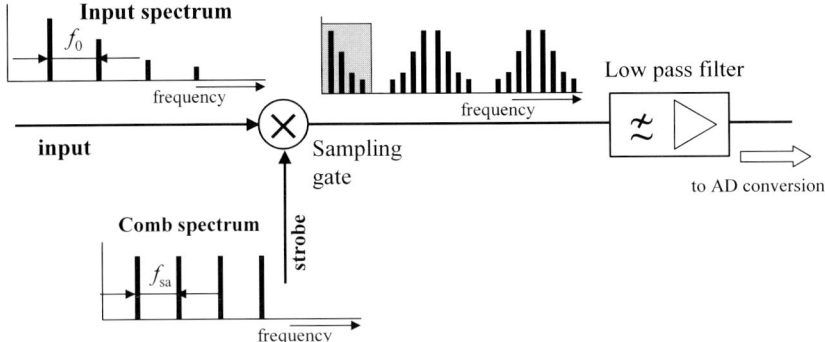

Fig. 17.13. Bandwidth compression by convolving the input spectrum with a comb spectrum. (A comb spectrum is generated by a train of short pulses)

However, the original order of spectral lines and the samples will only be respected for sequential sampling, i.e.,

$$u_{sf} = n\,N + 1 \quad (n = 1, 2, 3, \cdots). \tag{17.28}$$

In all other cases, the samples must be reordered before further processing.

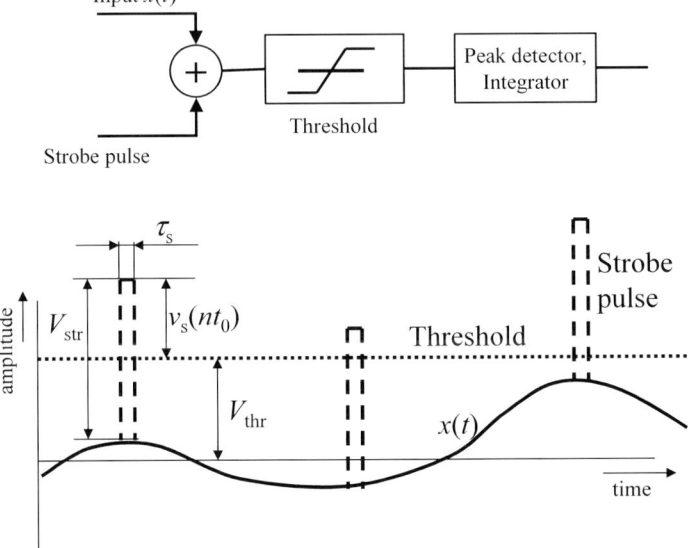

Fig. 17.14. Additive sampling. Either the peak value $v_s(nt_0)$ or the signal area (charge) beyond the threshold provides the value of interest

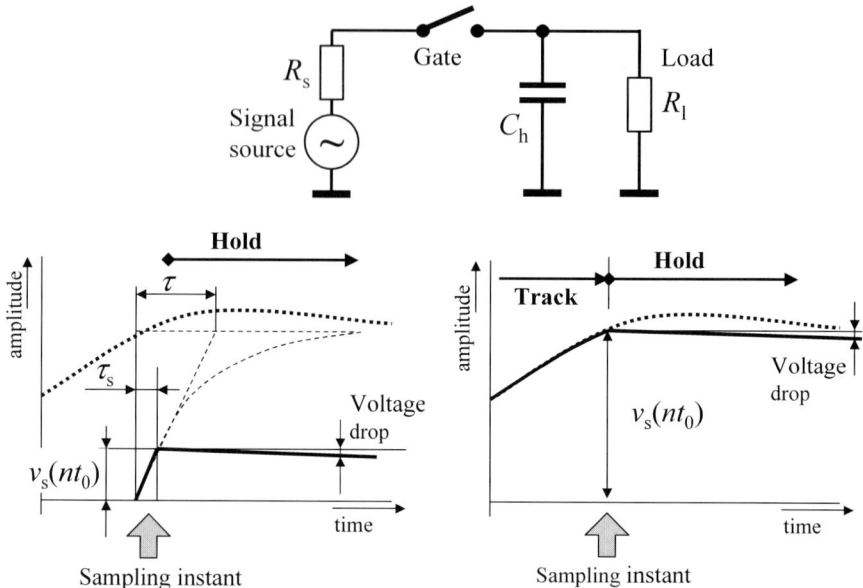

Fig. 17.15. Multiplicative sampling. Circuit schematic (top) and typical signals in the sample and hold mode (bottom left) and in the track and hold mode (bottom right)

The probing itself is done by two approaches – additive or multiplicative sampling. For additive sampling, the measurement value results only from that signal part which goes beyond a threshold (see Fig. 17.14). The threshold is chosen in such a way that the pure input signal can never pass it. An additive sampler can be implemented by very simple and power-efficient circuits [6]. However, they are strongly sensitive to threshold drift and variations of the strobe pulse magnitude.

Multiplicative sampling (Fig. 17.15) provides the measurement value by charging a hold capacitor via the sampling gate, which is controlled by the strobe pulse. Two modes of charging the capacitors are in use. In the S&H mode, the gate is only closed for a very short time τ_s. The capacitor will only be charged to a small fraction of the input amplitude. Thus the sampling efficiency will be low (only a few percent) but the operational bandwidth will be high. The opposite behavior is to be found for the T&H mode. The gate is closed over a long time so that the voltage across the capacitor follows the input signal. This will result in a higher measurement value (higher efficiency). But with increasing frequency, the signal source is no longer able to drive the capacity even if a drive amplifier is inserted. If the gate is open, the circuit is in the hold state. The capacitor voltage drops only slowly and a simple ADC is able to capture the value.

Multiplicative sampling circuits are more stable than additive sampler ones but they are also more power consuming. Sampling circuits can be built for a bandwidth from DC up to 100 GHz approximately.

17.5.4 UWB Principles for Volume Application

Costs, size, robustness, and power consumption of the sensor electronics will finally decide against or in favor of a volume application of UWB principles in moisture sensing. With the exception of laboratory tasks, sophisticated classical wideband systems such as network analyzers or TDR-sampling scopes will hardly find use in field applications. However, pushed by the progress in semiconductor technology, there exist some UWB principles which will have the power to transfer UWB moisture sensors into large-scale applications. Three of them will be discussed shortly in what follows.

FMCW principle: The FMCW method permits a simple and cost-effective system implementation. It uses a frequency-modulated continuous wave (FMCW) signal to stimulate the test objects. For that purpose a VCO is swept linearly over the frequency band of interest. The system response is transformed by the homodyne concept (see Fig. 17.11) into the low-frequency range where the signals can be easily captured by general purpose digital hardware. The commercial availability of different VCOs provides a great deal of different operational frequency bands. Due to the large bandwidth, the FMCW method has to waive the quadrature demodulation, so only the real part of the FRF can be measured. This greatly limits the signal processing since only part of the information about the DUT is available.

Let us consider for simplicity the wave transmission between two frequency-independent antennas. The complex FRF of such an arrangement looks like a screw thread line ($\underline{H}(f) \sim \exp(j2\pi f \tau)$) if visualized in 3D coordinates (real axes – imaginary axes – frequency axes). The "thread lead" depends on the traveling time τ which can be a measure of moisture, for example. The projection of such a line into the real-axes – frequency-axes plane results in a sinusoid as function of f having a period proportional to τ. Since in the case of FMCW the VCO sweeps linearly in time, frequency and time can be interchanged. Thus the measured signal is finally a "time-dependent" sinusoid whose frequency gives access to the traveling time τ of the wave. However, each deviation from a linear frequency sweep causes remarkable errors which must be suppressed by additional means [7, 8].

Pulse method: The pulse method represents another principle which profits from the simplicity of the electronic circuits. The DUT is excited by a short, i.e., wideband, pulse and the response of the DUT is usually obtained by sequential sampling (see Fig. 17.16). The pulse generation is typically based on avalanche principles, step recovery diodes, or tunnel diodes. The peak power of the stimulation pulse largely determines the *SNR* value. Very simple, power, and cost-efficient pulse modules can be built up to a few gigahertz bandwidth, particularly if they are equipped with additive samplers [6]. But such simple solutions usually suffer from some drawbacks. As a matter of principle, the strobe pulse generation tends to drift and produces jitter (compare Fig. 17.9 and Eq. (17.26)) and a non-linear ramp results in timing errors. Thus additional and expensive measures must be taken to improve the behavior. Sophisticated pulse systems are available up to 100 GHz bandwidth.

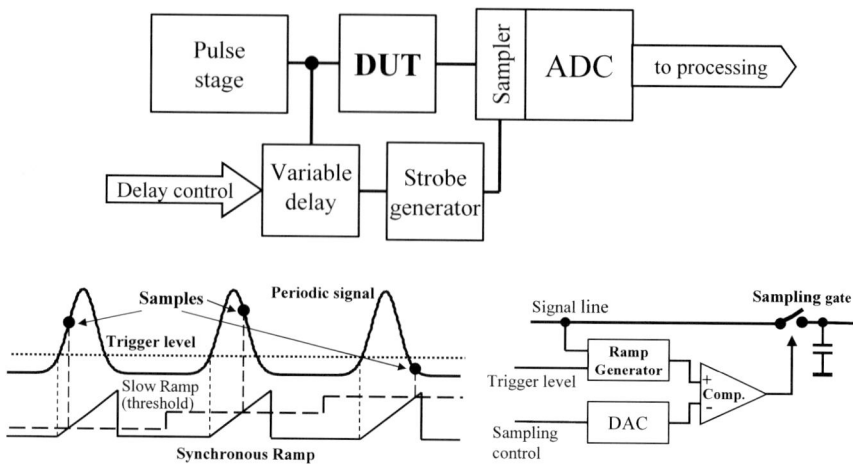

Fig. 17.16. Typical structure of a pulse system (top). The example shown uses sequential sampling for data gathering in which the strobe pulse is triggered from the stimulus via a variable delay. The variable delay is performed by sweeping a threshold (dashed line) over a voltage ramp which is synchronized with the stimulus (bottom)

MLBS approach: The direct use of a maximum length binary sequence (MLBS) with microwave frequencies in connection with under-sampling is a further approach promising superior performance at a bandwidth up to the 10 GHz range. Mainly its stable operation gives access to sophisticated data processing which is the key issue of all UWB measurement methods.

Figure 17.17 demonstrates the basic concept and a practically implemented RF part. A fast digital shift register with an appropriate feedback is able to provide an MLBS having a bandwidth up to several gigahertz. This signal stimulates the DUT. The MLBS is a periodic signal consisting of numerous pulses having a variable width. They are quasi-randomly distributed over the whole period. In contrast to the classical impulse excitation, the MLBS distributes the pulse energy over the complete measurement time. Thus the signal amplitude may be comparatively low even if a large energy is required in order to gain a certain *SNR*. Signals with low amplitudes are easy to handle and they promote monolithic circuit integration, resulting in an improved RF behavior. As in the pulse method the wideband measurement signal is captured by under-sampling. But now it is controlled by a binary divider. This is a very effective method since drift and jitter are largely suppressed due to the steep flanks of the divider and furthermore it absolutely avoids nonlinear sample spacing.

The period length of an MLBS corresponds to $N = 2^n - 1$ clock cycles in which n is the length of the shift register. If the equivalent sampling rate f_s is to be chosen equal to the clock rate f_c the usable bandwidth will cover the range from DC to $f_c/2$ and the measured system response functions will consist of N data points. Moreover it can be seen from Eq. (17.27) that any under-sampling factor of a power of 2, $u_{sf} = 2^m$, is applicable because $2^n - 1$ can never be divided by 2. See [9 – 12] for more information on the MLBS principle.

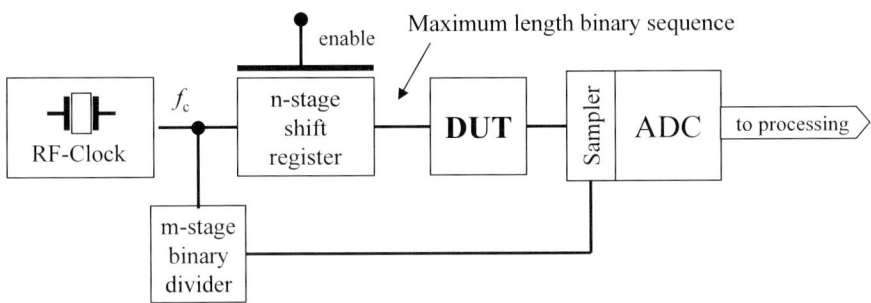

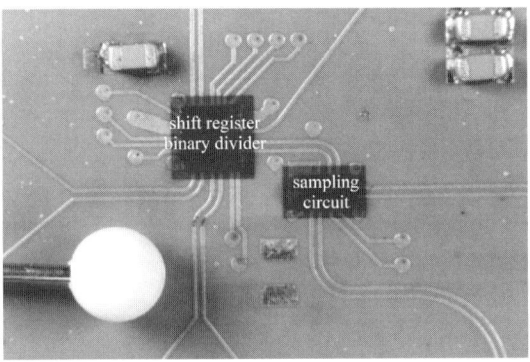

Fig. 17.17. Basic structure of MLBS baseband UWB system (top) and an example of the RF front end (except the clock generation) of a version for 10 GHz clock rate (bottom, courtesy of MEODAT GmbH). The pin head on the left is for comparison of dimensions

Admittedly the MLBS principle requires a more complex circuit structure compared to the basic versions of the above-mentioned pulse and FMCW methods. Nevertheless it can be economically manufactured thanks to modern low-cost RF and semiconductor technologies. In connection with its outstanding features the MLBS approach open up a great deal of future applications in UWB moisture sensors.

As expected, the power consumption of the MLBS RF part is higher than for simple pulse systems with additive samplers. However, concerning the total power consumption, it plays only a secondary role since the required signal processing usually has a greater power need. Furthermore it should be underlined that the MLBS approach enables in a simple way the building of fast-operating multi-channel arrangements [9]. It is much less expensive with this technique to equip every measurement channel with its own signal source (shift register) and receiver circuit (T&H or S&H) than to use RF switches. The shift register can be controlled by a simple TTL signal, so no additional RF hardware is required to build a multi-channel arrangement.

UWB-modulated carrier: It is often desirable to work in a certain frequency band. The FMCW method gives this possibility. The pulse- and MLBS approaches are restricted to the baseband. By combining them with the homodyne concept, the situation can be changed. Figure 17.18 shows a possible circuit structure using an MLBS to modulate the carrier. Note that it is more suitable to modulate by an MLBS than by a pulse, because it does not burden the modulator with short high-power shocks.

The advantage compared to the FMCW method is mainly to be seen in the possibility to gather completely the information content of measurement data due to quadrature demodulation. In contrast to the FMCW method a 90° phase shifter for the quadrature demodulator can now be built since the device is working with a fixed carrier frequency.

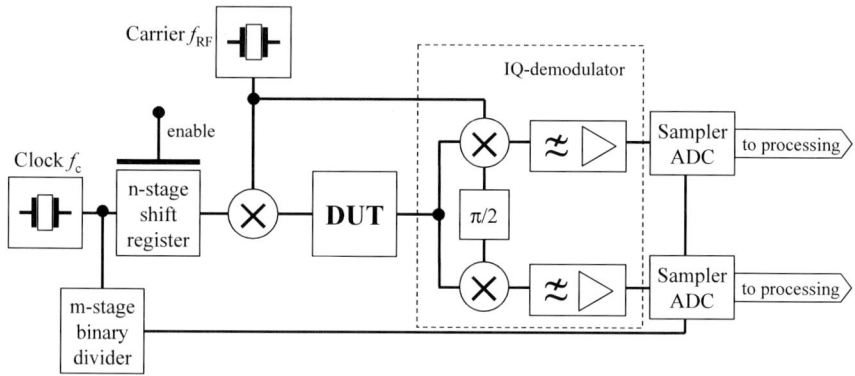

Fig. 17.18. Basic structure of a carrier frequency MLBS system. The measurement signal is divided into three parts: (1) the carrier which is known a priori; (2) a baseband signal describing the even part of the spectrum – detected by the upper channel; (3) a baseband signal describing the odd part of the spectrum – detected by the lower channel

17.6 Examples

This section refers to some simple laboratory experiments which demonstrate the versatility and the usability of the UWB technique. It is not the intention of the author to deeply interpret the measurement results with respect to quantitative statements. Furthermore, the demonstrations are limited for simplicity to

SiSo sensors (single-input – single-output) even if it is an easy task to upgrade the measurement head to a multi-channel system. All experiments are based on MLBS RF front end, as shown in Fig. 17.17, which was used in the measurement arrangements as indicated in Fig. 17.19.

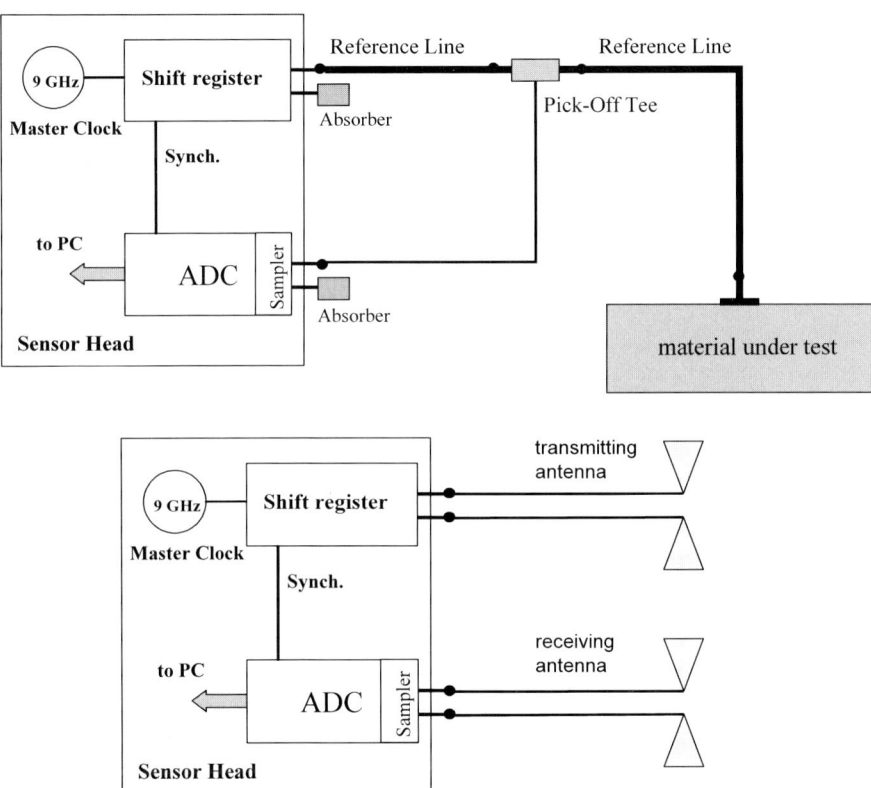

Fig. 17.19. Typical UWB measurement arrangement. Top: Time domain reflectometry (TDR) arrangement for measuring the reflection of an open coax line. Bottom: transmission arrangement for imaging purposes or if waves should cross the object under test

Water diffusion in a brick: Corresponding to the schematics on the top of Fig. 17.19, an open coaxial line was placed on the top side of a brick. Four minutes after the start of the measurement, the bottom side of the brick was exposed to fresh and in a second experiment to salty water. The temporal changes in the reflection coefficient of the probe were captured. They are presented for both experiments in Fig. 17.20. As expected, no effect appears when the water was added to the bottom side. But after approximately 20 min, the water arrived at the upper surface by diffusion and the reflection coefficient of the probe changed in both amplitude and phase. To make the presentation in Fig. 17.20 more clear, the same situation is pictured in Fig. 17.21 representing the data in the complex plane.

salty water   fresh water

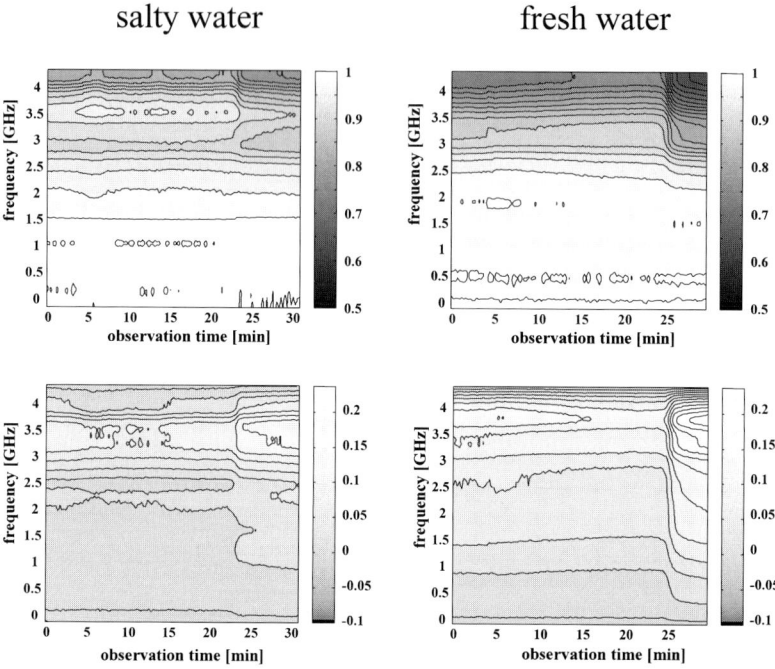

Fig. 17.20. Contour plots showing the temporal variation of the magnitude (top) and the phase (bottom) of the normalized reflection coefficient of the open coax line gained from TDR measurements. The considered frequency band covers DC to 4.5 GHz

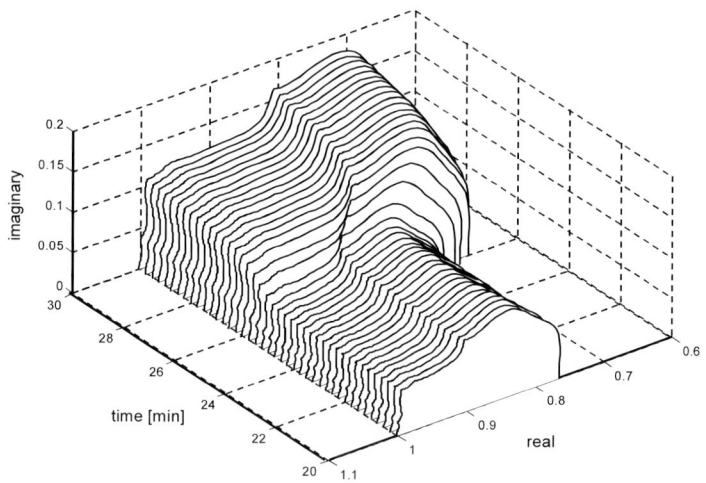

Fig. 17.21. Variation of the normalized complex reflection coefficient. The data shown refer to the fresh-water example

Fig. 17.22. Laboratory test arrangement using 3 by 5 bricks. The antennas used were of the Vivaldi-type

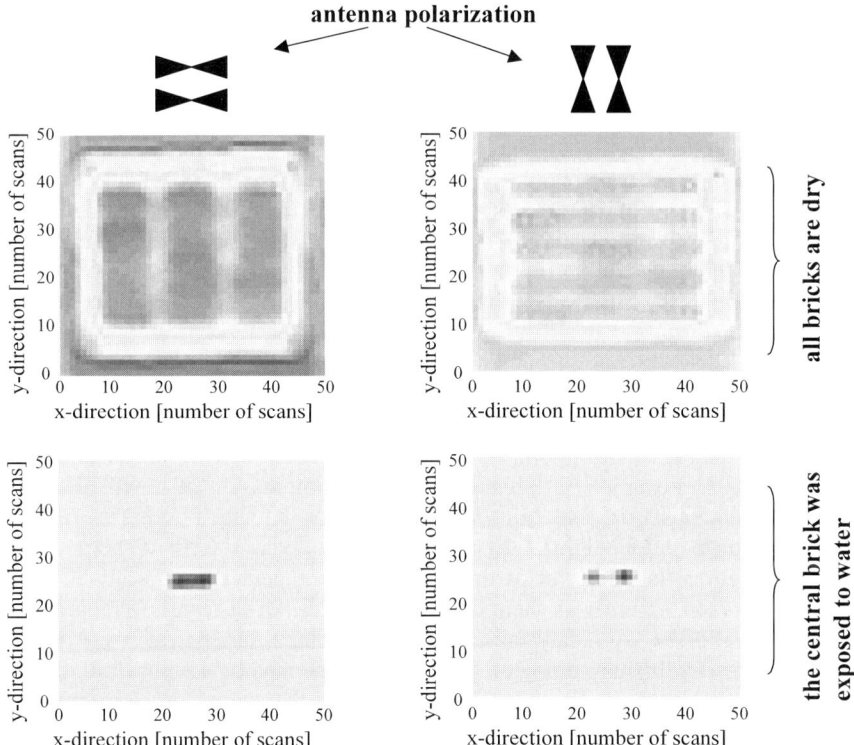

Fig. 17.23. C-scan representation of the scattered waves for different polarization. It should be noted that in the lower images the gaps between the brick are not visible because of a bigger scaling factor. By applying the same scaling factor to all images they will appear once again

Microwave imaging: The UWB technology offers interesting possibilities to build hand-held or automatic scanners in order to capture "moisture images" or moisture profiles respectively. The high bandwidth of the measurement electronics as well as of the antennas results in a range and cross-range resolution of a few centimeters. The following pictures demonstrate an example in which the signal processing is limited to a minimum in order to give an impression of the quality of the measurement data. It is interesting to note that by deploying array processing or synthetic aperture algorithms the image resolution can be greatly increased.

The goal of the experiment was to investigate the influence of dry and moist bricks on the scattering of electromagnetic waves. For that purpose, an arrangement of 3 by 5 bricks (see Fig. 17.22) was automatically scanned by a grid of 2 by 2 cm. Figure 17.23 shows some radar images. Obviously, dry and moist stones differ and it can also be seen that the polarization of antennas emphasizes either the lengthwise or cross-gaps between the bricks. A C-scan is also often called as time slice. It represents the reflectivity at a certain depth.

Time – frequency analysis [13, 14]: As mentioned above, all accessible information about the sensor behavior is included in the IRF or FRF. The problem which often arises is to extract it from these functions. Propagation effects such as scattering at boundaries or similar can be easily interpreted in the time domain. However, the frequency domain is preferred in connection with wave dispersion caused by a frequency-dependent permittivity as for water, for example. A new approach to data analysis should help to overcome these drawbacks. It is the so-called time – frequency representation (TFR) which joins the advantage of the pure IRF and FRF representation and makes it possible to analyze signals in the joint time – frequency domain.

One distinguishes between linear and non-linear TFRs. The well-known Wavelet transform and the short time Fourier transform count among the linear types. Linear TFRs have a time – frequency resolution which does not meet our requirements. Non-linear TFRs have a better resolution but they provoke cross-terms which complicate the data interpretation. One of the best known non-linear TFRs is the Wigner distribution.

The following example is intended to give an impression of how TFR works. The experiential set-up was built from a sandbox having a footprint of about 50 cm by 50 cm and a height of 20 cm. The sandbox was placed on sheet metal of the same size as the footprint of the box. This arrangement was scanned by the same UWB radar scanner as mentioned in Fig. 17.22. The resulting radargram is given in Fig. 17.24. A radargram or B-scan displays the IRF at every scanner position whereby the magnitude of the IRF is represented by a color. In the central region of the radargram, it can be clearly observed that signals arise by reflection at the surface and the ground (sheet metal) of the sandbox and by multiple reflections. Outside the sandbox, to the right and left, only the surface reflection of the laboratory floor appears. Since the sheet metal and floor are on the same level, the image shows well the lower propagation speed within the sand.

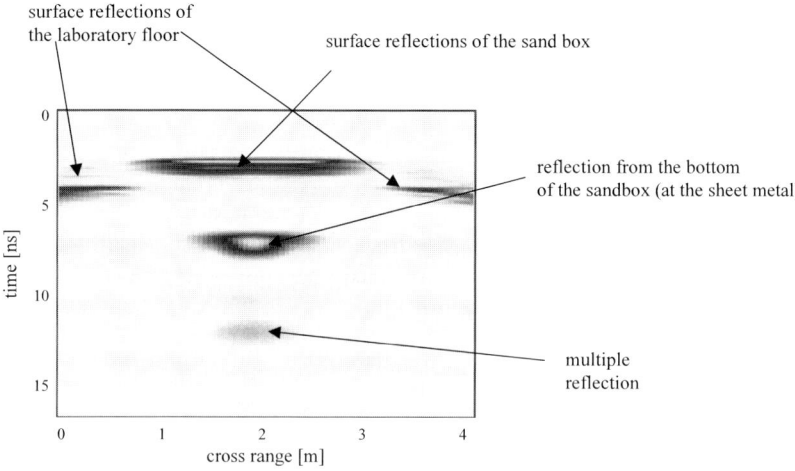

Fig. 17.24. Radargram of the sandbox (Courtesy of R. Zetik)

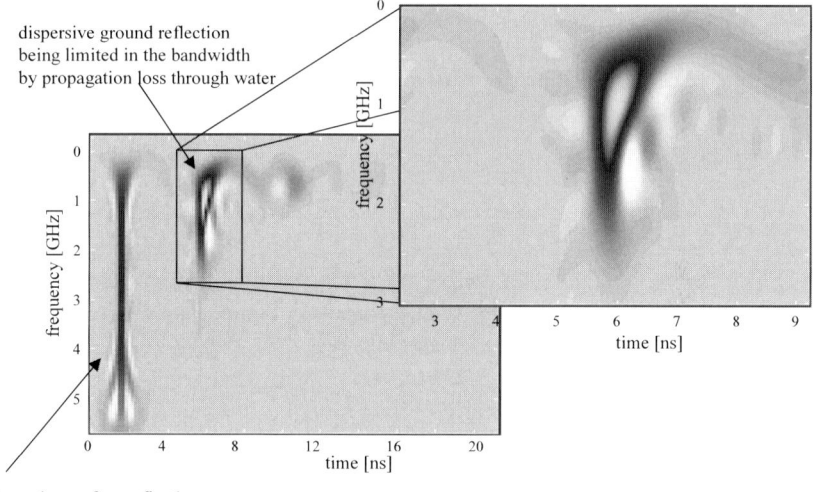

Fig. 17.25. Joint time – frequency representation of the central IRF of the radargram obtained by back scattering from a box containing moist sand. The ordinate of the images represents the propagation time of the sounding wave. The abscissa corresponds to the frequency and the color (gray scale) is a measure of energy. (Courtesy of R. Zetik)

In order to deal with the TRF, one IRF was isolated from the central part of the radargram in Fig. 17.24 and considered separately. Figure 17.25 shows the joint time frequency representation of this IRF.

How do we interpret this image? The first impulse is caused by reflection at the surface of the sandbox. The sounding wave propagates only through air, thus no dispersion arises within the TFR. That means all spectral components of the pulse are delayed by the same time. This will give rise to a trace parallel to the frequency axis. The length of the trace reflects the bandwidth of the pulse.

The pulse which travels through the sandbox and back is subjected to a dispersion the strength of which depends upon the moisture content. That means the spectral components of the low-frequency part are delayed more strongly than the higher spectral components. Furthermore, higher frequencies will be attenuated. In the TFR, this results in a skewing of the impulse trace and a reduction of its length. For simplicity, we used only the skewing for the first trials to demonstrate the moisture effect. The results are presented in Fig. 17.26. But the joint time – frequency approach offers many more possibilities to separate different effects.

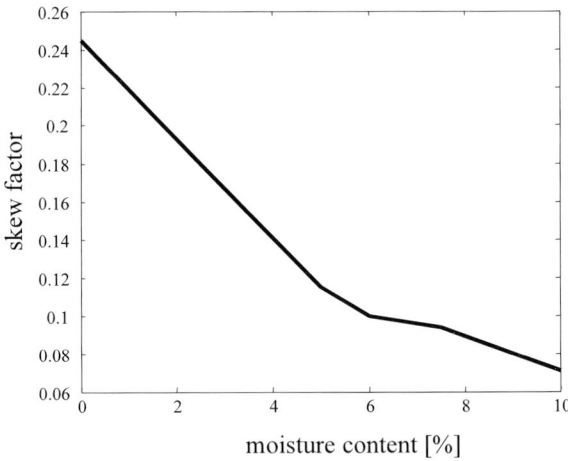

Fig. 17.26. Skew factor in dependence on the humidity of the sand. (Courtesy of R. Zetik)

17.7 Summary

UWB methods will open a promising perspective for further improvements in microwave moisture sensors and electromagnetic sensors in general. Measurements over a large frequency range are able to provide more information about the test object due to more and better data.

A successful application of UWB sensor principles requires:

- appropriate UWB applicators (antennas, electrode configurations, etc.),

- adequate UWB electronics, and

- a suitable model of the test arrangement in connection with (usually digital) data processing for extraction of the required information.

The greatest hindrance to the application of UWB sensor principles has been the lack of cost-effective, small-sized, power-saving, and robust UWB electronics in the past. But this situation is now changing. Different UWB electronics are available and under development. These electronics are supported by permanent improvements in size, power consumption, costs, and numeric power of the digital hardware which is inevitable in UWB sensors.

This chapter gave a short view of the philosophy of UWB measurements in order to enable the readers to estimate the potential of the UWB technique and its problems. Furthermore, some hints were given on how to choose the right measurement principle for the users purpose. But finally attention will be drawn the to the fact that the success of a UWB measurement strongly depends also on the quality of the signal processing.

References

1. Federal Communications Commission (2002) FCC 02-48, 14 Feb
2. Mirabbasi Sh, Martin K (2000) Classical and modern receiver architectures. IEEE Commun Mag Nov: 132-139
3. Stephan R, Loele H (1999) Ansätze zur technischen Realisierung einer Geschwindigkeitsmessung mit einem Breitband-Rausch-Radar. In: Proceedings of the workshop of the German IEEE/AP chapter on short range radars, Technische Universität Ilmenau, July 1999, pp 65-70
4. Narayanan RM, Xu Y, Hoffmeyer PD, Curtis JO (1995) Design and performance of a polarimetric random noise radar for detection of shallow buried targets. Proc SPIE 2496:20-30
5. Zollinger E (1993) Eigenschaften von Funkübertragungsstrecken in Gebäuden. Ph.D. thesis, ETH no 10064, Swiss Federal Institute of Technology, Zurich
6. Barrett TW (2001) History of ultra wideband communications and radar: Part II, UWB radars and sensors. Microwave J Feb
7. Musch T (2002) A high precision 24 GHz FMCW-radar using a phase-slope signal processing algorithm. In: Proceedings of the 32nd European microwave conference, Milan, vol 3, pp 945-948
8. Heide P, Vossiek M, Nalezinski M, Oréans L, Schubert R, Kunert M (1999) 24 GHz short-range microwave sensors for industrial and vehicular applications. In: Proceedings of the workshop of the German IEEE/AP chapter on short range radars, Technische Universität Ilmenau, July 1999, pp 4-9
9. Sachs J, Peyerl P (1999) A new principle for sensor-array-application. In: Proceedings of 16th IEEE instrumentation and measurement technology conference, IMTC/99, Venice, 24-26 May 1999, pp 1390-1395
10. Rossberg M, Sachs J, Rauschenbach P, Peyerl P, Pressel K, Winkler W, Knoll D (2000) 11 GHz SiGe circuits for ultra wideband radar. In: Bipolar/BiCMOS circuits and technology meeting, BCTM-2000, 25-26 Sept, Minneapolis
11. Sachs J, Peyerl P (2001) Integrated network analyser module for microwave moisture sensors. In: Proceedings of fourth international conference on "electromagnetic wave interaction with water and moist substances", 13-16 May 2001, Weimar, p 165ff

12. Sachs J, Peyerl P, Kmec M, Tkac F (2002) Digital ultra-wideband-sensor electronics integrated in SiGe-technology. Proceedings of 32nd European microwave conference, Milan, vol 2, pp 539-542, Milan

13. Zetik R, Sachs J (2002) Moisture determination of solid material by means of ultra-wideband radar and time-frequency signal representations. Acta Electron Inf 2(1):15

14. Zetik R (2000) Dual L-Wigner distribution and applications of time-frequency signal representations in ultra-wideband radar systems. Dissertation, University of Technology in Košice

Methods and Sensors for Quality Assesment to Products of Agriculture, Food, and Forestry

18 Permittivity Measurements and Agricultural Applications

Stuart O. Nelson, Samir Trabelsi

U. S. Department of Agriculture, Agricultural Research Service,
Athens, GA, USA

18.1 Introduction

The electrical characteristics of agricultural materials have been of interest for many years. One of the earliest applications was the correlation noted between the electrical resistance, or conductance, of wheat grain and its moisture content [10]. This relationship was eventually utilized for the development of electrical moisture meters that became commonly used in the grain trade [42]. Later, studies on the use of radio-frequency (RF) instruments for rapidly determining moisture content in wheat and rye were reported [8], and grain moisture meters were developed that were based on the relationships between grain moisture content and the capacitance of parallel-plate or coaxial grain sample holders [42]. Moisture measuring instruments based on both principles of DC conductance and RF capacitance measurements were available for many years and both principles are still in use today, though the convenience and sophistication of such moisture meters are much improved in modern instruments.

With the advent of RF dielectric heating, and later microwave heating, potential agricultural applications were explored, and the need was recognized for data on the dielectric properties, or permittivities, of the agricultural products that were being considered for these applications. In connection with research on RF dielectric heating exposures of grain, a technique was developed for reliable measurement of the dielectric properties of grain, and the method and the first data for grain were reported in 1953 [62]. A few years later, dielectric properties were reported for several different kinds of grain and seed at frequencies between 50 kHz and 50 MHz [24, 34].

Upon recognition that dielectric properties data would be of reference value for moisture meter design and dielectric heating applications, such data for grain and seed were organized and first published by the American Society of Agricultural Engineers in 1966 [2]. These data have been published annually and updated on several occasions. They currently appear in the ASAE Standards 2002 as ASAE D293.2 Dielectric Properties of Grain and Seed [4].

Effectively using the dielectric properties of grain for moisture sensing requires that information be obtained on the influence of other variables on these properties as well. Measurements have been taken over ranges of frequency on a number of types of grain and seed at different moisture levels, bulk densities, and temperatures. Dependence of the dielectric properties of grain on these variables was also summarized for reference purposes [64, 45, 46]. In another study, dielectric properties of corn, wheat, and soybeans were measured at 1 to 200 MHz with respect to factors affecting their use for moisture sensing [20].

Another study of the dielectric properties of wheat, barley, and rice was conducted for purposes of evaluating moisture sensing instrumentation for control of grain dryers [5]. Measurements included frequencies of 0.1, 1, and 10 MHz, and effects of temperature and kernel moisture distribution.

The dielectric properties of grain as well as those of grain-infesting insects were needed in research on RF dielectric heating for potential control of the insects through selective heating of the insects [66], and permittivity data for hard red winter wheat and adult rice weevils were reported for a wide range of frequencies, from 250 Hz to 12 GHz [54].

Permittivity measurements for a wide range of different kinds of grain and seed were also made in connection with studies of dielectric heating exposures for improving the germination of these seeds [65]. These dielectric properties data were also summarized for reference [4].

The potential use of permittivities for quality sensing in fruits and vegetables prompted the measurement of microwave permittivities of several fruits and vegetables for background information on their dielectric properties [43, 47, 85, 57]. Some measurements were taken on fruits of different maturities over the frequency range from 200 MHz to 20 GHz [58].

Of course, many measurements have been reported for food materials in connection with microwave cooking applications, but the scope of this chapter is limited to agricultural products prior to processing for food purposes. Some of the permittivity measurement techniques will be reviewed, and some of the newer dielectric properties data will be presented.

In addition, applications of permittivity or dielectric properties data will be discussed with respect to applications such as RF dielectric or microwave heating for product drying, insect control, seed treatment, and product conditioning. The use of permittivities for quality sensing and grain and seed moisture measurement will be discussed further, and other potential applications will also be mentioned.

18.2 Permittivity Measurements

The permittivity and dielectric properties as used in this chapter are defined as follows:

$$\varepsilon = \varepsilon' - j\varepsilon'' = |\varepsilon|e^{-j\delta} \tag{18.1}$$

where ε is the complex permittivity relative to free space, ε' is the dielectric constant, ε'' is the dielectric loss factor, and δ is the loss angle of the dielectric where $\tan\delta = \varepsilon''/\varepsilon'$. The AC conductivity $\sigma = \omega\varepsilon_0\varepsilon''$ S/m, where $\omega = 2\pi f$ is the angular frequency, with f in Hz, and ε_0 is the permittivity of free space, 8.854×10^{-12} F/m.

Hereafter in this chapter, "permittivity" will imply the complex relative permittivity. The dielectric constant is associated with the capability for energy storage in the electric field in the material and the loss factor is associated with energy dissipation in the material or the conversion from electric energy to heat energy. Here, all loss mechanisms, due to both dipole relaxation and ionic conduction, are included in the dielectric loss factor ε''.

18.2.1 Measurement Techniques

The first permittivity measurements on grain samples were obtained by adaptation of the reactance variation technique [19] for use with the Boonton 160A Q-Meter[1] [87], and development of a suitable new sample holder for use with grain and seed samples [62]. The measurement was based on a series resonant circuit and employed the replacement of capacitance lost when a sample was removed from the sample holder by a calibrated variable capacitor in the sample holder, connected in parallel with the sample holder capacitance and the main tuning capacitor of the Q-Meter. The LC resonant circuit was tuned to resonance when the sample was present in the coaxial sample holder. The sample was then removed, and resonance was restored by adjusting the calibrated variable capacitor in the sample holder for this measurement. The Q of the circuit changed also when the sample was removed, and this was taken into account by measuring the change in capacitance (reactance variation) necessary to lower the peak voltage V_m reading across the capacitive portion of the resonant circuit to a chosen reference voltage V on both sides of the resonant peak, both with the sample in the sample holder and with the sample removed. These two capacitance values were then used with the voltage ratios V_m /V to calculate the loss tangent, $\tan\delta$ [62]. Measurements on many grain and seed samples were taken in the frequency range from 1 to 50 MHz with this measurement system and reported [34, 4].

For measurements in the 50 – 250 MHz range, a coaxial-electrode sample holder was designed for use with the Boonton 250-A RX Meter, which consisted of variable frequency oscillators and a modified Schering impedance bridge circuit. The sample holder, with a built-in open-circuit termination, was analyzed as several transmission line sections with lumped-circuit element values determined from measurements on standard materials. By measuring the impedance at the input to the sample holder, the dielectric properties of the grain sample, which filled the sample-holding section of the sample holder, were then calculated directly from the complex impedance [21].

The frequency for permittivity measurement capability for grain samples was then extended to the 200 – 500 MHz range through use of a newly designed coaxial sample holder for use with the General Radio Type 1602-B Admittance Meter.

This coaxial sample holder with built-in open-circuit termination to enclose the sample was also modeled as transmission line sections, and lumped-circuit values

---

[1] Mention of company or trade names is for the purpose of description only and does not imply endorsement by the U S Department of Agriculture.

were determined from measurements on standard materials. The permittivity of the sample in the sample-holding portion was then determined by calculation from the admittance measured at the input to the sample holder [72].

To extend the measurement capability to higher frequencies, the short-circuited transmission line technique of Roberts and von Hippel [71] was utilized with several measurement systems, including the X-band rectangular waveguide [37], 21 mm coaxial line [40], 25.4 mm coaxial line (1 and 3 GHz) and circular waveguide (8.5 GHz) [54], and K-band rectangular waveguide [47]. A general computer program for precise calculation of dielectric properties of high- or low-loss materials from short-circuited waveguide measurements was also developed for use with all of these and other measurement systems [67, 68], which was widely used for many dielectric properties measurements and studies at microwave frequencies.

Free-space measurement systems, consisting of transmitting and receiving antennas connected to a vector network analyzer, with grain and seed samples confined in sample holders between the antennas, were used for permittivity measurements in the 5 – 18 GHz range [27, 80, 83]. Measurements of attenuation and phase shift as the wave traversed the grain or seed layer provided the information needed for calculation of the dielectric properties. The problem of phase ambiguity for samples in which the phase shifted more than 180° was resolved [81], a one-way attenuation of 10 dB in the sample rendered the effects of internal reflections and reflections between the antennas negligible, approximations requiring that $\varepsilon'' \ll \varepsilon'$ were satisfied, and other techniques were used to improve the accuracy of the measurements [83].

For permittivity measurements on fresh fruit and vegetable tissue at 2.45, 11.7, and 22 GHz, core samples were cut to fit short-circuited coaxial line and rectangular waveguide sample holders [43, 47] and the Roberts and von Hippel short-circuited line technique was used [71] with computation of permittivity by the general computer program for this technique [68]. For broadband permittivity measurements on fruits and vegetables in the 10 MHz - 20 GHz range, open-ended coaxial line measurements with network and impedance analyzers were used [56, 51].

18.2.2 Permittivity Data

Available reliable data for the dielectric properties of agricultural products were tabulated 30 years ago [38]. Included were data for animal tissues, foods, plant material, fruits and vegetables, grain and seed, wood and textiles. This tabulation also listed the frequency for which the dielectric constant, loss factor, loss tangent, and conductivity were listed and, when known, the moisture content, temperature, and specific gravity, or density, of the materials, because these factors all influence the permittivity of agricultural materials. Some of these data were also included in later tabulations [75, 22].

The dielectric properties data for grain and seed, which were deemed useful for reference, as already mentioned, currently provide permittivity data for a reasonably wide range of grain and seed types at frequencies from 1 to 50 MHz [4]. Data

are also given for several crops in the audio-frequency range, and models for estimating the dielectric constant at 24°C of several major grain crops and soybeans
are given as functions of frequency (20 MHz - 2.45 GHz), moisture content and
bulk density of the grain and seed. The most comprehensive data are given for
hard red winter wheat in the form of contour plots for ε' and ε'' as functions of
frequency and moisture content [45], as shown in Fig. 18.1. Here, it is noted that
the behavior of ε' is much more regular with frequency and moisture content than
is that of ε''.

Some new data on the frequency and temperature dependence of the dielectric
properties of fresh fruits and vegetables have recently been obtained [51]. The
typical variation of the dielectric constant and loss factor of fresh avocado is
shown in Fig. 18.2. The very high values for ε' at the lower end of the frequency
range are no doubt attributable to the polarization contributed by ionic conduction,
while the behavior of ε' at the higher end of the frequency range is characteristic
of dipolar relaxation. It is evident in Fig. 18.2 that at about 90 MHz the temperature dependence of ε' disappears, and the ionic conduction becomes the dominant
mechanism influencing the value of ε' below that frequency. This phenomenon
was noted for all the fruits and vegetables at some frequency in the 10 - 100-MHz
range.

Data for the permittivity of nine fruits and vegetables are shown for a temperature of 25°C at frequencies of 10 and 100 MHz and 1 GHz in Table 18.1.

Table 18.1. Permittivities of fresh fruits and vegetables at indicated frequencies

Fruit or vegetable	10 MHz		100 MHz		1 GHz	
	ε'	ε''	ε'	ε''	ε'	ε''
Apple	109	281	71	33	64	10
Avocado	245	759	66	89	56	14
Banana	166	834	76	91	65	18
Cantaloupe	260	629	70	72	63	14
Carrot	598	1291	87	157	72	23
Cucumber	123	361	80	39	77	9
Grape	122	570	78	60	73	13
Orange	197	617	78	69	72	14
Potato	183	679	73	77	62	16

The data in Table 18.1 show considerable variation among the different fruits
and vegetables. Both real and imaginary parts of the permittivity are particularly
large for carrot tissue at the lower frequencies, but these differences largely diminish at microwave frequencies.

Dielectric properties data showing further behavior of these properties with
respect to frequency, moisture content, and other variables are summarized elsewhere for grains, fruits, vegetables, and other food materials [48, 55].

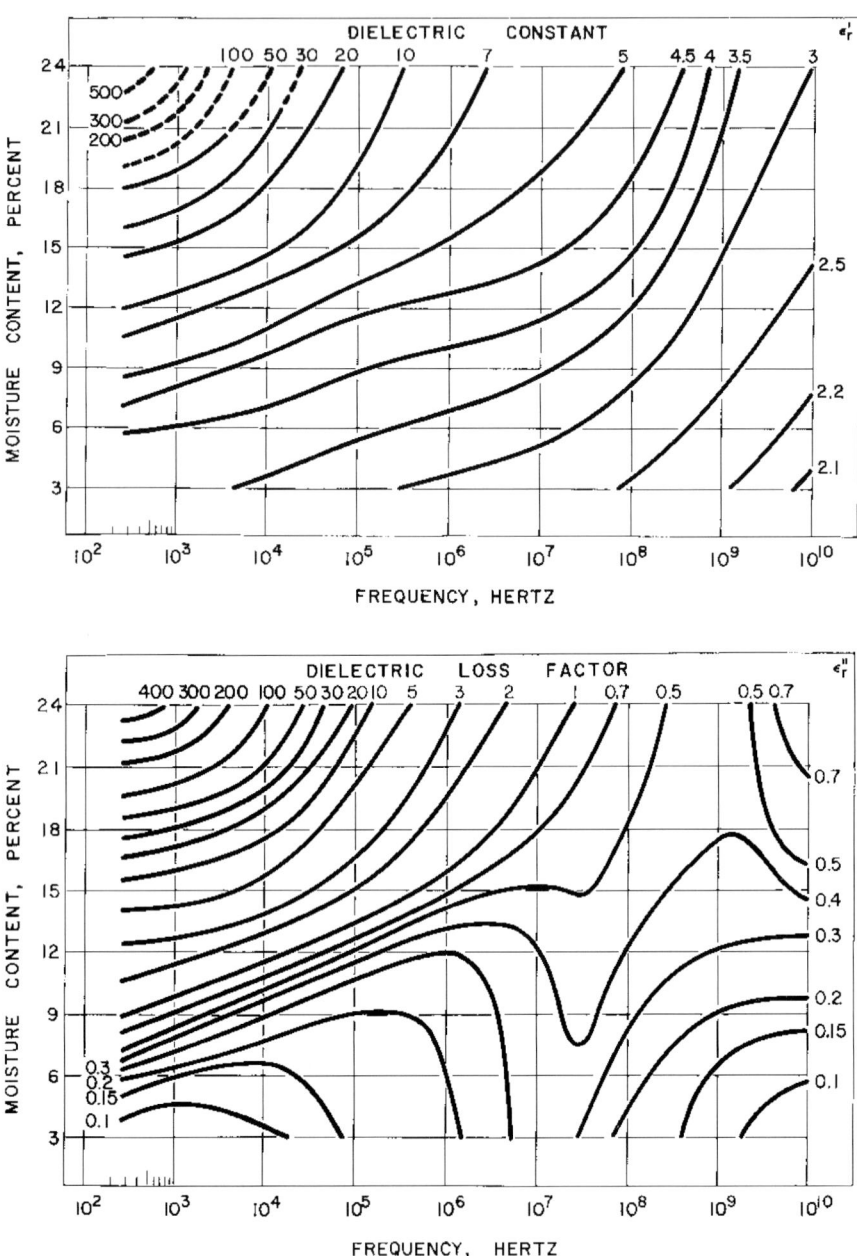

Fig 18.1. Dielectric properties of hard red winter wheat as a function of frequency and moisture content, wet basis, at 24°C [46].

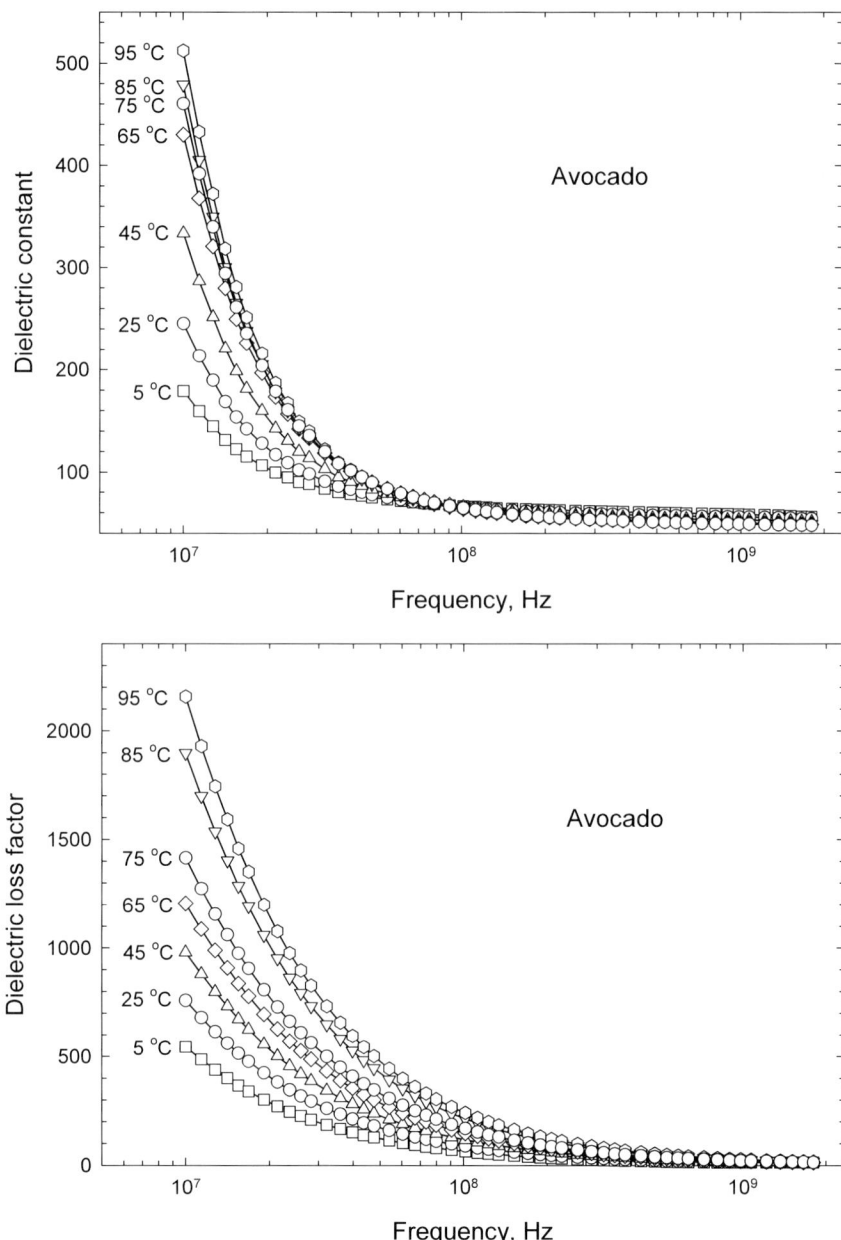

Fig. 18.2. Frequency and temperature dependence of avocado permittivity [51]

18.3 Dielectric Heating Applications

For most agricultural products, the unit value is too small to justify the economic costs of dielectric heating, either at high frequencies or at microwave frequencies. However, if either the speed of heating or an achieved improvement in product quality is sufficiently important, dielectric heating applications might be considered. A few applications will be cited here, but there are probably many more that have been given some consideration.

18.3.1 Product Drying

Drying plant products for safe storage is a common practice in agriculture. As soon as they are sufficiently mature, many must be harvested to avoid potential loss from inclement weather or other undesirable conditions that can develop. If the moisture content of products, cereal grain for example, is too high for short-term storage without spoiling, it must be dried, by circulation either of heated or unheated air through the material. Depending on the product, drying facilities, and the economic environment, the speed of drying can be important. For these reasons studies have been conducted to answer fundamental questions concerning the use of dielectric heating for drying.

Dielectric heating of materials involves the absorption of energy from the high-frequency or microwave electric fields created in the materials and its conversion to heat energy. The power dissipated per unit volume in the material, P, can be expressed as

$$P = 55.63 f E^2 \varepsilon'' \times 10^{-12} \quad \text{W/m}^3 \tag{18.2}$$

where f is frequency in Hz, and E is the electric field intensity in the material in V/m. The time rate of temperature, T, rise in the material as a consequence of the power absorption is then

$$dT / dt = 0.239 P / (c\rho) \quad °\text{C/s} \tag{18.3}$$

where c is the specific heat of the material and ρ is its specific gravity.

Early research on drying rice with 27 MHz dielectric heating showed promise for continuous drying of rice from field conditions to safe storage levels without damaging milling quality [90]. Experimental high-frequency drying of grain and grass seed at frequencies from 1 to 12 MHz showed improvements in the drying process over conventional methods and that drying could be accomplished at lower temperatures with dielectric heating in combination with heated air drying [24].

Experiments with corn showed that it could be dried much more rapidly by microwave heating at 2.45 GHz than by conventional hot-air drying, but that physical damage to kernels could result if corn was dried too rapidly [17].

Other studies at 2.45 GHz and 915 MHz also showed that speed of drying had to be limited to prevent physical damage to kernels and that this might discourage the use of microwave energy for this purpose. Costs of electricity and equipment for microwave drying in the field were considered too high for practical use.

Microwave heating was investigated for drying seed cotton prior to ginning, but was found impractical [89]. Other studies on microwave and vacuum drying of cotton reduced drying time and was found to improve cottonseed oil marketing properties [1]. Microwave heating under partial vacuum for drying rice and soybeans resulted in products of improved quality, but higher costs have precluded practical use [15].

RF dielectric heating at 43 MHz, with moisture removal by circulation of unheated air, provided rapid drying of chopped alfalfa forage, and resulted in much improved carotene retention without affecting crude protein content [73]. The method could not be recommended for field-drying alfalfa, but did show promise for blanching samples to inactivate carotene-destroying enzymes as an aid in research. Combined microwave and unheated air drying at 2.45 GHz provided rapid drying of laboratory samples of forages and improved quality as well [14]. Such uses of microwave drying have been limited to laboratory and research applications, with costs being too high for consideration of practical-scale use.

18.3.2 Pest Control

Many pests must be dealt with in the production of agricultural crops and in the preservation of agricultural and food products. These include insects, nematodes, fungi, and bacteria that attack growing plants and insects and fungi that infest and infect products in storage. Weeds are also serious pests in the production of field crops. New methods for coping with pests are always being sought, and it is not surprising that high-frequency and microwave energy have been explored for these purposes.

Interest in controlling insects with high-frequency radio waves was recorded in the scientific literature more than 70 years ago. These reports, along with many since then on RF and microwave treatments for insect control, have been analyzed in previously published reviews [36, 39, 50, 49]. Most studies were related to control of stored-product insects and wood-infesting insects. Cereal grains and their products are among those most susceptible to insect infestation, particularly in the tropical and temperate regions of the world. Increasing concern about chemical pesticide residues in the 1950s and 1960s enhanced the interest in non-chemical methods for controlling such insects. Dielectric heating offered one possible alternative method.

Because the dielectric properties of insects and their host materials may differ, there is the possibility of selectively heating the insects to lethal temperatures through dielectric heating [74, 69, 36, 50]. The principle of selective dielectric heating is best explained by examining equations (18.2) and (18.3). Because the heating rate depends on the power dissipation P and the specific heat and specific gravity of the materials, those variables that influence P will be important in determining differential heating of components of a mixture. Thus, if a mixture of materials, such as insects and grain kernels, is subjected to dielectric heating by high-frequency or microwave electric fields, the relative power absorption will depend upon the relative values of the electric field intensity E and the dielectric loss factor ε'' for each of the materials in the mixture. The values of ε'' are characteristics of the materials and can be measured. The values for E in the different materials are more difficult to determine, because the field intensity distribution depends upon the dielectric constants ε' of the two materials and geometric factors [54]. Furthermore, both dielectric properties ε' and ε'' can vary with the frequency of the applied electric field.

Measured values for these properties of bulk samples of hard red winter wheat, *Triticum aestivum* L., and adult rice weevils, *Sitophilus oryzae* (L.), are shown in Fig. 18.3 [54]. Both materials exhibit broad dielectric relaxations between 1 MHz and 1 GHz, probably associated with bound water. It is interesting to note that the dielectric loss factor of the insects from 5 to 70 MHz is more than five times larger than that of the wheat. Based on these measurements, relative values for the electric field intensity E in the insect and in the grain were calculated according to a simple mathematical model, and power dissipation ratios for the insects and the grain were estimated. The most advantageous frequency range for selectively heating the insects was from about 10 to 100 MHz, where 3 to 3.5 times greater power dissipation could be expected in the insects than in the grain. Results of the analysis indicated that little selective heating of the insects could be expected at microwave frequencies (1 GHz or higher). Experimental treatment of wheat infested with adult rice weevils at frequencies of 39 MHz and 2.45 GHz confirmed this prediction [63]. Complete mortality of the insects was achieved by 39 MHz treatments at grain temperatures below 40°C, whereas exposures at 2.45 GHz had to produce temperatures in the grain above 80°C for comparable insect mortalities.

Laboratory and pilot-scale tests in Switzerland of high-frequency insect control equipment operating at 12.6 MHz [7] confirmed earlier findings, in work at about 40 MHz [66], that final grain temperatures of about 60°C were necessary for control of the most resistant stored-grain insects.

Dielectric properties data on stored-grain insects were recently extended to higher frequencies and their dependence on temperature as well as frequency was determined [52, 53]. The new data provided no evidence for more efficient treatment of stored-grain insects at higher frequencies as suggested recently [18]. Some studies with pulsed 47.5 MHz electric fields and high-frequency and vacuum treatments with plasma formation have been reported [33], but conclusive work remains to be completed. No economically feasible RF or microwave treatments for controlling stored-grain insects have yet been demonstrated.

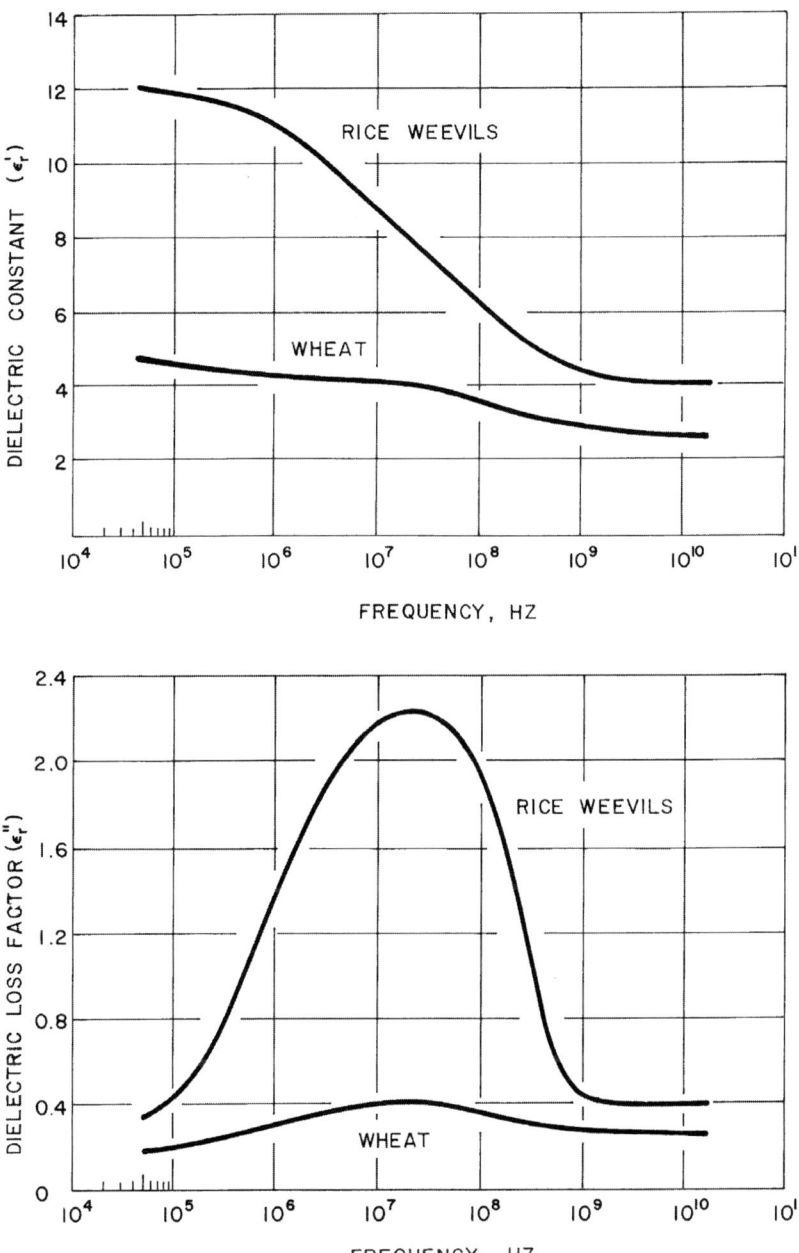

Fig. 18.3. Dielectric properties of bulk samples of hard red winter wheat (10.6% moisture content) and adult rice weevils (49% moisture content) [54].

Use of microwave energy has also been considered for soil treatment to control weed seeds, insects, nematodes, and soil-borne micro-organisms [12, 88]. However, consideration of the dielectric properties of soils and the attenuation to be expected render the possibilities of success extremely remote as explained in a review and assessment of microwave energy for soil treatment to control pests [49].

Very recently, tests were made to determine how effective RF dielectric heating of alfalfa seed, inoculated with human pathogenic bacteria, might be for controlling such infections of seed used for production of edible sprouts [59]. Dielectric heating treatments at 39 MHz provided significant reductions in bacterial populations, but did not control the bacteria, *Salmonella, E. coli* O157:H7, and *Listeria monocytogenes*, without serious reductions in alfalfa seed viability. However, the treatments did provide significant increases in germination while at the same time providing moderate reduction in bacterial populations.

18.3.3 Seed Treatment

Studies reported more than 60 years ago on high frequency or RF treatment of seeds to improve germination are cited in a previous review [35]. Since then a large number of experiments concerning RF and microwave treatments of various kinds of seed have been described. Such studies on alfalfa, a species that has always responded favorably to RF and microwave dielectric heating treatments, were summarized and cited in an earlier review [41].

Results of experimental treatment of seed of more than 80 plant species with RF and microwave dielectric heating have been summarized [65]. Some small-seeded legumes, such as alfalfa, red clover, and arrowleaf clover, which often have naturally impermeable seed coats, responded consistently to dielectric heating treatments with marked increases in germination. Others, such as sweetclover, did not respond nearly so well. Benefits from RF treatment have been retained in alfalfa seed for up to 21 years in storage after treatment with no evidence of any detrimental effects. Several vegetable and ornamental species did not show any improvement as a result of RF seed treatment, but some, such as okra, garden peas, and garden beans, responded favorably. Germination of spinach was consistently accelerated [60].

In general, seeds of grasses, woody plants, and tree species did not respond very well. However, acceleration of germination was noted in some tests, and increases in germination were sometimes noted. Tests with field crops such as corn, cotton, and wheat showed acceleration of germination in some lots, as a result of treatment, but not in others. In those kinds of seed that did respond favorably, moisture content of the seed at the time of treatment influenced the degree of response. Generally, seeds of lower moisture content responded better to treatment than high-moisture seeds. The frequency of the fields used in seed treatment did not appear to be important. Instead, the final temperature of the seeds treated at any given moisture content seemed to be a good indicator of the degree of response. These findings still remain to be implemented for practical use.

18.3.4 Product Conditioning

Although microwave dielectric heating has found widespread application in the food industry and in microwave ovens in homes, it has found little use in the processing of other agricultural products. Microwave heating at 2.45 GHz was investigated for purposes of grain sorghum eversion for use in cattle feeding rations [6]. Kernel eversion was less than 50%, and very large equipment investments would have been required for practical-scale application.

RF and microwave heating have been studied for treatment of soybeans. Raw soybeans have a trypsin inhibitor that must be inactivated for efficient nutritional use by humans and monogastric farm animals. The enzyme, trypsin, is important in the digestion of protein; so trypsin inhibitor is normally inactivated by a moist heating process. Inactivation of the trypsin inhibitor can be accomplished by RF or microwave dielectric heating of soybeans with the natural moisture present in the soybeans [9, 44, 70]. Lipoxygenase, an enzyme associated with off-flavors, was also inactivated by the treatments, and peroxidase, a desirable enzyme, remained active in the treated soybeans. Protein digestibility and body weight gains of laboratory rats were improved by 2.45 GHz microwave heating of soybeans in other studies also [16, 11].

Tests have been conducted to determine whether dielectric heating treatments of pecans might prevent or delay development of rancidity during storage. Pecans must be stored at reduced temperatures to preserve their quality, and if holding periods are more than a few months, refrigerated storage at 0°C or lower temperatures is required. Limited studies showed that 43 MHz dielectric heating exposures of 1 to 2 min and steam treatments of 4 min were effective in stabilizing flavor quality during accelerated storage tests, but further studies would be needed to evaluate the processes before practical recommendations could be offered [61].

Product quality is not always improved by microwave heating. Adverse effects for milling purposes were noted in durum wheat after 2.45 GHz exposures [13].

18.4 Product Quality Sensing

When the dielectric properties of agricultural products can be well correlated with other physical or chemical characteristics that determine their suitability for particular uses or their value to consumers, there is the possibility of developing instruments for rapidly sensing these characteristics. The use of grain moisture meters for sensing the moisture content of grain is the best-known example of such applications. However, other possible applications have been considered, and some of these will be discussed.

18.4.1 Fruit and Vegetable Quality Sensing

Techniques for rapidly sensing quality factors of fresh fruit and vegetable produce are of great value in sorting such products for grading and marketing operations. Optical equipment for color sorting and surface defect detection has been in use for some time. Because high-frequency and microwave electric fields penetrate materials much better than visible, infrared, or ultraviolet radiation, they offer possibilities for detection of internal quality factors.

For these reasons, the dielectric properties of a few fruits and vegetables have been measured. In connection with quality sensing in fruits and vegetables, the dielectric properties of mature-green and full-ripe peaches at 2.45 GHz were examined to see whether these properties might be useful in distinguishing degree of maturity [43]. The same kind of measurements were taken on normal sweet potatoes and those that had a hard-core condition induced by chilling injury in storage [43]. Permittivity measurements at the single frequency of 2.45 GHz did not appear to offer promise for detecting either of these quality factors. Following permittivity characterization measurements for 23 kinds of common fresh fruits and vegetables over the frequency range from 200 MHz to 20 GHz at 23°C [56, 57], similar measurements were taken over a narrow range of peach maturity, and evidence for possible distinction of degree of maturity was obtained [58]. A permittivity-based maturity index was suggested, based on differences in both components of the permittivity, the dielectric constants at the low end of the frequency range, and the loss factors at 10 GHz near the higher end of this frequency range. More research and development are needed for determining the potential for practical use of the technique, including measurements at frequencies lower than 200 MHz, since the curves for the dielectric constants of the different maturities appeared to be diverging as they approached the lower end of the frequency range.

18.4.2 Grain and Seed Moisture Sensing

Moisture content is the most important characteristic of cereal grain affecting its suitability for harvesting, storage, transport, and processing. It is also an important factor affecting the price of grain. Therefore, moisture content must be determined whenever grain is traded. If moisture content is too high at the time of harvest, the grain kernels can be damaged in the mechanical harvesting process, leaving them more susceptible to infection by fungi. If they are stored at moisture contents too high for the prevailing environment, they can spoil because of the action of microorganisms, and the value is degraded or completely lost for human and animal consumption.

Reference methods for determining moisture in grain generally require oven drying at specified temperatures following prescribed laboratory procedures [86, 3], or chemical titration methods, which are also laboratory procedures. Therefore, these methods are too slow and tedious for practical use in the grain trade. Electrical measurement methods have been developed that depend on correlations between the electrical properties of the grain and moisture content [34, 42].

Electrical moisture meters for grain moisture determination have evolved over the past century [42, 48], and grain moisture meters today are predominantly those operating at frequencies in the range from 1 to 20 MHz that sense the dielectric properties, or permittivity, of the grain samples. These instruments, although troubled with inconsistency at moisture contents above 20% to 25% moisture content, perform reasonably well, and calibrations are maintained by the manufacturers for many different grain and seed commodities.

Moisture meters used in the trade require static samples, and corrections are made for variations in temperature and bulk density of the grain samples. Needs have long been recognized for moisture sensing instruments for applications with moving grain, and efforts have been devoted to developing RF dielectric-type moisture monitoring instruments. The need for moisture monitoring on combines as grain is harvested has spurred such development. Modern agriculture, involving "precision farming," which generally implies yield mapping with the application of global positioning systems and grain mass flow monitoring, requires reliable moisture monitoring also, because yield data need to be based on a specific moisture content. Fluctuation in bulk density, when grain is flowing, causes errors in moisture readings unless some compensation is provided for bulk density changes. Thus, moisture monitoring system design must provide some means for minimizing bulk density variation.

Research on sensing moisture content in grain by microwave measurements has indicated two important advantages for microwave frequencies. The inconsistency of moisture measurements by instruments operating in the HF range may be due, in part, to the influence of ionic conduction on the measured dielectric properties at high moisture levels. At microwave frequencies, the influence of ionic conduction is negligible, and better correlations between permittivity and moisture content can be expected. In addition, techniques for density-independent moisture sensing in granular materials have been reported for measurements at microwave frequencies [25, 26, 28, 29, 76, 77, 82]. Therefore, the principles for sensing moisture content in grain by microwave measurements are reviewed briefly and newer developments are summarized here.

18.4.3 General Principles

From a dielectrics viewpoint, a mass of grain can be characterized by the effective complex permittivity of the material, relative to free space, $\varepsilon = \varepsilon' - j\varepsilon''$, where ε' is the dielectric constant, and ε'' is the dielectric loss factor. For cereal grains, the permittivity is not only a function of the moisture content M, but also of the frequency f of the applied electric field, the temperature T of the grain, and the bulk density ρ of the grain. Thus, a plane wave traversing a layer of grain of thickness d will interact with the granular material as depicted in Fig. 18.4, where E_i represents the incident wave electric field, E_r is the reflected wave electric field, E_t is the electric field of the transmitted wave, Γ is the reflection coefficient, and τ is the transmission coefficient.

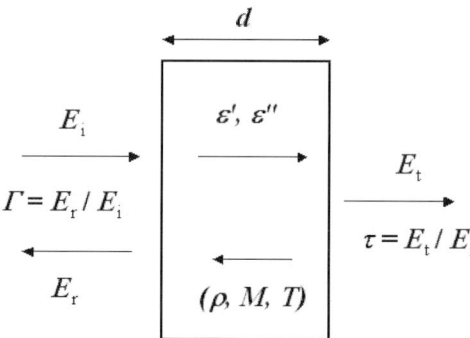

Fig. 18.4. Diagram of wave-dielectric material interaction

The components of the permittivity, ε' and ε'', can be obtained by measurement of the attenuation A and phase shift ϕ of the wave as it traverses the dielectric layer, because for a plane wave, when $\varepsilon'' << \varepsilon'$,

$$\varepsilon' = \left(\frac{\beta}{\beta_0}\right)^2 \tag{18.4}$$

$$\varepsilon'' = \frac{2\alpha\beta}{\beta_0} \tag{18.5}$$

where $\alpha = A/d$ is the attenuation constant, $\beta = \phi/d + \beta_0$ is the phase constant, and $\beta_0 = 2\pi/\lambda_0$ is the phase constant for free-space wavelength λ_0. The disturbance caused by any reflected waves within the grain layer can be made negligible by using a layer thickness that provides at least 10 dB of attenuation for waves traveling one way through the layer.

Measurements of attenuation and phase shift for grain have shown that both are relatively linear with moisture content [25]. Therefore the ratio of phase shift and attenuation, ϕ / A, has been considered for providing density-independent deter-minations for moisture content of grain [25]. For plane wave propagation through low-loss materials, this ratio can be expressed in terms of the permittivity of the grain as [23, 30]

$$\frac{\phi}{A} = \left(\frac{\varepsilon' - 1}{\varepsilon''} \right) \left(\frac{2\sqrt{\varepsilon'}}{\sqrt{\varepsilon'} + 1} \right) \tag{18.6}$$

The first term of the right-hand side of Eq. (18.6), $(\varepsilon' - 1)/\varepsilon''$, was considered earlier as a density-independent function in calibration equations for microwave measurement of moisture content of a number of particulate dielectrics [31, 32], because the second term had little significance for low-loss materials. However, work by Kress-Rogers and Kent [30] in food powders revealed that this term could be too important to neglect.

A newer density-independent function of the permittivity for moisture calibra-tion in microwave measurements was reported by Trabelsi et al. [78]. This function was based on an observation of the complex-plane plot of ε'/ρ vs. ε''/ρ for a large set of measurements on hard red winter wheat at several frequencies, moisture contents, temperatures, and bulk densities. It was noted that, for permit-tivities determined from attenuation and phase measurements at a given frequency, all of the points fell along a straight line and that differences in either moisture content or temperature amounted to translations along that same line (Fig. 18.5). The lines for each frequency intersected the $\varepsilon''/\rho = 0$ axis at a common point, $\varepsilon'/\rho = k$, which represents the value of ε''/ρ for 0% moisture content or the value at very low temperatures. Any change in frequency amounted to a rotation of the straight line about that intersection point. Thus, for a given frequency, the equation of the line is expressed as

$$\varepsilon''/\rho = a_f(\varepsilon'/\rho - k) \tag{18.7}$$

where a_f is the slope at a given frequency. It was determined that the slope varied linearly with frequency. Solving Eq. (18.7) for ρ, we have

$$\rho = \frac{a_f \varepsilon' - \varepsilon''}{a_f k} \tag{18.8}$$

For a given frequency, a_f is a constant, and for a given material, k is a constant. Thus, the bulk density is provided by Eq. (18.8) in terms of the permittivity alone, without regard for temperature or moisture content. Considering that $\tan\delta = \varepsilon''/\varepsilon'$, where δ is the loss angle of the dielectric, expresses the distribution between dissi-pated and stored energy in a dielectric, and that $\tan\delta$ varies with bulk density, it was divided by bulk density.

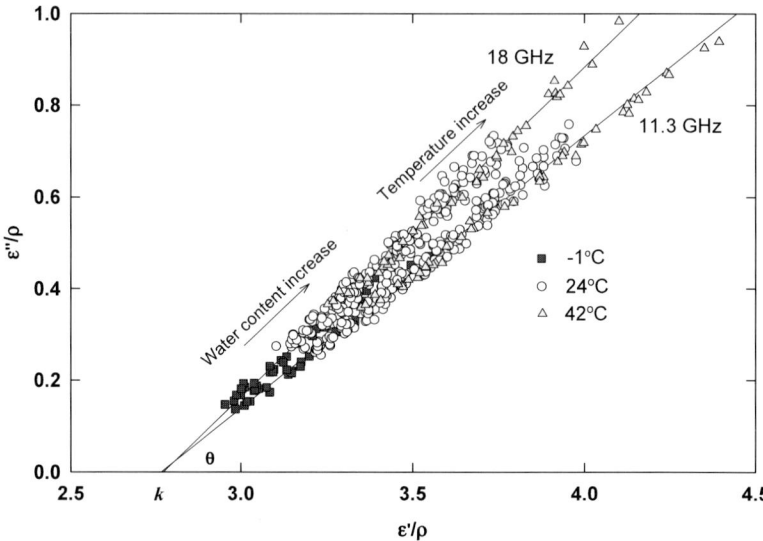

Fig. 18.5. Complex-plane plot of the dielectric constant and loss factor, divided by bulk density, for hard red winter wheat of various moisture contents and bulk densities at indicated temperatures for two frequencies, 11.3 and 18.0 GHz [78].

Using Eq. (18.8) for ρ, we can write

$$\frac{\tan\delta}{\rho} = ka_f\left(\frac{\varepsilon''}{\varepsilon'(a_f\varepsilon' - \varepsilon'')}\right) \qquad (18.9)$$

For a given frequency and particular kind of material, ka_f is a constant, and a new density-independent moisture calibration function can be defined as follows:

$$\psi = \sqrt{\frac{\varepsilon''}{\varepsilon'(a_f\varepsilon' - \varepsilon'')}} \qquad (18.10)$$

The quadratic relationship between the calibration function ψ and the permittivity function of Eq. (18.9) for changes in moisture content was determined empirically [78].

The new calibration function has been studied for a large set of measurements on hard red winter wheat over practical ranges of moisture content, bulk density, and temperature. Figure 18.6 shows the variation of ψ with moisture content M and temperature T at one frequency, 14.2 GHz. Because ψ is linear with both M and T, all of the data points fall on a plane for which the equation can be obtained by regression analysis, providing the values for the constants, a, b and c in the following equation:

$$\psi = aM + bT + c \qquad (18.11)$$

The resulting equation for moisture content,

$$M = (\psi - bT - c)/a \qquad (18.12)$$

is then given in terms of the density-independent calibration function ψ, which, at any given frequency, depends only on the grain permittivity, as shown in Eq. (18.10). The dielectric constant and loss factor can be determined by any suitable microwave measurement.

Further research with this new density-independent moisture calibration function has shown that very similar values of regression constants were obtained for kinds of grain as different as wheat and corn [79]. In another comparison, the same constants performed very well for wheat, oats, and soybeans, which have very different characteristics with respect to kernel shape, size, and composition [84]. These findings support the idea of a universal calibration for grain and seed.

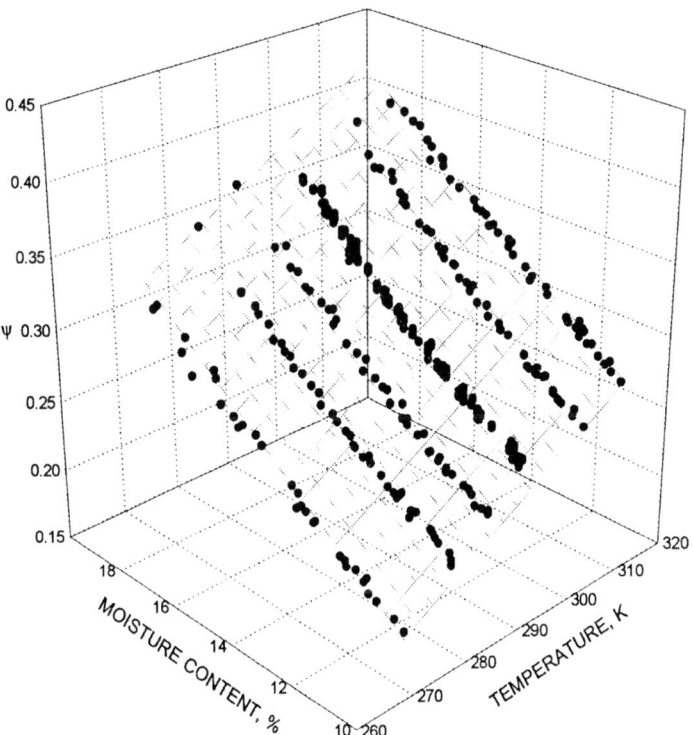

Fig. 18.6. Moisture and temperature dependence of density-independent function ψ at 14.2 GHz for hard red winter wheat [79]

The rapid, non-destructive, reliable sensing of moisture content in cereal grains and other agricultural crops is essential in modern agriculture for prevention of losses and improvements in efficiency of production of these sources of food. Measurement or sensing of the relative complex permittivity, or dielectric properties, at microwave frequencies offers advantages that include density-independent sensing of moisture content and the reduction of variations that arise from ionic conductivity of high moisture grain samples at lower radio frequencies. These advantages, along with the promise for a universal moisture calibration, should encourage the development of microwave measurement systems for on-line moisture sensing for cereal grains and granular materials in other industries as well as in agriculture.

18.5 Potential for Further Applications

Predicting further applications for dielectric properties information entails the risk of being mistaken, but in view of past applications, some observations can be offered concerning likely developments.

The need for rapid quality determination for many agricultural products will inspire further research on correlations between the quality factors of products and their dielectric properties. No doubt there will be quality factors other than moisture content that can be sensed through these correlations, particularly when broad frequency band data are considered. The availability of broadband permittivity measurement techniques will permit relatively efficient collection of such data for developing useful correlations. When such potentially useful correlations are verified, the engineering challenges of designing equipment to utilize the new information must be met for economically feasible applications.

With respect to grain and seed moisture sensing, and the sensing of bulk density also where it may be of value, the advantages outlined for microwave frequencies should encourage the development of instruments operating at these frequencies. Because of the advances in the development of communications equipment, and the scale of these applications, the costs of many components for such circuits have become very reasonable. This should improve the chances for developing economically practical instruments for sensing the dielectric properties of interest in agricultural products. The development of applicable microstrip antennas, which can be produced at low cost, is another factor that would contribute substantially to the development of microwave moisture meters and other instruments that can utilize the dielectric properties for rapid assessment of quality in agricultural products.

Of course, further research may open entirely new applications, where the dielectric properties of products will provide the means for sensing internal quality, conditions, or other information of practical interest.

References

1. Anthony WS (1983) Vacuum microwave drying of cotton: effect on cottonseed. Trans ASAE 26:275-278
2. ASAE (1966) ASAE Data: ASAE D293. Dielectric properties of grain and seeds Agricultural Engineers Yearbook. American Society of Agricultural Engineers, St. Joseph, MI
3. ASAE (2000) ASAE S352.2 Moisture measurement - Unground grain and seeds ASAE Standards 2000. American Society of Agricultural Engineers, St. Joseph, MI, 563 pp
4. ASAE (2002) ASAE D293.2 Dielectric properties of grain and seed. In: ASAE Standards 2002. American Society of Agricultural Engineers, St. Joseph, MI, pp 575-584
5. Ban T, Suzuki M (1977) Studies on electrical detection of grain moisture content in artificial drying (Japanese) Technical Report No. 11, Institute of Agricultural Machinery, Omiya, Japan
6. Beerwinkle KR, McCune WE (1969) Factors affecting eversion of sorghum grain using energy in 2450 MHz range. Trans ASAE 12:295-297
7. Benz G (1975) Entomologishe Untersuchungen zur Entwesung von Getreide mittels Hochfrequenz. Alimenta 14:11-15
8. Berliner E, Ruter R (1929) Über Feuchtigkeitsbestimmungen in Weizen und Roggen mit dem D K-Apparat. Z Gesamt Mühlenwes 6:1-4
9. Borchers R, Manage LD, Nelson SO, Stetson LE (1972) Rapid improvement in nutritional quality of soybeans by dielectric heating. J Food Sci 37:333-334
10. Briggs LJ (1908) An electrical resistance method for the rapid determination of the moisture content of grain. Bureau of Plant Industry Circular No. 20, Washington, DC
11. Chen XJ, Bau HM, Giannangeli F, Debry G (1986) Evaluation de l'influence de la cuisson par les micro-ondes sur les proprietes physico-chimiques et nutritionnelles de la farine entiere de soja. Sci Aliment 6:257-272
12. Davis FS, Wayland JR, Merkel MG (1971) Ultrahigh-frequency electromagnetic fields for weed control: phytotoxicity and selectivity. Science 173:535-537
13. Doty NC, Baker CW (1977) Microwave conditioning of durum wheat. 1. Effects of wide power range on semolina and spaghetti quality. J Agric Food Chem 25:815-819
14. Fanslow GE, Gittins LL, Wedin WF, Martin NP (1971) Power absorption and drying patterns of forage crops dried with microwave power. J Microwave Power 6:229-235
15. Gardner DR, Butler JL (1981) Preparing crops for storage with a microwave vacuum (MIVAC) drying system. Paper presented at Second International Symposium on Drying. Montreal, Canada
16. Hafez US, Mohamed AI, Hewedy FM, Singh G (1985) Effects of microwave heating on solubility, digestibility and metabolism of soy protein. J Food Sci 50:415-417
17. Hall GE (1963) Preliminary investigation of shelled corn drying with high-frequency energy. Ohio Agricultural Experiment Station, Wooster, OH
18. Halverson SL, Burkholder WE, Bigelow TS, Nordheim EV, Misenheimer ME (1996) High-power microwave radiation as an alternative insect control method for stored products. J Econ Entomol 89:1638-1648
19. Hartshorn L, Ward WH (1936) The measurement of permittivity and power factor of dielectrics at frequencies from 10^4 to 10^8 c. p. s. J IEE 79:597-609
20. Jones RN, Bussey HE, Little WE, Metzker RF (1978) Electrical characteristics of corn, wheat, and soya in the 1-200 MHz range. NBSIR 78-897 National Bureau of Standards, Boulder, CO, 67 pp
21. Jorgensen JL, Edison AR, Nelson SO, Stetson LE (1970) A bridge method for dielectric measurements of grain and seed in the 50 to 250MHz range. Trans ASAE 13:18-20, 24

22. Kent M (1987) Electrical and Dielectric Properties of Food Materials. Science and Technology Publishers, Hornchurch, Essex, England
23. Kent M, Kress-Rogers E (1986) Microwave moisture and density measurements in particulate solids. Trans Instrum, Meas Control 8:161-168
24. Knipper NV (1959) Use of high-frequency currents for grain drying. J Agric Eng Res 4:349-360
25. Kraszewski AW (1988) Microwave monitoring of moisture content in grain further considerations. J Microwave Power 23:236-246
26. Kraszewski AW, Kulinski S (1976) An improved microwave method of moisture content measurement and control. IEEE Trans Ind Electron Control Instrum 23:364-370
27. Kraszewski AW, Trabelsi S, Nelson SO (1996) Wheat permittivity measurements in free space. J Microwave Power Electromag Energy 31:135-141
28. Kraszewski AW, Trabelsi S, Nelson SO (1998) Comparison of density-independent expressions for moisture content determination in wheat at microwave frequencies. J Agric Eng Res 71:227-237
29. Kraszewski AW, Trabelsi S, Nelson SO (1999) Temperature-compensated and density-independent moisture content determination in shelled maize by microwave measurements. J Agric Eng Res 72:27-35
30. Kress-Rogers E, Kent M (1987) Microwave measurement of powder moisture and density. J Food Eng 6:345-376
31. Meyer W, Schilz W (1980) A microwave method for density independent determination of the moisture content of solids. J Phys D: Appl Phys 13:1823-1830
32. Meyer W, Schilz W (1981) Feasibility study of density-independent moisture measurement with microwaves. IEEE Trans Microwave Theory Tech 29:732-739
33. Mishenko AA, Malinin OA, Rashkovan VM, Basteev AV, Basyma LA, Mazalov YP, Kutovoy VA (2000) Complex high-frequency technology for protection of grain against pests. J Microwave Power Electrom Energy 35:179-184
34. Nelson SO (1965) Dielectric properties of grain and seed in the 1 to 50 MC range. Transactions of the ASAE 8:38-48
35. Nelson SO (1965) Electromagnetic Radiation Effects on Seeds. Conf Proc - Electromagnetic Radiation in Agriculture:60-63
36. Nelson SO (1967) Electromagnetic energy. In: Kilgore WW, Doutt RL (eds) Pest Control - Biological, Physical and Selected Chemical Methods. Academic Press, New York and London
37. Nelson SO (1972) A system for measuring dielectric properties at frequencies from 8.2 to 12.4 GHz. Trans ASAE 15:1094-1098
38. Nelson SO (1973) Electrical properties of agricultural products - a critical review. Trans ASAE 16:384-400
39. Nelson SO (1973) Insect-control studies with microwave and other radiofrequency energy. Bull Entomol Soc Am 19:157-163
40. Nelson SO (1973) Microwave dielectric properties of grain and seed. Trans ASAE 16:902-905
41. Nelson SO (1976) Use of microwave and lower frequency RF energy for improving alfalfa seed germination. J Microwave Power 11:271-277
42. Nelson SO (1977) Use of electrical properties for grain moisture measurement. J Microwave Power 12:67-72
43. Nelson SO (1980) Microwave dielectric properties of fresh fruits and vegetables. Trans ASAE 23:1314-1317
44. Nelson SO (1981) Effects of 42 and 2450 MHz dielectric heating on nutrition-related properties of soybeans. J Microwave Power 16:313-318

45. Nelson SO (1981) Review of factors influencing the dielectric properties of cereal grains. Cereal Chem 58:487-492
46. Nelson SO (1982) Factors affecting the dielectric properties of grain. Trans ASAE 25:1045-1049, 1056
47. Nelson SO (1983) Dielectric properties of some fresh fruits and vegetables at frequencies of 2.45 to 22 GHz. Trans ASAE 26:613-616
48. Nelson SO (1991) Dielectric properties of agricultural products - measurements and applications. IEEE Trans Elec Insul 26:845-869
49. Nelson SO (1996) A review and assessment of microwave energy for soil treatment to control pests. Trans ASAE 39:281-289
50. Nelson SO (1996) Review and assessment of radio-frequency and microwave energy for stored-grain insect control. Trans ASAE 39:1475-1484
51. Nelson SO (2003) Frequency- and temperature-dependent permittivities of fresh fruits and vegetables from 0.0l to 1.8 GHz. Trans ASAE 46:567-574.
52. Nelson SO, Bartley PG, Jr., Lawrence KC (1997) Measuring RF and microwave permittivities of adult rice weevils. IEEE Trans Instrum Meas 46:941-946
53. Nelson SO, Bartley PG, Jr., Lawrence KC (1998) RF and microwave dielectric properties of stored-grain insects and their implications for potential insect control. Trans ASAE 41:685-692
54. Nelson SO, Charity LF (1972) Frequency dependence of energy absorption by insects and grain in electric fields. Trans ASAE 15:1099-1102
55. Nelson SO, Datta AK (2001) Dielectric properties of food materials and electric field interactions. In: Datta AK, Anantheswaran RC (eds) Handbook of Microwave Technology for Food Applications. Marcel Dekker, New York
56. Nelson SO, Forbus WR, Jr., Lawrence KC (1994) Microwave permittivities of fresh fruits and vegetables from 0.2 to 20 GHz. Trans ASAE 37:181-189
57. Nelson SO, Forbus WR, Jr., Lawrence KC (1994) Permittivities of fresh fruits and vegetables at 0.2 to 20 GHz. J Microwave Power Electromag Energy 29:81-93
58. Nelson SO, Forbus WR, Jr., Lawrence KC (1995) Assessment of microwave permittivity for sensing peach maturity. Trans ASAE 38:579-585
59. Nelson SO, Lu C-Y, Beuchat LR, Harrison MA (2003) Radio-frequency heating of alfalfa seed for reducing human pathogens. Trans ASAE 45:1937-1942.
60. Nelson SO, Nutile GE, Stetson LE (1970) Effects of radiofrequency electrical treatment on germination of vegetable seeds. J Am Soc Hortic Sci 95:359-366
61. Nelson SO, Senter SD, Forbus WR, Jr. (1985) Dielectric and steam heating treatments for quality maintenance in stored pecans. J Microwave Power 20:71-74
62. Nelson SO, Soderholm LH, Yung FD (1953) Determining the dielectric properties of grain. Agric Eng 34:608-610
63. Nelson SO, Stetson LE (1974) Comparative effectiveness of 39 and 2450 MHz electric fields for control of rice weevils in wheat. J Econ Entomol 67:592-595
64. Nelson SO, Stetson LE (1976) Frequency and moisture dependence of the dielectric properties of hard red winter wheat. J Agric Eng Res 21:181-192
65. Nelson SO, Stetson LE (1985) Germination responses of selected plant species to RF electrical seed treatment. Trans ASAE 28:2051-2058
66. Nelson SO, Stetson LE, Rhine JJ (1966) Factors influencing effectiveness of radio-frequency electric fields for stored-grain insect control. Trans ASAE 9:809-815
67. Nelson SO, Stetson LE, Schlaphoff CW (1972) Computer program for calculation dielectric properties of low- or high-loss materials from short-circuited waveguide measurements. ARS-NC-4 Agricultural Research Service, US Department of Agriculture, 30 pp

68. Nelson SO, Stetson LE, Schlaphoff CW (1974) A general computer program for precise calculation of dielectric properties from short-circuited waveguide measurements. IEEE Trans Instrum Meas 23:455-460
69. Nelson SO, Whitney WK (1960) Radio-frequency electric fields for stored-grain insect control. Trans ASAE 3:133-137
70. Pour-El A, Nelson SO, Peck EE, Tjiho B (1981) Biological properties of VHF- and microwave-heated soybeans. J Food Sci 46:880-885, 895
71. Roberts S, von Hippel A (1946) A new method for measuring dielectric constant and loss in the range of centimeter waves. J Appl Phys 17:610-616
72. Stetson LE, Nelson SO (1970) A method for determining dielectric properties of grain and seed in the 200- to 500-MHz Range. Trans ASAE 13:491-495
73. Stetson LE, Ogden RL, Nelson SO (1969) Effects of radiofrequency electric fields on drying and carotene retention of chopped alfalfa. Trans ASAE 12:407-410
74. Thomas AM (1952) Pest control by high-frequency electric fields - critical resume. British Electrical and Allied Industries Research Association, Leatherhead, Surrey, 40 pp
75. Tinga WR, Nelson SO (1973) Dielectric properties of materials for microwave processing - tabulated. J Microwave Power 8:23-65
76. Trabelsi S, Kraszewski A, Nelson SO (1997) Simultaneous determination of density and water content of particulate materials by microwave sensors. Electron Lett 33:874-876
77. Trabelsi S, Kraszewski A, Nelson SO (1998) A microwave method for on-line determination of bulk density and moisture content of particulate materials. IEEE Trans Instrum Meas 47:127-132
78. Trabelsi S, Kraszewski A, Nelson SO (1998) New density-independent calibration function for microwave sensing of moisture content in particulate materials. IEEE Trans Instrum Meas 47:613-622
79. Trabelsi S, Kraszewski A, Nelson SO (2001) New calibration technique for microwave moisture sensors. IEEE Trans Instrum Meas 50:877-881
80. Trabelsi S, Kraszewski AW, Nelson SO (1997) Microwave dielectric properties of shelled yellow-dent field corn. J Microwave Power Electromag Energy 32:188-194
81. Trabelsi S, Kraszewski AW, Nelson SO (2000) Phase-shift ambiguity in microwave dielectric properties measurements. IEEE Trans Instrum Meas 49:56-60
82. Trabelsi S, Nelson SO (1998) Density-independent functions for on-line microwave moisture meters: a general discussion. Meas Sci Technol 9:570-578
83. Trabelsi S, Nelson SO (2003) Free-space measurement of dielectric properties of cereal grain and oilseed at microwave frequencies. Meas Sci Techn 14:589-599
84. Trabelsi S, Nelson SO, Kraszewski AW (2001) Universal calibration for microwave moisture sensors for granular materials. Proc 18. IEEE Conf Instrum and Meas Techn, 1808 - 1813. Budapest, Hungary
85. Tran VN, Stuchly SS, Kraszewski A (1984) Dielectric properties of selected vegetables and fruits. J Microwave Power 19:251-258
86. USDA (1986) Chapter 4. Air-Oven Methods. In: Service FGI (ed) Moisture Handbook. US Department of Agriculture, Washington, DC
87. Wangsgard AP, Hazen T (1946) The Q-Meter for dielectric measurements on polyethylene and other plastics at frequencies up to 50 megacycles. Trans Electrochem Soc 90:361-375
88. Wayland JR, Merkle MG, Davis FS, Menges RM, Robinson R (1975) Control of seeds with UHF electromagnetic fields. Weed Res 15:1-5
89. Wesley RA, Lyons DW, Garner TH, Garner WE (1974) Some effects of microwave drying on cottonseed. J Microwave Power 9:329-340
90. Wratten FT (1950) The application of dielectric heat to the processing of rice. M.Sc Thesis, Louisiana State University, Baton Rouge

19 Determination of the Composition of Foodstuffs Using Microwave Dielectric Spectra

Frank Daschner, Reinhard Knöchel

Christian Albrechts University of Kiel
Kaiserstrasse 2; 24143 Kiel; Germany

19.1 Introduction

Water is often added to foodstuffs during processing. This is inevitable when the raw materials have to be cleaned. Furthermore the nutritiousness is retained and oxidation of the surface is reduced when the raw materials are wet. Because water is inexpensive it is sometimes added deliberately in order to increase the financial profit. For quality control purposes there is a need to measure the water content of the foodstuffs. Furthermore the determination of other constituents like protein, fat, and salt is needed. There are many measurement systems available to measure the composition of foodstuffs, but they have the disadvantage that they can only measure one constituent and they are relatively expensive. In addition they are very time consuming. A measurement in real time in order to control the production process is not possible. Generally these methods alter the materials or they are destructive. In short there is a need for inexpensive, fast, non-destructive, and compact systems for measuring the composition of foodstuffs.

The method followed here to realize such a system uses the fact that a change in the composition of the foodstuffs also alters dielectric spectra in the microwave region. The problem is to relate changes in the dielectric spectrum to the composition of the material. Kent et al have already carried out investigations in this area. They used principal components analysis and regression [1, 2]. Other methods to described here are used for comparison purposes. At first a dielectric modeling is applied. Due to the high complexity of foodstuffs it is difficult to get precise results with this approach. Therefore multivariate calibration methods are applied. The measured dielectric data are used in a multiple linear regression. An orthogonalization of the measured data is carried out with partial least squares regression to improve the performance. Non-linear processing of the data is described using artificial neural networks. Finally all these methods are compared.

19.2 Experiments

The water content of flesh can be increased by dipping it in polyphosphate solution and saline solution. The two treatment parameters are the dipping time and the concentration of the solution. Because they have a different influence on the dielectric spectra of the samples, both parameters have to be taken into account. For example, the ionic conductivity of the samples will increase with the concentration of the solution.

19.2.1 Preparation of the Samples

In the framework of a project financed by the European Commission (FAIR CT97-3020) many experiments with poultry, pork, and fish samples were carried out. Two test series were selected to demonstrate the methods which were utilized. For the first experiment prawns (*Pandalus borealis*) were peeled and treated in a polyphosphate solution with different concentrations (0%, 0.5%, 1%, and 2%) and for different times (0 h, 2.5 h, 24 h, and 48 h). The temperature of the solution was 4–5°C. After the treatment the prawns were sieved and gently dried by dabbing with a paper towel. The amount of added liquid was determined by weighing the sample before and after the treatment. The samples were minced and halved. One part was used for the dielectric measurements while the other part was sent to an external laboratory which determined the composition of the samples using the conventional methods mentioned above. These measurements served as reference for the methods discussed here.

In the second experiment herring (*Clupea harengus*) was treated with a solution with constant polyphosphate content (2%) while the NACl concentration was varied (0%, 0.5%, 2%, and 3%). The treatment time was 0 h, 2.5 h, 24 h, and 48 h again. In comparison to the prawn samples the herring samples had a naturally high variation of fat content.

19.2.2 Dielectric Measurements

The dielectric measurements were carried out using an open-ended coaxial line with a diameter of 3 mm. This sensor has been well investigated and has become the standard for measuring dielectric spectra with liquid and soft materials [3–5]. The dielectric probe kit HP85070 in combination with an HP8510 automatic network analyzer was used for the measurements. The dielectric spectra of the samples were measured in the frequency range from 200 MHz to 12 GHz at 31 frequency points with logarithmic separation. Each measurement was repeated five times. Outliers caused by bad contacts between the probe and the sample were removed. The means of the permittivity values were calculated and used subsequently in order to remove the influence of any remaining inhomogeneities in the samples.

Generally the permittivity of a sample depends on its temperature. Therefore the dielectric measurements were taken at a sample temperature of 3°C, 8°C, and 20°C and 30°C. A total number of 91 data sets were obtained in the prawn test series and 52 in the herring test series.

19.3 Qualitative View on the Influence of the Sample Treatment

Due to the fact that the investigated foodstuffs have relatively high losses in the measured frequency range, the dielectric spectra are complex. Before the processing of the measured data is discussed in detail a qualitative view on the influence of the treatment of the samples is presented. The dielectric spectra of some prawn samples are shown in Fig. 19.1. On the left hand side is the Cole–Cole diagram while on the right hand side the curves of both the real and imaginary part (losses) of the permittivity are plotted vs. frequency. The selected samples are:

1. The untreated sample (solid line), liquid uptake: 0%.
2. Short treatment with weak solution (dashed line), liquid uptake: 5.5%.
3. Long treatment with pure water (dot-dashed line), liquid uptake: 10.9%.
4. Long treatment with strong solution (dotted line), liquid uptake: 11%.

The sample which was treated in pure water has lower losses at low frequencies in comparison to the untreated sample. This is explained by the fact that ions were diffused into the water. Hence the ionic conductivity decreases. With the highly treated sample (4) the opposite effect is observable: the losses increase at lower

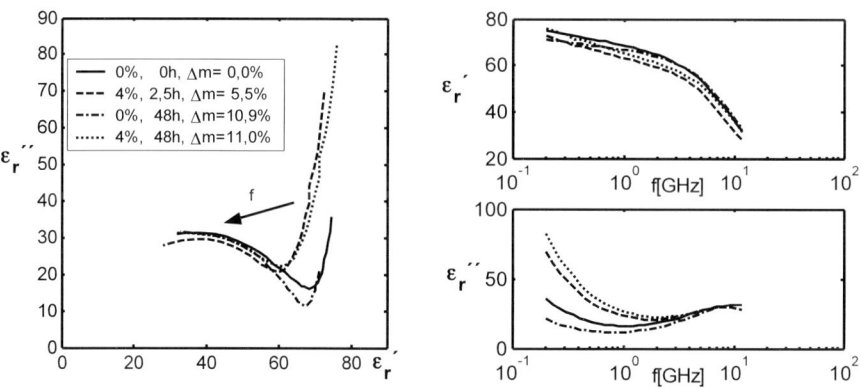

Fig. 19.1. Effects of the treatment of the samples on the dielectric spectra of prawns. The measurement temperature was 3°C

frequencies because ions diffuse from the solution into the tissue. With sample 2 it is observable that the polyphosphate accelerates the liquid uptake. With the real part of the permittivity it is difficult to detect any evident effects of the treatment.

However, the treatment has an influence on the shape of the dielectric spectrum. Signal processing for the determination of the composition of the samples has to extract the hidden information from the changes in the shape. Remarkably the samples which were treated for 48 h, (3) and (4), have nearly the same amount of liquid uptake although their dielectric spectra have a different shape. The signal processing has to tolerate such relatively great variations in the same prediction value.

19.4 Dielectric Modeling

Foods are generally complex stuffs. First of all flesh is inhomogeneous. It consists of different kinds of tissues like muscles, blood vessels, connective and fatty tissue. The main constituent of flesh is water. Part of it lies within the cells and part occurs in the extras cellular space. But flesh also consists of many other organic and inorganic substances, some of which have a complex interaction. A Parts of the water is bound by hydrophilic groups of organic molecules. The free water is rotationally hindered by other molecules. The dispersion of water lies within the range of microwave frequencies depending on its temperature. Smaller peptides and amino acids also have its relaxation frequencies in the microwave region [6].

The microwave dielectric spectrum is mainly formed by water. However, due to the complex interaction it is difficult to create a sufficiently accurate model for the dielectric behavior of flesh. Nevertheless, it should be possible to determine the existence of constituents which do not have its relaxation frequency in the measured frequency range – provided that they have an interaction with water. As mentioned above, salts increase the ionic conductivity. Hence at lower frequencies the losses also increase. In spite of this direct effect one cannot distinguish between the kinds of diluted salts.

The first method discussed here tries to apply a dielectric model of the flesh. The model parameters are taken from the measurements. A correlation between these parameters and the ingredients is used to determine the composition of the samples. The model applied in the following text consists only of three components: the free water (γ-dispersion), the bound water (δ-dispersion), and the salt content:

$$\varepsilon_r(\omega) = \varepsilon_\infty + \frac{\varepsilon_{s\gamma} - \varepsilon_\infty}{1 + (j\omega\tau_\gamma)^{(1-\alpha_\gamma)}} + \frac{\varepsilon_{s\delta} - \varepsilon_\infty}{1 + (j\omega\tau_\delta)^{(1-\alpha_\delta)}} - j\frac{\sigma}{\varepsilon_0\omega}$$

$$(19.1)$$

Where ε_∞ is the permittivity at infinite frequency, ε_s is the static permittivity, ω is the angular frequency, τ is the relaxation time, α describes the distribution of the relaxation time, and σ is the conductivity. Both the free and bound water are described by a Cole–Cole-relaxation and the salt content is considered in the third term. ε_∞ is fixed at the value of 4.6 for both dispersions..

The seven model parameters are calculated from the measured dielectric spectra using a non-linear curve fitting procedure. This procedure is described in [6] and based upon the Gauss–Newton method. The computational effort of the curve fitting is relatively high ($\approx 2\times10^6$ floating point operations for one curve fitting).

After the calculation of the model parameters, the samples were divided randomly, two-thirds into the calibration group, and one third into the validation group. The model parameters of the calibration group are used as regressors in a multiple linear regression (MLR). A composition value (z_c, e.g. water content) is described as the linear combination of the model parameters:

$$\vec{z}_c = A_c\vec{\beta} + \vec{e}_c . \tag{19.2}$$

The error vector is $\vec{e}_k$ while the regressor matrix is built in the following way:

$$A_c = \begin{bmatrix} 1 & \varepsilon_{s\gamma 1} & \tau_{\gamma 1} & \alpha_{\gamma 1} & \varepsilon_{s\delta 1} & \tau_{\delta 1} & \alpha_{\delta 1} & \sigma_1 & T_1 \\ & & & & \vdots & & & & \\ 1 & \varepsilon_{s\gamma t} & \tau_{\gamma t} & \alpha_{\gamma t} & \varepsilon_{s\delta t} & \tau_{\delta t} & \alpha_{\delta t} & \sigma_t & T_t \end{bmatrix} . \tag{19.3}$$

Each row consists of the model parameters of one sample of the calibration group. The number of samples in the calibration group is t. A least squares estimation for the vector of coefficients can be estimated by [7]

$$\hat{\vec{\beta}} = (A_c^T A_c)^{-1} A_c\vec{z}_c . \tag{19.4}$$

The composition value of the samples in the calibration and validation groups is estimated by

$$\vec{\hat{z}}_{c,v} = A_{c,v}\vec{\hat{\beta}} . \tag{19.5}$$

The calibration procedure described above has to be performed for each composition value[1]. The quality of the prediction is evaluated by the root mean square error[2]

$$RMSE_{c,v} = \sqrt{\frac{\vec{e}_{c,v}^{T} \cdot \vec{e}_{c,v}}{t}}$$ (19.6)

and the coefficient of determination R^2

$$R^2 = 1 - \frac{\vec{e}_{c}^{T} \cdot \vec{e}_{c}}{\sum_{i=1}^{t} (z_{c,i} - \bar{z}_c)^2}.$$ (19.7)

The results obtained from the dielectric modeling are shown in Table 19.1. By examining R^2 one can see that the performance of this method is disappointing. Only for the prediction of the salt content of the herring samples is $R^2 > 90\%$ obtained. With the prawn samples only the water and protein content is predicted with moderate accuracy.

Table 19.1. Results of the dielectric modelling procedure.

Prawns	R^2	$RMSE_c$	$RMSE_v$
Liquid uptake	0.745	2.75% of m_0	3.37 % of m_0
Fat	0.616	0.064 % abs.	0.077 % abs.
Protein	0.889	0.59 % abs.	0.73 % abs.
Water	0.852	0.61 % abs.	0.70 % abs.
Salt	0.771	0.047 % abs.	0.061 % abs.
Herring	R^2	$RMSE_c$	$RMSE_v$
Liquid uptake	0.634	6.50 % of m_0	5.89 of m_0
Fat	0.696	1.35 % abs.	1.67 % abs.
Protein	0.650	1.28 % abs.	1.06 % abs.
Water	0.623	2.19 % abs.	2.29 % abs.
Salt	0.955	0.125 % abs.	0.264 % abs.

[1] It is also possible to expand Eqs. 19.2, 19.3, and 19.5 to matrix equations. But for reasons of clarity this was not carried out here.
[2] The RMSE and R^2 for the method described later are determined in the same way.

Obviously the poor performance can be explained by the simplicity of the model used, which considers only three components. Now one could expand the model to improve the accuracy. But there are many reasons why this strategy does not succeed:

- Water has the most intense influence on the dielectric spectrum of flesh. All other polar molecules have only a minor influence on it.
- The dispersions of the diverse small protein molecules are widely spread[3]. Hence there is no clear relaxation frequency despite its distribution is considered with the parameter α.
- The interactions between the components are complex, possibly unknown as yet, and it is not possible to consider all of them.
- The measurement accuracy is not high enough to acquire all the details of the model. The dominant source of error with the open-ended coaxial line is the contact between the probe and the sample. Furthermore the material is not homogeneous enough.

For these reasons the determination of the model parameters of the bounded water is critical. In Fig. 19.2 the confidence intervals of all the model parameters of all the prawn samples are shown. They were calculated with a 95% level as described in [6]. This means the probability that the true model parameter is within the calculated confidence interval is 95%. In the figure one can see for some samples that the distributions of the relaxation times (α_γ and α_δ) are not determined reliably. Due to the numerical instability of the curve fitting procedure some of the determined relaxations times of the bounded water (τ_δ) samples are useless.

Another method to check the model parameters is to calculate the correlation between the them[4]. If the correlation is high, although they are physically independent this is a warning signal of the low reliability of the determination of the model parameters. For example, the correlation factor between the dielectric increment of the free water ($\Delta_\gamma = \varepsilon_{s\gamma} - \varepsilon_\infty$) and the distribution factor of relaxation time of the bounded water (α_δ) is unrealistic for one sample: $\rho = -0.99$. In comparison to other polar molecules the bounded water i.e. has nevertheless a determinable influence on the dielectric spectrum. But it is already difficult to measure its model parameters.

In short, it is not reasonable to expand the model with other components. Although the three-component model is primitive not all its parameters can be determined reliably. On the other hand, the use of a very primitive model can lead to a loss of information during the curve fitting procedure.

---

[3] Leucine: 56 MHz; glycine: 3.23 GHz [6].
[4] The calculation procedure is also described in [6].

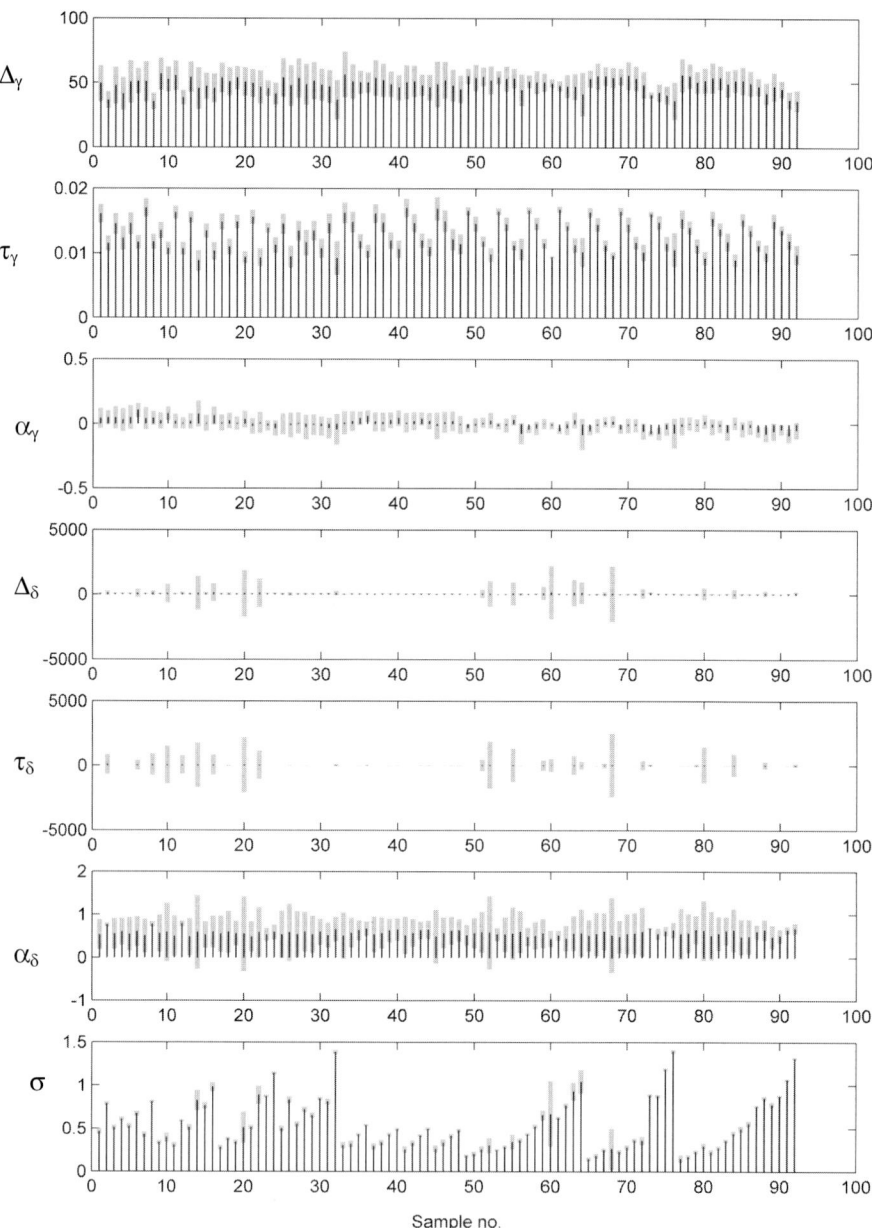

Fig. 19.2. Confidence intervals (95%) of all model parameters of all prawn samples. The determined parameters are shown as thin black lines while the confidence intervals are plotted as thick gray lines

19.5 Direct Processing of the Dielectric Data in a Multiple Linear Regression

The measurement problem is generalized in the following text. It is necessary to find a function that describes the relationship between the dielectric spectrum and the composition values:

$$z = f(\varepsilon(\omega, T)) . \tag{19.8}$$

It is preferable to use the dielectric data directly as regressors in a MLR instead of the tediously determined, but inaccurate, model parameters. The procedure is nearly the same as that described above (Eqs. (19.2–19.5)). Only the matrix of regressors has to be changed. It is now built from the real and imaginary parts of the dielectric data and the temperature:

$$A_{k,v} = \begin{bmatrix} 1 & \varepsilon'_{r1(\omega_1)} & \cdots & \varepsilon'_{r1(\omega_n)} & \varepsilon''_{r1(\omega_1)} & \cdots & \varepsilon''_{r1(\omega_n)} & T_1 \\ & & & \vdots & & & & \\ 1 & \varepsilon'_{rt(\omega_1)} & \cdots & \varepsilon'_{rt(\omega_n)} & \varepsilon''_{rt(\omega_1)} & \cdots & \varepsilon''_{rt(\omega_n)} & T_t \end{bmatrix} \tag{19.9}$$

The calibration equation Eq. (19.5) estimates a composition value by a linear combination of the dielectric data and the temperature. The problem here is that the dielectric data are too collinear, i.e. the adjacent frequency points are relatively

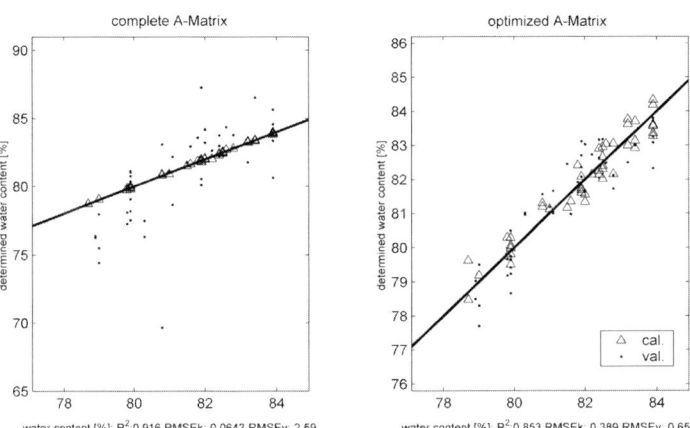

Fig. 19.3. Determination of the water content of prawns using the dielectric data as regressors in MLR. <u>Left hand side:</u> overfitting caused by collinear data. <u>Right hand side:</u> reduced matrix

highly correlated. For this reason the matrix inversion in Eq. (19.4) is numerically unstable [8]. In this case overfitting occurs. (Overfitting means that the calibration equation only treats the calibration group and the performance is bad for the validation group.) The effect of the collinearity is shown in Fig. 19.3.

A primitive method to reduce the collinearity is to thin out the matrix of regressors (19.9). That means only every second, third, etc. column is used. The optimum configuration is found when $RMSE_v$ has a minimum. This procedure is carried out for every composition value and the results are shown in Table 19.2. In comparison to the performance of the dielectric modeling a definite improvement results. However, the reduction of the collinearity has been taken arbitrarily. A more fashionable and precise method is used in the next section.

Table 19.2. Results obtained from the optimized regressor matrix composed of dielectric data

Prawns	R^2	$RMSE_c$	$RMSE_v$
Liquid uptake	0.827	2.26 % of m_0	3.18 % of m_0
Fat	0.394	0.080 % abs.	0.086 % abs.
Protein	0.910	0.53 % abs.	0.65 % abs.
Water	0.905	0.48 % abs.	0.58 % abs.
Salt	0.772	0.047 % abs.	0.052 % abs.
Herring	R^2	$RMSE_c$	$RMSE_v$
Liquid uptake	0.738	5.49 % of m_0	4.31 % of m_0
Fat	0.817	1.05% abs.	1.13 % abs.
Protein	0.586	1.39% abs.	0.73% abs.
Water	0.792	1.63% abs.	1.44% abs.
Salt	0.938	0.147% abs.	0.224% abs.

19.6 Elimination of the Collinearity using Partial Least Squares Regression

Kent et al [1, 9] proposed the use of principal components analysis (PCA) and principal components regression (PCR) for the processing of dielectric data. The principal components are an orthogonal transformation of the measured data. Hence the collinearity is removed completely. The criterion of the orthogonal transformation is to maximize the variance of the principal components, where the first one has the greatest variance, sorted in descending order. Normally the first few principal components have nearly the total variance of the data and the last one contains only the influence of noise. Hence PCA can be used for data reduction purposes. After the transformation some of the first components are used as regressors in MLR. One problem is to make a good selection for this purpose. If too many components are used, overfitting can also occur. Another problem is that the transformation is done completely independently of the composition values.

Partial least squares regression (PLSR) performs the calibration more directly. The composition values are considered during the orthogonalization. Originally PLSR was developed to process economic data, and it was more or less developed intuitively. But Martens and Naes have used this technique also for near infrared spectroscopy [8]. In [10] PLSR is also used to determine the water content of wheat using microwave transmission measurements.

PLSR reduces the data to a set of data what is called "hidden path variables." The PLSR algorithm used here is called "PLS1" and is described in [8]. The structure of this method is as follows:

1. After subtracting the means the measured data are weighted in such a way that the covariance between them and the composition values is maximal.
2. A factor is defined as the projection of the data onto the vector of weights.
3. There then follows a regression analysis between the input variables and the composition values and the factor. The parts described by this linear regression are subtracted from the input data and composition value. These new data are used in the next iteration.
4. The algorithm is repeated until a specific number of factors are calculated.

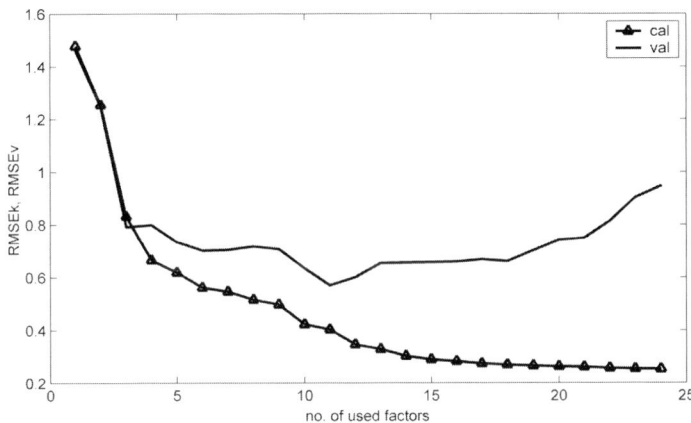

Fig. 19.4. Relationship between the RMSE of the calibration and validation groups for the determination of the water content of the prawn samples and the number of factors used in the PLSR.

F. Daschner, R. Knöchel

Hence the only degree of freedom is the number of factors to be calculated. If too many are used overfitting occurs again. This effect is shown in Fig. 19.4. The RMSE of the calibration group decreases with the number of factors used. First, the RMSE of the validation group also decreases, and a minimum is reached at 12. But if more than 12 factors are used the $RMSE_v$ increases. That means overfitting appears. The optimal number of factors has to be determined empirically for each composition value. However, in comparison to the more or less arbitrary deletion of columns of the regressor matrix described in the previous section, the determination of a minimum is much easier. This optimum can be determined more reliably the greater is the number of available data sets in the calibration and the validation groups.

The results obtained with PLSR are shown in Table 19.3. They are better than those obtained with the dielectric modeling but comparable to those of the dielectric data used directly as regressors described in the previous section. For the herring test series high values of R^2 are observable with PLSR, but this is only because of a higher overfitting.

Table 19.3. Comparison of the results obtained with PLSR

Prawns	R^2	$RMSE_c$	$RMSE_v$
Liquid uptake	0.744	2.76 % of m_0	3.27 % of m_0
Fat	0.436	0.078 % abs.	0.086 % abs.
Protein	0.936	0.45 % abs.	0.66 % abs.
Water	0.942	0.38 % abs.	0.57 % abs.
Salt	0.845	0.039 % abs.	0.052 % abs.
Herring	R^2	$RMSE_c$	$RMSE_v$
Liquid uptake	0.968	1.92 % of m_0	3.79 % of m_0
Fat	0.914	0.72 % abs.	1.19 % abs.
Protein	0.955	0.46 % abs.	0.80 % abs.
Water	0.981	0.49 % abs.	1.48 % abs.
Salt	0.945	0.139 % abs.	0.228 % abs.

19.7 Non-linear Data Processing Using Artificial Neural Networks

Apart from the curve fitting procedure of the dielectric modeling all calibration equations are a linear combination of the input variables. But the unknown function $z = f(\varepsilon(\omega, T))$ may be non-linear. One method to approximate an unknown function is the use of artificial neural networks (ANNs). Bartley et al used ANNs for the determination of the water content of wheat using free-space transmission measurements from 10 to 18 GHz [11]. Before such networks can be used successfully, a suitable architecture has to be chosen. Even if such a fundamental architecture is found many degrees of freedom still remain. For a functional approximation the use of multi-layer feed forward (MLFF) networks is recommended in the literature [12].

The architecture of the MLFF ANN used is shown in figure 19.5. The configuration of the displayed network proves very well suited for the application discussed here. It has one hidden layer which contains 5–10 neurons. The activation functions of the neurons in the hidden layer are non-linear (*tansig* function) while those of the output layer are linear.

It has been shown by Kreinovich and other authors that such types of ANNs are able to estimate unknown functions with a limited number of discontinuities at any desired accuracy [14].

If this architecture is chosen various degrees of freedom are left:

- The number of hidden layers.
- The number of neurons of the hidden layer.
- The kind of activation function of the neurons of each layer.

A more complex function to be approximated requires a greater number of neurons, hidden layers, and training data sets. Unfortunately the only guidelines

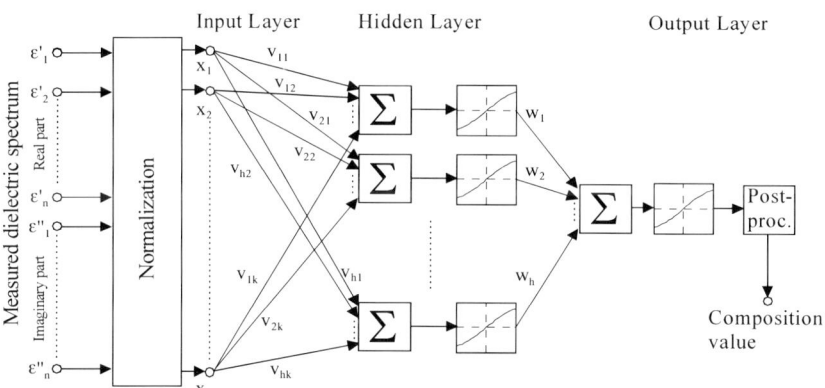

Fig. 19.5. Architecture of the MLFF network

known are to choose all these parameters. Therefore one has to evolve intuition in designing such a network by using trial and error.

19.7.1 Training of the Artificial Neural Network

After a fundamental architecture has been chosen the ANN has to be trained. That requires the determination of the weights (w and v) of the connections between the inputs and neurons in order to minimize the error of the function approximation. The ANNs discussed here are trained with back-propagation, which is a descent gradient algorithm. It uses a performance function which delivers the sum of the squared residuals as output values. The absolute minimum of this performance function has to be found by moving the network weights along the negative gradient of the performance function. The new calculated weights are used in the next iteration.

If, for example, five complex values and the temperature are used as input variables and a network consists of one hidden layer with eight neurons and one output variable is used, the performance function is a surface in a space with 88+1 dimensions. The surface may have a complex curvature which generally increases with the size of the network. Therefore the gradient algorithm may converge only to a local minimum or it may not converge at all.

Another problem is the requirement of random numbers to be used as starting values of the weights. Hence the success of a training exercise depends on those starting values. From that point of view this kind of ANN is not strictly deterministic. A further disadvantage is that ANNs have the tendency of overfitting. That means the network imprints only the training data and loses its required properties of generalization. The more complex the network, the higher is the risk of overfitting. To avoid overfitting the method of "Early Stopping" [13] was used. With this method the network output is observed if the validation set is used as the input. If its performance decreases while that of the calibration data output still increases the training is stopped.

The training was performed using the efficient Levenberg–Marquardt algorithm which is implemented within the "Neural Network Toolbox" of MATLAB. The different network parameters were varied in order to find an optimal configuration. Because of the sensitivity of the training with respect to the randomly selected starting values, the training was repeated 60 times in order to find a good pass.

19.7.2 Optimal Architecture Found

Using an oversized network is punished by a bad convergence of the training. For the application discussed here one hidden layer is sufficient. The number of neurons in the hidden layer should be between 5 and 10. If the number is below 5 the performance is not optimal. If it is increased above 10 the training becomes unstable again, the calculation effort rises, and the risk of overfitting increases. Quoting a range rather than a precise number is deliberate. The reason is that there is not a clear optimum. As mentioned above because of the random starting values of the weights the results also have variations. Apart from this, small changes in the number of neurons in the hidden layer have no dramatic effects on the results. Such a general robustness is also an appreciated quality of the ANN.

The use of a non-linear activation function is necessary if an MLFF network is required to approximate non-linear functions. Predominantly the *tansig* functions are used as non-linear activation functions [13].

The experiment results in the experience that a linear activation function is the best choice for the neurons of the output layer. If the *tansig* function is used for the output layer the convergence of training and performance is worse in comparison to the application of the linear activation function.

19.7.3 Results Obtained with ANN

The results obtained using the best found configuration are given in the diagrams of Fig. 19.6 (prawn), Fig. 19.7 (herring), and Table 19.4. On the abscissa of the diagrams the composition values determined by the reference methods are plotted while the predicted values by the ANN are shown on the ordinate. Hence a perfect prediction would hit the quality line.

With the prawn test series an improvement of the $RMSE_v$ is observable. Especially with the estimated liquid uptake the accuracy is conspicuously higher. But with the herring test series the training of the ANN was not stable. Therefore the coefficient of determination R^2 is relatively poor. This is explainable by the low number of calibration data sets of this test series (herring: only 52; prawn: 92). In short the training of an MLFF ANN needs more effort and more calibration data sets. But the higher effort leads to a better prediction of the composition values.

Table 19.4. Results obtained with ANNs

Prawns	R^2	$RMSE_c$	$RMSE_v$
Liquid uptake	0.919	1.55 % of m_0	2.59 % of m_0
Fat	0.800	0.046 % abs.	0.076 % abs.
Protein	0.939	0.43 % abs.	0.53 % abs.
Water	0.933	0.41 % abs.	0.46 % abs.
Salt	0.792	0.045 % abs.	0.047 % abs.
Herring	R^2	$RMSE_c$	$RMSE_v$
Liquid uptake	0.503	6.45 % of m_0	5.04 % of m_0
Fat	0.304	1.61 % abs.	1.15 % abs.
Protein	0.366	1.42 % abs.	0.92 % abs.
Water	0.605	1.64 % abs.	1.62 % abs.

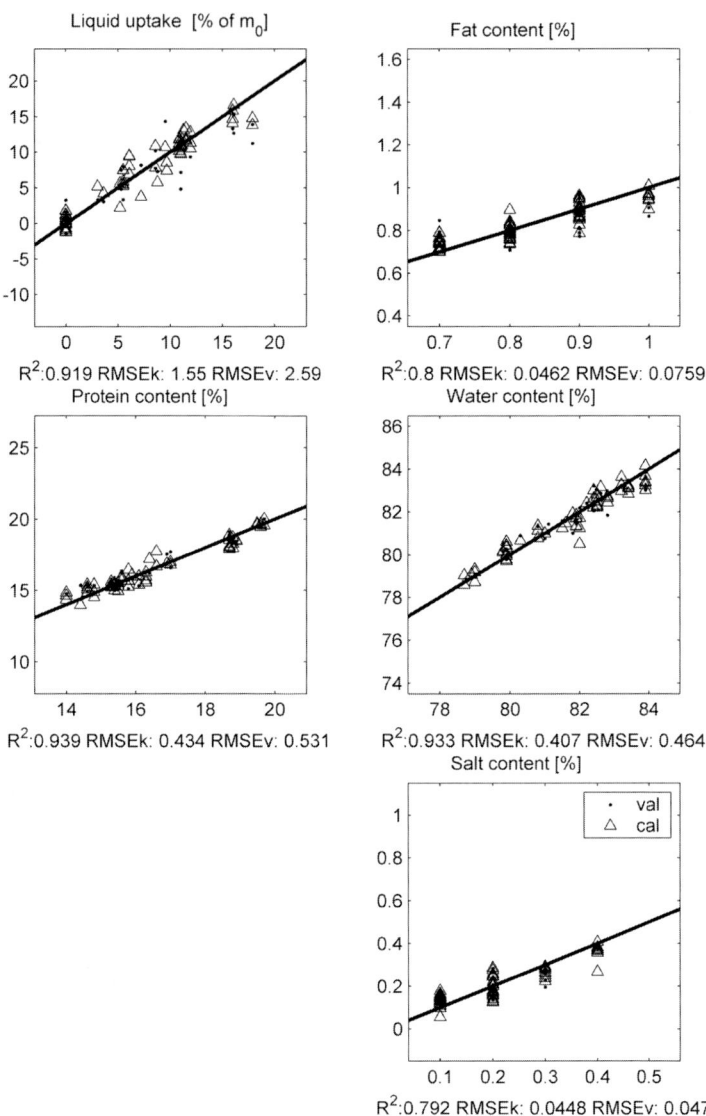

Fig. 19.6. Results obtained with ANNs for the prediction of the composition of prawn

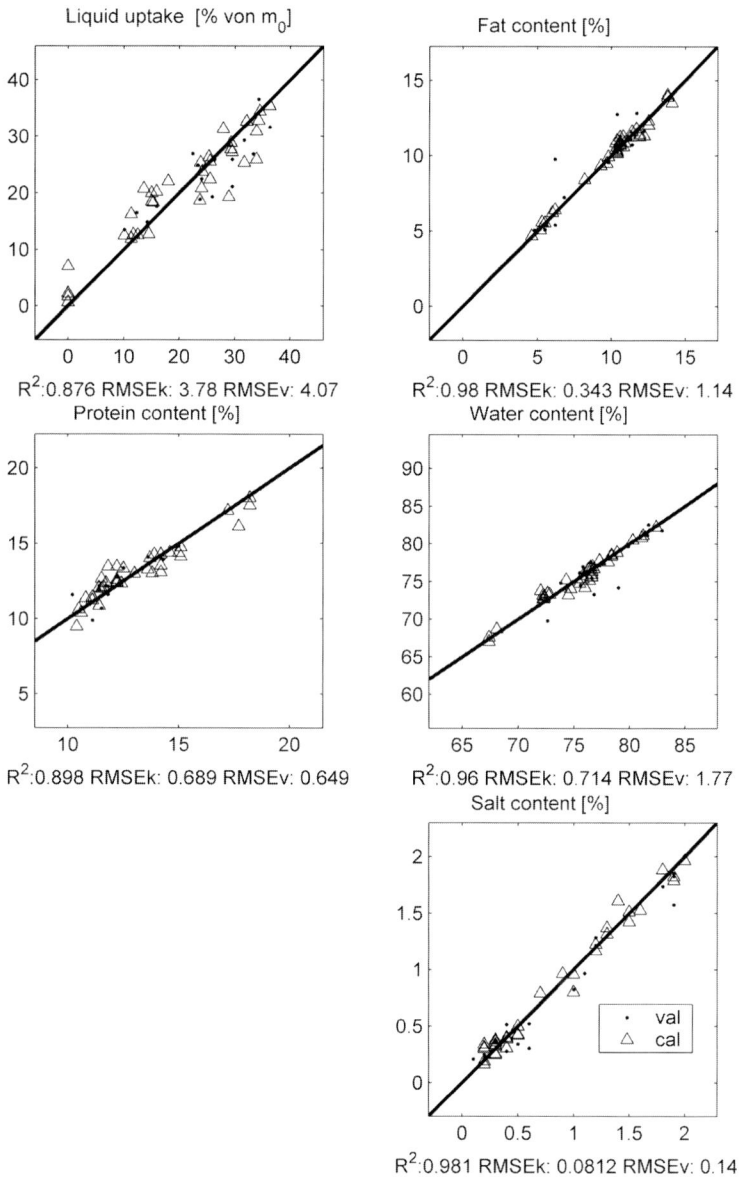

Fig. 19.7. Results obtained with ANNs for the prediction of the composition of herring

19.8 Comparison of Methods

For the comparison of methods for the prediction of the composition of foodstuffs using dielectric spectra three criteria are used:

1. Accuracy.
2. Calculation effort for calibration and validation.
3. Number of data sets needed for calibration and training.

In order to measure the calculation effort the *flops* function of MATLAB was used. In figure 19.8 the calculation effort for the complete calibration procedure of all methods is displayed versus the $RMSE_c$. One can see that the dielectric modeling has the worst performance although the calculation effort is relatively high. This high effort is caused by the curve fitting procedure. The highest calculation effort has to be done during the training of the ANN. But here the effort pays because the performance has the best results. In comparison, the direct processing of the dielectric data in a MLR is relatively effortless.

However, for the practice it is more important to know how costly the prediction of the composition values of a sample is. While the calibration does not have to be done in real time, a realized system may be controlled by a relatively simple microcontroller which has less computational power. In Fig. 19.9 the calculation effort for the estimation of the water content of one prawn sample is plotted versus the $RMSE_v$. Here the dielectric modeling also has the highest computational effort while the performance is relatively low. The performance of the direct processing of the dielectric data in a MLR and PLSR is similar. The effort of the MLR is lower. But with PLSR the orthogonalization proceeds more directly. The best results are taken with the ANN. The computational effort is higher but lower than with the dielectric modeling. With nearly 2,000 floating point operations it should be possible to use ANN in real time.

With all methods expect the ANN it is possible to make a successful calibration with fewer calibration data sets. If only a relatively small number of calibration data sets are available the internal cross-validation may be used [8]. But ANNs need a relatively high number of calibration data sets. However, the higher effort leads to better results.

19.8.1 General Evaluation of the Methods Discussed

The enforced experiments and discussed data processing methods showed that it is in principle possible to determine important composition values of foodstuffs using microwave dielectric spectra in combination with a multivariate processing of the measurement data. But because this method works in an indirect way two requirements have to be fulfilled:

1. The range of a composition value has to be wide enough.
2. The reference methods have to be accurate and should have a high resolution.

For example, these requirements are not given for the prediction of the fat content of the prawn samples. The range was only from 0.7% to 1.0% and the resolution of the reference method[5] was 0.1%. The results are accordingly moderate. This is the same case for the prediction of the salt content of prawns. The range was only 0.1% to 0.4% and the resolution of the reference method[6] was 0.1% too. But with the herring test series these requirements are fulfilled for the prediction of the fat and salt content. The fat content of the herring samples varied between 6% and 14.1%. Due to the treatment with different saline solutions the salt content varied between 0.1% and 2%. Hence the performance is comparatively better.

The quality of the reference methods has a key role for the methods discussed here. An indirect measurement method cannot be better than its reference method. But for complex items like food it is also difficult to make accurate measurements with the reference method. For example, the reference method for the prediction of the protein content may be disputed. Proteins are a whole class of constituents with multifaceted properties. The determination of the protein content by the reference method works with the measurement of the nitrogen content[7]. The protein content is then estimated using a more or less arbitrary factor (the nitrogen factor). The prediction of the water content is also critical[8]. Not only does water vaporize during the procedure, but so too do other volatile substances. The results obtained with ANN reach the accuracy of the reference method but they cannot better it.

The performance could be improved with another sensor. Because it is not easy to make a good contact between the open-ended coaxial line and the sample, and

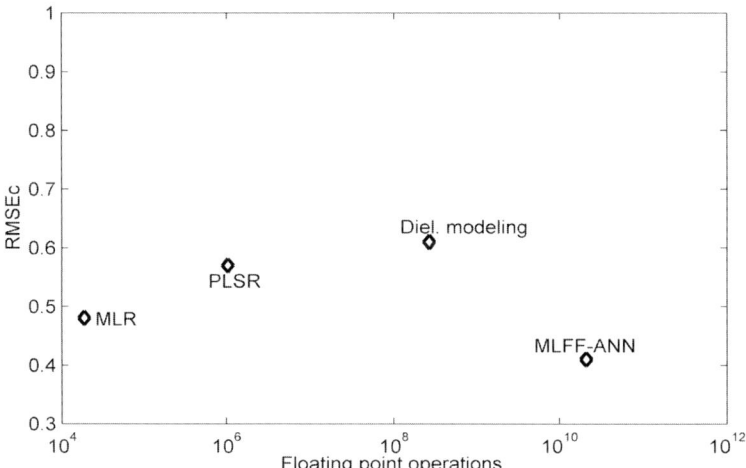

Fig. 19.8. Comparison of the complete calculation effort of the calibration (water content of prawn)

---

[5] Fat: NMR (SLV 195:9).
[6] Salt: conductometry.
[7] Protein: AOAC No. 999.15 and 990.03.
[8] Water: oven drying (104°C, 104 h).

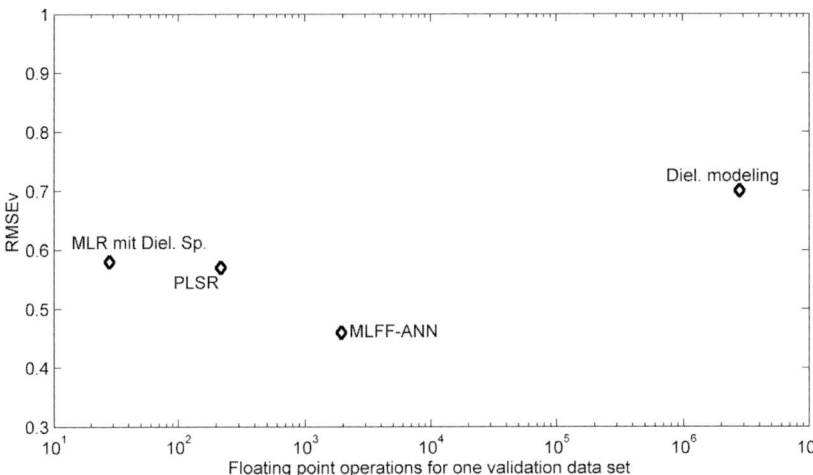

Fig. 19.9. Comparison of the calculation effort of the validation respectively the application of the methods for one dataset (water content of prawn).

because flesh is an inhomogeneous material, it is desirable to have a sensor with a higher sensitive area. The multivariate methods do not necessarily need the permittivity as an input variable because various other arrangements are possible as sensors. Because the measured S-parameters could be used as input variables directly, there is no need to create a model of the arrangement. It is only important that the sensor is unambiguously sensitive enough to changes in the composition.

Another aspect is the question of how far the prediction of the protein and salt content is independent of the prediction of the water content. The treatment of the samples leads to the addition of water. This means that the other components are diluted. Water has a dominant influence on the microwave dielectric spectrum of flesh. Hence there is a risk that the protein and fat content is only apparently predicted by the use of the negative correlation between the water content and the fat and protein content. Furthermore, there is a complex interaction between the water and the other constituents as mentioned above. This complex interaction could make it possible to detect other components if they modify the dielectric behavior of the water.

In Fig. 19.10 the correlation between water and protein and between water and fat is shown. The correlation factors are $\rho = -0.75$ and $\rho = -0.9$ approximately. In order to test how far the prediction of the protein and fat content depends on the negative correlation they were determined using the predicted water content (PLSR) and a linear regression only. In this case the coefficient of correlation decreases to $R^2 = 0.64$ (protein) and $R^2 = 0.53$ (fat). That means not only does the negative correlation deliver information about the fraction of protein and fat, but the hidden information in the dielectric spectrum is also effective.

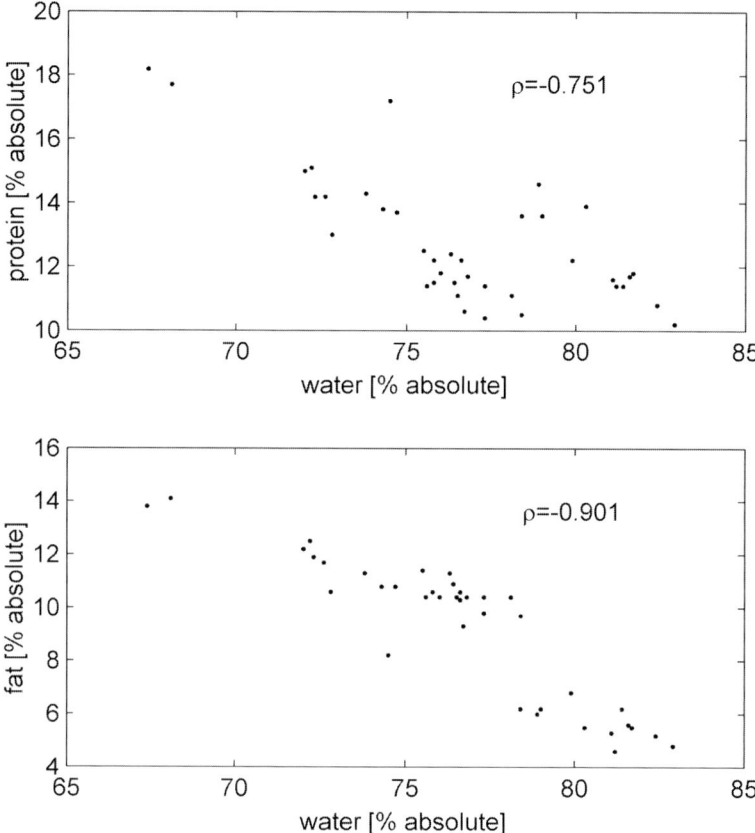

Fig. 19.10. Correlation between the water and the protein and fat content (herring).

Because the salt content causes higher losses at lower frequencies it has a direct and strong influence on the dielectric spectra. Hence the problem discussed above does not exist with the prediction of the salt content.

19.9 Conclusions

The determination of the composition of foodstuffs using conventional methods is time consuming, alters or destroys the samples, and needs many expensive devices. Therefore these methods are not applicable on the whole for the control of production processes. The approach suggested here is based on the influence of the ingredients on the microwave dielectric spectrum. The samples of two test series were diluted with water and their dielectric spectra were measured in the frequency range between 200 MHz and 12 GHz using an open-ended coaxial line.

A qualitative view on the measurements data showed that the dielectric spectra change with the amount of added water. Several methods were investigated to extract the hidden information about the composition of the samples from the dielectric spectrum.

The first method used a simple dielectric model. However, the performance of this method was not satisfactory and the computational effort was relatively high. Statistical analyses result in the perception that it is not reasonable to expand the dielectric model because the measurement accuracy is not good enough to detect small details of a more complex model. It was easier and more accurate to process the measured dielectric data directly with multiple linear regression (MLR). Due to the fact that the data have a high collinearity there is the problem that numerical instability occurs.

This problem can be avoided by the use of principal components analysis (PCA) or partial least squares regression (PLSR). If it is assumed that the unknown function which describes the relation between the permittivity and the composition of the samples might be non-linear the use of multi-layer feed forward artificial neural networks (MLFF ANNs) can be used to approximate it. The computational effort during calibration and training is relatively high and more calibration data are needed. But the best results were obtained with this method. If the number of calibration data sets is below approximately 50 the use of PLSR should be preferred.

It was also shown that besides water other important ingredients like protein, fat, and salt can be determined. In fact the reference methods cannot be exceeded. But all constituents can be measured in one step and in a short time by the multivariate processing of the microwave dielectric spectra.

Acknowledgments

The contributions of and many discussions with Dr. Mike Kent are greatly appreciated. This study has been carried out with financial support from the Commission of the European Communities, Agriculture and Fisheries (FAIR) specific RTD program FAIR CT97-3020 ''Added Water in Foods''. This study does not necessarily reflect the Commission's views and in no way anticipates its future policy in this area.

References

1. Kent M, Anderson D (1996) Dielectric studies of added water in poultry meat and scallops, J Food Eng 28: 239–259
2. Kent M (1999) Simultaneous determination of composition and other material properties by using microwave sensors. In: Sensors update, Wiley-VCH, Weinheim, vol 7, pp 4–25

3. Gajda G, Stuchly S (1983) An equivalent circuit of an open-ended coaxial line. IEEE Trans Instrum Meas IM-32(4): 506–508
4. Stuchly M, Brady M, Stuchly S, Gajda G (1982) Equivalent circuit of an open-ended coaxial Line in a lossy dielectric. IEEE Trans Instrum Meas IM-31(2): 116–119
5. Marsland TP, Evans S (1987) Dielectric measurement with an open-ended coaxial probe, IEE Proc 134(4): 341–349
6   Grant EH, Sheppard RJ, South GP (1978) Dielectric behaviour of biological molecules in solution. Oxford University Press, Oxford
7. Hartung J, Epelt B (1995) Multivariate Statistik, Lehr- und Handbuch der angewandten Statistik, 5. Auflage, R. Oldenbourg Verlag, Munich
8. Martens H, Naes T (1989): Multivariate Calibration. Wiley, Chichester
9. Kent M, Knöchel R, Daschner F, Berger UK (2000): Composition of foods using microwave dielectric spectra. Eur Food Res Technol no 210: 359–366
10. Archibald DD, Trabelsi S, Kraszewski AW, Nelson SO (1998): Regression analysis of microwave spectra for temperature-compensated and density-independent determination of wheat moisture content. Appl Spectrosc 52(11): 1435–1446
11. Bartley PG, Nelson SO, McClendon RW, Trabelsi S (1999): Determination of moisture content in wheat using an artificial neural network. In: 3rd workshop on electromagnetic interaction with water and moist substances, Athens, GA, pp 74–78
12. Patterson D (997) Künstliche neuronale Netze, Prentice Hall Verlag, Haar
13. Neural Network Toolbox, User's Guide, Version 4, The Mathworks Inc., 2000
14. Kreinovich VY(1991): Arbitrary nonlinearity is sufficient to represent all functions by neural networks: a theorem. Neural Networks 4: 381–383

20 Microwave Dielectric Properties of Hevea Rubber Latex, Oil Palm Fruit and Timber and Their Application for Quality Assessment

Kaida bin Khalid, Jumiah Hassan, Zulkifly Abbas and Mohd Hamami[1]

Physics Department, Univ. Putra Malaysia, 43400 Serdang Selangor, Malaysia
[1] Faculty of Forestry, Univ. Putra Malaysia, 43400 Serdang Selangor, Malaysia

20.1 Introduction

Hevea rubber, palm oil, and timber are three important commodities for Southeast Asia. In 2001, Thailand, Indonesia, and Malaysia remained the three leading world rubber producers contributing about 68% of total world production (7.17 million tonnes) [1]. Approximately half of the production comes from rubber smallholdings. Latex collected by smallholders is sold to a collector who pays according to the **D**ry **R**ubber **C**ontent (DRC) or **T**otal **S**olid **C**ontent (TSC) or **M**oisture **C**ontent (MC). It is important that the true DRC or MC of the latex is determined correctly to ensure a fair price is paid to the tapper.

Field latex is a biological product with a complex composition which varies widely according to clone type, seasonal variations, tapping system, soil conditions, adulteration etc. It consists of 55–80% water or MC, 15–40% of rubber hydrocarbon (DRC), and 3–5% of non-rubber solid [2]. The basic components of the non-rubber solid (NRS) are proteins, lipids, quebrachitol, and inorganic salts. The total concentration of organic salts is approximately 0.5% which consists of phosphate ions (~0.25%), potassium (0.12–0.25%), and small amounts of copper, iron, calcium, sodium, and magnesium [3].

Palm oil has become one of the important edible oils and besides its nutritional and medicinal value, it is increasingly popular as a raw material for the oleochemical industries and a fuel for automobiles. The world production of palm oil in 2001 was about 23.3 million tonnes and a major production is from Southeast Asia region with 20 million tonnes [4]. Palm oil is obtained from the mesocarp of the oil palm fruits and oil quantity is determined by the quality of fruits during harvesting or amount of oil, water and free fatty acid (FFA) in the fruit. Normally the amount of MC in fresh mesocarp is about 85% at 14–15 weeks after anthesis and decreases rapidly to about 30% in ripe fruits at about 20–24 weeks after anthesis [5]. At 20 weeks after anthesis, only a small amount of oil increases and at the same time the percentage of FFA in oil increases, which reduce the quality of oil. The fiber content in mesocarp is almost constant after a certain stage of maturity whereas water and oil vary with ripeness. Therefore, there is an optimum time by which the bunch should be harvested. The close relationship between MC and stage of ripeness gives a possibility of using this parameter to gauge the ripeness of the fruit.

The third commodity is timber or wood, which is also an important product from Southeast Asia. Malaysia alone produces various kinds of timber products with a total market of about US$ 3.5 billion for 2001 [6]. Density and stage of decay are two main parameters which determine the quality of wood.

This chapter will review some of the previous work and current findings on the dielectric properties of hevea rubber latex, oil palm fruit, and timber or wood. The information on these properties is useful to the state of the art for the development of a latexometer for measurement of rubber content, a microstrip and coplanar moisture sensor for determination of ripeness level of oil palm fruit, and a wood densitometer for density and decay level assessment.

20.2 Dielectric Properties

The interaction of electromagnetic waves with the non-magnetic material is determined by the complex permittivity or dielectric properties of the material, $\varepsilon = \varepsilon' - j\varepsilon''$. The real part, ε', which is known as the dielectric constant, expresses the ability of the material to store energy and the imaginary part, ε'', known as the dielectric loss factor, is a measure of the energy absorbed from the applied field. The dielectric properties of the material depend on the physical properties of the material, especially MC.

20.2.1 Hevea Rubber Latex

20.2.1.1 Dielectric Properties

Field latex is a highly perishable material that will coagulate a few hours after tapping, unless a preservative is added. Normally, 0.3 to 0.6% of ammonia is added as a preservative. Field latex can be concentrated to a higher rubber content to make it more uniform and economically more attractive. The present standard of latex concentrate has a dry rubber content of about 60% and approximately 39% MC. This chapter will review some of previous work on the dielectric phenomena of rubber latex at various MCs, temperatures from −30 to 50°C, and frequencies from 10^{-3} Hz to 20 GHz. This study aims to find out some important aspect of these phenomena, such as operating frequency with respect to the development of a latexometer and estimating the power absorption during microwave drying or curing. Measurements of dielectric properties ranging from 200 MHz to 20 GHz are carried out by using an open-ended coaxial probe (HP 85070B) and at 10^{-3} Hz to 10^{6}Hz by using a capacitance method (Chelsea Dielectric Spectrometer).

Figure 20.1 shows the dielectric spectrum of hevea latex from various samples as fresh field latex, diluted fresh latex, latex concentrate, and diluted latex concentrate. The dielectric constant ε' for almost all samples follows the trend of deionized water and their magnitudes are found decrease as the MC in the latex decreases. For the frequency range between 0.2 and 20 GHz, the loss factor ε'' can be divided into two regions. For frequency less than 2.5 GHz, ε'' is dominated by conductive loss while in the upper region (> 2.5 GHz) the loss mechanism is dominated by the dipole orientation of water molecules in the latex.

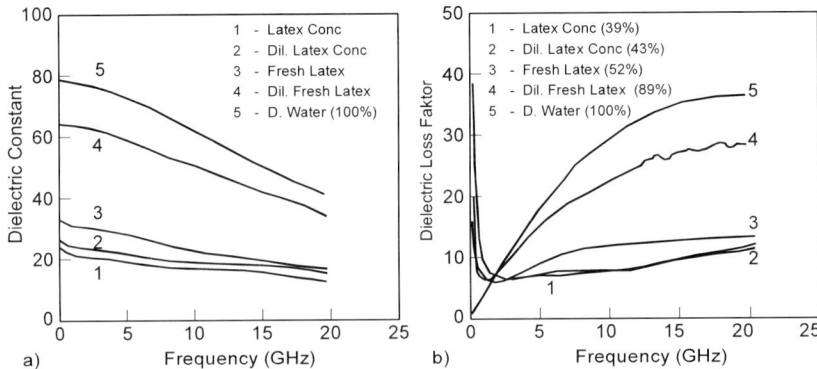

Fig. 20.1. Dielectric properties of hevea rubber latex as a function of frequency. **a** Dielectric constant. **b** Dielectric loss factor. (See [7])

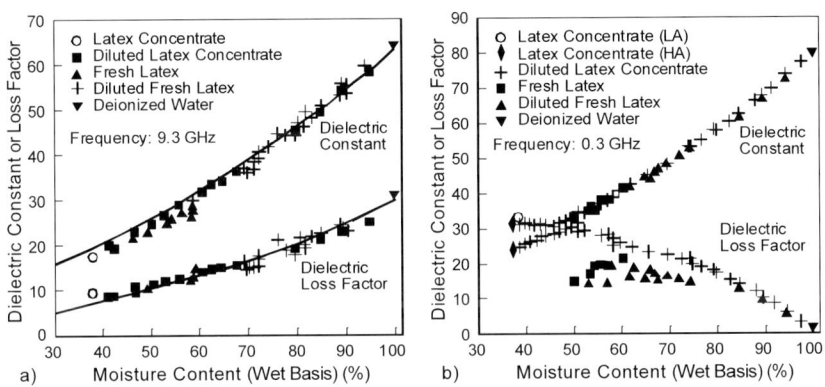

Fig. 20.2. Dielectric properties of hevea rubber latex versus MC (wet basis) at 26°C: **a** 9.3 GHz; **b** 0.3 GHz (See [7, 8])

The effect of conductive loss can be seen clearly in Fig. 20.2. At 0.3 GHz, ε'' spreads depending on the strength of ionic species in the latex. For latex concentrate, the higher value of ε'' at a particular MC is due to ammonium ions associated with ammonium added to the latex concentrate acting as a preservative.

However, at 9.3 GHz, no spreading of ε'' was observed. This implies that the contribution from the conductive loss is small and ε'' is merely dominated by the dipolar orientation of water molecules. A close relationship between dielectric properties and MC at this frequency, regardless of the types of latex, gives important information on the state of the art of the development of microwave moisture meters for latex.

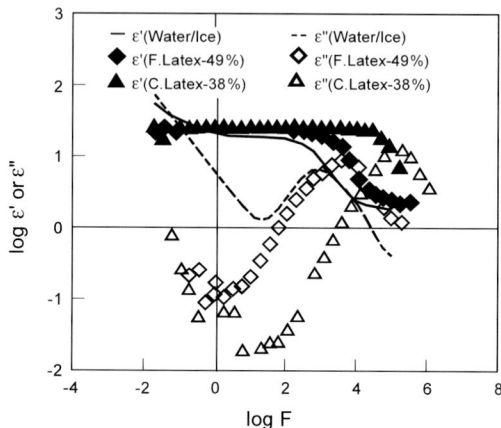

Fig. 20.3. Variation of dielectric properties of solidified latex at lower frequencies ranging from 0.01 Hz to 1 MHz and a temperature of −30°C. (See [7])

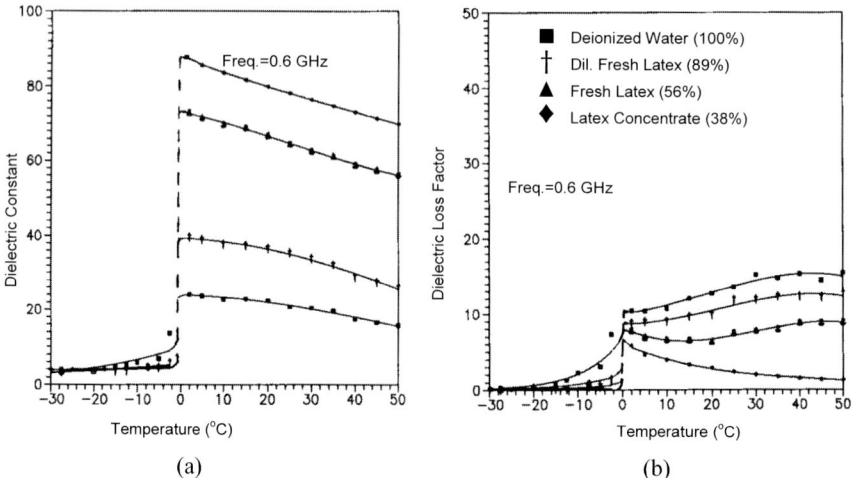

Fig. 20.4. Dielectric properties of hevea rubber latex with respect to temperature. **a** Dielectric constant at 0.6 GHz. **b** Dielectric loss at 0.6 GHz.

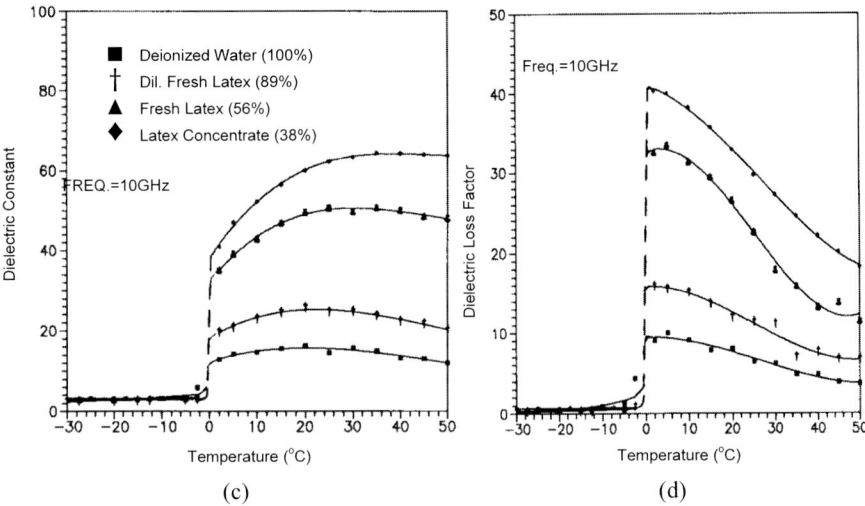

Fig. 20.4. (cont.) Dielectric properties of hevea rubber latex with respect to temperature. **c** Dielectric constant at 10 GHz. **d** Dielectric loss at 10 GHz. (See [9])

It was stated earlier that the hevea rubber latex is conductive for frequencies less than about 2 GHz. However, solidified latex at temperatures less than 1°C shows a dielectric phenomenon, which is almost similar to ice [7]. Figure 20.3 shows the dielectric spectrum of fresh latex and latex concentrate with frequency ranging from 10^{-2} Hz to 10^{6} Hz and at a temperature of –30°C. The relaxation peak is shifted to a higher value as the water content in the latex decreases.

The temperature dependence of ε' and ε'' of hevea latex at 0.6 and 10 GHz is shown in Fig. 20.4 [9]. Three different states exist in the temperature range of –30 to 50°C. These are the frozen (solid) state (–30 to –3°C), the transition state (–3 to 3°C), and the liquid state (above 3°C). In the transition region both ε' and ε'' show a steep increase as phase of the latex changes from solid to liquid. This may be due to a change in the physical state of water molecules from bound water to free water. In the frozen state, both ε' and ε'' at 0.6 GHz show a spreading in their values with temperature and this may be due to the effect of conducting phases in the samples. In the liquid state region and at 0.6 GHz, the increase in temperature raises the mobility of ions in the solution resulting in an increase in ε''. At 10 GHz, ε'' decreases as temperature increases, which is similar to the trend of deionized water. The shapes of the latex curves show a depression, which may be due to the water binding by dissolved ions. The details of the study on hevea latex in the frequency range of 0.2 to 20 GHz can be found in the previous works [8-10].

20.2.1.2 Mixture Model

The prediction of dielectric properties of latex from the dielectric mixture model has been carried out using the formulas from Wiener's upper bound [11], Bruggeman [12], and Kraszewski et al [13]. In these formulas, hevea latex is treated as a biphase liquid, consisting of water and solid rubber. In Wiener's upper bound formula the relative dielectric permittivity of the mixture is written as

$$\varepsilon = (1-\delta)\varepsilon_s + \delta\varepsilon_w \tag{20.1}$$

where ε_w and ε_s are the relative dielectric permittivity for water and solid materials respectively, and δ is the water volume fraction.

In this expression, it is assumed that the water molecule (dipole) is an ellipsoid with the major axis parallel to the direction of the applied field and the dipole is free to orientate. A previous study has shown that for the frequency range 2 to 20 GHz, the dielectric properties of a liquid are strongly dependent upon the geometrical shape of the ellipsoid. For a spheroid with three main axes a, b, and c with $a \neq b = c$ the equation for ε which are based on the Bruggemann model may be written as [14]

$$(1-\delta) = \left[\frac{\varepsilon_s}{\varepsilon}\right]^{3d} \frac{(\varepsilon_w - \varepsilon)}{(\varepsilon_w - \varepsilon_s)} \left[\frac{\varepsilon_s(1-3A) + \varepsilon_w(2-3A)}{\varepsilon(1-3A) + \varepsilon_w(2-3A)}\right]^K \tag{20.2}$$

where

$$d = \frac{A(1-2A)}{(2-3A)} \text{ and } K = \frac{2(3A-1)^2}{(2-3A)(1-3A)}$$

A is known as the depolarization factor, δ is the volume fraction, and the value of ε may be obtained from Eq. (20.2) by the numerical root-seeking method. Kraszewski et al [13] have developed a simple model based on the relation between the propagation constants and relative dielectric permittivity and is written as

$$\varepsilon^{1/2} = \delta\varepsilon_w^{1/2} + (1-\delta)\varepsilon_s^{1/2} \tag{20.3}$$

The volume fraction δ is related to the MC (wet basis) by

$$\delta = M_w[M_w + (D_s / D_w)] \tag{20.4}$$

M_w is the MC, D_w and D_s are the relative density of the water and solid rubber respectively and are considered to be constant with $D_w = 1.0$ and $D_s = 0.94$.

Figure 20.5 shows the variation of ε' and ε'' with MC at 10.9 GHz and at 26°C. The experimental data are shown by the point symbols and the lines are the theoretical values predicted from mixture equations given by Eq. (20.1, 20.2) (with $a/b = 0.01$), and Eq. (20.3). Throughout these figures ε' and ε'' demonstrate a good relationship with the MC and are almost independent of the type of solution.

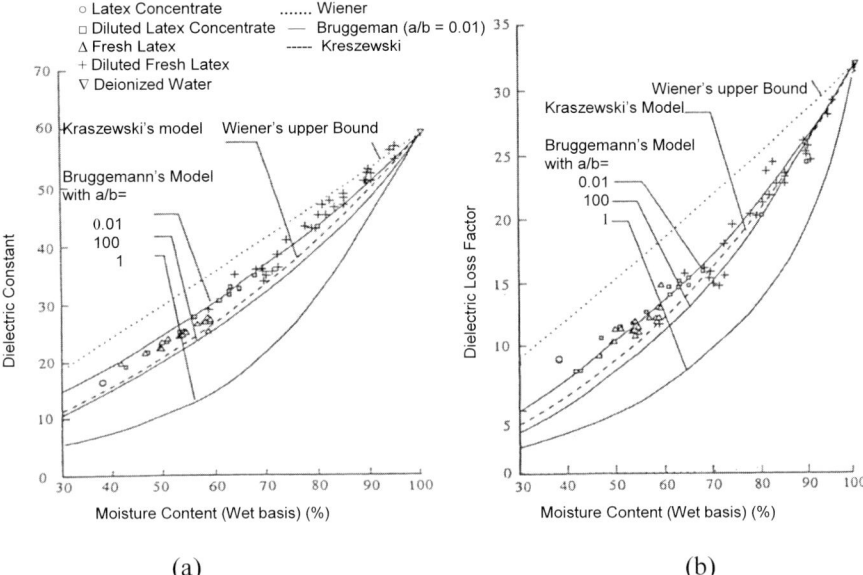

(a)   (b)

Fig. 20.5. Comparison between experimental dielectric data for hevea latex with theoretical data calculated from the mixture model at 10.9 GHz and 26°C. **a** Dielectric constant **b** Dielectric loss factor (See [8])

The experimental results of Fig. 20.5 are very close to the prediction values of Bruggemann's model with $a/b = 0.01$ (prolate spheroid) and well below and close to the upper limit of Wiener's model. These results suggest that the water molecules in hevea latex are loosely bound and are easily aligned by the electric field, and the shape of the water molecules is probably close to that of prolate spheroid.

The model from Kraszewski et. al. is also suitable to predict the variation of ε' and ε'' with MC with the average deviation from measured values of within 5–7% as compared with that of Bruggemann's model of about 3%.

20.2.2 Oil Palm Fruit

20.2.2.1 Dielectric Properties

The rate of ripening of a single fruit is not uniform with ripening, but usually starts from the upper to the lower region and from the outer to the inner region. There is a wide variation of MC, especially in the under-ripe and nearly ripe fruit. Therefore, it is better to use the MC of the mashed mesocarp of the whole fruit in order to give a reliable relationship between MC or permittivity and level of maturity [16].

Figure 20.6 shows the dielectric spectrum of mashed mesocarp at various stages of maturity and mesocarp constituents such as water, fiber and oil. The ε' for most samples follows the trend of water and their magnitudes decrease as the MC in the mesocarp decreases. In the frequency range of 0.2 to 20 GHz, the loss factor can be divided into two regions. For frequencies less than 3 GHz, dielectric loss is dominated by conductive loss, while in the upper region the loss mechanism is dominated by the dipole orientation of the water molecule. Dielectric properties of the fiber and oil are about the same and their values are about 2.6 - j0.02 and almost constant throughout the frequency of measurement.

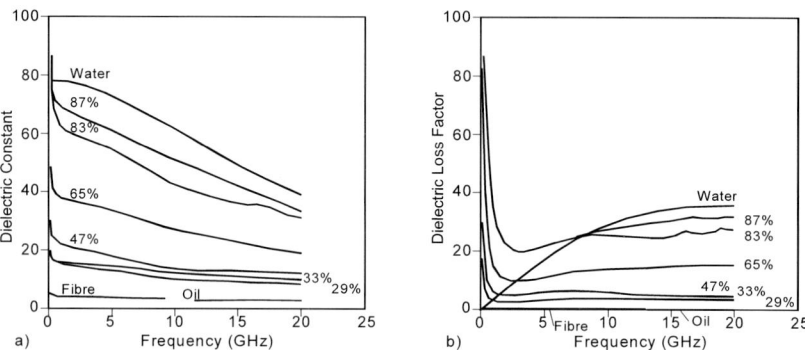

Fig. 20.6. Dielectric spectrum of mashed mesocarp from 0.2 GHz to 20 GHz at various moisture contents. **a** Dielectric constant. **b** Dielectric loss. (See [16])

Previous workers (Nelson et al) [17] introduced a permittivity maturity index based on the real part ε' at 0.2 GHz and dielectric loss, ε'' at 10 GHz, which was written as follows:

$$M_p = \frac{(\varepsilon')_{0.2} + (\varepsilon'')_{10}}{100}.$$

(20.5)

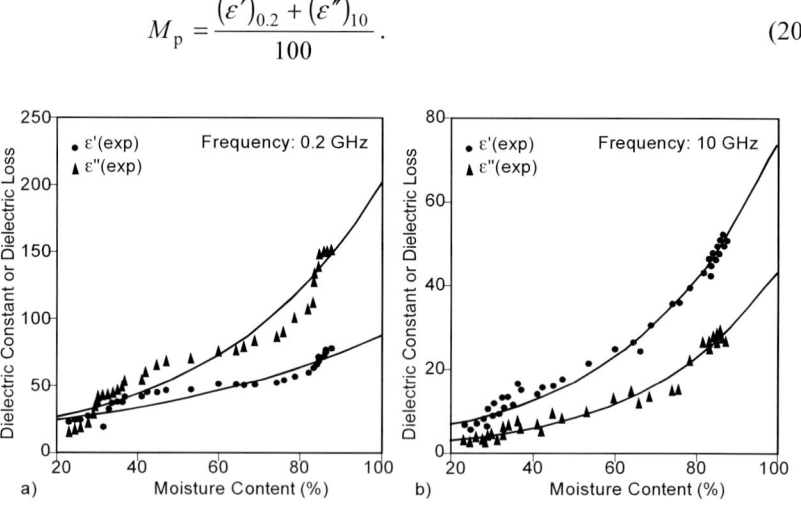

Fig. 20.7. Variation of dielectric properties of mashed mesocarp with respect to MC at **a** 0.2 GHz and **b** 10 GHz (See [16])

The variation of ε' and ε'' with MC at 0.2 and 10 GHz is shown in Fig. 20.7. Throughout these figures ε' and ε'' demonstrate a good relationship with MC or maturity of the fruits. Based on the dielectric properties of oil palm mesocarp at 0.2 and 10 GHz with respect to maturity, the index is found suitable for oil palm fruits. From Table 20.1 the maturity index for oil palm fruit varies from about 1.0 for young fruit to 0.3 for fully ripe fruit.

Table 20.1. Dielectric properties and maturity index at 26°C of mashed mesocarp at four stages of maturity. (See [16])

Stage of maturity Maturity index	0.2 GHz	10 GHz
Under–ripe	ε' : 35–70	ε' : 13.5–48.3
0.4–0.97		
40–85% MC	ε'' : 42.5–140	ε'' : 6–26.7
Nearly ripe and ripe	ε' : 29–35	ε' : 10.8–13.3
0.3–0.4		
33–40% MC	ε'' : 36.5–42.5	ε'' : 5–6
Fully–ripe	$\varepsilon' < 29$	$\varepsilon' < 10$
<0.3		
< 33% MC	$\varepsilon'' < 36.5$	$\varepsilon'' < 5$

20.2.2.2 Mixture Model

The results of measurements are compared with a simple mixture model [13] where the relative dielectric permittivities of the mesocarp can be expressed as follows:

$$\sqrt{\varepsilon_m} = v_\omega \sqrt{\varepsilon_\omega} + v_1 \sqrt{\varepsilon_1} + v_f \sqrt{\varepsilon_f} \qquad (20.6)$$

where v_ω, v_1, and v_f are the volume fractions of water, oil and fiber respectively and ε_ω, ε_1, and ε_f are the corresponding relative permittivities. The volume fraction for fiber is about 16% [15] and v_1 and v_ω can be written as follows:

$$v_1 = 1 - v_\omega - v_f \qquad (20.7)$$

and

$$v_\omega = \frac{M_w \left(\rho_f v_f + \rho_1 - \rho_1 v_f \right)}{\rho_\omega - M_w \rho_\omega + M_w \rho_1} \qquad (20.8)$$

where M_w is the MC (wet basis), and ρ_f, ρ_1 and ρ_ω are the densities of fiber, oil, and water respectively.

Generally, the mixture equation is suitable for predicting the dielectric properties of oil palm mesocarp at 10 GHz and with MC ranging from 3 to 85%.

The deviation between predicted values and measured values is within 5%. In Fig. 20.8 (a), the experimental results for dielectric loss are higher than the predicted values, which are based on the dipolar mechanism; these properties may be contributed by conduction loss influenced by conducting phases in mesocarp.

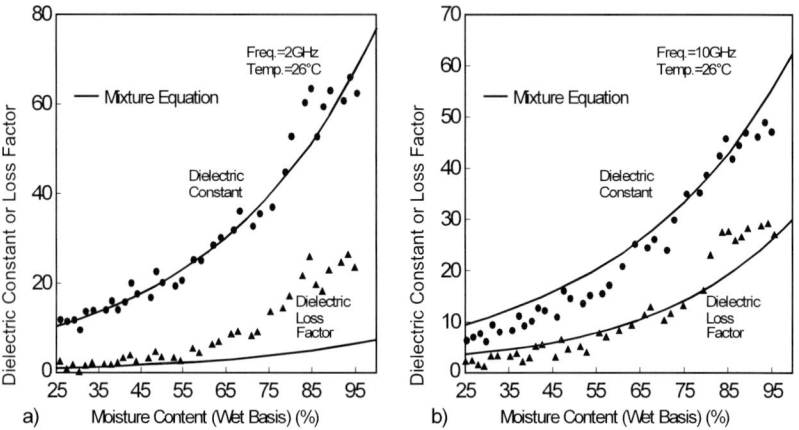

Fig. 20.8. Relationship between dielectric properties of mashed mesocarp with MC at **a** 2 GHz and **b** 10 GHz. Solid lines show the theoretical data from the mixture model (See [16])

20.2.3 Timber

20.2.3.1 Dielectric Properties

Various microwave techniques have been proposed for the density measurement of wood, which are based on the variation of attenuation and phase shift with the density [18] [19]. Most of these methods are expensive to develop, quite delicate to run, and only suitable for laboratory work. There is a need to develop a simple, portable, easy and accurate wood densitometer, which might be useful for field application. In this project, a simple microwave wood densitometer will be developed based on the reflection method. This meter will be used to measure the wood density ranging from 300 kg/m^3 to 1200 kg/m^3. The range covers most of the wood species originating from tropical countries.

The design involves the determination of the functional dependence between material permittivity, ε with MC, and ε with density. This scheme enables us to find the accuracy, sensitivity, and dynamic range of the measuring system in terms of geometrical and electrical properties of the sensor before the actual design is done. Fig. 20.9 shows the variation of dielectric properties with density of the various species of tropical woods [20]. The close relationship between these parameters enables us to design a wood densitometer.

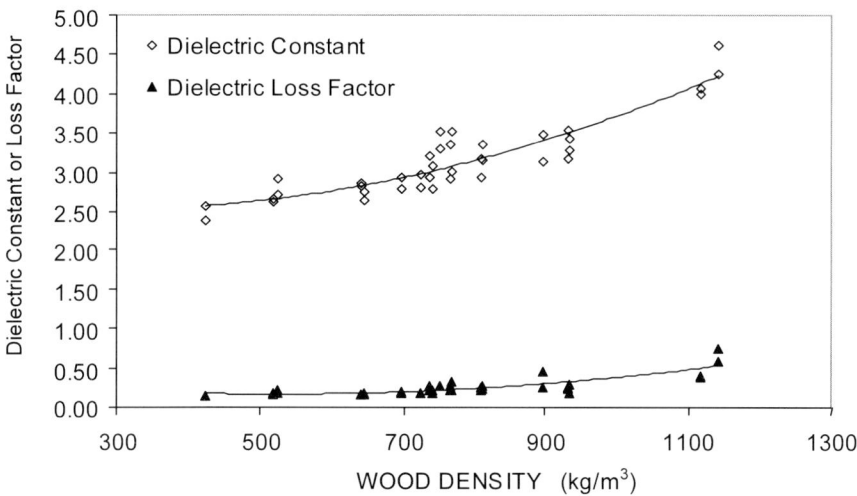

Fig. 20.9. Variation of dielectric properties of wood at 10.9 GHz and its density. (See [20])

In certain countries, some power lines are supported with wooden poles and cross-arms. Normally, local hardwoods such as Chengal are used extensively for this purpose. When wood is exposed to the weather, its surface undergoes changes. Initially, the dark-colored woods tend to fade, and as weathering continues, all woods assume a silver-gray color, with the gray layer extending 0.007 to 0.025 mm in depth. Other changes due to natural weathering include warping, loss of some surface fibers, surface roughening, and end checkings. As a result of weathering, unprotected surface tends to absorb more moisture compared to the surface of normal wood. The moisture will be sustained in the weathered wood for some time before it dries out. Moisture removal normally occurs at the ends and on the surface. There is a possibility to relate the quality of the wood to the amount of its moisture at a particular time. For example, during the rainy season, weathered or degraded wood will have more moisture at the surface than normal wood and replacement should be made when the level of MC has passed a certain limit.

The weathered wooden cross-arms can be classified into three categories, namely severely decayed wood, incipient wood, and sound wood. The descriptions of each category are as follows:

Severely decayed wood (type-A) shows very advanced level of decay. The color of the samples is brown to dark brown. There are some signs of crevices and small valleys on the surface. These crevices may result from severe decay and severe weathering on the sample surfaces. The samples are very soft and can be easily broken. The sample can achieve up to 70% MC at saturation level.

The obvious indicator of **partly (incipient) decay wood (type-B)** is the change in its color. The samples are brown to dark brown in color with white lines (axial resin canals) evenly spaced in between. The samples appear to be attacked by brown rot fungi. The samples are quite soft and can be broken using little extra force. Some remnants of fungal hypae occur in the vessels. The MC of the sample at saturation level is about 50%.

Sound wood (type-C) appears to be very sound without any visible attack from fungi. The fibers are still intact and maintain Chengal's natural color even after being used for more than 10 years. Small cracks do appear but the wood still shows no sign of visible decay and is still hard. The MC of this sample at saturation level is about 35%.

20.2.3.2 Mixture Model

Water is present in wood in two forms, one as free water in the cell cavities and pores and the other as bound water in cell walls. The free water is held by capillary forces whereas bound water is chemically bonded to the cell walls. When the wood is dried, the first water to be removed is the free water, and eventually a stage is reached when the cell cavities and pores are empty but the cell wall is still saturated. This is the fiber saturation point (FSP), generally ranging from 10 to 35% depending on the species. This means that the moisture in wood contributes to the dielectric properties in two ways: one below FSP and the other above FSP (Fig. 20.10). The MC below the FSP constitutes bound water whereas the MC above the FSP constitutes as free water plus bound water.

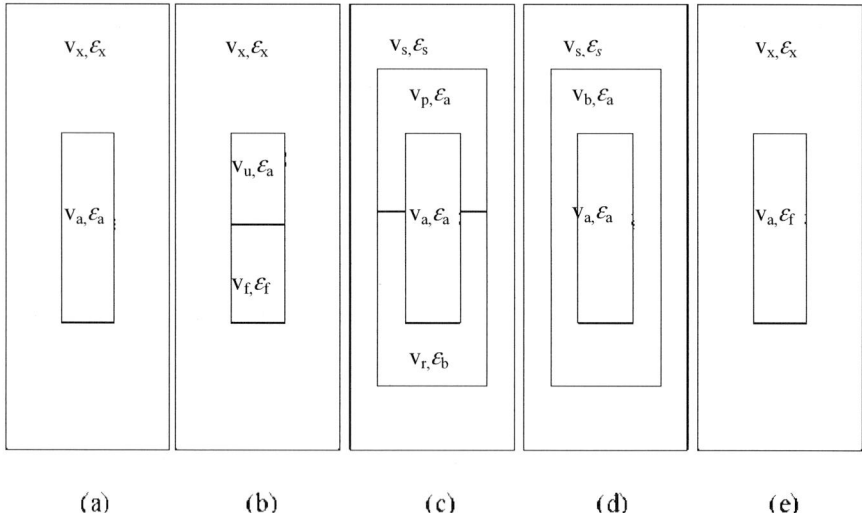

(a)   (b)   (c)   (d)   (e)

Fig. 20.10. Different types of water in wood: **a** at FSP; **b** above FSP; **c** Below FSP; **d** fully dried; **e** fully saturated. (See [21])

At FSP the dielectric properties of the wood sample, ε_w may be written using the known mixture equation as

$$\varepsilon_w = V_x \left(\varepsilon_x\right)^k + V_a \left(\varepsilon_a\right)^k \tag{20.9}$$

where ε_x, ε_a, V_x, V_a are the dielectric constant and volume fraction of a mixture of bound water and cell wall substance and air respectively. k is a coefficient which varies from 0 to 1 and is determined experimentally or from theoretical predictions [11].

ε_x can be written in terms of the dielectric properties of bound water, ε_b, and cell wall substance, ε_s, as

$$\varepsilon_x = V_b \left(\varepsilon_b\right)^k + V_s \left(\varepsilon_s\right)^k \tag{20.10}$$

Above FSP, the dielectric properties of wood can be written as the combined dielectric properties from three phases, i.e. ε_x, the dielectric constant of free water, ε_f, and air,

$$\varepsilon_w = V_x \left(\varepsilon_x\right)^k + V_f \left(\varepsilon_f\right)^k + V_u \left(\varepsilon_u\right)^k \tag{20.11}$$

where

$$V_a = V_f + V_u \tag{20.12}$$

Below FSP, the dielectric properties of wood may be written as

$$\varepsilon_w = V_s \left(\varepsilon_s\right)^k + V_r \left(\varepsilon_b\right)^k + V_p \left(\varepsilon_a\right)^k + V_a \left(\varepsilon_a\right)^k \tag{20.13}$$

where V_r and V_p are the volume fractions for bound water and air in the cell wall region respectively.

When the wood is fully dry dielectric properties can be written simply as

$$\varepsilon_w = V_s \left(\varepsilon_s\right)^k + \left(V_b + V_a\right)\left(\varepsilon_a\right)^k \tag{20.14}$$

where

$$V_b = V_r + V_p \tag{20.15}$$

And lastly, when the wood is fully saturated with water its dielectric properties are written as

$$\varepsilon_w = V_x \left(\varepsilon_x\right)^k + V_a \left(\varepsilon_f\right)^k \tag{20.16}$$

The volume fractions V_b and V_a can be determined experimentally when the wood is totally dry and fully saturated with water, while V_f and V_r can be calculated from the known MC and density of the wood.

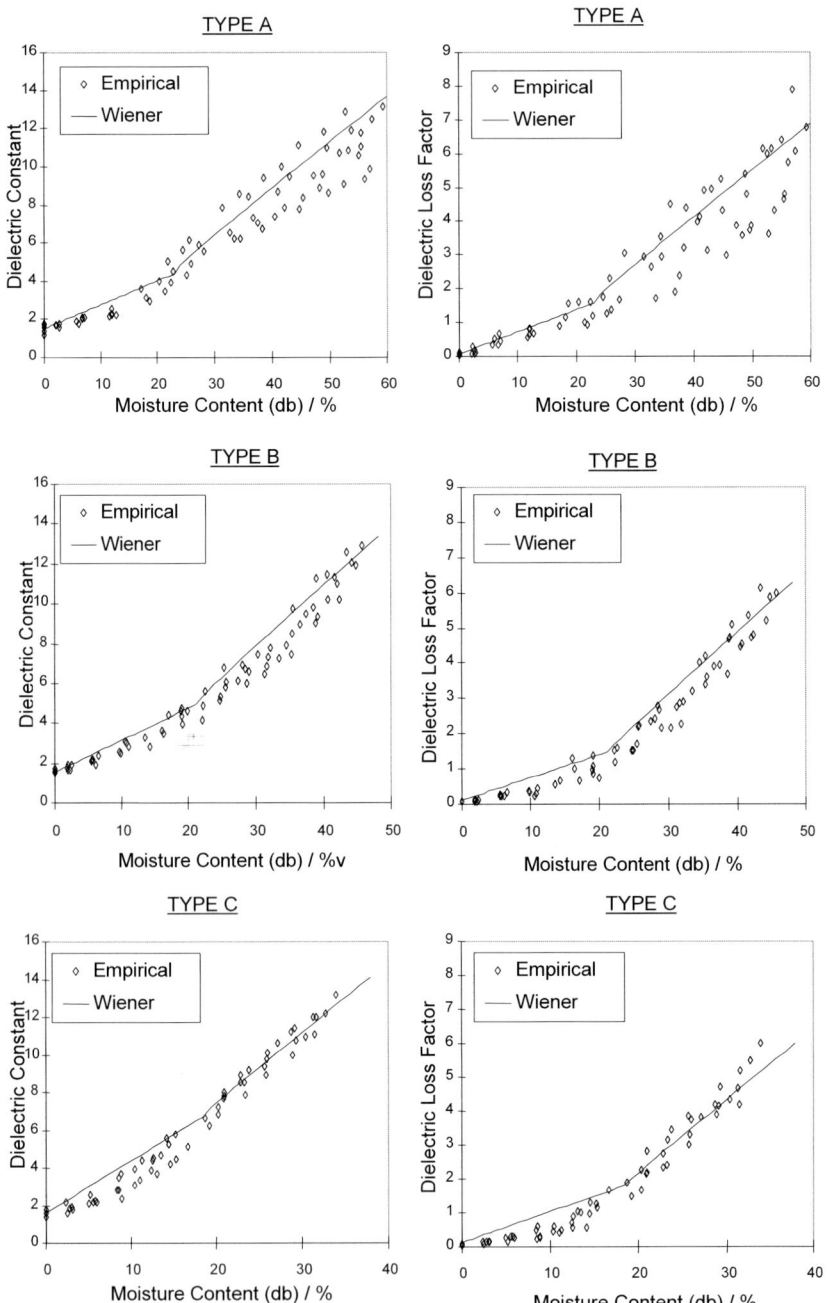

Fig. 20.11. Comparison of the experimental dielectric data for wooden cross-arms with mixture model data at 10.7 GHz and 26°C for types A, B, and C (See [21])

For example, V_f can be calculated in terms of MC, M_w density of wood, γ_x at FSP, density of water, γ_w, and density of air, γ_a as follows:

$$V_f = M_w \left(\gamma_a V_a + \gamma_s - \gamma_x V_a\right)/\left(\gamma_w - \gamma_w M_w + \gamma_a M_w\right) \tag{20.17}$$

Since γ_a is very small, then Eq. (20.17) can be simply expressed as

$$V_f = M_w \gamma_s \left(1 - V_a\right)/\left[\gamma_w \left(1 - M_w\right)\right] \tag{20.18}$$

ε_s, ε_f and ε_a can be obtained from the well-established data while ε_b can be determined from Eqs. (20.10 and 20.11) by knowing the value of ε_w at FSP.

The comparison between the dielectric mixture model of Wiener ($k = 1$) [13] with experimental results for the dielectric properties of wooden cross-arms versus MC, at 10.7 GHz, is shown in Fig. 20.11. The FSP is about 17% MC and the dielectric parameters at FSP are 3.0–j0.89, 3.5–j0.7, and 5.2–j1.1 for types A, B, and C respectively. As the MC increases above 17%, both ε' and ε'' increase abruptly. At this range of MC which is due to free water, high values of dielectric constant and loss factor are obtained. For example, in type A with the MC ranging from 30 to 55%, ε' increases from 5 to about 15 while ε'' increases from 2 to 6.

20.3 Application of Dielectric Properties

20.3.1 Latexometer

Based on the basic dielectric properties of hevea rubber latex as mentioned earlier, we have developed a simple, portable, easy to use, and cheap latexometer which can be used for MC or DRC measurement of hevea rubber latex. A photograph of the meter is shown in Fig. 20.12. Basically, the sensor consists of a microwave transmitter and receiver, non-lossy protective cover, detector, signal conditioning and display unit.

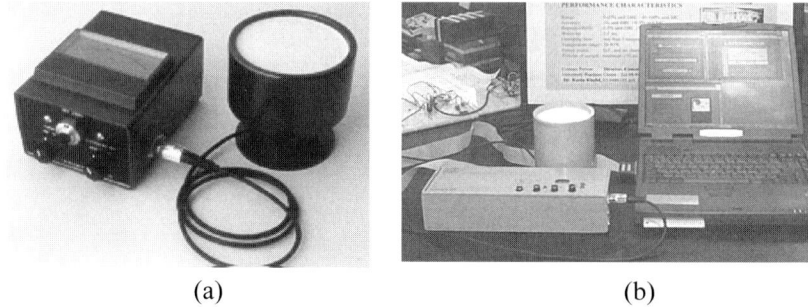

(a)   (b)

Fig. 20.12. Latexometer **a)** Analog model. **b)** Digital model. (See [24])

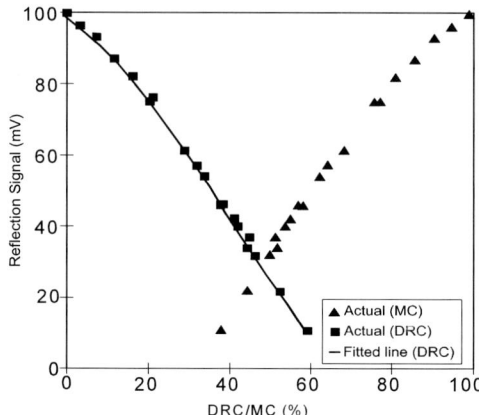

Fig. 20.13. Variation of reflection signal in the form of detected current versus DRC and MC of hevea latex. (See [24])

The analysis of the propagation of a wave through this multi-layer system is very complex and it can be simplified by using a signal flow graph and Mason's non-touching loop rules [22]. The details of the analysis, structure of the sensor, and prediction of reflection power can be found in the previous work [23]. The variation of the detected current of the meter with MC and TSC is shown in Fig. 20.13. It is clearly seen that the relationship between MC and detected current is good. Over the temperature range of 25 to 45°C the variation in MC is less than 1%. The performance characteristic of the latexometer is shown in Table 20.2. This kind of meter is suitable for use at the latex collecting center and for process control in the latex dipping industries.

Table 20.2. Performance characteristic of Latexometer (See [24])

Range	Accuracy	Reproducibility	Warm-up time	Operating time	Temperature range	Volume of sample
0–60% unit DRC	1% unit DRC	0.5% unit DRC and MC	2–5 s	< 5 min	20–45°C	150 ml
40–100% unit MC	0.5% unit MC					

20.3.2 Microstrip and Conductor-Backed Coplanar Waveguide (CBCPW) Moisture Sensors for Oil Palm Fruit

Microstrip and CBCPW moisture sensors have been developed to measure the variation of MC in oil palm fruit and subsequently predict the level of maturity. Both structures as shown in Fig. 20.14 consist of an input and output, a sensing area, and a stripline or coplanar line. The sensing area consists of a substrate, a protective layer, and a sample with semi-infinite thickness (see Fig. 20.15). The transmission and reflection phenomena in the structure can be analyzed using a signal flow graph and Mason's non-touching loop rules. The scattering parameter S_{21} or attenuation can be predicted from the dielectric properties of the sample. The detailed analysis of this structure can be found in the previous work [25-27].

The variation of MC of fresh mesocarp as a function of S_{21} or attenuation at 10.7 GHz is shown in Fig. 20.16. For the microstripline sensor, the curve is almost linear and the dynamic range of S_{21} is about 10 to 40 dB, which corresponds to the MC between 25 and 80%. The performance of the CBCPW sensor is shown in Fig. 20.16 with a sensing area length of about 1.6 cm, small gap size ($(b-a)/h = 0.3$, and s/h ratio (or SPH) equal to 0.0, 0.04, 0.08, 0.13, 0.18, and 0.22. The sensitivity for SPH=0.08 is about 0.15 dB/% MC.

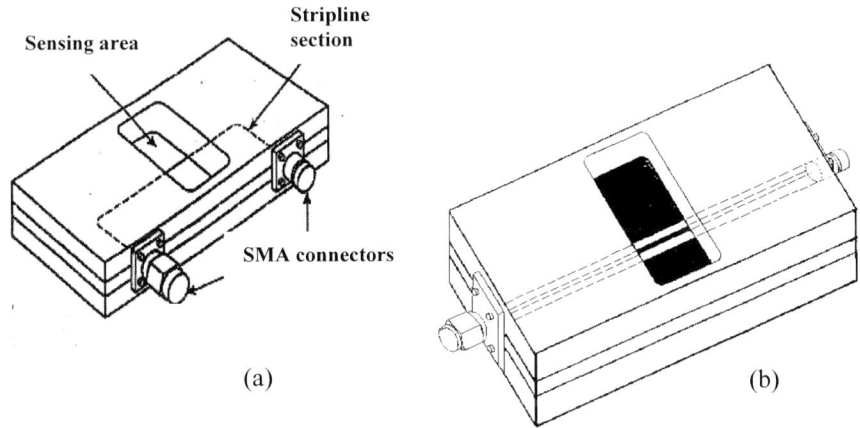

Fig. 20.14. Moisture sensors for oil palm fruit: **a** Microstrip (See [25]); **b** CBCPW (See [27]).

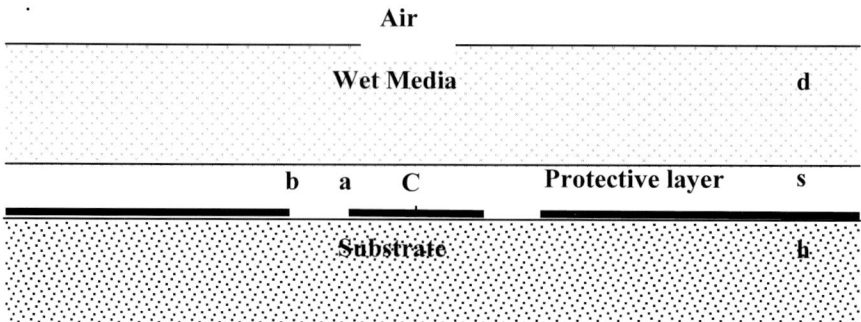

Fig. 20.15. Cross-section of CBCPW sensing area. (See [27])

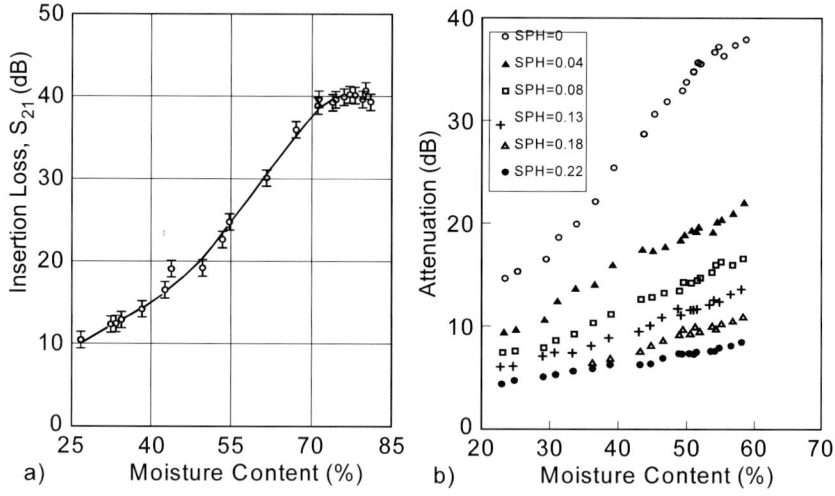

Fig. 20.16. Variation of insertion loss or attenuation with respect to MC in mesocarp: **a** microstrip sensor (See [25]. **b** CBCPW sensor for different s/h (or SPH) ratio, i.e. 0, 0.04, 0.08, 0.13, 0.18, and 0.22. (See [27])

The relationship between MC in fresh mesocarp and the magnitude of insertion loss with time after anthesis for six bunches is shown in Fig. 20.17. It is clear that water rapidly decreases in the mesocarp from approximately 15 to 20 weeks after anthesis. After the 20th week, there is a further small decrease in MC until full ripeness. Although the optimum time to harvest the fruit is at the 21 to 22 weeks after anthesis, time must be allowed for most of the fruits of the bunch to ripen. A single test of the sample is adequate for predicting the optimum time of harvesting by applying the profile of ripeness. This method is found to be suitable for assessing the quality of the fruit that reaches the factory.

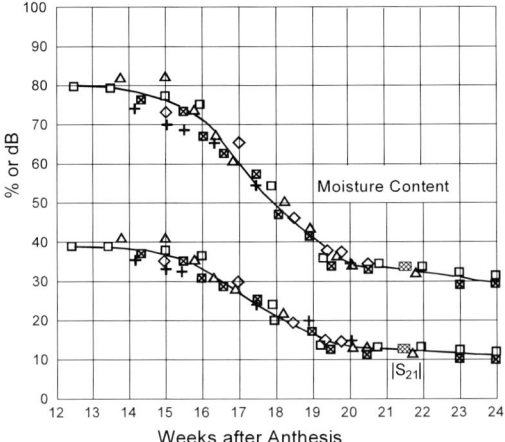

Fig. 20.17. Moisture Content and insertion loss of microstrip sensor as a function of development time of oil palm for six different bunches. (See [25])

20.3.3 Wood Meter

This chapter describes the development of a simple microwave wood meter based on the variation of dielectric properties of wood with moisture, density, and stage of decay as explained earlier.

The experimental results of the variation of reflection coefficient with MC at various stages of decay for Chengal wood and variation of reflection coefficient with stages of decay have already been described in detail in previous studies by the same workers [21, 28]. These studies provide a basis for the development of a simple instrument for the detection of decay in wooden cross-arms, which is based on the extent of MC absorbed by the decayed wood and also a small reflection at the equilibrium MC (~ 10–15% (dry basis)) from the decayed wood.

The characteristic of detected current at receiver with MC for decayed and sound wood is shown in Fig. 20.18. In actual conditions, especially during the rainy season, the maximum MC that can be absorbed by sound wood is about 15%, while for decayed wood it is about 35%. There is a possibility of using MC and its corresponding detected current as the indicator of decay. For example, if the detected current for a particular sensor is more than 50 mA, the wood is considered already decayed. However, the measurement has to be done immediately after the rain and it is not recommended as far as safety of the operator is concerned.

In Fig. 20.18, it is seen that at environmental moisture content (EMC) of about 10% there is quite a reasonable difference in the detected current between decayed wood and sound wood. This property gives the possibility for the detection of decay while the wood is dried or at EMC.

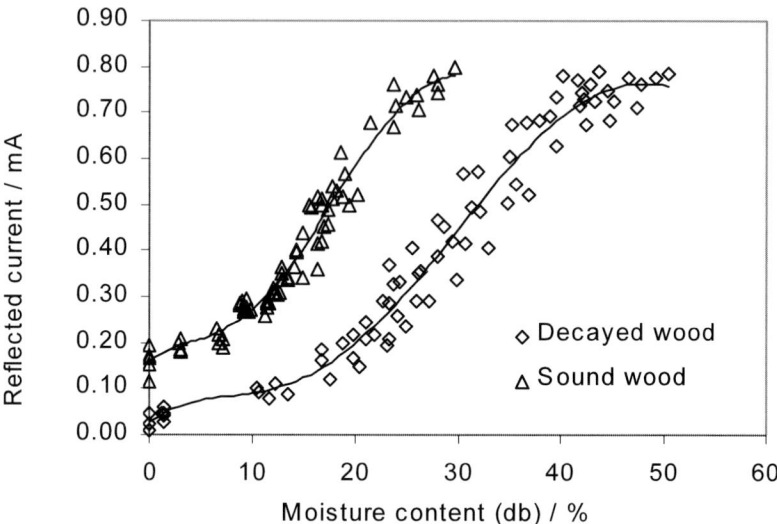

Fig. 20.18. Microwave detector current versus moisture content (dry basis) for decayed wood and sound wood of Chengal cross-arms. (See [20, 28]).

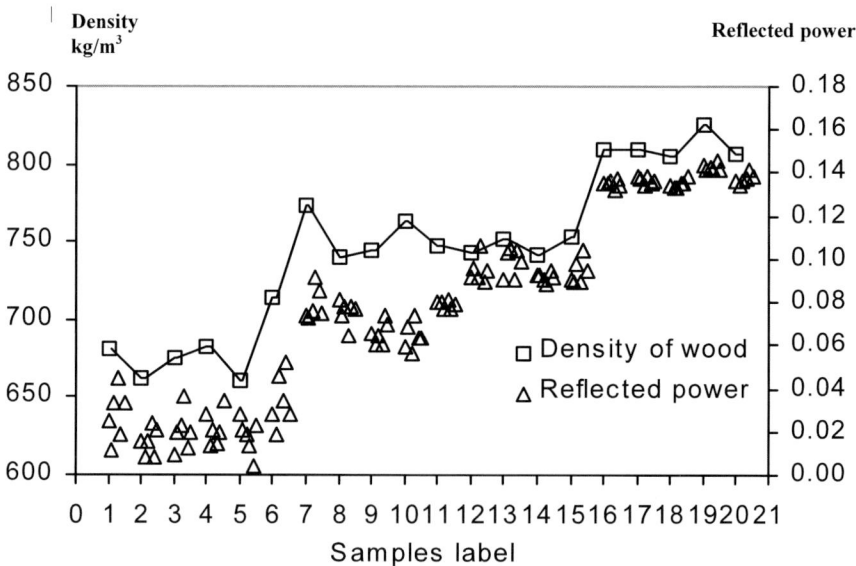

Fig. 20.19. Variation of density and reflected power (mW) for 20 samples of dried wood at different stages of decay at EMC level. (See [20])

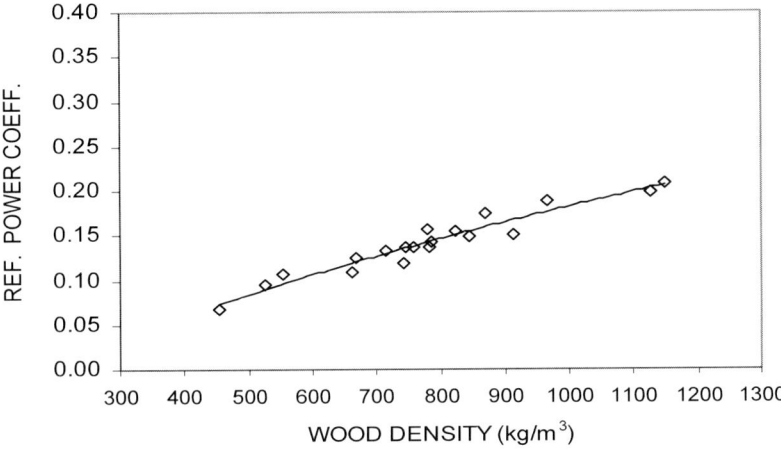

Fig. 20.20. Dependence of reflected power coefficient on wood density (See [20])

Shown in Fig. 20.19 is the variation of density and reflection power for 20 specimens of wood including sound wood, partially decayed and decayed wood (the measurement of each specimen is repeated five times). The reflection power from decayed wood is quite low (close to zero) while sound wood gives a reflection power of about 0.14. Their corresponding density varies from 670 kg/m^3 for decayed wood to about 800 kg/m^3 for sound wood.

The measurement results of the variation of the microwave reflection power coefficient with density of the wood with respect to a protective cover of 1cm thickness are shown in Fig. 20.20. It is found that there is a good relationship between reflection coefficient and wood density and the sensitivity is about 2.7×10^{-4} kg/m^3.

Various microwave wood meters (see Fig. 20.21) have been developed and all designs are taken into consideration for portability, ease of use, price, and acceptable accuracy. Further information on each version is described as follows:

Wood moisture meter and decayed level detection: This instrument is suitable for use in routine inspection of the stage of decay of wooden cross-arms based on the MC in the sample. The dynamic range of MC is about 0 to 40% for decayed wood and 0 to 30% for sound wood. The accuracy of the MC is about 2 to 5% (dry basis).

Wood densitometer: This instrument is used to measure the density of wood with a thickness greater than 5 cm and a density ranging from 400 kg/m^3 up to 1200 kg/m^3. The accuracy of the meter is about ±5 kg/m^3 and the measurement is operated at EMC condition.

Computer-assisted wood meter: This instrument is suitable for the measurement of density, moisture, and stage of decay of the wood. It is designed for multi-purpose measurements and the averaging technique base on the Lab View package is applied in order to reduce the error due to inhomogeneity of the samples. The possible species of the wood corresponding to the measured density can be displayed on the screen.

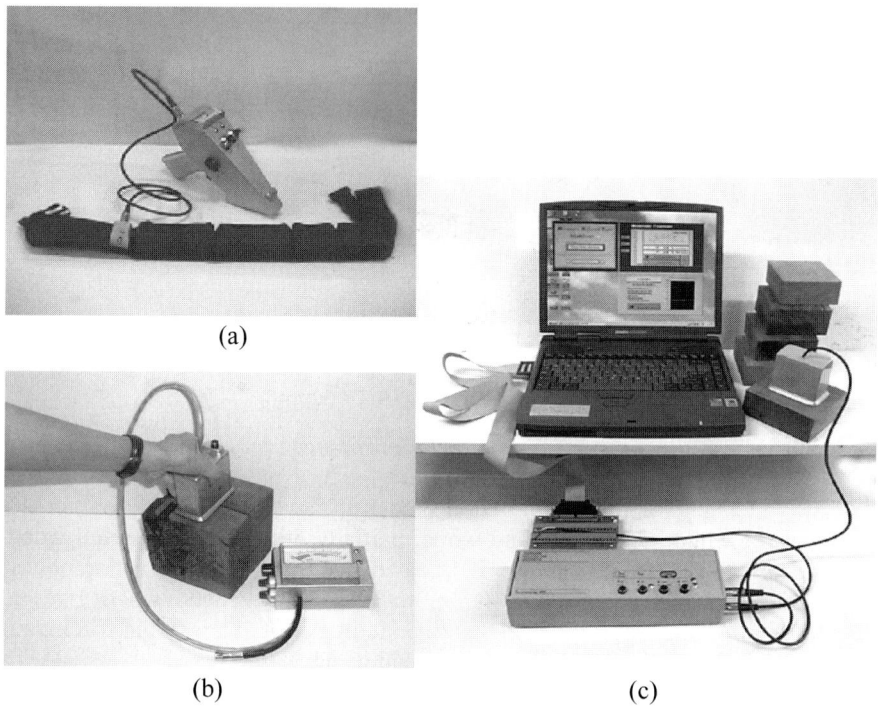

(a)

(b)   (c)

Fig. 20.21. Various types of wood meter. **a)** Wood moisture meter and decayed level detection in wooden cross-arms. **b)** Wood densitometer. **c)** Computer-assisted wood densitometer (See [20])

20.4 Conclusion

The dielectric properties of important commodities from Southeast Asia such as hevea rubber, oil palm fruit, and timber have been described. The second part of the chapter described the features and performance of microwave sensors for the determination of MC, density, stage of maturity, and stage of decay based on the dielectric properties of these materials. It was found that the most important features of the instrument that can be developed from these studies are the accepted accuracy, price, ease of use, portability, and non-destructive nature. This kind of work also provides a basis for a computer-assisted instrument for multi-tasking measurement of various parameters related to the assessment of the quality of the commodities.

References

1. International Rubber Study Group (2002) Rubber Statistical Bulletin, Wembley, UK
2. Chin HC (1979) RRIM training manual on analytical chemistry. Research Institute Malaysia, Kuala Lumpur p67
3. Cook AS, Sekhar BC (1953) Fraction brown Hevea brasilinsis latex centrifuged at 59,000g. J Rubber Inst Malaya 14:163
4. Oil World Annual (2001) World production of 17 oils and fats, Hamburg
5. Thomas RL, Phang S, Mok CK, Chan KW, Easau PT, Ng SC (1971) Fruit ripening in the oil palm Elaies guineensis. Ann Box 35:1219–1225
6. Year book of statistics (2001) Statistics Department of Malaysia, Kuala Lumpur
7. Khalid KB, Hassan J, Daud WM (1997) Dielectric phenomena in hevea rubber latex and its applications. In: Proceedings of the 5th international conference on the properties and applications of dielectric materials, Seoul, pp 25–30
8. Khalid KB, Daud WM (1992) Dielectric properties of natural rubber latex at frequencies from 200 MHz to 2500 MHz. J Natl Rubber Res 7(4): 281–289
9. Khalid KB, Hassan J, Daud WM (1996) The effect of ionic conductivity and dipole orientation on the dielectric loss of the hevea rubber latex. In: Proceedings of IEEE-MTTS international microwave symposium, San Francisco, pp 23–26
10. Khalid KB, Hassan J, Daud WM (1994) Dielectric properties of hevea latex at various moisture content. J Natl Rubber Res 9(3): 172–189
11. Suresh N, Calloghan JC, Creelman AE (1967) Microwave measurement of the degree of binding of water absorbed in soils. J Microwave Power 24:129
12. Bruggeman DAG (1935) Dielektrizitatskonstanten und leitfahigkeiten der mischkorper aus isotropen substanzen. Ann Phys Lpz 24(5): 636
13. Kraszewski A, Kulinski S, Matuszewski M (1976) Dielectric properties and model of biphase water suspension at 9.4 GHz. J Appl Phys 47:1275
14. Boned C, Peyrelasse J(1983) Some comments on the complex permittivity of ellipsoids   dispersed in continuum media. J Phys D: Appl Phys 16: 1777
15. Hartley CWS (1977) The Oil Palm, 2nd Edition Longman Group Limited, London pp 222-223
16. Khalid KB, Zakaria Z, Daud WM (1996) Variation of dielectric properties of oil palm mesocarp with moisture content and fruit maturity at microwave frequencies. Elaeis 8(2): 83–91
17. Nelson SO, Forbus WR, Lawrence KC (1995) Assessment of microwave permittivity for sensing peach maturity. Trans ASAE 3812:579–585
18. Peysken E, de Poureq M, Stevens M, Schalck J (1984) Dielectric properties of softwood species at microwaves frequencies. Wood Sci Technol 18:267–280
19. Nyfors E, Vainikainen P (1989) Industrial microwave sensors. Artech House, Norwood, MA, pp 216–224
20. Khalid K, Sahri MH, Cheong NK, Fuad SA (2001) Microwave reflection technique for determination of density, moisture and stage of decay in wood. In: Kupfer K (ed) Proceedings of the 4th international conference on electromagnetic wave interaction with water and moist substances, Weimar, pp 79–87
21. Khalid K, Sahri MH, Cheong NK, Fuad SA (1999) Microwave dielectric properties of wooden cross-arms. Proc SPIE, pp 146–156
22. Warner FL (1977) Microwave attenuation measurement, IEE monograph series no 19. Peter Peregrinus, Stevenage, Herts, pp 272–277

23. Khalid KB (1994) Portable microwave moisture meter for lossy liquids. In: Proceedings of the Asia Pacific microwave conference, Tokyo, pp 477–481
24. Khalid KB, Mohd R (2002) Development of microwave moisture sensors for hevea rubber latex and its application for latex dipping industries. In: Proceedings of the 4th ISHM, Taipei, pp 241–247
25. Khalid KB, Abbas Z, (1992) A microstrip sensor for determination of harvesting time for oil palm fruits (Genera: Elaeis guineensis). J Microwave Power EM Energy 27(1): 3–10
26. Khalid KB, Abbas Z (1996) Development of microstrip sensor for oil palm fruit. In: Kraszweski A (ed) Microwave aquametry, IEEE Press book series, New York, pp 239–248
27. Khalid KB, Hua TL (1998) Development of conductor-backed coplanar waveguide moisture sensor for oil palm fruit. Meas Sci Technol 9:1191–1195
28. Khalid K, Sahri MH, Cheong NK, Fuad SA (2000) Microwave reflection sensor for determination of decay in wooden cross-arms. In: Proceedings of the 6th conference on properties and applications of dielectric materials, Xian, pp 595–598

Application of Nuclear Magnetic Resonance

21 Moisture Measuring with Nuclear Magnetic Resonance (NMR)

Bernd Wolter[1], Martin Krus[2]

[1]Fraunhofer-Institut für Zerstörungsfreie Prüfverfahren (IZFP), Universität Gebäude 37, 66123 Saarbrücken, Germany

[2]Fraunhofer-Institut für Bauphysik (IBP), Fraunhofer Straße 10, 83626 Valley, Germany

21.1 Introduction

The nuclear magnetic resonance (NMR) phenomenon in bulk matter was first demonstrated in 1946 independently by two separate groups, Bloch's group from Stanford and Purcells group at Harvard. Purcell and Bloch were jointly awarded the Nobel Prize for their discoveries in 1952 [1, 2]. Over the next 50 years NMR has evolved from a scientific curiosity into one of the most powerful spectroscopic techniques. Today it is routinely used as an analytical tool in chemical and biochemical research but also in physics, material science and even in geochemistry. NMR spectroscopy probes the physical state of matter, determines the structure of complex molecules and helps to characterize intra- and intermolecular interactions and molecular dynamics. But NMR does not solely provide physical and chemical information. In magnetic resonance imaging (MRI) this information is combined with spatial resolution in the microscopic range.

Even though the NMR phenomenon can be observed in roughly two-thirds of all stable atomic nuclei, the majority of NMR applications use the hydrogen (^{1}H) nucleus. This is due to the high natural abundance and the large nuclear magnetic moment of this isotope. Perhaps the most famous application of ^{1}H-NMR is found in medicine, where it is used as a non-invasive imaging (MRI) method. In contrast to traditional x-ray imaging it enables the radiologist to diagnose cancer and diseases of the central nervous system without the need for employing any ionizing radiation [3].

Of course, ^{1}H-NMR is also known to be one of the most versatile methods for determining the moisture content within a solid material [4]. It is highly precise, non-destructive, non-contacting, fast, easy to use and provides additional information about the binding state of water. Nowadays, the technology has reached a level of high sophistication and maturity.

More and more, the potentialities of NMR and MRI are appreciated in industrial quality control. Moisture content determination in foodstuffs, powders, and other bulk materials is a routine task for low-cost NMR benchtop systems.

Hence NMR is far from being limited to application in research laboratories and it has already been successfully used for on-line monitoring of industrial processes [5].

Especially for building materials there is a considerable demand for reliable moisture measuring methods. Special interest is expressed in methods like NMR allowing for measurement of water content distributions, in order to determine material properties of water storage and transport. These material properties can be used as a base for a computation model to predict complex situations of moisture and heat transfer in a building construction.

On-site determination of moisture distribution can be used to describe the drying process in a building component and to predict its sensitivity for pollutant absorption in order to prevent fatal damage during service-life. With the advent of one-sided access (OSA) instrumentation, these "outdoor" applications of NMR have been rendered possible.

21.2 Physical Background

21.2.1 Basics of Hydrogen NMR

A full treatment of the physical background is beyond the scope of this text. Interested readers are referred to the exhaustive literature [6, 7]. The basic physical concept underlying NMR is the simple fact that a moving electrical charge produces a magnetic field and vice versa. The angular momentum or spin is a property of all atomic nuclei having an odd number of protons or neutrons. As atomic nuclei are charged, the spinning motion causes a magnetic dipole in the direction of the spin axis, and the intrinsic magnitude of this dipole is a fundamental nuclear property called the nuclear magnetic moment, μ. Hydrogen nuclei (^1H) possessing the strongest magnetic moment, fortunately are in high abundance in moist material. Consequently this isotope is used for NMR moisture measurements. Furthermore a variety of other nuclei are routinely observed in scientific research such as ^2H, ^7Li, ^{11}B, ^{13}C, ^{14}N, ^{15}N, ^{17}O, ^{19}F, ^{23}Na, ^{27}Al, ^{29}Si, ^{31}P, etc.

Consider a collection of a large number N_0 of ^1H nuclei as found in a specimen of a moist material (see Fig. 21.1a). In the absence of an external magnetic field, the magnetic moments have random orientations. The first step in an NMR experiment is to supply an external magnetic field, B_0. Now the magnetic moments tend to align with the external field like compass needles as described in Fig. 21.1b. However, unlike compass needles the ^1H spins may align either with the field or against it according to the laws of quantum mechanics.

Spins with anti-parallel orientation are at a higher energy state than spins aligned with the field, leading to a slightly overbalanced population of the latter energy state in thermal equilibrium. This excess population ΔN provides a detectable macroscopic magnetization (equilibrium magnetization M_0). The magnetization vector is aligned with the external field (z direction) and its modulus is proportional to the overall number of ^1H nuclei in the specimen.

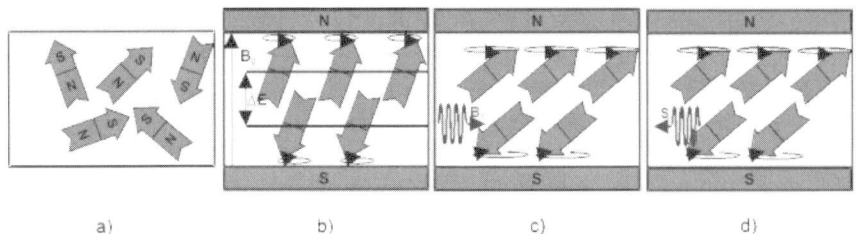

Fig. 21.1. a-d Schematic description of the physical principle of Nuclear Magnetic Resonance, NMR

Generally, reducing the specimen temperature and increasing the magnetic field strength leads to higher values of ΔN and with it higher magnitudes of M_0 ($\sim\Delta N$). That is why NMR spectrometers operate with extremely strong magnets (superconducting magnets). At finite temperatures $\Delta N/N_0$ is very small. In a magnetic field with a flux density of 1 T (1 tesla) it is only about 0.00005 % at room temperature. On the other hand, 1 cm^3 of H_2O contains $N_0 = 6.7 \cdot 10^{22}$ hydrogen nuclei, leading to an excess population ΔN of $3.7 \cdot 10^{16}$ nuclei, which is more than enough to generate a detectable NMR signal

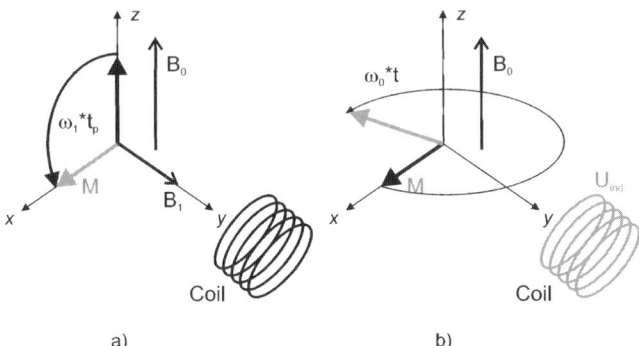

Fig. 21.2. a-b Magnetization M precessing under the influence of the pulsed RF magnetic field B_1 with pulse duration t_p **a** and free precession of M after termination of the pulse **b**

Another important distinction to compass needles is that instead of simply aligning themselves with the external magnetic field, the spins rotate around it, like a spinning top precessing in the earths gravitational field. The spins precess with a characteristic frequency, called the Larmor frequency ω_0. It depends on the type of nuclei and it is directly proportional to the strength of the magnetic field, resulting in

$$\omega_0 = \gamma \cdot B_0. \tag{21.1}$$

The proportional constant γ is called the gyromagnetic ratio (2π 42.58 MHz/T for ^{1}H). In order to detect an NMR signal, radio-frequency (RF) energy must be applied exactly at the Larmor frequency. This RF field is indicated by its amplitude magnetic field vector B_1 as it is shown in Fig. 21.1c. The nuclear dipoles will be induced to resonantly absorb and afterwards emit energy (Fig. 21.1d).

In modern NMR spectrometers, a coil of a resonant circuit generates one or more short pulses of the RF field. The RF pulse transfers longitudinal magnetization into phase coherent transversal magnetization, i.e., it rotates the magnetization M from its original z direction towards the xy plane (see Fig. 21.2a). The angle of deflection is proportional to B_1 and the duration of the pulse. A 90° pulse rotates the magnetization completely in the xy plane, whereas a 180° pulse inverts it. Phase coherent magnetization in the xy plane will induce an electromotive force in a receiver coil often identical to the transmitter coil according to Faradays law of magnetic induction (Fig. 21.2b). This voltage is proportional to the number of hydrogen nuclei (~ number of water molecules) in the test volume and thus it is a measure of the moisture content.

21.2.2 NMR Relaxation and Signals

Following the termination of the pulse, internal processes cause a loss of xy magnetization and a gradual increase of z magnetization. Nuclei will dissipate their excess energy as heat to the surrounding environment (called the lattice) and revert to their equilibrium position. This recovery process is called longitudinal relaxation or spin lattice relaxation and it is usually described by an exponential equation with the time constant T_1 (relaxation time T_1). The individual nuclear magnetic moments also interact with each other. As a result, the phase coherence immediately following the pulse gets lost and the magnetization in the xy plane disappears. This decay process does not involve the emission of energy and is called transverse relaxation or spin-spin relaxation. The rate of decay is described by a further time constant T_2 (relaxation time T_2).

Both time constants, T_1 and T_2 are very sensitive to molecular mobility. Pure water at room temperature shows large values of both T_1 and T_2, typically in the range of seconds. These large values are related to the high degree of Brownian motion in the low-viscous liquid. Decreasing the molecular mobility, as it results from solidification or increasing viscosity, leads to a monotonous decrease in T_2 up to values between 10^{-5} and 10^{-4} s in rigid solids. In contrast T_1 first decreases and afterwards increases again, if molecular mobility is reduced.

The simplest NMR experiment is the application of a single 90° pulse. Subsequently an NMR signal is detected, which is called the free induction decay, abbreviated to FID. This signal is shown in Fig. 21.3a. Usually, the FID will last only a few microseconds or milliseconds. Besides relaxation, a major reason for the fast decaying character of the signal is the magnetic field inhomogeneity, because it is virtually impossible to construct an NMR magnet with perfectly uniform magnetic field strength B_0. For that reason, the signal decays according to a time constant $T_2^* \leq T_2$. For single-exponential relaxation, the FID is described by

$$S = S_0 \left[1-\exp(-t_r/T_1)\right] \exp(-t/T_2^*).$$ (21.2)

The signal amplitude S_0 is a measure for the equilibrium magnetization M_0 ($\sim$ hydrogen/water content). In order to enhance the signal-to-noise ratio (SNR), usually a large number of NMR signals are registered and subsequently averaged. The time between the detection of two successive signals is called repetition time t_r. As can be concluded from Eq. (21.2), the relation between t_r and T_1 determines the maximum achievable signal amplitude. On the other hand, varying the repetition time t_r provides a possibility to determine the relaxation time T_1.

A further NMR experiment is the so-called spin echo pulse sequence, which is described in Fig. 21.3b. This signal can be written as

$$S = S_0 \, [1\text{-}\exp(\text{-}t_r/T_1)] \, \exp(\text{-}t_e/T_2) \, \exp(\text{-}t/T_2^*). \tag{21.3}$$

Here, the exponential term in the middle is independent of magnetic field inhomogeneity and decays with the "true" T_2 relaxation time. Therefore, T_2 may be determined by varying the echo delay time t_e.

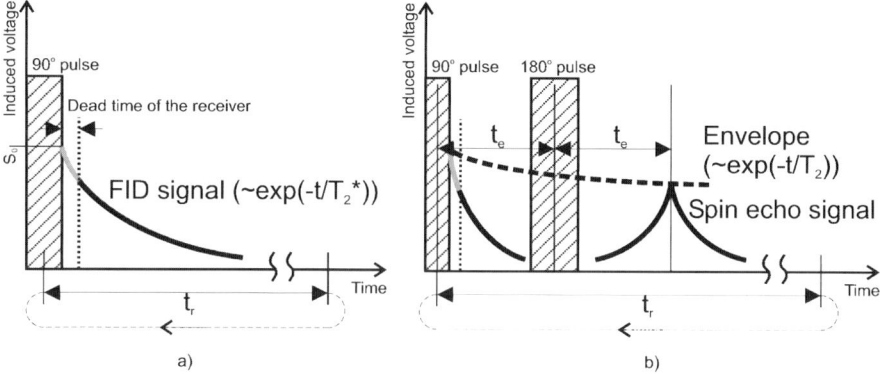

Fig. 21.3. a-b NMR signals: FID (Free Induction Decay) a and spin echo b

21.2.3 Spatially Resolved NMR and MRI

Spatially resolved NMR is based on the resonance condition of Eq. (21.1). When a non-uniform magnetic field is applied, the constant magnetic field B_0 has to be replaced by a spatially varying magnetic field B. In the one-dimensional case this is

$$B(x) = B_0 + gx. \tag{21.4}$$

Here g is the magnetic field gradient and x the position. Hence each position x in the specimen will be characterized by a specific frequency offset $\Delta\omega_0(x) = \gamma gx$. Therefore the frequency spectrum of the received signal can be directly translated into a corresponding spatial signal distribution, i.e., now the position x is frequency encoded. This approach is used in the second type of IBP instrumentation (see next section). Limiting the frequency range of the detected NMR signal allows selecting the signal contribution from the corresponding range of positions. The IZFP instrumentation (one-sided access NMR) profits by this procedure of spatial selection.

An additional possibility for selecting positions of interest within the specimen arises from the limited size of the coil. Significant signal contributions are restricted only to the part of the specimen which is in the sensitive range of the coil. IZFP as well as IBP instrumentation use this simple fact for spatially resolved measurements too.

21.3 NMR Hardware

21.3.1 Basic Components of NMR Instrumentation

The main parts of an NMR spectrometer are the magnet, the probe coil, and the electronic control unit. The NMR magnet is one of the most expensive components. Electromagnets made of superconducting wire, resistive electromagnets, and permanent magnets are commonly used. In conventional NMR the sample is placed in a probe coil inside the magnet geometry (within the magnet coil or between the poles of the magnet) in order to ensure the sample sits in the homogeneous focal point of the magnetic field. In case of MRI additional gradient coils serve for applying the magnetic field gradient g.

The specimen, for instance a human patient, has to be inserted into the enclosing apparatus generally a large hollow cylinder. For this reason the maximum volume of the specimen is restricted and the apparatus has huge dimensions and weight. In the case of a very large specimen, e.g., building constructions, a sampling is required. Therefore, traditional NMR instrumentation, even though in principle non-destructive, does not allow application without damaging the specimen in this case.

21.3.2 IBP Instrumentation

The NMR instrumentations used at IBP consists of an electronic control unit and a permanent magnet with measuring coil. The "small" measuring unit generates a magnetic field strength of 0.23 T, meaning that the resonance frequency for hydrogen nuclei is about 10 MHz. This equipment operates with open coils, through which the sample can be passed stepwise providing scanning of the specimen (see Fig. 21.4).

The specimen itself is suspended horizontally and hooked up to a glass liquid supply system to provide free water absorption without gravitational influences. Capillary liquid flow is determined by continuous weighing of the reservoir container for its mass loss, corresponding to determination of the water absorption coefficient (A value) from a suction test. Weighing provides a check of the liquid distributions recorded in the specimen. A photographic view is shown in Fig. 21.5 (left).

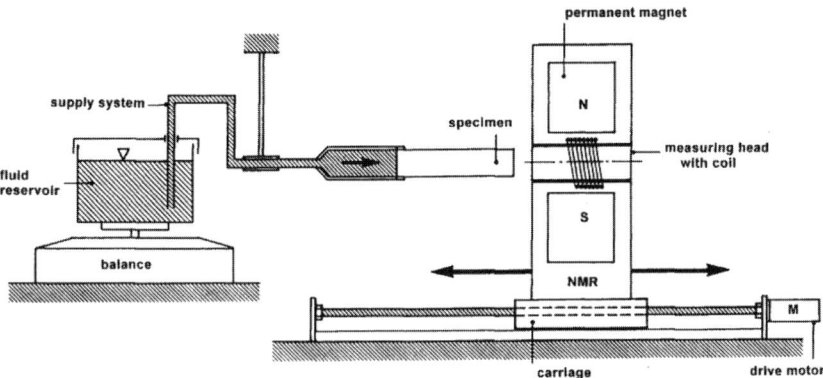

Fig. 21.4. Schematic layout of the NMR unit from IBP for determining liquid distributions in prismatic specimens during capillary liquid absorption

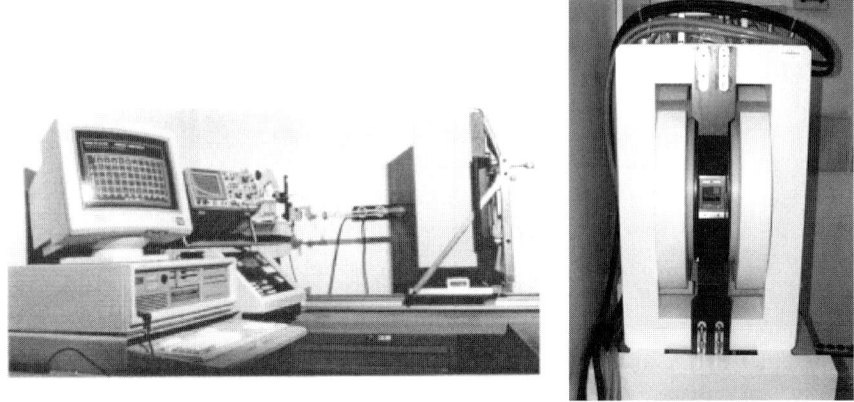

Fig. 21.5. "Small" NMR unit (*left*) allowing measurements on specimens up to 2×2 cm^2 in sectional area and "large" NMR unit (*right*) with 5×5 cm^2 maximum cross-sectional area

To enable the measurement on larger specimens a second ("large") measuring unit was set up, working with an electromagnet instead of a permanent magnet (see Fig. 21.5, right). Its magnetic field strength is 0.46 T, meaning that the resonance frequency for hydrogen nuclei is about 20 MHz. In comparison to the small unit this one works with a pulsed field gradient superimposing the static magnetic field, providing frequency encoding of position as described in Sect. 21.2.3.

21.3.3 IZFP Instrumentation One-Sided Access NMR (OSA-NMR)

In OSA-NMR the stray field of a U-shaped magnet assembly is used to obtain a B_0 distribution outside the NMR apparatus as described in Fig. 21.6. The required RF field is generated by a flat surface coil. At a predefined carrier frequency ω only nuclei in a small sensitive volume will be excited, in which B_0 fulfills the resonance condition of Eq. (21.1). This sensitive volume is located at a definite distance to the probe surface (depth x). Varying the measuring depth x can be accomplished, either by changing the frequency ω at constant B_0 distribution or changing $B_0(x)$ with a variable (electro) magnet at constant frequency ω. If none of them can be varied, the measuring depth in the specimen can be modified only by changing the lift-off of the probe.

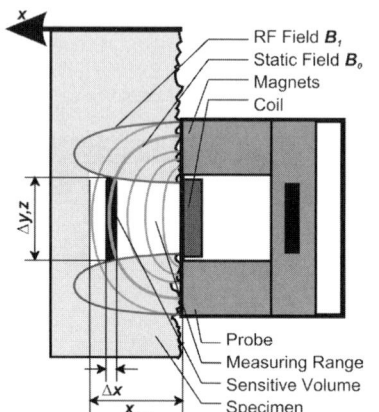 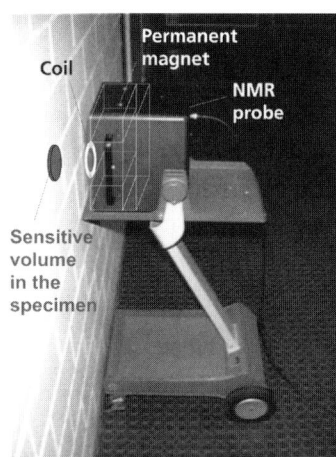

Fig. 21.6. Schematic layout of the OSA NMR unit

In order to find and characterize deposits of oil and gas underground in beds of sedimentary or other porous rock, in the early 1980s an instrumentation had already been developed for measuring NMR signals in objects which are outside the NMR apparatus [8]. Nowadays several commercial manufacturers offer NMR logging tools, which can be lowered into a well to measure the physical properties of oil and water saturated earth formation in situ. Employees of SwRI (Southwest Research Institute, San Antonio, USA) have developed the first prototype of the OSA-NMR technology [9]. With a view to using NMR for moisture measurements in concrete bridge decks and near-surface soils, SwRI utilized an electromagnet for B_0 generation. Even though very large maximum measuring depths could be achieved with this equipment, its serviceability was limited due to the heavy probe weight and the need for an auxiliary electromagnet power supply.

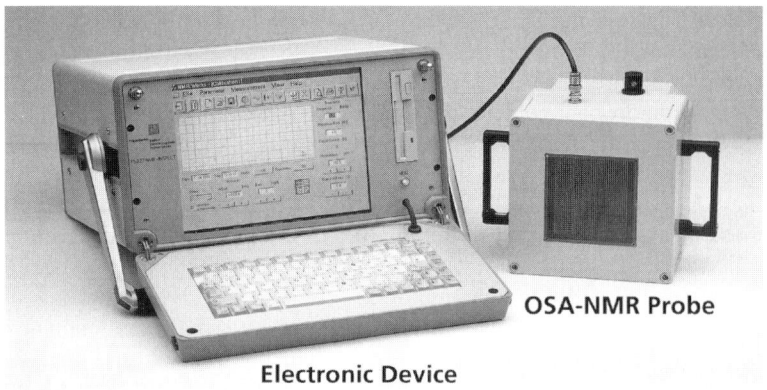

OSA-NMR Probe

Electronic Device

Fig. 21.7. Photograph of "NMR-INSPECT", the OSA-NMR unit developed at IZFP

A novel OSA-NMR approach was developed at IZFP (see Fig. NMR-INSPECT is the first completely portable, battery-powered measuring unit in the world based on the OSA-NMR principle. This equipment was intended especially for on-site application in terms of diagnosing the moisture situation in porous building materials. Its compact, robust design, its easy handling, and its optional battery power supply allows application on building sites [10].

21.4 Potential and Limits

21.4.1 Potential of the Method

NMR is one of the most successful methods for moisture content determination [11]. In contrast to nearly all other methods for measuring material moisture, NMR does not sense any changes of physical properties, which are influenced among other things by moisture, but it directly detects the hydrogen content and with it the moisture content of the material. The functional dependence between NMR signal and moisture content is linear. Often, the same calibration curve can be used for different materials (e.g., one calibration for most mineral building materials).

Moisture determination with NMR is rapid and accurate and, most importantly, non-destructive to the sample [12]. This is in contrast to conventional, highly precise measuring methods like gravimetry and Karl Fischer titrimetry, which modify the sample and require an expensive sample preparation. Nevertheless, the NMR method is far from being exclusively appropriate for application in laboratory environments. An example for on-line application is the measurement of moisture and calorific value in flowing coal [5].

By the use of imaging methods (MRI) it is possible to evaluate the internal moisture distribution within the material in order to analyze the actual moisture situation in detail. Due to the fact that these images can be acquired very rapidly, it is possible to monitor time-dependent processes as the absorption/desorption and migration of moisture.

In addition to hydrogen content the hydrogen relaxation times T_1 and T_2 represent a further information source concerning the characterization of a material's moisture. As described in Sect. 21.2.2, hydrogen in a solid state environment can be distinguished from hydrogen in a liquid state environment by means of these measuring quantities. In a mixture of components having different values of T_1 (T_2), the relaxation signal curve is composed of overlapping exponential curves representing the individual components. By separating the measuring curve into single exponential curves the relative hydrogen content of each component can be determined. As an example, this relaxation analysis is routinely used in the food industry to determine the fat/water content of margarine [13]. Later it will be shown, that in the same way density as well as moisture can be determined in organic materials like wood.

If a water molecule is in close vicinity to a surface, its NMR hydrogen relaxation will be accelerated with respect to that of pure water. Water within a confining geometry as a pore shows a characteristic value for T_1 (T_2), depending on the relation between the inner surface and the volume (surface-to-volume ratio, SVR), which in turn is a property of the pore geometry. Water-filled porous materials are often characterized by a broad distribution of T_1 (T_2) values, corresponding to the pore-size distribution. Hence, analysis of NMR relaxation times is a precise method for the quantitative determination of the pore-size distribution [14].

21.4.2 Limits of the Method

Problems can occur in the measurement of materials containing ferromagnetic inclusions, like the steel reinforcement in concrete. As a result the magnetic field distribution B_0 can be altered, preventing reasonable measurements in extreme cases. For OSA-NMR, the considerable effect of a reinforcement steel bar was noticed, if the bar was closer than 30 mm to the concrete's surface, which is not unusual for near-surface reinforced concrete. This application restriction can be overcome by using an adapted magnet assembly.

Materials with very small T_2 relaxation times (e.g., fired clay brick, blast furnace cement concrete) can cause another problem. If T_2 is very short, the hydrogen nuclei already completely relax within the dead time of the NMR measuring device (see Fig. 21.3). Hence these nuclei cannot be detected. Additionally the resolvable increment of MRI is inversely proportional to T_2. Therefore small T_2 values prevent high spatial resolution.

21.4.3 Accuracy and Spatial Resolution

Generally, the inaccuracy of moisture determination in solid materials with NMR is well below 1% by weight. Water detection down to moisture levels of 0.012% by weight has been reported [15]. IBP instrumentation, described in Sect. 21.3.2 is able to achieve accuracy better than 0.2% by volume in natural sandstones. The OSA-NMR from IZFP is slightly less precise. In concrete, an accuracy of about 0.7% for volumetric moisture can be achieved.

Conventional MRI equipment was used for the application described in Sect. 21.5.1. It provides two-dimensional images with pixel sizes of about 100×100 mm^2. But distinct higher spatial resolution down to the 10^{-6} m range can be realized by using special techniques of NMR hardware and methodology [16]. The instrumentations developed at IBP and IZFP provide high spatial resolution in one direction. In the case of the "small" IBP unit as well as for the OSA-NMR apparatus from IZFP, the sensitivity profile in this direction can be approximated by a Gauss function, i.e., it is smeared out instead of being a sharp function. Half-width at half-maximum of relative sensitivity is a measure for the resolvable linear increment. In the case of the IBP apparatus this is about 6 mm. For the IZFP apparatus the resolvable increment is the thickness of the sensitive volume (see Fig. 21.6). Depending on the sensor configuration, this depth increment Δx is typically between 0.2 and 1.5 mm. By employing switchable gradients (see Sect. 21.2.3) the "large" IBP unit provides a sharp rectangular sensitivity profile with 1 mm thickness.

21.4.4 Maximum Specimen Size and Maximum Measuring Depth

Because the specimen material has to be placed in the probe coil of the IBP instrumentation, the maximum cross-section of the specimen is restricted. In the case of the "small" apparatus prismatic bodies having maximum sectional dimensions of 20×20 mm^2 or cylindrical sample bodies 30 mm in diameter can be investigated. Specimens with a cross-section up to 50×50 mm^2 can be measured with the "large" apparatus.

For OSA-NMR (IZFP) the maximum specimen size is arbitrary. Here the coil diameter determines the lateral dimension of the sensitive volume ($\Delta y,z$), which is about a few centimeters (see Fig. 21.6). The field strengths B_0 and B_1 and with them the measuring sensitivity rapidly decrease with increasing distance x to the OSA-NMR sensor. Hence there is a threshold level the maximum measuring depth x_{max} which characterizes the measuring range in which the observed nuclei can be detected with an acceptable sensitivity. With today's available equipment this x_{max} is between 5 and 30 mm.

21.5 Application

21.5.1 Foods, Pharmaceuticals and Consumer Products

For random sample testing of granular materials and mass production articles the necessity to place the sample into the interior of the conventional NMR apparatus is less obstructive as in the case of non-destructive testing of valuable components. Today, NMR is one of the most successful moisture measuring methods for foods, pharmaceuticals, and cosmetics. Several suppliers offer low-cost benchtop NMR systems[1,2] for routine applications in industrial quality control. These systems measure the moisture content in starch, potatoes, seed, tobacco, sugar, rice, and pharmaceutical catalysts [17-19]. Simultaneously they determine oil/fat and moisture in nuts, chocolate, marzipan, flesh, milk powders, margarine, oilseeds, emulsions and even in whole bodies of animals [20-22].

Today, NMR is used of both as a research tool and as a quality control (off- and on-line) method for food components and food systems. A comprehensive review of NMR water determination in foods and biological materials was given by Ruan et al [4]. Imaging methods (MRI) enable food researchers to monitor water penetration into the material of interest, for instance into extruded pasta or even into a single rice grain during cooking [23, 24].

The imaging of the time-dependent moisture distribution also finds applications in pharmaceutical research, e.g. for on-line monitoring of the disintegration process of starch tablets in water, as shown by Köller and co-workers [25]. Many drugs are formulated as tablets for oral administration. The availability of drugs depends on the rate at which the tablet disintegrates in the gastrointestinal tract. Therefore the mechanisms of water uptake and swelling of compressed tablets in contact with water are of interest in pharmaceutical research. Tablet disintegration testing is used as a quality assurance measure. Ordinary disintegration testing, conforming to the current requirements as laid down in the German DAB Pharmacopoeia, are limited to the observation of the disintegration time. NMR microscopy offers the possibility to study diffusion of the solvent into the tablet and erosion of the tablet simultaneously, hence providing a detailed picture of the dynamics of disintegration.

In the present study, a special preparation technique was used in order to monitor tablet disintegration within conventional MRI apparatus. Immediately after starting the NMR experiment a drop of water was applied to the top surface of the cylindrical tablet (see Fig. 21.8a). Following one-dimensional distributions of hydrogen density and relaxation times (T_1 and T_2) of water upon and within the tablet have been observed as a function of time. The results of these experiments are represented in an imaginary coordinate system of position (X) and time (Y), as shown in Fig. 21.8b. The position X shows the one-dimensional distribution of NMR quantities in the direction of tablet thickness (diffusion direction). The NMR

---

[1] Bruker Minispec, Bruker BioSpin GmbH, Rheinstetten, Germany
[2] MQA7005 and MQA7020, Oxford Instruments, Abingdon, UK

measuring quantities are grey level encoded, showing a change of grey levels from white to dark grey for decreasing values.

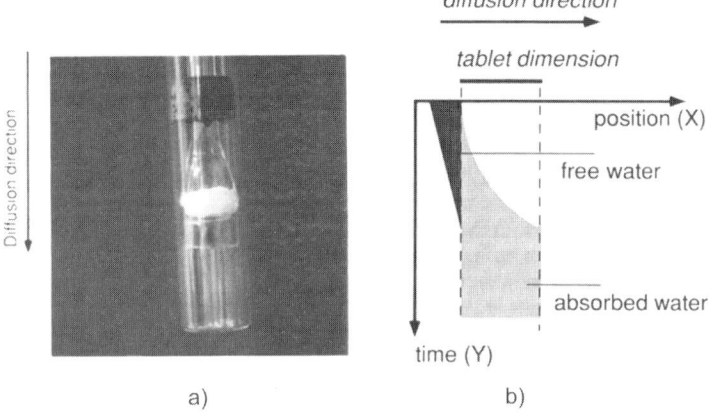

a)   b)

Fig. 21.8. a-b Imaging of time-dependent moisture distribution during tablet disintegration **a**; a drop of water is applied directly to the top of the tablet, which is placed inside the NMR apparatus **b**

Figure 21.9 shows the disintegration of a tablet made from insoluble calcium phosphate dihydrate with 10% of native maize starch. The tablet disintegrates within 60 seconds. T_2 relaxation time is plotted over 4 minutes.

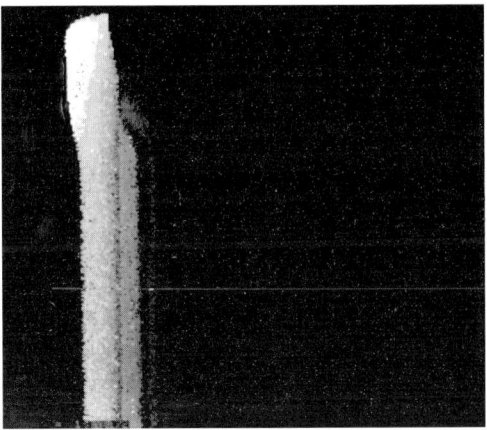

Fig. 21.9. Disintegration of a tablet observed by T_2 relaxation time images; T_2 limits: 3 ms (dark grey) to 110 ms (white); spatial resolution: ~ 80 μm; recording time: 0-4 min

After the water drop has contacted the tablets surface one can distinguish water of high mobility on the upper side (white) and less mobile water inside the tablet (light and dark grey). During disintegration, the part with less mobile water moves to the left, indicating an increase in the tablet thickness due to the swelling of the starch components. The moment, when all water is absorbed by the tablet can be exactly determined (vanishing of the white component). Simultaneously, the water diffusion front proceeding through the tablets thickness can be observed.

21.5.2 Liquid Transport Coefficients in Building Materials

Laboratory NMR equipment (e.g., IBP instrumentation) is routinely used to investigate the absorption and redistribution of water in different building materials. In this way, it was possible to determine the liquid transport coefficients and storage parameters in an exhaustive variety of materials [26, 27]. These coefficients come into general use in the basic data set for numerical computer programs allowing precise one- and two-dimensional calculation of simultaneous heat and moisture transport in building components even under complex conditions [28].

21.5.2.1 Measuring Procedure

Moisture distribution measurements were carried out using the "small" NMR unit from IBP. Each point in the moisture profile was measured within 5 s. Therefore the total measuring time for an incremental width of 4 mm in a specimen 15 cm in length takes less than 5 min. The specimen is connected to the glass liquid supply system by way of shrinkable PVC plastic tubing, some sealing compound being placed between the tubing and the specimen.

Immediately after the liquid supply is opened and the face of the specimen is fully moistened, the test routine can start. The specimen is scanned in the direction of absorption. Recording of moisture profiles is repeated at constant root-time intervals to obtain constant spacing of the profiles despite the absorption process becoming slower and slower ($\sqrt{t}$ law). Once the liquid has penetrated deeply enough (6 to 12 cm), sampling can be stopped. Depending on the absorption capacity of the material, a test period up to several days may be necessary.

After interruption of the water supply, the process of subsequent moisture redistribution starts. This transport phenomenon, which is considerably slower, may also be observed through subsequent determination of moisture profiles. However, this requires a test period from roughly a day to as much as several weeks. Observation time is limited as a result of the simultaneous process of diffusion which causes drying of the specimen through the open sectional surface on the end of the specimen. Once the total moisture content of the specimen has dropped by more than 5%, the sampling operation is discontinued.

21.5.2.2 Results and Discussion

In the following discussion four types of natural stone and four different artificial building materials are combined into two composite graphs. This means of depiction is also used to show the capillary transport coefficients. In addition to the measuring data the calculated moisture distributions are also plotted on the graphs.

Figures 21.10 and 21.11 show the test results of water absorption experiments. The moisture profiles measured at different times during absorption clearly show that water does not penetrate in the form of a "definite moisture front." Fully in accord with theoretical approaches describing capillary liquid transport in porous building materials, water penetrates in a complex way with gradients changing over space and time. All natural sandstones show a similar behaviour. n the absorbing face of the specimen a constant moisture content corresponding to capillary saturation quickly establishes itself. Moisture content declines continuously in the direction of absorption. All measured profiles have a convex shape. As depth of penetration increases, the curves become flatter. The individual types of sandstone differ in rate of penetration and steepness of the resultant curves.

Of the artificial building materials, gypsum and sand-lime brick behave in a manner similar to the natural sandstones (see Fig. 21.11). But water penetration presents a different picture for aerated concrete. Once the absorption times exceed 24 hours, this material shows an uncommon absorption behavior. The moisture content of the absorbing surface climbs above capillary saturation to 375 kg/m^3. Moisture plateaus develop at a water content of about 200 kg/m^3 as well as at capillary saturation. After an absorption period of 8 days (192 hours), the moisture content of the absorbing surface rises further to 450 kg/m^3 with a third moisture plateau beginning to form at 375 kg/m^3. In the case of brick a moisture plateau develops too, albeit after only roughly an hour of absorption. This plateau appears at about half the capillary saturation. Here no change in water content on the absorbing surface is noticed.

An explanation for the peculiar behavior of aerated concrete and clay brickcould not be found in the course of this research work. Studies of the pore radii distribution in clay brick using mercury porosimetry produced two pore radii frequency maxima. This could be an explanation for the appearance of water content plateaus.

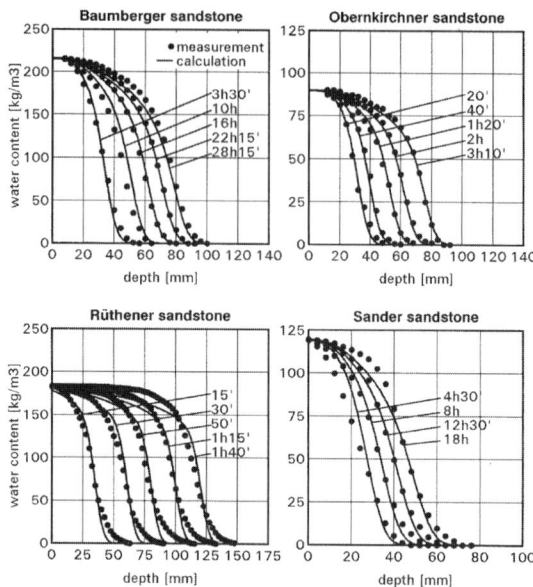

Fig. 21.10. Moisture distributions over depth of several natural stones during absorption

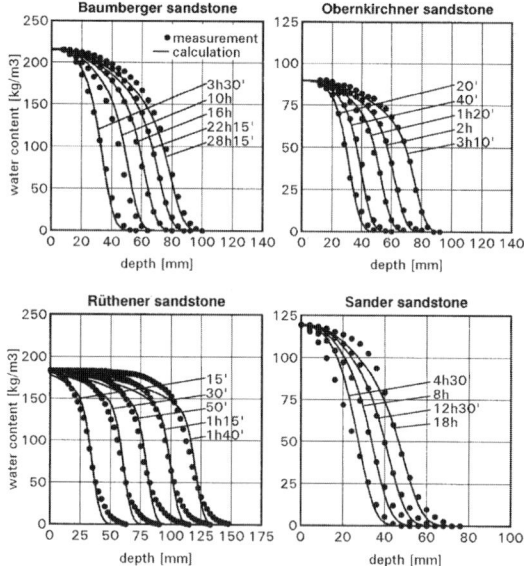

Fig. 21.11. Moisture distributions over depth of several building materials during absorption

Since the operative tractive forces of capillary action in the pore space cannot be measured directly, but since their measurable effects (gradient in water content) must have a functional relationship with them, Krischer at al formally introduced water content as motive potential [29]. This yields the following diffusion equation:

$$g_w = -D_w(w)\frac{dw}{dx},$$
(21.5)

with g_w (kg/m²s) as the liquid transport flux density, $D_w(w)$ (m²/s) as the liquid transport coefficient, and w (kg/m³) as the water content. The mass transport density $g_{w(x,t)}$ may be evaluated from profiles determined at two different points at times t_1 and t_2,

$$g_{w\left(x,t-\frac{t_1+t_2}{2}\right)} = \frac{1}{(t_2-t_1)}\int\limits_{a=x}^{\infty}(w(a,t_2)-w(a,t_1))da.$$
(21.6)

The upper limit of integration may be replaced by the location of the moisture front since a further integration cannot result in any change.

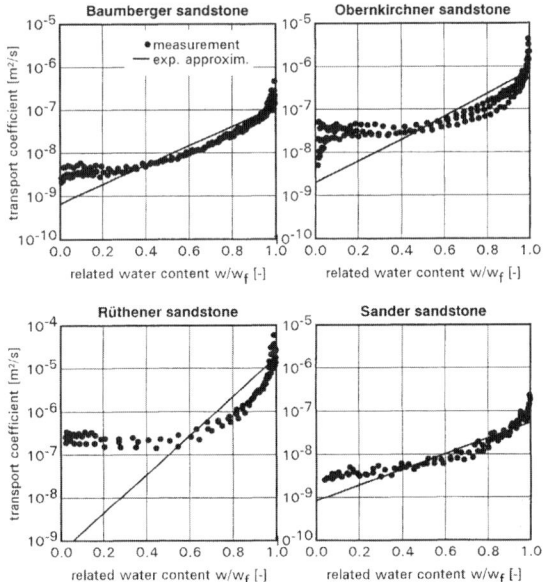

Fig. 21.12. Transport coefficients calculated from moisture profiles of Fig. 21.11 (sandstones)

Figure 21.12 shows the transport coefficients determined in this fashion for the sandstone. In all cases the transport coefficient is heavily dependent on water content. At capillary saturation the transport coefficient is above that in the hygroscopic moisture region by roughly two orders of magnitude. The transport coefficients of the artificial building materials are shown in Fig. 21.13. Sand-lime brick and gypsum show transport coefficients comparable to the natural sandstones. As anticipated, aerated concrete by contrast presents a different image. For the moisture contents in the plateau range, very high transport coefficients result. Here the moisture transport function is a function not only of water content but also of time, since the plateaus only form with time and the water content of the absorbing surface changes constantly. Behavior of this nature can no longer be described by a diffusion equation. For clay brick as well, the plateau which forms at half saturation, yields an increase of the correspondent transport coefficient. But the time-dependence is only slight.

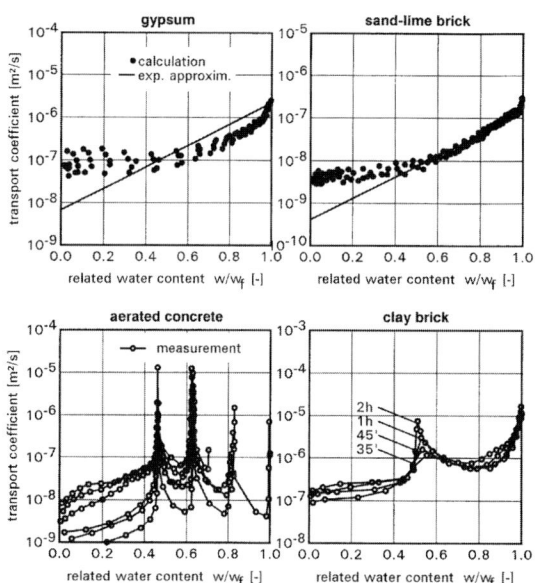

Fig. 21.13. Transport coefficients calculated from moisture profiles of Fig. 21.12 (building materials)

According to Kießl [30] the capillary transport coefficients for most building materials can be approximated very well by an exponential function. To verify this, an exponential approximation was included in Fig. 21.12 and 21.13 for all building materials except aerated concrete and clay brick. The exponential function is chosen in order to minimize the deviations between the measured moisture profiles and the profiles, which have been calculated by using this exponential function. The calculated distributions are also shown in Fig. 21.10 and 21.11.

21.5.3 Application of OSA-NMR

OSA-NMR, as used at IZFP, offers the possibility for on-site and on-line inspection. Comparable to an ultrasound reflection sensor, it can be applied from one side to the specimen (e.g., a building wall) and provides depth-resolved information about its internal structure. The equipment is portable and it can be used even in the harsh environment of a construction site. From this and in combination with the extensive information content of NMR, a variety of novel applications arise.

21.5.3.1 On-Site Determination of Moisture Profiles in Buildings

With OSA-NMR the current moisture situation and variations in moisture distribution during wetting and drying can be observed directly on the building component. Entire building constructions can be inspected in a complete non-destructive manner, without the need for impairing their integrity or their appearance by taking a sample. By determining the water tightness of building structures, environmental durability problems can be detected at an early stage, preventing unexpected fatal damage due to moisture ingress. A further application is to monitor the drying process in fresh concrete or cement floorings [31, 32].

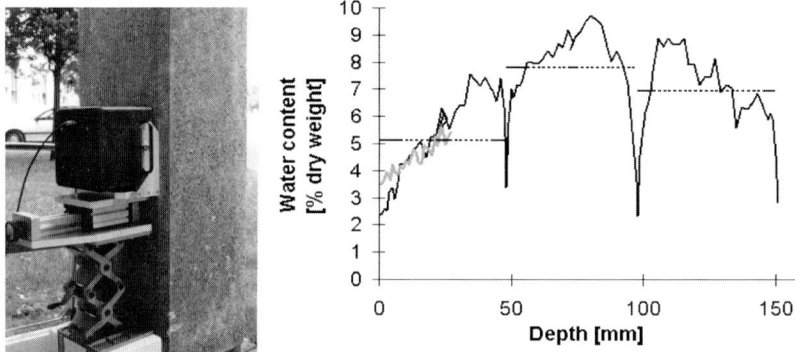

Fig. 21.14. Moisture profile measurement on a lightweight concrete pillar; the diagram on the left shows the on-site measuring result (thick gray curve) as well as the moisture profiles (black curve) and the integral moisture values (dotted straight lines) from laboratory measurements

In Fig. 21.14 application and results of a moisture profile measurement on a lightweight concrete pillar are presented. The thick grey curve in the diagram shows the moisture profile as it was determined on-site. The profile could be determined up to the maximum measuring depth of the used OSA-NMR device, which was 26 mm. This result indicates a water content increasing strongly in depth. The strength of this result was not clear, because the concrete was reinforced with embedded steel bars. As described in Sect. 21.4.2, this ferromagnetic steel could eventually result in false readings of the moisture content.

In order to verify the measurement result, a drilling core 150 mm in length was drawn leaving out the steel reinforcement bars. Later this drilling core was investigated in the laboratory. For the purpose of measuring the moisture distribution across the entire length of this sample, it was cut into three pieces of about 50 mm in length. By measuring from the upper side as well as from the bottom side of each piece, it was possible to determine the moisture content at each point of the drilling core despite the limited maximum measuring depth of 26 mm. This "laboratory" moisture profile is shown as the black curve in the diagram of Fig. 21.14. Apart from small differences at the beginning, probably as a result of the one-day delay between on-site measurement and sampling, a good agreement between the "on-site" profile and the "laboratory" profile is obvious. Therefore it can be concluded that in the present case the steel reinforcement does not have a significant influence on the NMR measurement. Note that the strongly decreased moisture contents at about 50 and 100 mm are a consequence of the cutting process. Finally, the integral moisture contents of the cut pieces have been determined by gravimetry (weighing, drying, and re-weighing). The moisture contents determined by this method are shown as the dotted straight lines in the diagram.

21.5.3.2 On-Line Determination of Density and Moisture in Wood

The profiles of moisture and density across the panel thickness are the most important measuring parameters for characterizing and controlling the product quality in the fabrication of chip (and fibre) wood panels. Usually, the density profile of small samples is determined by laboratory methods. Nowadays, the raw density profile in wood panels can be measured on-line with devices based on Compton backscattering [33]. But due to the similar scattering and absorption properties of solid wood and moisture, these methods do not accurately determine the dry wood density, which is by far more interesting than the raw density.

Based on the analysis of the T_2 relaxation curve, solid and liquid wood components can be quantitatively distinguished by NMR. As mentioned in Sect. 21.4.1, the relaxation curve is composed of two overlapping exponential curves representing hydrogen in solid wood and hydrogen in wood moisture. By analyzing the measuring curve the relative hydrogen content of both components can be determined. In this approach wood moisture as well as the dry wood density can be measured in a single procedure (see Fig. 21.15). By using OSA-NMR, profiles of moisture and density can be determined on-line during panel fabrication.

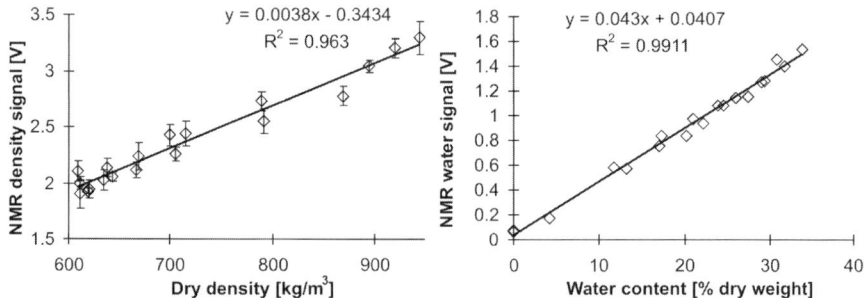

Fig. 21.15. Simultaneous density and moisture determination in wood panels; by analyzing the T_2 relaxation curve the signal contribution from hydrogen in solid wood (*left*) can be separated from the signal contribution of moisture (*right*)

21.5.3.3 Monitoring of Concrete Hardening

Fresh concrete develops in strength over time. Initially, after casting, the concrete is supported within formwork until it gains sufficient strength to support its own weight. Early age damage as well as increased susceptibility to detrimental influences from the environment (reduced durability) can often be traced to insufficient strength development prior to loading of the structure.

The development of mechanical properties in concrete is closely related to the chemical reaction between cement and water a process called hydration. During hydration the matrix phase of concrete, the so-called cement-stone is formed. The macroscopic strength of concrete is a consequence of an irregular structure of nano-crystallites in the cement-stone. Water is an important structural component of cement-stone. On one hand it is chemically bound in the crystallites and on the other hand it is physically adsorbed to the huge internal surface as a part of the gel pore phase. Furthermore, water is also bound in larger capillary pores, which are responsible for the moisture transport and the permeation of substances (e.g. water, aggressive ions) from outside into the internal structure of concrete.

Event though a variety of methods to measure the properties of fresh concrete are already available, monitoring the strength development in early-age concrete is still an unsolved testing problem. Conventional methods, such as the Vicat needle test, the slump test, the flow table test, etc., are not objective because the results are highly dependent on the measuring device and the measuring procedure. None of these methods allows continuous monitoring of material properties from the fresh to the hardened state, because they are destructive.

During hydration a part of the mixing water is chemically combined in hydration products and the residual water is confined in pores, which gradually decrease in size as cement hydration proceeds. These processes gradually diminish the molecular mobility of hydrogen in the concrete's water. As described in Sect. 21.4.1 this strongly affects the T_1 and T_2 relaxation times. On the other hand, the same changes of cement-stone's microstructure affecting the relaxation times are also responsible for the development of the physical properties, as strength and tightness. Therefore, measures of T_1 (T_2) allow for monitoring the strength development in hardening concrete. By use of OSA-NMR instrumentation this method is a new approach to continuously monitor the hardening of concrete in a building component.

The hardening behavior of five different concrete mixtures has been investigated in the study, which is described in the following. The specimen pure cement pastes without any aggregates as well as "real" concrete mixtures were prepared by using conventional Portland cement and gravel with 5 to 8 mm maximum grain size. Different water-to-cement ratios (w/c) as well as enrichment of some specimen with a retarder (tetra potassium pyrophosphate) should provide a wide range of hardening behavior (see Table 21.1).

In order to measure the T_2 relaxation curve, the mixtures were poured into plastic containers, which were placed on the top of the OSA-NMR unit. Detection of a depth profile was not of interest in this study, hence the measuring depth within the material was fixed. The hardening behavior was monitored over 3 days.

Starting with the preparation, the T_2 relaxation curve for each specimen was measured after hardening times t_h of approximately 1, 2, 3, 4, 5, 6, 8, 9, 23, 25, 27, 29, 31, 59, and 60 hours.

Table 21.1. Characteristics of the concrete specimen

Specimen	Cement (g)	H$_2$O (g)	w/c	Gravel (g)	Retarder (g)
S1	200	120	0.6		
S2	200	120	0.6		4
S3	200	70	0.35		
S4	200	120	0.6	200	
S5	200	120	0.6	200	4

The left diagram of Fig. 21.16 shows the evolution of the T_2 relaxation curve of specimen S1 for four different hardening times. It is evident that the decay is accelerating as hardening, i.e., cement hydration, proceeds. Fitting the experimental data to a one-exponential approximation function provides the time constant of the decay curves, which is the relaxation time T_2. From the beginning up to the end of hardening, this relaxation time shows typically a strong decrease (right diagram of Fig. 21.16).

In the case of cement paste or concrete without retarder (S1, S3 and S4), this decrease is moderate in the first 2–3 hours after mixing water and cement. This "Induction Period" is followed by an accelerated decrease ("Acceleration Period") up to 20-30 hours and finally hydration and with it T_2 decrease are slowing down ("Decay Period"). Hence the T_2 evolution follows the hydration progress and with it the qualitative behavior of strength development in concrete. The influence of a water-to-cement ratio (w/c) which is only 0.4 instead of 0.6 can be observed by comparing the hydration behaviour of specimen S1 and S3. Specimens with retarder (S2 and S5) show the most peculiar behaviour. As expected, the "Acceleration Period" is strongly delayed.

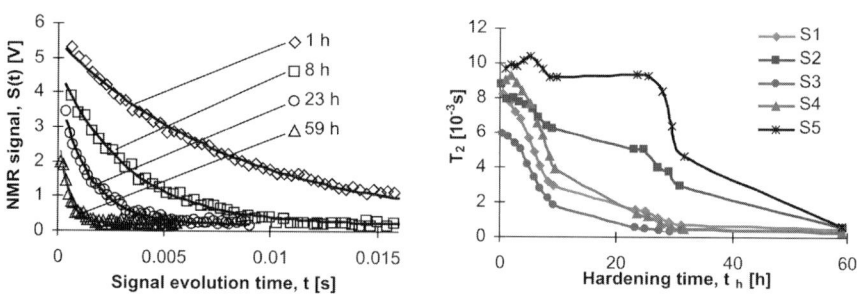

Fig. 21.16. Monitoring of concrete hardening by T_2 relaxation curve measurements; as hardening proceeds the decay of the T_2 relaxation gets faster (*left*), which is equivalent to the decrease of the relaxation time T_2 as a function of hardening time (*right*)

The strong dependence between T_2 (and also T_1) and the hardening state in cementitious materials is a well-known fact [34]. In the literature it has been shown, that this correlation can be used to predict the strength development of early-age concrete [35]. The concern of the present study was to demonstrate the feasibility of monitoring the hardening in concrete by conducting T_2 measurements not only with laboratory NMR equipment but also with OSA-NMR, offering the possibility to inspect early-age concrete on a building site.

21.6 Summary and Conclusions

Today hydrogen nuclear magnetic resonance (NMR) is a standard method for determining the water content of solid materials. As a non-destructive, non-contacting and direct moisture measuring technique it combines a variety of benefits. Highly sophisticated magnetic resonance imaging (MRI) instrumentation enable researchers to get a detailed look into water absorption and transport processes by taking a fast series of two-dimensional pictures of internal moisture distribution with microscopic spatial resolution.

Less expensive and more rugged NMR equipment, as developed at the Fraunhofer-Institut fuer Bauphysik (IBP), allows the determination of one-dimensional moisture distribution within the test object. A resolvable depth increment down to 1 mm can be achieved and the measurement inaccuracy is well below 1% of volume. This equipment is routinely applied to monitor the absorption and redistribution of water in building materials such as sandstones, brick, gypsum, and concrete in order to determine the liquid transport coefficients and storage parameters of these materials. These coefficients come into general use as the basic data set for numerical computer programs allowing precise one- and two-dimensional calculation of simultaneous heat and moisture transport in building components even under complex conditions.

Traditionally, NMR cannot be used for in situ industrial application. This disadvantage is mainly due to the need for sample extraction required by these enclosing systems, but also to their large size and their fragility. These limitations have been overcome with the advent of one-sided access (OSA) NMR. At the Fraunhofer-Institut fuer zerstoerungsfreie Pruefverfahren (IZFP) a portable, rugged, and easy to use OSA-NMR measuring device was developed. It provides a variety of new NMR application possibilities. As a tool for on-site building inspection, it can be used to record moisture depth profiles (preventive diagnostics) or to monitor the hardening degree of fresh concrete (quality control). As an on-line monitoring system, it can help to control and optimize the wood panel fabrication process by continuously measuring the moisture as well as the density depth profile.

References

1. Bloch F (1952) The principle of nuclear induction. In: Nobel Lectures, Physics 1942-1962
2. Purcell EM (1952) Research in nuclear magnetism. In: Nobel Lectures, Physics 1942-1962
3. Gilles RJ (1994) NMR in Physiology and Biomedicine. Academic Press, New York
4. Ruan RR, Paul L, Chen PL (1998) Water in foods and biological materials - A Nuclear Magnetic Resonance Approach. Technomic Publishing Company, Lancaster
5. King JD, Rollwitz WL (1982) Magnetic resonance measurement of flowing coal. In: Proc 10th Annual Mining and Metallurgy Industries Symp and Exhib, May 05-07, Denver, pp 145-157
6. Abragam A (1996) Principles of Nuclear Magnetism. Oxford Science Publications University Press, Oxford
7. Farrar TC, Becker ED (1971) Pulse and Fourier Transform NMR. Academic Press, New York
8. Jackson JA, Cooper RK (1980) Magnetic resonance apparatus. United States Patent No. 4,350,955, Oct. 10, 1980
9. Matzkanin GA, King JD, Rollwitz WL (1981) Nondestructive Measurement of Moisture in Concrete Bridge Decks Using Pulsed NMR. In: Proc 13th Symp on NDE, April 21-23, San Antonio, p 454
10. Wolter B, Dobmann G, Surkowa N, Kohl F (2002) Kernresonanz in Aufsatztechnik. In: TM - Technisches Messen 69:43-48
11. Schmidt SJ (1991) Determination of moisture content by pulsed nuclear magnetic resonance spectroscopy. Adv Exp Med Biol 302:599-613
12. Krus M (1997) Kernresonanzverfahren. In: Kupfer K (ed) Materialfeuchtemessung, Expert-Verlag, Renningen-Malmsheim, pp 173-190.
13. McCarten J (1991) Fat Analysis in Margarine Base Products. In: Minispec Application Note 1, Bruker Physik AG, Karlsruhe
14. Pel L, Kopinga K, Hazrati K (1997) Water distribution and pore structure in concrete as determined by NMR. Proc 9. Feuchtetag, Sept. 17-18, Weimar, pp. 294-300
15. Sobottka J, Kalähne R, Rössling G, Harde C (2001) Feuchtebestimmung in Pharmaka. In: Labor Praxis, 25(2):30-32
16. Kinchesh P, Samoilenko AA, Preston AR, Randall EW (2002) Stray Field Nuclear Magnetic Resonance of Soil Water: Development of a New, Large Probe and Preliminary Results. In: J Environ Qual 31: 494-499
17. Harz HP, Weisser H (1986) Einsatz von Kernresonanzspektrometern in der Lebensmittelindustrie. In: ZFL - Internationale Zeitschrift für Lebensmittel-Technik 4:278-281
18. Brosio E, Conti F, Lintas C, Sykora S (1978) Moisture determination in starch-rich food products by pulsed nuclear magnetic resonance. In: J Food Technol 13: 107-116
19. Kuhn K (1986) Kernresonanzmessungen und Wasseraktivitätsbestimmungen zur Aufklärung des Trocknungsverhaltens von Kartoffeln. Ph.D. thesis, Universität Hamburg
20. Koch A (1993) Schnellbestimmung des Wassergehaltes mit NMR. In: ZFL - Internationale Zeitschrift für Lebensmittel-Technik 44:122-124
21. Chang K, Ruan R, Chen PL, Ning A (1997) Quantification of moisture distribution in cheese block during cooling using MRI. In: Proc ASAE Annual International Meeting, Aug 10-14, Minneapolis

22. Lewis DS, Rollwitz WL, Bertrand HA, Masoro EJ (1986) Use of NMR for measurement of total body water and estimation of body fat. In: J Appl Physiol 60 (3): 836-840
23. Hills BP, Babonneau F, Quantin VM, Gaudet F, Belton PS (1996) Radial NMR microimaging studies of the rehydration of extruded pasta. In: J Food Eng 27:71-86
24. Takeuchi S, Fukuoka M, Gomi Y, Maeda M, Watanabe H (1997) An application of magnetic resonance imaging to the real time measurement of the change of moisture profile in a rice grain during boiling. In: J Food Eng 33:181-192
25. Köller G, Köller E, Kuhn W, Moll F (1991) Protonenresonanz-Mikro-Imaging der Diffusions- und Quellungsvorgänge von Poly(L-, DL-)Lactid-Tabletten. In: Pharm. Ind. 53:955-958
26. Krus M (1993) Determination of Dw from A-value. In: IEA-Annex 25 Projekt, Report T3-D93/01
27. Krus M, Künzel HM, Klier M (1993) Liquid Transport Over the Boundary Layers of Two Different Hygroscopic Capillary Active Materials. In: IEA-Annex 24 Projekt, Report T1-D93/08
28. Künzel HM (1994) Verfahren zur ein- und zweidimensionalen Berechnung des gekoppelten Wärme- und Feuchtetransports in Bauteilen mit einfachen Kennwerten. Ph.D. thesis, Universität Stuttgart
29. Krischer O, Kast W (1978) Die wissenschaftlichen Grundlagen der Trocknungstechnik. Springer Verlag, Berlin.
30. Kießl, K (1983) Kapillarer und dampfförmiger Feuchtetransport in mehrschichtigen Bauteilen. Ph.D. thesis, Universität Gesamthochschule Essen
31. Wolter B, Netzelmann U, Dobmann G, Lorentz OK, Greubel D (1997) Kontrastierende 1H-NMR-Messungen in Aufsatztechnik zur Bestimmung von Feuchteverteilungen in Zementestrichen und Holz. In: Proc 9. Feuchtetag, Sept 17-18, MFPA Weimar, pp 217-228
32. Wolter B (2001) Development and Application of a Portable NMR Moisture Measuring System in One-Sided Access Technique. In: Kupfer K, Hübner C (eds) Proc 4th Intern. Conf. Electromagnetic Wave Interaction with Water and Moist Substances, May 13-16, Materialforschungs- und -prüfanstalt an der Bauhaus-Universität (MFPA), Weimar, pp. 500-508
33. Dueholm S (1995) Bestimmung des Dichteprofils von Holzspanplatten. Holz- und Kunststoffverarbeitung 11:1394
34. Colombet P, Grimmer AR (1994) Application of NMR Spectroscopy to Cement Science. Proc 1st Intern Conf NMR Spectroscopy of Cement-Based Materials, Guerville, France 1992
35. Barbic L, Kocuvan I, Ursic J, Blinc R, Zupancic I, Rozmarin M (1979) Recherches de l'hydratation et des reistances des ciments par la resonance magnetique nucleaire. Ciments-Betons-Platres-Chaux 718:172-174

Index

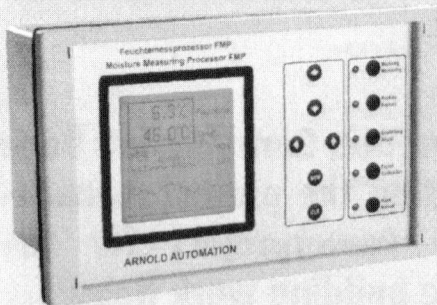
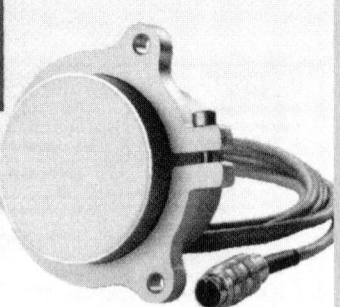
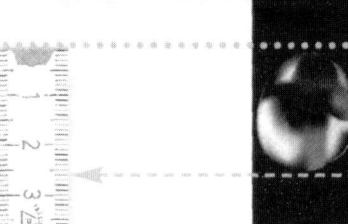